Frontiers of Organosilicon Chemistry

Frontiers of Organosilicon Chemistry

Edited by

A.R. Bassindale
The Open University, Milton Keynes, UK

P.P. Gaspar
Washington University, St Louis, Missouri, USA

ISBN 0-85186-097-4
A catalogue record for this book is available from the British Library

Proceedings of the IXth International Symposium on Organosilicon Chemistry
organised by the Royal Society of Chemistry, 16-20 July, 1990, University of Edinburgh

© The Royal Society of Chemistry 1991

All Rights Reserved
No part of this book may be reproduced or transmitted in any form
or by any means — graphic, electronic, including photocopying,
recording, taping, or information storage and retrieval systems —
without prior permission from the Royal Society of Chemistry

Published by The Royal Society of Chemistry,
Thomas Graham House, Science Park, Cambridge
CB4 4WF

Printed and bound in Great Britain by Bookcraft (Bath) Ltd

Preface

Organosilicon chemistry has a considerable, and growing, importance both in industry and in the academic world. It is a subject that draws on and contributes to the traditional areas of organic, inorganic and physical chemistry. At the same time it is fast becoming a mainstay of the interdisciplinary area of materials science. This multidisciplinary approach gives a unique strength to organosilicon chemistry that we believe is well-reflected in this volume.

This book has arisen from the IXth International Symposium on Organosilicon Chemistry held in Edinburgh in July 1990. There were over five hundred participants from all over the world and it was a particular pleasure to welcome many colleagues and friends from Eastern Europe following the momentous events of 1989 and 1990. There were about one hundred papers and two hundred posters presented. The aim here is to present an account of the frontiers of organosilicon chemistry by a selection of its leading exponents. In inviting contributions the Organizing Committee, (chairman Bernard Aylett), aimed to achieve a mix of some established experts with some of the exciting younger scientists active in organosilicon chemistry.

The most active and fruitful areas of organosilicon chemistry are all represented. The first part of the book comprises seven chapters concerned with perhaps the most dominant area at present; organosilicon-based polymeric materials. The range of such materials is impressive, with polyacetylenes, polysilazanes, polycarbosilanes, boron-containing materials and polysilaethers being amongst those described in the first section.

Much of the impetus for the development of silicon based materials grew from mechanistic studies, particularly of high temperature thermolyses. Those mechanistic studies were, for many years, driven only by a desire to understand the nature of what were then considered to be fleeting intermediates such as silylenes and silenes. Subsequent developments in ceramic and other technologies are another example of the rewards of supporting (and pursuing) fundamental research in areas of apparently purely academic interest. Accordingly, mechanistic organosilicon chemistry is well-represented and comprises the second section which is divided into two parts. The first is concerned with gas phase and photochemical studies, usually involving uncharged, low-coordination-number reactive intermediates. The second part of the mechanistic section relates to nucleophilic substitution at silicon with particular emphasis on hypervalent silicon species - currently an area of great interest and activity. Bioorganic silicon chemistry is also covered in this section.

The third part of the volume covers the structural chemistry of organic, inorganic and organometallic silicon compounds. The range of unusually coordinated and substituted silicon compounds that are described here was unthinkable only a few years ago. It is interesting to note that in five of the chapters the strategy of employing especially bulky ligands around silicon has been deployed with great success. In that way previously unknown or unstable species have been isolated and characterized. A remaining challenge is to produce structural data for silicenium ions.

The final section shows how mature the art of using silicon in organic synthesis has become. The emphasis is no longer simply on synthetic methods; 'control' is now the theme. The special activating and directing effects of silicon are shown to advantage in allylsilane and other reactions. Enantioselective reactions, once the weak links in synthetic organosilicon methodology are now a reality.

This book is a balanced account of current organosilicon chemistry. It should appeal to chemists in industry and in academic life. It will provide specialist information as well as being a useful overview for those chemists or materials scientists who are new to the subject.

Thanks are due to the Royal Society of Chemistry for supporting this volume and the IXth Organosilicon Symposium. Miss Catherine Lyall of the RSC deserves particular mention for patiently guiding the editors towards the completed volume. We also thank Mrs Jennifer Burrage for able assistance and typing.

Contents

Section I: Silicon-based Polymeric Materials

The Wonderful World of Silicon on Unsaturated Carbon. From Gas
Phase Rearrangements to Ceramic Fibers   3
   Thomas J. Barton

Silicon Ceramics with a Dash of Boron   15
   Dietmar Seyferth, Herbert Plenio, William S. Rees, Jr.,and Klaus Büchner

Organosilicon Preceramic Polymer Technology   28
   William H. Atwell, Gary T. Burns, and Gregg A. Zank

New Convergent Strategies for the Synthesis of Silicon Carbide
and Carbide Nitride Precursors   40
   J.-P. Pillot, M. Birot, F. Duboudin, M. Bordeau, C. Biran, and J. Dunoguès

Heterochain Polysilahydrocarbons containing Silicon-Silicon Bonds
in the Main Chain   50
   L. E. Gusel'nikov and Yu. P. Polyakov

Small Polysilylalkane Molecules: Feedstock Gases for Chemical Vapour
Deposition of Silicon and Silicon Alloys   62
   H. Schmidbaur, R. Hager, and J. Zech

Synthesis and Properties of Some Polysilaethers -
Polyoxymultisilylenes   70
   J. Chojnowski, J. Kurjata, S. Rubinsztajn, and M. Scibiorek

Recycling - Magic or Tragic?   81
   A. Stroh

Section II: Mechanistic Organosilicon Chemistry
a) Gas Phase and Photochemical Reactions

Mechanisms of Pyrolysis of Some Polysilanes   89
   Iain M.T. Davidson, Terry Simpson, and Richard Taylor

Recent Results in the Chemistry of Silylenes and Germylenes   100
   *P.P. Gaspar, K.L. Bobbitt, M.E. Lee, D. Lei, V.M. Maloney, D.H. Pae, and
   M. Xiao*

Direct Kinetic Studies of Silicon Hydride Radicals in the Gas Phase 112
Joseph M. Jasinski

Silylene and Disilene Addition Reactions 122
Manfred Weidenbruch

Reactive Intermediates and Mechanistic Aspects of Organosilicon
Photoreactions 134
M. Kira, T. Taki, T. Maruyama, and H. Sakurai

Structure and Reactivity of 7-Sila- and 7-Germanorbornadienes as
Precursors of Silylenes and Germylenes 145
O.M. Nefedov and M.P. Egorov

Reactions of Silane with Some Halogen and Oxygen-containing
Organic Compounds Induced by Infrared Laser Radiation 159
J. Pola

Reactions of Functionalized Silenes and Cyclosilanes 170
Wataru Ando and Yoshio Kabe

Section III: Mechanistic Organosilicon Chemistry
b) Hypervalent Silicon, Nucleophilic Substitution, and Biotransformations

Hypervalent Species of Silicon: Structure and Reactivity 185
R.J.P. Corriu

Hydrogen Peroxide Oxidation of the Silicon-Carbon Bond:
Mechanistic Studies 197
Kohei Tamao, Takashi Hayashi, and Yoshihiko Ito

Nucleophilic Substitution Reactions of Functional Siloxanes 208
K. Rühlmann, U. Scheim, K. Käppler, and R. Gewald

Bioorganosilicon Chemistry - Recent Results 218
*R. Tacke, S. Brakmann, M. Kropfgans, C. Strohmann, F. Wuttke,
G. Lambrecht, E. Mutschler, P. Proksch, H.-M. Schiebel, and L. Witte*

Section IV: Structural Organosilicon Chemistry and New Organosilicon Compounds

The Role of Trimethylsilyl-substituted Ligands in Coordination
Chemistry; Bis(Trimethylsilyl)methylmetal Complexes 231
Michael F. Lappert

Structure and Reactivity - The Role of Bulky Silyl Groups 253
David W.H. Rankin.

Sterically Overloaded Organosilicon Compounds 263
N. Wiberg

Structure Systematics of Di- and Oligosilanes 271
  L. Párkányi

Unusual Low Valent and Multiply Bonded Silicon Compounds from
Cyclopropenyl Silanes 285
  Mark J. Fink

Transition-metal Complexes of Reactive Silicon Intermediates 295
  T.D. Tilley, B.K. Campion, S.D. Grumbine, D.A. Straus, and R.H. Heyn

Decamethylsilicocene: Synthesis, Structure, and Some Chemistry 307
  P. Jutzi

Section V: Organic Synthesis using Silicon

Stereocontrol in Organic Synthesis using Silicon Compounds 321
  Ian Fleming

Stereocontrol by Means of Organosilanes. Some Observations on
Chemoselectivity, Stereoselectivity, and Enantioselectivity in
Allylsilane Reactions 332
  Cristina Nativi, Alfredo Ricci, and Maurizio Taddei

Enantioselective Synthesis using Organosilicon Compounds 344
  T.H. Chan, D. Wang, P.Pellon, S. Lamothe, Z.Y. Wie, L.H. Li, and
  L.M. Chen

β-Lactams from Allylsilanes 356
  E.W. Colvin, M.A. Loreto, M. Monteith, and I. Tommasini.

Some Synthetic Applications of Organosilicon Reagents to
Organic Synthesis 366
  K. Utimoto

Alkenylsilanes in the Synthesis of Nitrogen-Containing Heterocycles 378
  E. Lukevics and V. Dirnens

Palladium-catalyzed Insertion Reactions of Unsaturated Carbon
Compounds into Silicon-Silicon Bonds of Polysilanes. 391
  Y. Ito.

Subject Index 399

Section I: Silicon-based Polymeric Materials

The Wonderful World of Silicon on Unsaturated Carbon. From Gas-phase Rearrangements to Ceramic Fibers

Thomas J. Barton
AMES LABORATORY,* IOWA STATE UNIVERSITY, AMES, IOWA 50011, USA

This chapter may seem a bit disjointed since it will range from gas-phase rearrangements to preceramic and "optical" polymers. However, I believe that this spectrum is fitting for this volume since it is tracking a trend that is very much a part of organosilicon chemistry - a shift of emphasis from reactive intermediates to the materials science of silicon.

For quite a few years we have been examining the thermochemistry of organosilicon compounds containing silicon directly bonded to sp^2- or sp-hybridized carbon. One of the questions we repeatedly ask is, how does the presence of a silicon change the thermal behavior from that of the all-carbon analog? Not a particularly profound question, but one which sometimes yields profound answers. For example, although cyclobutene is well known to undergo thermally induced ring opening to afford butadiene, we have recently found that silacyclobutene **1** is, above 650°C, in equilibrium with allene **2**!

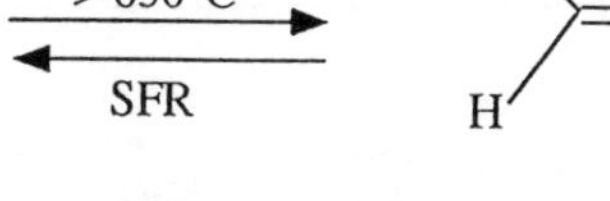

but:

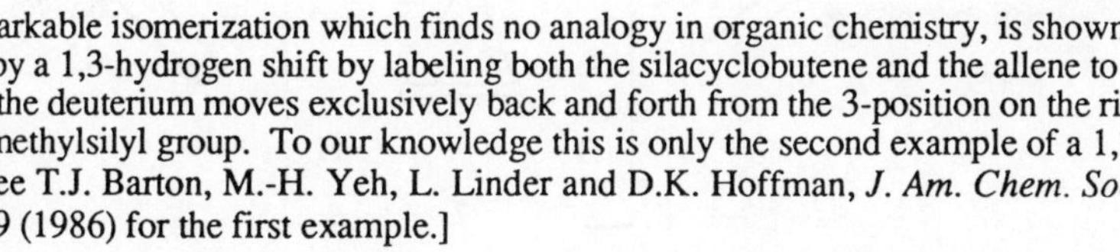

This remarkable isomerization which finds no analogy in organic chemistry, is shown to proceed by a 1,3-hydrogen shift by labeling both the silacyclobutene and the allene to find that the deuterium moves exclusively back and forth from the 3-position on the ring to the dimethylsilyl group. To our knowledge this is only the second example of a 1,3-H shift. [See T.J. Barton, M.-H. Yeh, L. Linder and D.K. Hoffman, *J. Am. Chem. Soc.* *108*, 7849 (1986) for the first example.]

---

* Ames Laboratory is operated for the U.S. Department of Energy by Iowa State University under Contract No. W-7405-ENG-82.

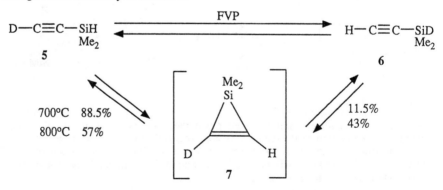

Another question we have recently asked is, since the allene-propyne equilibrium is now generally believed to proceed through the intermediacy of cyclopropene, does the sila-analog, **3**, isomerize to silirene **4**.

If $H_2C=C=CH_2 \rightleftarrows H_3C-C\equiv CH$ proceeds via   [cyclopropene structure]

Does $H_3Si-C\equiv CH$ isomerize to   [silirene structure]   ?

3   **4**

Evidence that this is indeed the case was obtained by finding that the deuterated molecules **5** and **6** are in equilibrium in a flash vacuum pyrolysis system above 700°C. Certainly, the most economical rationalization for this is that the isomerizations do indeed proceed through the intermediacy of silirene **7**.

$$D-C\equiv C-SiHMe_2 \xrightleftharpoons{FVP} H-C\equiv C-SiDMe_2$$

5   **6**

700°C 88.5%   11.5%
800°C 57%   43%

[silirene **7** structure with Me₂Si, D, H]

One of the more perplexing questions we have recently pondered has to do with the thermal isomerization of olefins to carbenes. The very few literature examples of this extremely difficult reaction involve twisted, strained alkenes with the exception of methylenesilacyclobutene **8**. Conlin [*J. Am. Chem. Soc.* (1986)] recognized that a

1,2-silicon shift to produce carbene **9** was the only reasonable way to rationalize ring expansion to the observed silacyclopentenes **10**.

$$\text{Me}_2\text{Si} \xrightarrow[\text{1,2-Si}]{621°\text{C}} \text{Me}_2\text{Si} \longrightarrow \text{Me}_2\text{Si} + \text{3-ene}$$

$$\quad\quad 8 \quad\quad\quad\quad\quad\quad 9 \quad\quad\quad\quad\quad\quad 10$$

If silicon as a migrating group facilitates olefin-carbene isomerization, we should be able to rationalize this. One approach would be to note that silicon has a unique potential to assist heterolytic rupture of a carbon-carbon π bond since it is well known that Si dramatically stabilizes both α-carbanions and β-carbocations.

Si stabilizes α-carbanions: [R$_3$Si–C:⁻ structure]   and β-carbocations: [R$_3$Si–C–C⁺ structure]

Thus twisting a silylolefin so as to heterolytically break the π-bond produces a zwitterion that only silicon can stabilize at both ends. As silicon facilely migrates to β-carbocations, we might expect that zwitterion **11** would isomerize to primary carbene **12** which we note would have the same hyperconjugative stabilization as a β-silyl carbocation.

$$\text{R}_3\text{Si} \diagup\!=\!\diagdown \longrightarrow [\text{zwitterion}] \longrightarrow [\text{carbene with SiR}_3]$$

$$\quad\quad\quad\quad\quad\quad 11 \quad\quad\quad\quad\quad\quad 12$$

We have spent considerable effort searching for evidence that silicon substitution actually does facilitate olefin → carbene isomerization. I won't review those efforts but suffice it to say they have been, without exception, negative. Thus, one must question the initial premise, namely, does silicon electronically lower π-bond rotation energies? To answer this question we have carried out numerous gas-phase kinetic studies on the thermally-induced <u>cis</u>-<u>trans</u> isomerization of substituted olefins, some of which are presented in Table 1.

One notes that replacing a methyl of <u>cis</u>-2-butene with a trimethylsilyl group reduced the rotation barrier by ca. 6 kcal. This is far too small a reduction to cause olefin isomerization by heterolytic cleavage rather than the usually lower energy diradical pathway. Indeed, if one examines the entire table, it is seen that the effect of a Me$_3$Si group is almost the same as a *t*-butyl, and we must conclude that the effect of Me$_3$Si is almost solely steric.

In years of examining the thermal chemistry of vinyl- and ethynylsilanes, our efforts have often been thwarted by the extreme stability of these systems. Eventually we asked whether we might utilize this stability by incorporating such units in polymer chains

TABLE 1
Gas-Phase, Thermally-Induced Cis-Trans Isomerization

Reactant (cis)	Product (trans)	E_{act} (kcal/mole)	(log A)
Me, Me / H, H	Me, H / H, Me	62.3 ± 0.8	(13.5)
Me, SiMe₃ / H, H	Me, H / H, SiMe₃	56.0 ± 0.3	(13.0)
Ph, Me / H, H	Ph, H / H, Me	60.2 ± 0.7	(15.1)
Ph, SiMe₃ / H, H	Ph, H / H, SiMe₃	53.0 ± 0.3	(14.3)
Ph, CMe₃ / H, H	Ph, H / H, CMe₃	54.9 ± 0.2	(14.8)
Me₃Si, SiMe₃ / H, H	Me₃Si, H / H, SiMe₃	52.7 ± 0.4	(13.9)
Me₃C, CMe₃ / H, H	Me₃C, H / H, CMe₃	54.4	

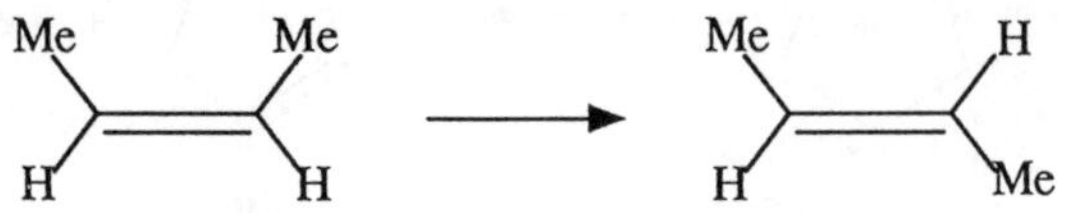

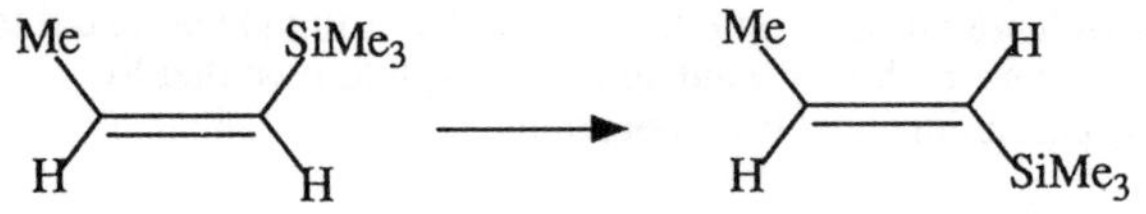

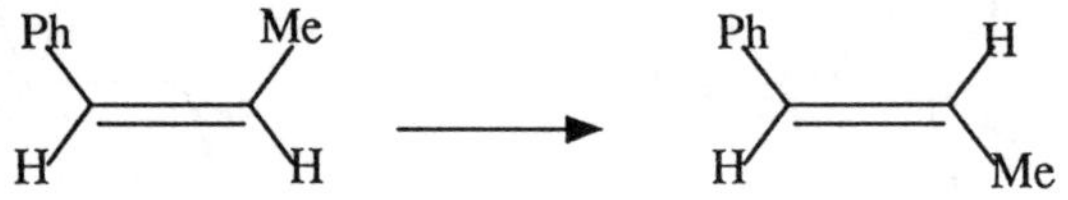

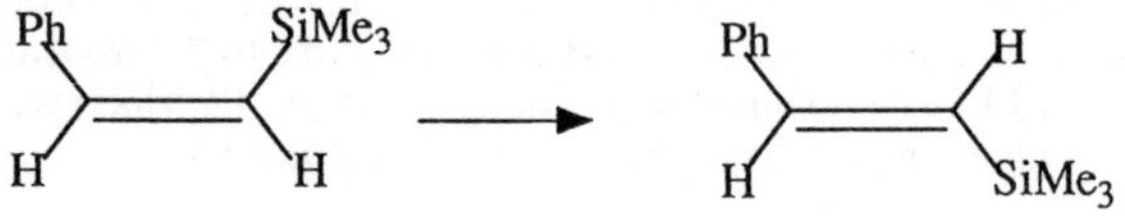

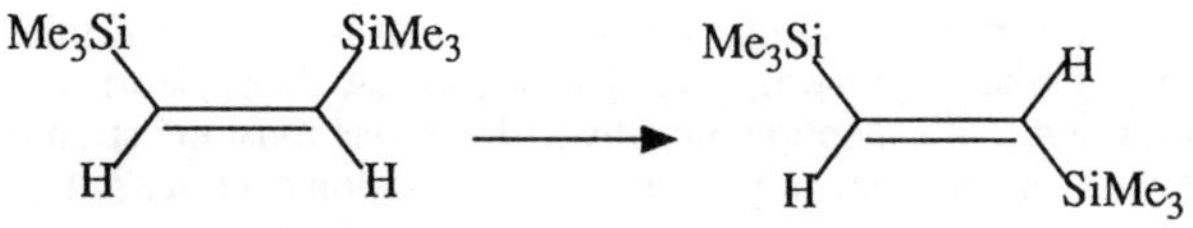

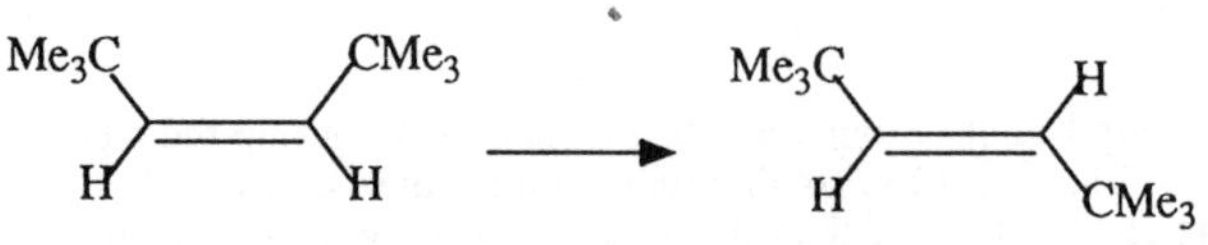

W. R. Roth *J. Org. Chem.*, **52**, 5636 (1987)

from which the substituents on silicon could be thermally eliminated to yield, for example, fibers of silicon carbide? Probably this was an idea whose time had come since we now know that at approximately the same time such polymers were also synthetic targets in the laboratories of Corriu, Henge, Ishikawa, Jones, Sakurai, Seyferth, West and probably many more of whom I don't yet know. Hardly the sort of competition I would have hoped for but certainly indicative of a major directional change in organosilicon chemistry.

Although we have worked less with silicon-olefin polymers, $\{R_2Si-CH=CH\}_{\overline{n}}$, in our hands they have been the easiest to prepare. We have simply taken advantage of the familiar chloroplatinic acid (CPA) catalyzed hydrosilylation of acetylenes, recognizing that if both functionalities, hydride and acetylene, are attached to the same silicon polymerization can occur as shown here.

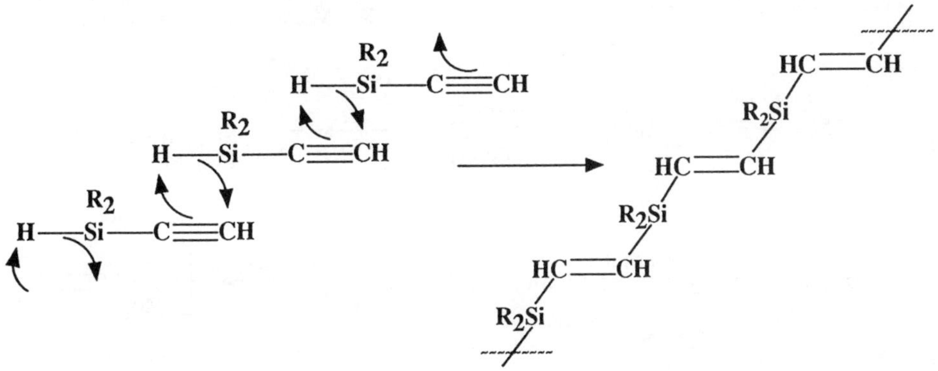

The polymerization of ethynylsilyl hydrides can be run either neat or in THF solution and quickly (too quickly sometimes) affords > 90% yields of high molecular weight silicon-olefin polymers which are soluble in a variety of organic solvents, melt reversibly, form excellent fibers and films, and fire to a mixture of silicon carbide and carbon. For example, neat silane **13** affords polymer **14**, with molecular weights controlled by the reaction temperature, which upon heating to 1,100°C yields 46% of ceramic char which is completely stable at 1,100°C in air.

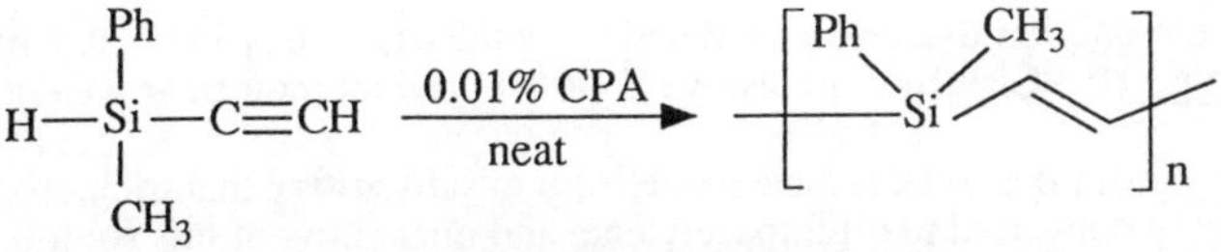

13   80° M_w 11,400 (PD 2.5)
         120° M_w 120,000 (PD 4.3)

14

The combination of synthetic ease, high yields, processing properties, and high ceramic yields makes these polymers excellent candidates for commercial silicon carbide precursor polymers.

Silicon-acetylene polymers were first reported by Korshak (*Izv. Akad. Nauk SSSR*, 1962) to be formed by the reaction of the acetylenic diGrignard reagent, $BrMg-C\equiv C-MgBr$, and Ph_2SiCl_2. The polymers, actually oligomers, were described as black resins and certainly didn't sound particularly appealing We spent considerable effort improving and modifying this process and were able to notably raise the molecular weights of these white powders, but not to values which we considered satisfactory.

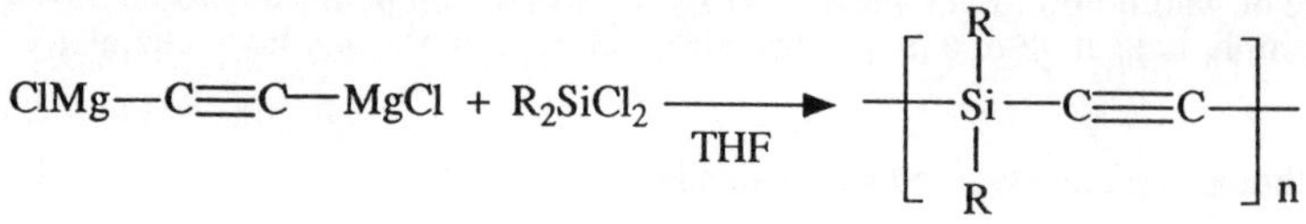

$$ClMg-C\equiv C-MgCl + R_2SiCl_2 \xrightarrow{THF} {\left[\begin{array}{c} R \\ | \\ Si-C\equiv C \\ | \\ R \end{array} \right]}_n$$

$R = R = Me$ M_w 2,500

$R = R = Ph$ M_w 3,000

$R = Me, R = Ph$ M_w 4,200

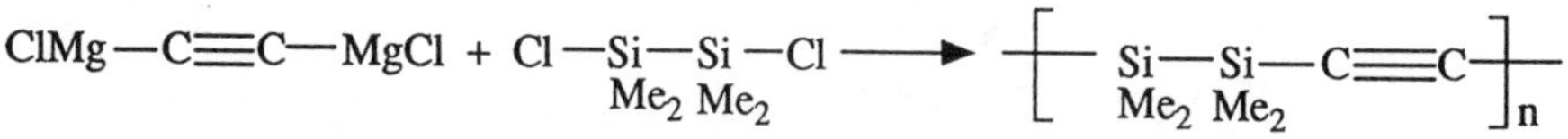

$$ClMg-C\equiv C-MgCl + Cl-\underset{Me_2}{Si}-\underset{Me_2}{Si}-Cl \longrightarrow {\left[\underset{Me_2}{Si}-\underset{Me_2}{Si}-C\equiv C \right]}_n$$

M_w 5,000

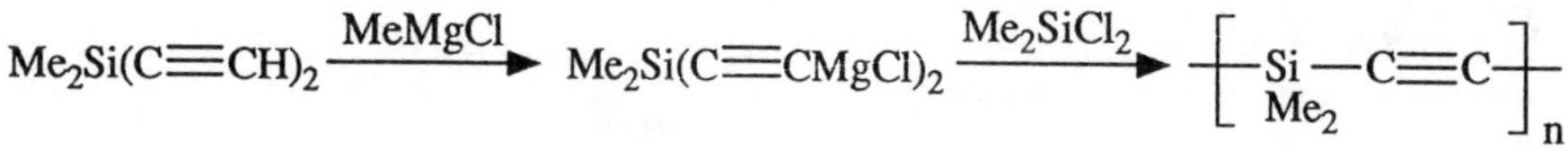

$$Me_2Si(C\equiv CH)_2 \xrightarrow{MeMgCl} Me_2Si(C\equiv CMgCl)_2 \xrightarrow{Me_2SiCl_2} {\left[\underset{Me_2}{Si}-C\equiv C \right]}_n$$

~ 50%, M_w ~ 5,000

$$HC\equiv C-\underset{Me_2}{Si}-\underset{Me_2}{Si}-C\equiv CH \xrightarrow[2)Cl-\underset{Me_2}{Si}-\underset{Me_2}{Si}-Cl]{1)\ MeMgCl} {\left[\underset{Me_2}{Si}-\underset{Me_2}{Si}-C\equiv C \right]}_n$$

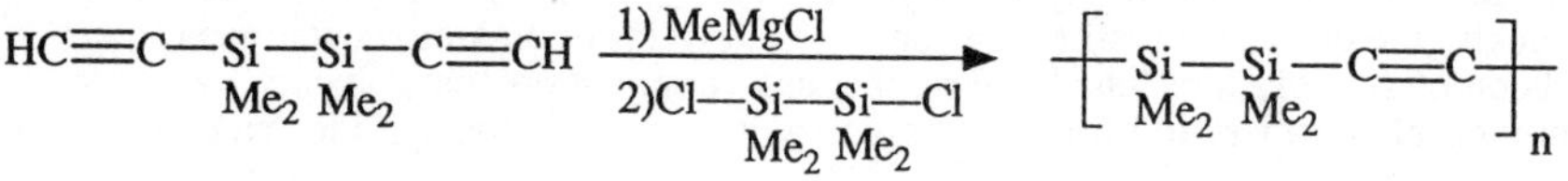

~ 50%, M_w ~ 6,000

We are of the opinion that chain termination by the essentially inevitable monoGrignard reagents, e.g., HC≡CMgBr, will always limit the product molecular weights in this route.

By far our best results have come from our discovery that trichloroethylene is quantitatively converted to dilithioacetylene, and quenching of this solution with dichlorosilanes affords the desired silicon-acetylene polymers in yields > 90% after purification.

Some of these polymers can be melt spun into fine (< 20 μm) continuous monofilaments, and all can be solvent-cast into coherent films and thermally converted into silicon carbide. Thermal decomposition is complete before 800°C, and total weight losses vary from 18 to 41% which compares very favorably with polysilastyrene (70% weight loss by 800°C) and $(Me_2Si)_n$ (90% by 400°C).

Wonderful World of Silicon on Unsaturated Carbon

$$Cl_2C=CHCl + 3\ n\text{-BuLi} \longrightarrow Li-C\equiv C-Li \xrightarrow{\underset{Me_2\ Me_2}{Cl-Si-Si-Cl}} \left[\begin{array}{c} Si-Si-C\equiv C \\ Me_2\ Me_2 \end{array}\right]_n$$

M_w 70,000 (dispersity, PD = 2.6)

Via Cl_2SiPh_2:
$$\left[\begin{array}{c} Si-C\equiv C \\ Ph_2 \end{array}\right]_n$$
M_w 30,000 (PD ~ 2.9)

Via $Cl_2SiPhMe$:
$$\left[\begin{array}{c} Ph \\ | \\ Si-C\equiv C \\ | \\ Me \end{array}\right]_n$$
M_w 20,000 (PD = 2.2)

Via Cl_2SiMe_2:
$$\left[\begin{array}{c} Si-C\equiv C \\ Me_2 \end{array}\right]_n$$
M_w 20,000 (PD = 1.7)

Direct thermal conversion of the melt-spun polymer fibers into silicon carbide fibers is, of course, impossible since they remelt to lose their shape. However, exposure of the fibers to triflic acid vapor results in replacement of surface phenyl groups by triflate. Submerging these surface-modified fibers into water hydrolyzes the silyl triflates to afford silanols which are apparently present in sufficient surface density to allow crosslinking via siloxane formation. Fibers thus treated could be fired to 1,300°C without losing their shape.

$$\underset{\text{(fiber)}}{\left[\begin{array}{c} Ph \\ | \\ Si-C\equiv C \\ | \end{array}\right]_n} + F_3CSO_3H \xrightarrow{-C_6H_6} \left[\begin{array}{c} OSO_2CF_3 \\ | \\ Si-C\equiv C \\ | \end{array}\right]_n \xrightarrow{H_2O}$$

$$SiC\ fibers \xleftarrow{1300°C} \left[\begin{array}{c} \left[\begin{array}{c} Si-C\equiv C \\ | \end{array}\right]_n \\ O \\ \left[\begin{array}{c} Si-C\equiv C \\ | \end{array}\right]_n \end{array}\right] \xleftarrow{-H_2O} \left[\begin{array}{c} OH \\ | \\ Si-C\equiv C \\ | \end{array}\right]_n$$

More recently, we have found that the "green" polymer fibers can be cured by exposure to UV irradiation, after which firing produces excellent ceramic fibers.

We have extrapolated to silicon-diacetylene polymers through the discovery that hexachlorobutadiene is quantitatively converted to dilithiobutadiyne by reaction with n-BuLi. The resulting solution can be titrated with dichlorosilanes to afford the desired polymers in yields > 90% and molecular weights (M_w) generally in the 20,000 range.

$$Cl_2C=\underset{\underset{Cl}{|}}{C}-\underset{\underset{Cl}{|}}{C}=CCl_2$$

$$\downarrow n\text{-BuLi}$$

$$Li-C\equiv C-C\equiv C-Li$$

$$\left[\begin{array}{c} \text{Me} \\ | \\ -\text{Si}-\text{C}\equiv\text{C}-\text{C}\equiv\text{C}- \\ | \\ \text{Me} \end{array}\right]_n \quad \xleftarrow{Me_2SiCl_2}$$

94%, M_w 20,000 (PD 1.7)

$$\xrightarrow{\substack{ClSi-SiCl \\ Me_2\ Me_2}} \left[\begin{array}{c} \text{Me Me} \\ |\quad | \\ -\text{Si}-\text{Si}-\text{C}\equiv\text{C}-\text{C}\equiv\text{C}- \\ |\quad | \\ \text{Me Me} \end{array}\right]_n$$

96%, M_w 20,000 (PD 2.0)

PhMeSiCl$_2$ ↙ ↘ Ph$_2$SiCl$_2$

$$\left[\begin{array}{c} \text{Ph} \\ | \\ -\text{Si}-\text{C}\equiv\text{C}-\text{C}\equiv\text{C}- \\ | \\ \text{Me} \end{array}\right]_n \qquad \left[\begin{array}{c} \text{Ph} \\ | \\ -\text{Si}-\text{C}\equiv\text{C}-\text{C}\equiv\text{C}- \\ | \\ \text{Ph} \end{array}\right]_n$$

92%, M_w 20,000 (PD 1.6) 91%, M_w 10,000 (PD 2.2)

Unlike the silicon-acetylene polymers, these silicon-diacetylene polymers are not pure white powders but always have at least a reddish tinge. In that form they are soluble in many organic solvents, but if heated much above 100°C, they blacken and become totally insoluble. This thermal crosslinking is hardly unexpected for diacetylene polymers and allows for the production of strong monolithic bodies simply by pressing the polymer powder, slowly heating to crosslink and firing to the ceramic. This technique also works to produce ceramic fibers which are initially pulled from the thermally-softened polymers. Thermal decomposition is complete by 800°C and in each case thus far examined results in ca. 80% char yields. Presumably, the extensive crosslinking traps many of the substituents and thus incorporates them into the carbon-rich char.

Silicon-diacetylene polymers were first reported in 1968 [Luneva, Sladkov and Korshak, *Izv. Akad. Nauk SSSR* (1968)] to be formed simply by heating diethynyldiphenylsilane (**14**) at ca. 200°C. That the red, low molecular weight polymer is not the claimed $-\!\!+\!\!Ph_2Si-C\equiv C-C\equiv C\!\!+\!\!\frac{}{n}$ is easily established by ^{13}C NMR and, of course, our independent synthesis of the actual polymer of this structure. What did turn out to be interesting was that when we polymerized **14** catalytically with MoCl$_5$ or WCl$_6$, a beautiful violet, soluble polymer was obtained, with vastly improved molecular weights, whose solvent-cast films can be doped with I_2 vapor to electrical conductivities up to 10^{-3} S/cm.

Ph$_2$Ge(C≡CH)$_2$ affords a very similar violet polymer, but Me$_2$Si(C≡CH)$_2$ catalytically polymerizes to a violet insoluble gel under a variety of reaction conditions. However, more bulky alkyl substituents such as butyl, isopropyl or hexyl allow formation of soluble violet or blue polymers whose doped film conductivities are in the region of 10^{-1} S/cm. Although most of our work has been with MoCl$_5$ catalyst, we obtain similar but structurally more ordered polymers with WCl$_6$ catalysis, for example:

$$Ph_2Si(C\equiv CH)_2 \xrightarrow{\Delta} \text{red polymer}$$

$$\downarrow \begin{array}{c} MoCl_5 \\ C_6H_6 \end{array}$$

$M_w \sim 10{,}000$
λ_{max} 450 nm

violet polymer

$M_n \sim 50{,}000$ ($M_w/M_w = 1.91$)

λ_{max} 550 nm $\lambda_{sh} \sim 590$ nm

$$s\text{-}Bu_2Si(C\equiv C)_2 \xrightarrow[\substack{MoCl_5 \\ (WCl_6)}]{} \text{violet polymer}$$

$\lambda_{max} = 560$ nm
($\lambda_{max} = 550$ nm)

$M_n = 35{,}200$ $PD = 1.8$ ($M_n = 26{,}987$) $PD = 2.5$)
$M_w = 63{,}700$ ($M_w = 66{,}335$

Conductivity: undoped $\sim 2 \times 10^{-8}$ S/cm
I_2-doped $\sim 1 \times 10^{-1}$ S/cm

At this time we are not 100% certain of the "structures" of these polymers–in fact, we may never be. From a considerable effort involving quantitative ^{29}Si NMR and ^{13}C NMR, resonant Raman and theoretical calculations, we believe the fundamental repeat unit to be a methylenesilacyclobutene ring with acyclic olefin units interspersed in the chain.

Formation of repeating methylenesilacyclobutene units can be mechanistically rationalized by a metallocarbene process involving only head-to-tail additions, both intra- and intermolecular.

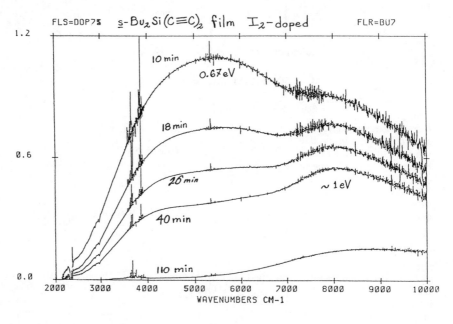
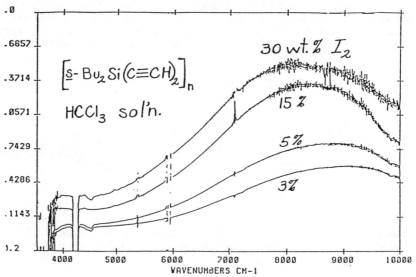

Iodine doping of these conjugated polymers is simply a redox reaction and would be expected to produce bipolarons (dications) as the charge carrying species on the chain. As a bipolaron has two inner gap states, one above and one below the Fermi level, upon doping we should see two new lower energy electronic transitions.

CONDUCTION BAND

FERMI LEVEL

VALENCE BAND

BIPOLARON (spinless dication)

observe two new transitions from the valence band

Indeed, after exposure of the polydiethynylsilanes (PDES) films to iodine vapor, two new absorptions are observed, at ca. 0.6 and 1.0 eV, along with a concomitant decrease in intensity of the violet band at ca. 2 eV. Thus for some time we were confident that bipolarons were the charge carriers in these polymer films. However, recently we have examined I_2-doping in solution and, quite to our surprise, only a single new band at ca. 1 eV is observed. My own favored explanation for this is that in both the film and solution this is a soliton band and thus the polymer has a degenerate ground state. The lower energy new band seen only in the films must be due to some unprecedented interchain interaction.

Of course, there is another way to produce these inner gap states and that is by photoexcitation. In this experiment the polymer film is pumped with a CW laser at 2.5 eV to reveal three new bands in addition to photobleaching. The higher energy band (HE) is also seen in <u>trans</u>-polyacetylene, and we believe it to be due to a neutral soliton. Of most interest are the two low energy bands, ω_1 and ω_2 (0.3 and 0.85 eV), which correlate with the two bands observed upon I_2 doping. These two bands are not due to a bipolaron since their intensities do not have the same dependences upon either temperature or excitation intensity.

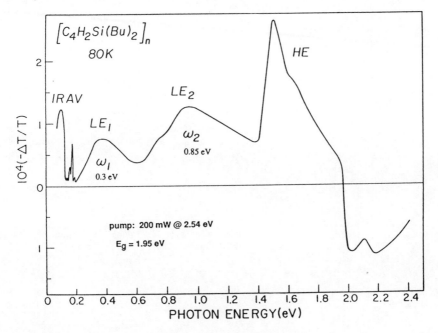

The potential of PDES's as optical switches is excellent since we find that the photobleaching recovery of the violet band at ~ 2 eV is less than one picosecond. In fact, after a 75 femtosecond pulse the 1/e decay time is about 320 femtoseconds–the fastest recovery time for an organic polymer ever reported. We are just beginning to examine the nonlinear optical properties and have measured one $\chi^{(3)}$ value by degenerate four-wave mixing. The $\chi^{(3)}$ value of 3×10^{-9} esu is to our knowledge the largest ever obtained for an unoriented organic polymer, but of equal importance is that the intensity of I_4, the phase-conjugated beams, drops to zero in less than 2 picoseconds. We attribute this important lack of tailing to energy dissipation by the long alkyl substituents on silicon.

So, in summary, what we have is a new polymer system which has all of the desirable properties of polyacetylene and substituted polyacetylenes and none of the undesirable properties. The PDES's are soluble, air-stable, dope to semiconductor conductivities and exhibit extremely exciting NLO properties.

I hope that this *potpourri* of wide ranging chemistry has served to convince you that much exciting and novel chemistry exists in systems containing silicon bound to sp^2- or sp-carbons.

Acknowledgments. Our gas-phase mechanistic studies were generously funded by the National Science Foundation. Our polymer research has been funded by the Director for Energy Research, Office of Basic Energy Sciences, United States Department of Energy. The polymer research has been truly a collaborative effort involving Dr. Sina Ijadi-Maghsoodi, Yi Pang and Professor Joseph Shinar of the Ames Laboratory/Iowa State University; Professor Zeev Vardeny of the University of Utah; and Drs. Stelian Grigoras and Bhukan Parbhoo of Dow Corning Corporation.

Silicon Ceramics with a Dash of Boron

Dietmar Seyferth, Herbert Plenio, William S. Rees, Jr., and Klaus Büchner

DEPARTMENT OF CHEMISTRY, MASSACHUSETTS INSTITUTE OF TECHNOLOGY, CAMBRIDGE, MA 02139, USA

1. INTRODUCTION

The pyrolysis of inorganic and organometallic polymers as a route to ceramics is a new area of polymer chemistry that has received much attention since the pioneering work of Yajima and his coworkers[1] and of Verbeek and Winter[2] in the mid-70's. Reviews that deal with preceramic polymers broadly[3] or in the context of silicon-containing ceramics[4] are available. Our own previous research has been concerned mainly with the development of useful polymers whose pyrolysis gives silicon-containing ceramics such as silicon carbide, silicon nitride, silicon carbonitride and silicon oxynitride.[4] An approach that has shown considerable promise in terms of useful applications involves the preparation of a cross-linked but still soluble polysilazane. The first step of this procedure, the ammonolysis of methyldichlorosilane, gives a silazane product that is a mixture mainly of cyclic oligomers, cyclo-$[CH_3Si(H)NH]_n$ ($n=3,4,5\cdots$). In order to obtain a useful preceramic material, this ammonolysis product must be converted to material of higher molecular weight. We effected such polymerization by treatment of the liquid mixture of cyclosilazanes with a catalytic quantity of a base such as KH (ca. 1 mol %), generally in an ethereal solvent such as diethyl ether or tetrahydrofuran.[5] Hydrogen was evolved, and a "living" polymer, one that contained potassium amide functions when KH was the catalyst used, was formed. Reaction of the "living" polymer with an electrophile such as CH_3I or a chlorosilane then gave the neutral polymer whose constitution, in terms of constituent groupings, was shown by proton NMR spectroscopy to be $[(CH_3Si(H)NH)_a(CH_3SiN)_b(CH_3Si(H)NCH_3)_c]_n$ ($a + b + c = 1$; $c \sim 0.01$, the KH catalyst concentration, a ~ 0.4, b ~ 0.6) when CH_3I was used. The treatment of the cyclo-$[CH_3Si(H)NH]_n$ mixture with a catalytic quantity of KH was based on the report of a catalytic cyclodisilazane synthesis in a Monsanto patent[6] (eq. 1). Our initial idea was that application of this catalytic process to the cyclo-$[CH_3Si(H)NH]_n$ mixture would lead to the formation of

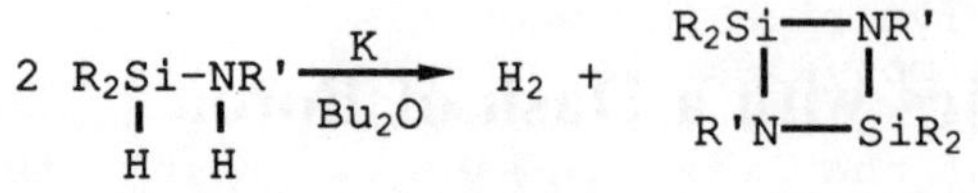

$$2\ R_2Si\text{-}NR' \xrightarrow[Bu_2O]{K} H_2 + \begin{array}{c} R_2Si\text{---}NR' \\ |\quad\quad | \\ R'N\text{---}SiR_2 \end{array} \quad (1)$$
    H H

a sheet polymer. Taking the 8-membered cyclosilazane as an example, the first step in the process would be the fusion of two such rings (eq. 2). The product formed contains six sets of Si(H)-N(H) units, and if all of these reacted in this sense, a sheet polymer would result.

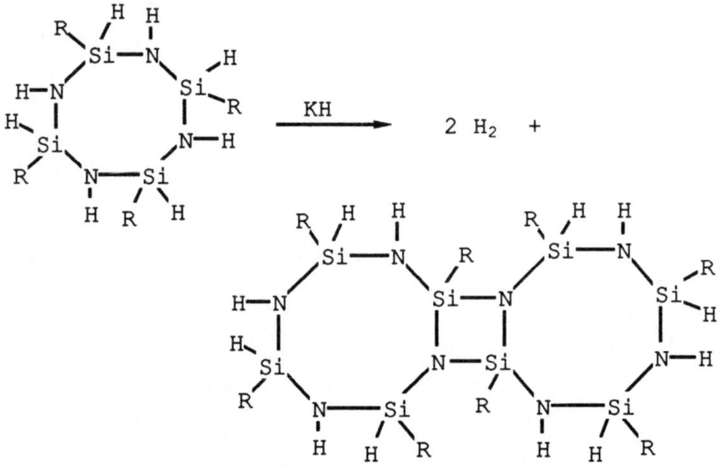

$$\text{[8-membered cyclosilazane]} \xrightarrow{KH} 2\ H_2 + \text{[fused bicyclic silazane]} \quad (2)$$

However, later studies have indicated that ring fusion polymerization via cyclodisilazane formation is not the principal polymerization process since ^{29}Si NMR studies of the products of the KH-catalyzed cyclo-$[CH_3Si(H)NH]_n$ polymerization have failed to provide evidence for the presence of Si_2N_2 rings. Instead the polymers produced appear to be composed in the main of larger silazane rings. It may be that the KH-catalyzed reaction occurs via an intermediate that contains an Si=N bond, **1** in the case of the 8-membered cyclosilazane, and that this then

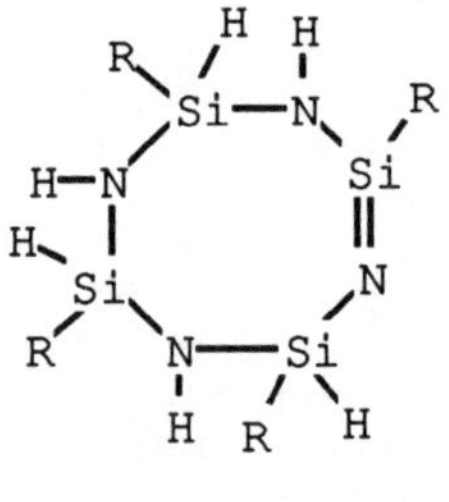

1 (R = CH_3)

inserts into an Si-N, Si-H and/or N-H bond of another cyclosilazane molecule. (The gas phase generation of >Si=N- intermediates and the study of their reactions by Sommer and his coworkers[7] are noted in this connection.)

Regardless of polymerization mechanism and the detailed structure of the polysilazane produced in the KH-catalyzed process, the polymer is highly cross-linked, about 60 mol % of the Si(H)-N(H) units having been lost. It is a soluble solid whose pyrolysis to 1000°C in an argon stream results in a _black_ ceramic residue (an amorphous silicon carbonitride) in about 85% yield. Pyrolysis to 1000°C in a stream of ammonia, on the other hand, gives a _white_ ceramic residue of silicon nitride in high yield, essentially all of the methyl groups on silicon having been lost during the middle stages of the pyrolysis. The polysilazane produced in the KH-catalyzed process serves excellently as a binder for SiC and Si_3N_4 powder in the fabrication of ceramic parts, serves with advantage as the main component in coating formulations, and may be dry-spun to give fibers whose pyrolysis in argon results in silicon carbonitride ceramic fibers.

In recent research we have been interested in composites that contain a silicon ceramic and a ceramic containing another element. It is known that such composites often have better physical and mechanical properties than do the pure components of the composite, but there are few examples of polymer pyrolysis having been used to prepare such composites. We report here concerning a novel polymer system whose pyrolysis gives ceramic composites that contain silicon and boron.

2. SILAZANE/BORAZINE POLYMERIC PRECURSORS FOR SILICON NITRIDE/BORON NITRIDE COMPOSITES

Boron nitride and silicon nitride by themselves are very useful ceramic materials. Thus boron nitride, BN, has high thermal stability (mp 2730°C), excellent high temperature strength, superior thermal shock resistance and it is chemically inert and resistant to molten metals.[8] Furthermore, it is a high resistance electrical insulator and it exhibits high thermal conductivity. Silicon nitride[9] also has very attractive properties, its low thermal coefficient of expansion and superior high temperature strength combining to make it one of the most thermally shock resistant ceramics known. To this can be added its high resistance to corrosion and its high temperature stability.

BN/Si_3N_4 blends are of interest. They have been prepared by chemical vapor deposition using $SiCl_4/B_2H_6/NH_3/H_2$[10] and $SiH_4/B_2H_6/N_2$[11] as source gases. It is noteworthy that the fracture toughness of the deposited thin films improved dramatically at Si contents >10 atom %, while their hardness gradually decreased as the Si content increased from 0 to 48 atom %. Mazdiyasni and Ruh[12] fabricated high density Si_3N_4 (+ 6% CeO_2) composites with 5 to 50% BN by hot pressing. These composites, it was found, had better thermal shock resistance than a sample of commercial hot-pressed Si_3N_4. In another approach,

Japanese workers[13] treated Si_3N_4 powder at high temperature with a gas containing BCl_3 and NH_3 to give BN/Si_3N_4 blends that could be used to make ceramic parts having high relative density and bending strength.

The entry to our synthesis of a polymer whose pyrolysis under appropriate conditions would give a Si_3N_4/BN composite was provided by a 1961 paper by Nöth that reported reactions of diborane(6) in diethyl ether with various silylamines and silazanes (eq. 3-6).[14]

$$Me_2HSiNMe_2 + \tfrac{1}{2} B_2H_6 \xrightarrow{Et_2O} Me_2HSiNMe_2 \cdot BH_3 \quad (3)$$

$$Me_2HSiNMe_2 \cdot BH_3 \longrightarrow Me_2SiH_2 + Me_2NBH_2 \quad (4)$$

$$(Me_3Si)_2NH + \tfrac{1}{2} B_2H_6 \xrightarrow{Et_2O} (Me_3Si)_2NH \cdot BH_3 \quad (5)$$

$$(Me_3Si)_2NH \cdot BH_3 \xrightarrow{120°C} H_2 + Me_3SiH + [Me_3SiNBH]_3 \quad (6)$$

Of particular interest was the observation that the $(Me_3Si)_2NH \cdot BH_3$ adduct formed a silyl-substituted borazine, **2**, when heated to 120°C, most likely by way of an intermediate borazane (Scheme 1). If this reaction could

SCHEME 1

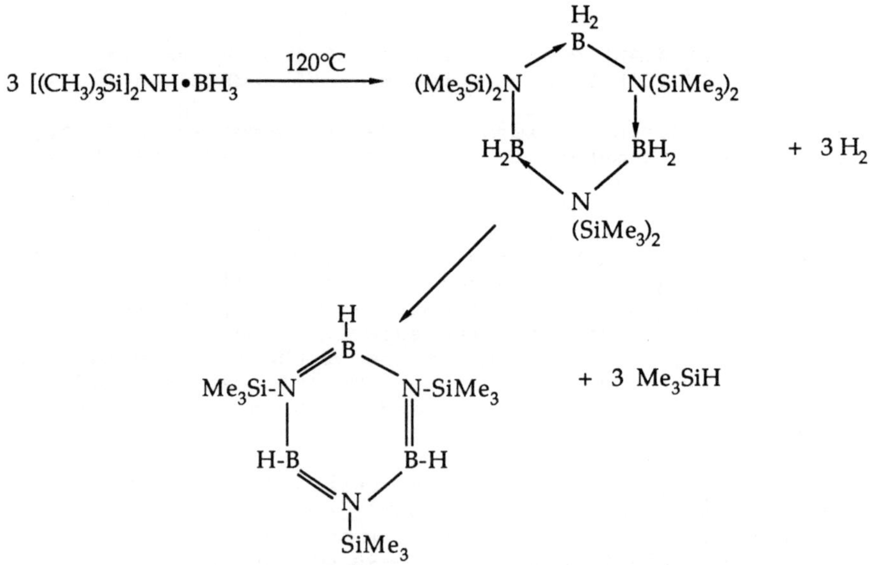

be applied to the mixture of cyclosilazanes obtained in the ammonolysis of methyldichlorosilane, then a network polymer involving borazine rings and silazane units should result since the initially formed borazine contains silazanyl side-chains on the boron atoms which can react further with $H_3B \cdot S(CH_3)_2$ (Scheme 2).

SCHEME 2

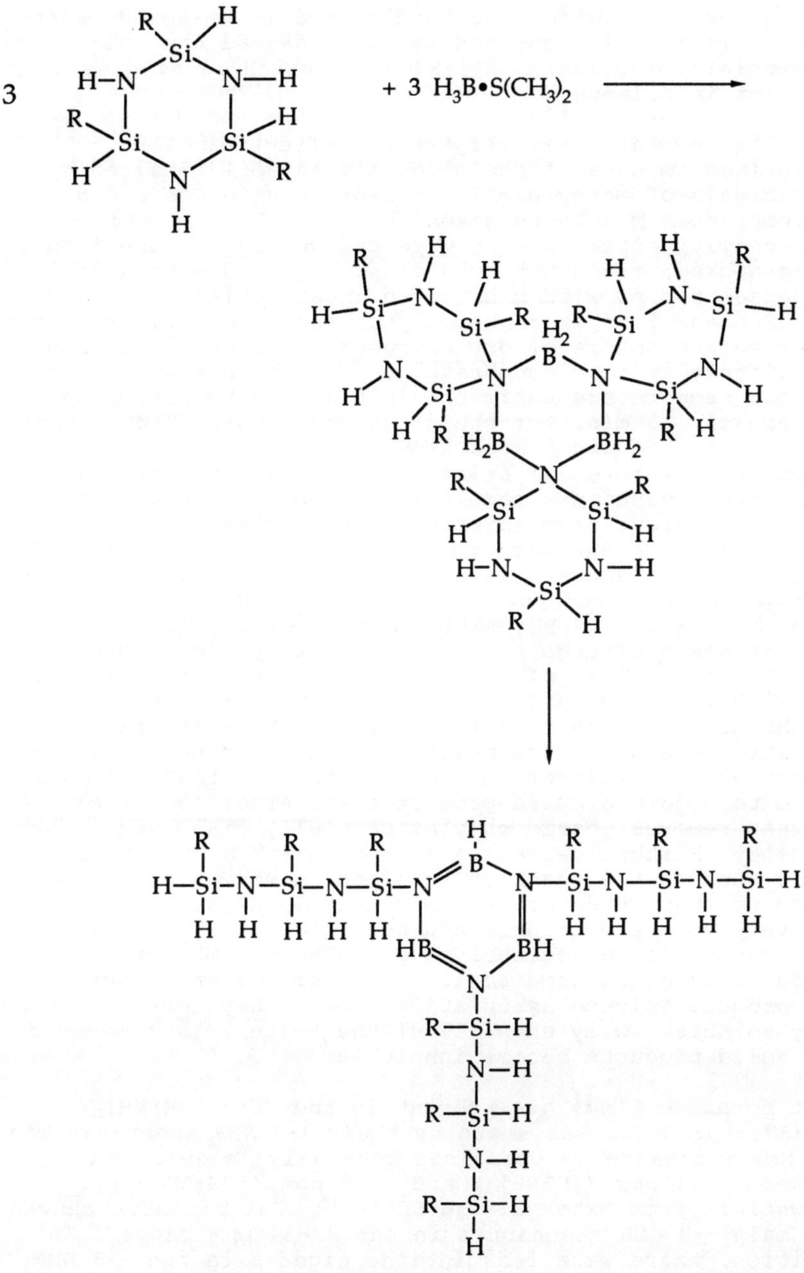

3 [cyclotrisilazane] + 3 H$_3$B•S(CH$_3$)$_2$ ⟶

↓

In principle, the reaction of interest would be that of cyclo-$[CH_3Si(H)NH]_n$ with diborane, B_2H_6. In practise, the latter is dangerous to handle on a large scale, even when it is in solution, and we used instead the safer, commercially available, $H_3B \cdot S(CH_3)_2$, which is sold as a 2M solution in toluene.

The reactions are very easily effected. The prescribed amounts of the oligosilazane oil obtained by ammonolysis of methyldichlorosilane in a suitable organic solvent, the 2M toluene solution of $H_3B \cdot S(CH_3)_2$ and (optionally) extra toluene were charged by syringe into a three-necked, round-bottomed flask equipped with a reflux condenser topped with a nitrogen inlet/outlet tube leading to a Schlenk line, a rubber septum and a magnetic stir-bar under an atmosphere of dry nitrogen. Gas evolution began immediately at room temperature. The mixture was stirred at room temperature until the initial gas evolution had ceased (ca. 15 min. for the 30-60 mmol $[CH_3Si(H)NH]_n$ scale used). It then was heated slowly (ca. 30 min.) to the reflux temperature and stirred at reflux for 1 hr. The resulting product was completely soluble in toluene and did not separate from solution as the released $(CH_3)_2S$ and the toluene solvent were removed at reduced pressure. The products are viscous oils or solids, depending on how much $H_3B \cdot S(CH_3)_2$ was used per molar unit of $CH_3Si(H)NH$. Thus, when the $CH_3Si(H)NH/BH_3$ molar ratio used was 20, the product was a viscous oil with a cryoscopic molecular weight (in benzene) of 390. When the ratio used was 10, the highly viscous oil produced had an average molecular weight of 520. Use of a ratio of 4 in this reaction resulted in a soft, low-melting solid, average molecular weight 800. A further decrease in the CH_3SiHNH/BH_3 molar ratio to 3 gave a solid product that, after complete solvent removal, could only be partially dissolved in hot toluene. Further decreases in the CH_3SiHNH/BH_3 molar ratio used in the reaction resulted in formation of solid products that were very soluble in toluene as formed. However, once the toluene had been completely removed, they could not be redissolved in toluene. When these products retained some small amount of toluene, such that the product/toluene syrup still flows, they continue to be very soluble. Only when all of the toluene is removed do the solid products become insoluble.

That borazine rings are present in the $[CH_3Si(H)NH]_n/H_3B \cdot SMe_2$ products was shown by their ^{11}B NMR spectra. The ^{11}B NMR resonance in borazines generally is observed between 29.1 ppm ($[HBNH]_3$) and 35.8 ppm ($[CH_3BNCH_3]_3$) (downfield from external $F_3B \cdot OEt_2$).[15] Our products showed the main ^{11}B NMR resonances in the 32-34 ppm range. In addition, there were less intense signals in the ^{11}B NMR spectra around 26 ppm, indicative of the presence of some species with BN_3 connectivity.

CH$_3$SiHNH/BH$_3$ molar ratios ranging between 20 and 1 were used in these experiments. As the ratio decreased, the yield of ceramic produced on pyrolysis in a stream of argon to 1000°C increased from 52% for a ratio of 20 to 91% for a ratio of 1. The ceramic residues were amorphous, black in color and contained boron, silicon, carbon and nitrogen, thus were borosilicon carbonitrides.

Pyrolysis of these [CH$_3$Si(H)NH]$_n$/H$_3$B•S(CH$_3$)$_2$ products in a stream of ammonia to 1000°C left a white ceramic residue that in most cases contained less than 0.5% by weight of carbon. Thus the pyrolysis product was a borosilicon nitride. As expected, as the CH$_3$Si(H)NH/BH$_3$ ratio that was used was decreased, the boron content of the ceramic product increased: from 1.72% when this ratio was 10 (equivalent to 2.5 Si$_3$N$_4$ + 1.0 BN) to 11.42% for a CH$_3$Si(H)NH/BH$_3$ ratio of 1 (equivalent to 0.42 Si$_3$N$_4$ + 1.0 BN). These are **nominal** compositions since the ceramic obtained is amorphous. No separation of phases into Si$_3$N$_4$ and BN had occurred. In fact, experiments in which such ceramic samples were heated up to 1700°C showed no crystallization of Si$_3$N$_4$; only broad features attributable to turbostratic BN were observed in the **X**-ray powder pattern. The amorphous ceramic products of the pyrolysis of our polysilazane alone in a stream of ammonia to 1000°C crystallize to form α-Si$_3$N$_4$ between 1450 and 1500°C, so it appears that BN retards the crystallization of Si$_3$N$_4$.

The simple chemistry described here also can be applied to mixed systems such as the coammonolysis products of CH$_3$SiHCl$_2$ and CH$_3$SiCl$_3$ and of CH$_3$SiHCl$_2$ and (CH$_3$)$_2$SiCl$_2$, as well as siloxazanes, such as the ammonolysis product of O[SiCH$_3$(H)Cl]$_2$. We note, however, that the reaction of H$_3$B•S(CH$_3$)$_2$ with cyclo-[(CH$_3$)$_2$SiNH]$_n$ (n=3,4) alone was much less facile, requiring long (ca. 72 hr) heating at reflux in toluene solution. Furthermore, the white solid product that was obtained in a reaction where the Si/B ratio was 2 left only a 2% yield of residual solid on pyrolysis to 1000°C in a stream of argon.

The use of [CH$_3$Si(H)NH]$_n$ is especially favorable for two reasons: (1) its reaction with H$_3$B•S(CH$_3$)$_2$ is facile and (2) if, after the reaction with the borane adduct, there is unreacted, latent Si(H)-N(H) functionality left, further elimination of hydrogen can take place during the initial stages of pyrolysis, leading to even further cross-linking, hence very high ceramic yields. The Si-H bond appears to be of critical importance in the chemistry we describe here.

To date, not much research on applications of these new polymers has been carried out. They serve well as binders for ceramic powders. In another experiment, the polymer from such a 2:1 Si/B preparation was pressed into a rectangular bar weighing 3.42 g (3.84 x 1.32 x 0.69 cm; density 0.98 g cm^{-3}). Pyrolysis to 1000°C in a stream of argon gave a black bar of the same shape (3.00 x 1.03 x

0.55 cm; 50% reduction in volume) that weighed 3.0 g and had a density of 1.76 cm^{-3}. From a toluene-containing thick syrup of the same polymer long thin fibers could be hand-drawn. When these were fired in a stream of ammonia to 1000°C, white ceramic fibers were obtained.

3. FROM BORON NITRIDE PRECURSORS TO AN o-SILABORANE

In research directed toward the preparation of polymers whose pyrolysis will give boron nitride, we prepared decaborane/diamine polymers of type **3** by the reaction of decaborane(14) with various diamines (eq. 7).[16]

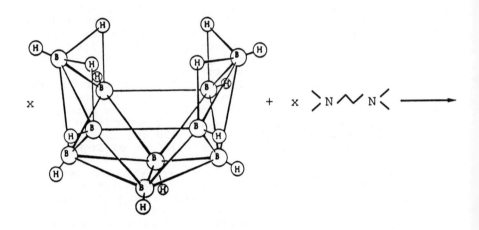

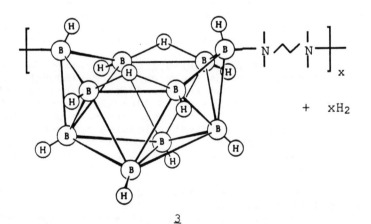

3

The $[B_{10}H_{12} \cdot \text{diamine}]_x$ polymers thus prepared are soluble in polar organic solvents such as dimethylformamide, dimethyl sulfoxide, hexamethylphosphoric triamide and acetone, but not in hydrocarbon solvents such as benzene, toluene or hexane. Their molecular weights appear to be in the 300,000+ range. Such $[B_{10}H_{12} \cdot \text{diamine}]_x$ polymers were prepared using $H_2NCH_2CH_2NH_2$, $(CH_3)_2NCH_2CH_2N(CH_3)_2$, $(CH_3)_2NCH_2CH_2NH_2$, and triethylenediamine. In the case of the $[B_{10}H_{12} \cdot H_2NCH_2CH_2NH_2]_x$ polymer, pyrolysis under argon to 1500°C (10°C/min) left a brown, crystalline ceramic residue, whose examination by powder $\underline{X}$-ray diffraction showed the presence of B_4C and BN. Examination of both the amorphous and crystalline pyrolysis products by diffuse reflection infrared Fourier transform spectroscopy showed absorptions due to B-C and B-N bonds. Similarly, pyrolysis of $\{B_{10}H_{12} \cdot (CH_3)_2NCH_2CH_2N(CH_3)_2\}_x$ gave an amorphous ceramic product of nominal composition $(B_4C)_1(BN)_1(C)_{0.53}$ (80% yield) at 1000°C and $(B_4C)_1(BN)_1(C)_{0.17}$ at 1500°C. High ceramic yields were observed in the pyrolysis to 1000°C under argon of the other diamine systems.

Usually, linear, noncross-linked polymers are not useful preceramic precursors, but in the case of the $[B_{10}H_{12} \cdot \text{diamine}]_x$ polymers a unique pyrolysis mechanism is operative. At temperatures above 150°C proton transfer from boron (the bridging H atoms) to nitrogen takes place, converting the covalent polymer to a diammonium salt of the $[B_{10}H_{10}]^{2-}$ anion. Such salts are not volatile, hence remain in the hot zone and are converted to a ceramic in high yield.

Ceramic monoliths could be produced by pyrolysis (under argon) of rectangular polymer bars. These $[B_{10}H_{12} \cdot \text{diamine}]_x$ polymers can serve as good to excellent binders for boron carbide powder.

The $[B_{10}H_{12} \cdot \text{diamine}]_x$ polymers also serve as boron nitride precursors. Their pyrolysis to 1000°C in a stream of ammonia (rather than argon) results in displacement of the diamine linker and reaction of ammonia with the B_{10} cage, leaving a <u>white</u> ceramic residue. The latter was spectroscopically and analytically indistinguishable from authentic boron nitride. For instance, the pyrolysis of $[B_{10}H_{12} \cdot N(CH_2CH_2)_3N]_x$ in a stream of ammonia gave a powdery ceramic residue in 70% yield which contained B and N in 0.99:1.0 ratio and only a slight amount of carbon. In a manner like that described above, a white ceramic bar was produced by pyrolysis (to 1000°C under ammonia) of a rectangular bar of BN powder/polymer binder that was of excellent strength and that exhibited shape retention in all dimensions.

Unfortunately, the $[B_{10}H_{12} \cdot \text{diamine}]_x$ polymers were not suitable for the preparation of ceramic fibers. They melted at higher temperatures with decomposition, so melt-spinning was not an option. In order to obtain [$B_{10}H_{12}$•diamine]$_x$ polymers with lower melting points, we replaced the organic diamines with dimethylbis(dimethylamino)silane. The latter reacted with $B_{10}H_{14}$ in refluxing benzene solution to give a pale yellow, resinous material that was soluble in polar organic solvents and melted in the range 80-110°C. Analytical and NMR spectroscopic data indicated that the moisture-sensitive product was the desired [$B_{10}H_{12}$•(CH_3)$_2$NSi(CH_3)$_2$N(CH_3)$_2$]$_x$. Pyrolysis of this polymer to 1000°C in a stream of argon left an amorphous black residue in 85% yield. Its analysis (18.0% C, 11.8% N, 6.4% Si, 57.9% B) established the formation of a borosilicon carbonitride, boron-rich, rather than silicon rich as was the case for the [CH_3Si(H)NH]$_n$/ H_3B•S(CH_3)$_2$ products. Pyrolysis to 1000°C in a stream of ammonia gave a white ceramic residue in 89% yield whose analysis (40% B, 55% N, 2% Si, 0.6% C) was fairly close to that required for BN (43.6% B, 56.4% N). Obviously, not all of the (CH_3)$_2$Si[N(CH_3)$_2$]$_2$ linker was displaced by the high temperature reaction with ammonia.

Long fibers could be drawn from a melt of this polymer. A moist air cure, followed by pyrolysis in a stream of ammonia to 1000°C gave white ceramic fibers. Since the T_g of the polymer was below room temperature, multifilament spinning proved to be impractical. In a search for a more suitable polymer (which is still in progress), we have varied the substituents on silicon in the bis(dimethylamino)silane linker. The results obtained when CH_3(H)Si[N(CH_3)$_2$]$_2$ was used are noteworthy.[17]

The reaction of this silane with $B_{10}H_{14}$ in refluxing benzene or toluene solution resulted in the formation of a mixture of white solid products. The major product was identified as $B_{10}H_{12}$•2(CH_3)$_2$NH (58% yield). The product formed in lesser (~15%) yield could be crystallized from benzene or sublimed at 90°C at 0.01 mm Hg. Very slow crystallization from dilute benzene gave diffraction-quality crystals which an X-ray diffraction study showed to be the 1:1 benzene solvate of 1,2-dimethyl-1,2-disila-closo-dodecaborane(12) (or 1,2-dimethyl-o-silaborane, in analogy to carborane nomenclature) whose structure is shown in Fig. 1. This is the first silicon analog of an ortho-carborane. Like the o-carboranes, it has a slightly distorted icosahedral structure. The Si-Si bond distance of 2.308(2)Å is slightly less than that of normal Si-Si single bonds (~2.33-2.34Å)[18] but significantly longer than those of known Si-Si double bonds.[19] In comparison, the C-C bond distances in o-carboranes (1.655Å, the mean value found in X-ray and gas phase electron diffraction studies[20]) are longer than the normal C(sp^3)-C(sp^3) distance of 1.54Å. This is understandable in terms of the larger size of a silicon atom (covalent radius 1.18Å), compared with a carbon atom (covalent radius 0.77Å). The Si-B bond distances in 1 [2.017(3), 2.018(3), 2.113(4) and 2.116(3)Å] are very close to the sum of the covalent radii of Si and B, 2.07Å (using 0.89Å as the average boron

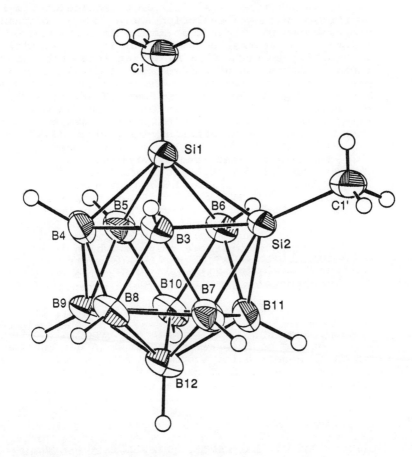

Fig. 1 ORTEP representation of 1,2-dimethyl-1,2-disila-_closo_-dodecaborane(12), drawn with 35% probability ellipsoids. Hydrogen atoms have been given arbitrary thermal parameters for clarity.

covalent radius in o-carboranes). These distances may be compared to those of the Si-B bonds in commo-3,3'-Si(3,1,2-SiC$_2$B$_9$H$_{11}$)$_2$, which contains a single silicon atom bonded within a 12-atom boron-containing cluster: 2.051(1) and 2.14(1)Å.[21]

1,2-Dimethyl-o-silaborane is less stable than the o-carboranes. It melts at 201-203°C (sealed capillary) and decomposes with gas evolution above 230°C. It remains undecomposed when heated at reflux in wet tetrahydrofuran solution for several days. It was unaffected when treated with an excess of CH$_3$CO$_2$H in THF at room temperature and remained largely undecomposed when this solution was heated at reflux for 14 hr. It was found to be stable toward AlCl$_3$ in refluxing benzene. It reacts with and is degraded by alcoholic KOH in THF at room temperature within minutes and it decolorizes a solution of bromine in carbon tetrachloride after several hours at reflux.

The o-carboranes are derivatives of acetylenes (eq. 8), and thus 1,2-dimethyl-o-silaborane may be regarded

$$B_{10}H_{12} \cdot 2Et_2S + RC \equiv CR' \longrightarrow 2Et_2S + \underset{B_{10}H_{10}}{\overset{R \diagdown \diagup R'}{C - C}} \quad (8)$$

as a _formal_ derivative of a disilyne, CH$_3$Si≡SiCH$_3$, although its formation in the reaction of B$_{10}$H$_{14}$ with CH$_3$(H)Si[N(CH$_3$)$_2$]$_2$ need not involve such an intermediate.

Acknowledgements. The authors are grateful to the Air Force Office of Scientific Research (AFSC) and the Office of Naval Research for support of this work. H. Plenio and K. Büchner thank the Alexander von Humboldt Foundation for the award of Feodor Lynen Fellowships.

REFERENCES

1. S. Yajima, Am. Ceram. Soc. Bull., 1983, 62, 893.
2. (a) W. Verbeek, U.S. Patent 3,853,567 (1974).
   (b) G. Winter and W. Verbeek, U.S. Patent 3,892,583 (1975).
3. (a) K.J. Wynne and R.W. Rice, Ann. Rev. Mater. Sci., 1984, 14, 297.
   (b) R.W. Rice, Am. Ceram. Soc. Bull., 1983, 62, 889.
4. D. Seyferth, in "Silicon-Based Polymer Science. A Comprehensive Resource", (Advances in Chemistry Series 224), J.M. Zeigler and F.W.G. Fearon, editors, American Chemical Society, Washington, DC, 1990, Chapter 31, p. 565.
5. D. Seyferth and G.H. Wiseman, J. Am. Ceram. Soc., 1984, 67, C-132; U.S. Patent 4,482,669 (1984).

6. Monsanto Co., Neth. Appl. 6,507,996 (1965); Chem. Abstr., 1966, 64, 19677d.
7. (a) C.M. Golino, R.D. Bush and L.H. Sommer, J. Am. Chem. Soc., 1974, 96, 614.
   (b) D.R. Parker and L.H. Sommer, J. Am. Chem. Soc., 1976, 98, 618.
   (c) D.R. Parker and L.H. Sommer, J. Organomet. Chem., 1976, 110, C1.
   (d) M. Elsheikh, N.R. Pearson and L.H. Sommer, J. Am. Chem. Soc., 1979, 101, 2491.
   (e) M. Elsheikh and L.H. Sommer, J. Organomet. Chem., 1980, 186, 301.
8. (a) "Gmelin Handbook of Inorganic Chemistry", 8th Edition, "Boron Compounds", Part 1, Springer-Verlag, Berlin, 1974, pp. 1-86.
   (b) R.T. Paine and C.K. Narula, Chem. Rev., 1990, 90, 73.
9. D.R. Messier and W.J. Croft, in "Preparation and Properties of Solid State Materials", Vol. 7, W.R. Wilcox, editor, Dekker, New York, 1982, pp. 132-212.
10. T. Hirai, T. Goto and T. Sakai, in "Emergent Process Methods for High Technology Ceramics", Material Science Research Vol. 17, R.F. Davis, H. Palmour III and R.L. Porter, editors, Plenum, New York, 1984, 347.
11. T. Nakahigashi, Y. Setoguchi, H. Kirimura, K. Ogata and E. Kamijo, Shinku, 1988, 31, 789; Chem. Abstr., 110, 81199m.
12. K.S. Mazdiyasni and R. Ruh, J. Am. Ceram. Soc., 1981, 64, 415.
13. Y. Nakamura and M. Nakajima, Jpn Kokai Tokkyo Koho JP 01 83,506 (89 83,506) and JP 01 83,507 (89 83,507) (1989), Chem. Abstr., 1989, 111, 44274g and 44275h.
14. H. Nöth, Z. Naturforsch. B, 1961, 16, 618.
15. H. Nöth and H. Vahrenkamp, Chem. Ber., 1966, 99, 1049.
16. (a) W.S. Rees, Jr. and D. Seyferth, J. Am. Ceram. Soc., 1988, 71, C-194; U.S. Patent 4,871,826 (1989).
    (b) W.S. Rees, Jr. and D. Seyferth, Ceram. Eng. Sci. Proc., 1989, 10, 837.
17. D. Seyferth, K. Büchner, W.S. Rees, Jr. and W.M. Davis, Angew. Chem., in press.
18. (a) B. Beagley, J.J. Monaghan and T.G. Hewis, J. Mol. Struct., 1971, 8, 401.
    (b) H.L. Carrell and J. Donahue, Acta Cryst., 1972, B28, 1566.
19. (a) M.J. Fink, M.J. Michalczyk, K.J. Haller, J. Michl and R. West, Organometallics, 1984, 3, 793.
    (b) S. Masamune, S. Murakami, J.T. Snow, H. Tobita and D.J. Williams, Organometallics, 1984, 3, 3330.
20. V.S. Matryukov, L.V. Vilkov and O.V. Dorofeeva, J. Mol. Struct., 1975, 24, 217.
21. W.S. Rees, Jr., D.M. Schubert, C.B. Knobler and M.F. Hawthorne, J. Am. Chem. Soc., 1986, 108, 5369.

Organosilicon Preceramic Polymer Technology

William H. Atwell, Gary T. Burns, and Gregg A. Zank
CENTRAL RESEARCH AND DEVELOPMENT, DOW CORNING CORPORATION, MIDLAND, MICHIGAN, USA

1. INTRODUCTION

Worldwide interest has emerged in the use of chemical precursors to produce improved, reliable ceramic materials. Interest has been especially active in the area of silicon-based systems[1]. The early work of Yajima and co-workers sparked an explosion of preceramic organosilicon polymer studies.

For the past fifteen years, Dow Corning researchers have actively pursued the development of this area of technology for the preparation of a variety of ceramic forms. Our program has focused on three major areas:

1. Ceramic fibers (Si_3N_4 and SiC).
2. Ceramic matrix composites (fiber and whisker reinforced).
3. Monolithic ceramics (primarily SiC).

We have previously described the preparation of silicon carbonitride ceramic fibers from polysilazane polymers[2]. This review will summarize our recent results on the preparation of dense silicon carbide monolithic components using organosilicon polymers as binders for crystalline silicon carbide powders.

2. POLYMER BINDER STUDIES

Objectives and Requirements

Numerous studies have been reported on the use of preceramic polymers for the preparation of ceramic materials[1]. Generally, these studies have not led to the preparation of dense, crystalline ceramic forms. Thus, the objective chosen for our studies was as follows: "Develop a preceramic organosilicon polymer binder for the fabrication and pressureless sintering of dense ($\geq 98\%$ theoretical density) beta-silicon carbide."

Silicon carbide was chosen for our studies because of its emerging position as a premium high performance ceramic material. Pressureless sintering was chosen as the densification method because of its low cost fabrication potential. The general approach chosen is given in Equation 1.

$$\begin{array}{c}\text{Organosilicon}\\ \text{Polymer}\\ +\\ \beta\text{-SiC Powder}\end{array} \xrightarrow{\text{Pressure}} \begin{array}{c}\text{Molded}\\ \text{Green}\\ \text{Body}\end{array} \xrightarrow{\text{Heat}} \begin{array}{c}\text{Silicon}\\ \text{Carbide}\\ \text{Body}\end{array} \quad (1)$$

Previous workers[3,4] have demonstrated with hydrocarbon binders that effective pressureless sintering of silicon carbide requires the following:

1. High purity, submicron silicon carbide powder.
2. Use of sintering aids (eg, 0.3-0.5 wt% boron).
3. Use of 1-2 wt% excess carbon (generally carbonaceous resin binders).
4. Temperatures >2000°C (inert atmosphere).

In state-of-the-art ceramic processing of silicon carbide, the binder provides particle lubrication for processing and must be removed prior to sintering.

In the case of organosilicon preceramic polymers, we concluded that these binders should provide the following: 1) silicon carbide to the sintered body; 2) excess carbon (1-2 wt%) required for sintering; 3) rheological control for improved fabrication; and 4) crosslinking to improve green body strength and dimensional stability.

Our studies with preceramic polymers demonstrated that effective densification is achieved under pressureless sintering conditions if 1.5 wt% of excess carbon is used (see Figure 1). Thus, to produce an effective binder it is important to control both char yield and composition for specific levels of polymer. The following procedures and calculations have been developed for this purpose.

1. Carry out the polymer pyrolysis to 1800°C (argon).
2. Determine the char yield and elemental (Si, C) composition.
3. For 100 g of polymer, calculate wt SiC and wt C_{excess} in the char (use rule-of-mixtures) using Equations 2 and 3.

Wt SiC = [wt polymer x char yield x wt% Si in char]
     + [wt polymer x char yield x wt% Si in char
        x 12/28] (2)

Wt C_{excess} = [wt polymer x char yield] - wt SiC (3)

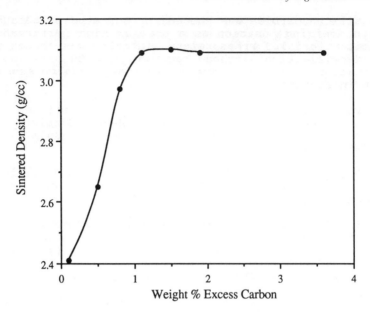

Figure 1. Sintered density of silicon carbide as a function of wt% excess carbon.

With a knowledge of the char yield and the wt SiC and wt C_{excess}, the appropriate levels of polymer binder and crystalline silicon carbide powder can be determined as follows:

4. For 100 g of SiC ceramic with 0.5 wt% boron additive and 1.5 wt% C_{excess},

$$\text{Polymer binder required} = \text{wt } P_b = \frac{1.5}{\text{wt } C_{excess}} \times 100 \quad (4)$$

$$\text{SiC powder required} = \text{wt } SiC_p = [100 \times 0.5 - (\text{wt } P_b) \times (\text{char yield})] \quad (5)$$

The use of these procedures and calculations will be illustrated later in the discussion.

Polymer Approaches

Several types of polycarbosilane[5,6] and polysilane[7,8] polymers have been reported to pyrolyze to mixtures of silicon carbide and excess carbon (Figure 2). We decided to examine the use of polycarbosilane (PCS) as a binder for beta-silicon carbide. Table 1 shows the results of the 1800°C PCS char study. As noted previously, our studies demonstrated that effective densification is achieved under pressureless sintering conditions if the appropriate polymer level was used. The amount of PCS to provide 1.5 wt% excess carbon was calculated using Equations 2, 3 and 4 (see Table 1). The required amount of SiC powder can be calculated using Equation 5.

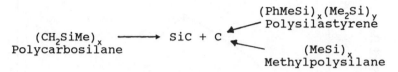

Figure 2. Known polymer approaches to silicon carbide.

The optimum PCS level was also experimentally determined by following the density of small test samples. The results are shown graphically in Figure 3. Note the poor correlation between the actual and calculated amount of polymer required for effective densification. Two main problems were encountered in the use of PCS as a binder for the preparation of high density silicon carbide. First, the polymer affords only 11.5 wt% excess C upon pyrolysis requiring binder levels of greater than 20% for effective sintering. Second, the lack of deep section cure for this polymer results in bloating and delamination during sintering and poor densification results at the higher polymer levels.

Following screening with PCS it became obvious that improved polymers that had respectable char yields (ca. greater than 40%), controllable carbon levels in their ceramic char and curability were required.

Table 1. Polymer Char Results

Polymer Type	1800°C Char Yield %	wt% Si	wt% C	Calculated wt P_b at 1.5 wt% C_{excess}
Polycarbosilane (PCS)	55.8	62.0	37.3	23.5 g
Polysilazane A[a] (SCBZ)	52.2	40.1	57.7	7.3 g
Polysilazane B[b] (SCBZ)	50.6	40.6	55.0	18.7 g[e]
Polysiloxane A[c]	42.5	47.5	50.4	11.0 g
Polysiloxane B[d]	43.0	55.0	44.8	22.0 g[e]

[a] $(C_3H_6SiNH)_{0.5}(Ph_2SiNH)_{0.25}(PhNH_{1.5})_{0.25}$
[b] $(C_3H_6SiNH)_{0.55}(Ph_2SiNH)_{0.15}(MeSiNH_{1.5})_{0.35}$
[c] $(MeSiO_{1.5})_{0.25}(PhSiO_{1.5})_{0.75}(Me_2ViSiO_{0.5})_{0.25}$
[d] $(MeSiO_{1.5})_{0.18}(PhSiO_{1.5})_{0.27}(Me_2SiO)_{0.18}(Ph_2SiO)_{0.12}(Me_2ViSiO)_{0.25}$
[e] Calculated wt P_b at 2.0 wt% C_{excess}

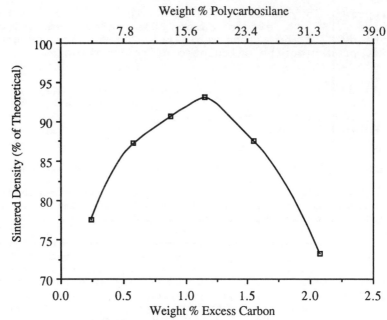

Figure 3. Sintered density of silicon carbide as a function of wt% excess carbon using polycarbosilane (PCS) binder.

Polysilazane Polymers

By combining portions of two previous studies (vide infra), we began investigation of silazane polymers as binders for SiC powders. A comprehensive study of the pyrolysis of alkyl and arylsilsesquiazanes was reported by Burns, et al[9]. Using a series of phenyl/methyl copolymers it was demonstrated that the carbon content of the char increased with increasing phenyl content (see Figure 4).

We prepared a high phenyl content silacyclobutasilazane (SCBZ) polymer[10] using the procedure in Figure 5. The silacyclobutane functionality provides rapid, deep section cure by ring-opening when the thermoplastic SCBZ polymer is heated above 240°C.

We examined the thermal stability of these silazane polymers and found that pyrolysis occurred in two distinct temperature ranges (see Figure 6). Pyrolysis to 1200°C gave a 70% char yield with an SiCN composition (wt% Si = 32.9, C = 48.1, N = 15.9). However, continued pyrolysis to 1800°C gave a char (52% yield) which contained no nitrogen (wt% Si = 40.2, C = 57.5, see Table 1). The results can be explained according to Equation 6.

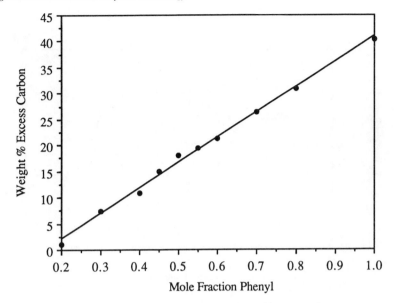

Figure 4. Calculated wt% excess carbon vs phenyl content from a series of $(PhSiNH_{1.5})_x(MeSiNH_{1.5})_{1-x}$ copolymers after pyrolysis to 1800°C.

$$\square\text{SiCl}_2 + PhSiCl_3 + Ph_2SiCl_2$$

$$\downarrow \quad NH_3/\text{toluene}/-78°C$$

$$\downarrow \quad \text{Filtration}$$

$$\downarrow \quad \text{Strip}$$

$$[C_3H_6SiNH]_{0.5}[PhSiNH_{1.5}]_{0.25}[Ph_2SiNH]_{0.25}$$

Silacyclobutasilazane Polymer (SCBZ)

(83%)

Figure 5. Reaction scheme for SCBZ polymer preparation.

$$(RSiN_{1.5})_n \xrightarrow{1200°C} C_xSi_yN_z$$
$$\xrightarrow{1800°C} ySiC + (x-y)C + (0.5)z\,N_2 \quad (6)$$

Experimentally, various levels of this polysilazane binder/SiC powder were screened and the sintered density of small test samples was determined. The results are shown in Figure 7.

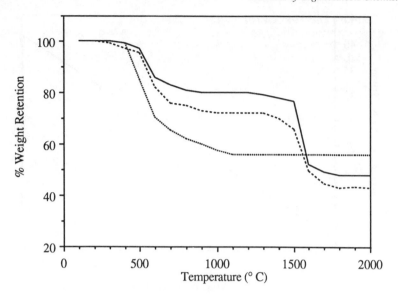

Figure 6. Representative TGA curves for the following polymers: ······ Polycarbosilane, ———— Polysilazane, ------ Polysiloxane

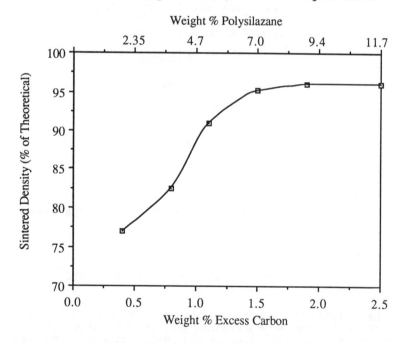

Figure 7. Sintered density of silicon carbide as a function of wt% excess carbon using polysilazane binder.

Note the excellent densities obtained at 1.5 wt% excess carbon and above. There is also a very good correlation between the calculated minimum binder level (see Table 1) and that observed experimentally (see Figure 7) for this silazane binder. This eliminated the need for experimental screening of each new polymer system.

The combination of polymer deep section cure and the ability to change the stoichiometry of the char allows these polymers to be used as binders in molding applications. For molding applications one needs to employ polymers that have suitably low melt viscosities and have suitable excess carbon at binder levels near 20 wt%. A typical moldable polysilazane (Polymer B) is shown in Table 1.

Polysiloxane Polymers

Silazanes function as binders for high density silicon carbide because they decompose, through loss of nitrogen, to silicon carbide and excess carbon below the required sintering temperature[11]. We discovered[12,13] that we could successfully employ siloxane polymers to give SiCO compositions upon pyrolysis to 1200°C and remove CO at higher temperatures to give the required SiC/C char composition (Equation 7).

$$(RSiO_{1.5})_n \xrightarrow{1200°C} C_xSi_yO_z \xrightarrow{1800°C} ySiC + (x-y-z)C + zCO \quad (7)$$

A number of reports have detailed the use of siloxanes as precursors to silicon carbide powders[14,15]; however, our studies are the first studies controlling the SiC to C content in the ceramic to prepare dense articles[12,13].

As is the case with silazanes, control of the carbon content of the ceramic is critical. Sufficient carbon must be present to allow CO evolution (Equation 7) while still providing the required SiC/C char composition. This is accomplished by controlling the phenyl content of the polymer. As the mole fraction of phenyl is increased within the $(MeSiO_{1.5})_{0.75-x}(PhSiO_{1.5})_x(Me_2ViSiO_{0.5})_{0.25}$ series of siloxanes, the wt% excess carbon in the 1800°C char increases (Figure 8).

In general, these polysiloxanes are made by co-hydrolysis of either the corresponding chloro- or methoxysilanes. The presence of the vinyl groups in the polymers provides reactive functionality for deep section cure using, for example, organic peroxides.

For powder pressing applications our work focused on Siloxane A. The TGA results for Siloxane A are given in Figure 6 and the structure and char results are presented in Table 1. Experimentally, various levels of Siloxane A binder/SiC powder were screened and the sintered density of small test samples was determined. The results are shown in Figure 9.

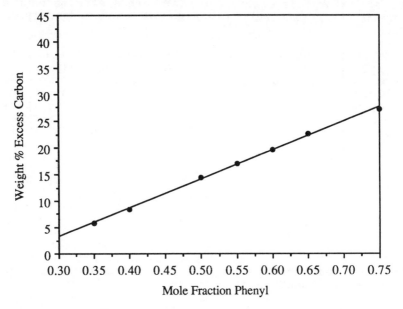

Figure 8. Weight % excess carbon as a function of phenyl content from a series of $(MeSiO_{1.5})_{0.75-x}$ $(PhSiO_{1.5})_x (Me_2ViSiO_{0.5})_{0.25}$ polymers after pyrolysis to 1800°C.

Excellent densities were obtained at 1.5 wt% excess carbon and above. Again, note the good correlation between the calculated minimum binder level (see Table 1) and that observed experimentally (see Figure 9) for this siloxane binder.

For molding applications our work focused on Siloxane B (see Table 1). This seemingly complicated formulation has a good viscosity profile so that stable, flowable mixes can be prepared even at 75 - 80 wt% silicon carbide powder loadings. These mixes flow under pressure at 100°C and can be cured by warming the mold cavity to 160°C. The molded materials have significant strength and are easily handled and removed from the cavity. After removal the articles can be directly sintered to high density without debinderizing. This technology affords complex shapes at significant reduction in cost.

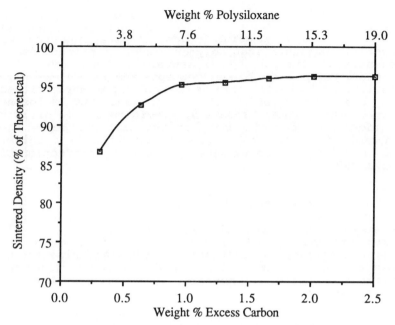

Figure 9. Sintered density of silicon carbide as a function of wt% excess carbon using Polysiloxane A binder.

3. POLYMER BINDER BENEFITS

Our studies have shown that the use of these newly developed silazane and siloxane preceramic polymers as binders in the fabrication of silicon carbide provides a unique combination of benefits. These include the following:

1. <u>More effective particle lubrication</u>. The polymer rheology and polymer level can be varied.

2. <u>Free carbon for sintering</u>. The polymer composition can be varied to provide the required carbon at different polymer levels.

3. <u>Elimination of debinderizing</u>. The polymer sinters to gaseous by-products, silicon carbide and carbon.

4. <u>High density, high strength green parts</u>. The polymer fills the void volume between particles and polymer cure provides strength.

5. <u>Lower shrinkage and excellent shrinkage control</u>. The polymer in the voids converts to silicon carbide during sintering, providing lower and more uniform shrinkage.

4. SUMMARY

This work has successfully demonstrated the use of preceramic organosilicon polymer binder to produce dense beta-silicon carbide by pressureless sintering. In addition, key polymer binder concepts and principles have been defined. Finally, two new types of polymer families have been developed. These silazane and siloxane polymers function as binders by providing stable silicon carbide/carbon ceramic chars at temperatures below the sintering temperatures required for silicon carbide. This technology is being developed for the production of fabricated parts.

5. ACKNOWLEDGEMENTS

The authors would like to thank Willard Hauth, Ronald Keller and Chandan Saha for their contributions to the materials science, understanding and characterization of these ceramic systems. The support of Dow Corning and contributions from all members of its Ceramic Group is also recognized.

REFERENCES

1. R. H. Baney, G. Chandra, "Encyclopedia of Polymer Science and Engineering," Wiley, New York, 1988, Vol. 13, p. 312.
2. W. H. Atwell, "Advances in Chemistry Series No. 224," J. M. Zeigler and F. W. G. Fearon Eds, 1990, pp. 593-606.
3. S. Prochazka, Am. Ceram. Soc. Bull. 1973, 52, 885.
4. See, W. vanRijswik and D. J. Shanefield, J. Am. Ceram. Soc. 1990, 73, 148 and references cited therein.
5. S. Yajima, J. Hayashi and M. Omuri, Chem. Lett. 1975, 931.
6. C. L. Schilling, J. P. Wesson, T. C. Wiliams, Am. Ceram. Soc. Bull. 1983, 62, 912.
7. R. H. Baney, J. H. Gaul, T. K. Hilty, Organometallics 1983, 2, 859.
8. "Ultrastructure Processing of Ceramics, Glasses and Composites," R. C. West, L. C. Hench, D. R. Ulrich Eds., Wiley, New York, 1984, p. 235.
9. G. T. Burns, T. P. Angelotti, L. F. Hanneman, G. Chandra and J. A. Moore, J. Mater. Sci. 1987, 22, 2609.
10. G. T. Burns, "Silane Modified Polysilacyclobutasilazanes," U.S. Patent 1990, No. 4,916,200.
11. G. T. Burns, C. K. Saha and R. J. Keller, "Highly Densified Bodies from Preceramic Polysilazanes Filled with Silicon Carbide," European Patent Application 1990, No. 89311235.9.

12. W. H. Atwell, G. T. Burns and C. K. Saha, "Organosiloxanes Filled with Silicon Carbide Powders and Highly Densified Bodies Therefrom," U.S. Patent 1990, No. 4,888,376.
13. W. H. Atwell, G. T. Burns and C. K. Saha, "Highly Densified Bodies from Organosiloxanes Filled with Silicon Carbide Powders," U.S. Patent 1990, No. 4,929,573.
14. D. A. White, S. M. Oleff, R. D. Boyer, P. A. Budinger and J. R. Fox, Adv. Ceram. Mater. 1987, 2, 45.
15. D. A. White, S. M. Oleff and J. R. Fox, Adv. Ceram. Mater. 1987, 2, 53.

New Convergent Strategies for the Synthesis of Silicon Carbide and Carbide Nitride Precursors

J.-P. Pillot, M. Birot, F. Duboudin, M. Bordeau, C. Biran, and J. Dunoguès*

LABORATOIRE DE CHIMIE ORGANIQUE ET ORGANOMÉTALLIQUE, URA 35 CNRS, UNIVERSITÉ BORDEAUX I, F-33405 TALENCE, FRANCE

1 INTRODUCTION

During the last fifteen years, silicon carbide and carbide nitride-based materials have created considerable scientific interest due to their potential utilization in the aeronautic and space industry and, more generally, in the production of a new generation of high performance engines. For this reason a number of investigations were devoted to the elaboration of ceramic fibers resistant to oxidation at very high temperatures (over 1000°C).

After the pioneering work of Verbeek and Winter[1,2] in the field of both SiC and SiCN precursors, Yajima et al.[3] have prepared SiC-based fibers commercialized by Nippon Carbon Co Ltd. (Nicalon®). These fibers were improved and modified[4] and this field has been widely investigated[5]. However, the presence of oxygen especially introduced for curing the polymeric fibers, limits the field of application of the materials to temp. lower than 1100°C to preserve their mechanical properties[6]. Considering the high thermal stability and resistance to oxidation as well as other advantageous properties of Si_3N_4, the presence of nitrogen seemed more desirable and, consequently, SiCN-based fibers were obtained from polysilazanes (see, for instance, the works of Seyferth[7], Laine[8], Rhône-Poulenc Co.[9]) or from polysilasilazanes (Dow Corning Corp.[10]).

In this context, our investigations have been focused on the synthesis of soluble, meltable and -a priori- spinnable, SiC or SiCN precursors. Although the objective of this research is the processing of inorganic materials, this report concerns only strategies involved in the synthesis of SiC and SiCN preceramic polymers.

2 RESULTS AND DISCUSSION

<u>Precursors of SiC-Based Ceramics</u>

We report here three different approaches of these

precursors : a) a chemical modification of the Yajima process ; b) an electrochemical route to polysilanes and polycarbosilanes and c) a synthesis of a linear well-defined polysilapropylene, a suitable precursor for investigating the transformation :

polydimethylsilane ⟶ polycarbosilane ⟶ SiC-based ceramic

    PDMS   PCS

in terms of mechanism studies, in order to better understand all the different steps of this conversion.

<u>Chemical Modification of the Yajima Process</u>. One work was directed to the use of new "catalysts" for the thermolysis of PDMS and another to the development of an approach involving copolymers.

Use of New "Catalysts". The conversion of PDMS into PCS can be carried out by heating PDMS in an autoclave (Marks I and II) or by using a non-recoverable "catalyst", under atmospheric pressure (Mark III). In the latter case Yajima and others[11,12] used a polyborodiphenylsiloxane resulting from the condensation of diphenyldichlorosilane on boric acid[11]. We have demonstrated that monomeric "catalysts" could assure a better reproducibility of the reaction and a better stability of the resulting PCS since the "catalyst" not incorporated would be evolved upon distillation under vacuum. So 3.3 % (weight) of $B(OR)_3$ (R = Me, iPrSiMe$_2$), $B(NEt_2)_3$ or $(Me_3Si)_2BNMe_2$ gave PCS similar to Mark III from IR or NMR data, but with SEC curves very much narrower to those of Mark I. The observed reproducible polydispersity D = M_w/M_n, D = M_w/M_n = 3.9 (with $B(NEt_2)_3$) or 5.8 (with $B(OMe)_3$) much lower than that of common Mark III, constitutes a noticeable advantage since the mechanical properties of SiC fibers depend on the dispersity of the precursor. Ceramic yields from these PCS are similar to those of Yajima (≈ 60 %) for spinnable precursors.

Strategy Involving Copolymers. The drastic conditions required to convert PDMS into PCS prompted us to modify its regular structure by the introduction of SiCSi sequences with reactive sites to facilitate this transformation. For this purpose we have proposed a route involving the co-condensation of Me_2SiCl_2 with bis(chlorosilyl)methanes such as $(ClMeHSi)_2CH_2$ in variable amount, in the presence of sodium[13] :

$$(Me_2SiCl_2)_{1-x} + \begin{matrix} Me & Me \\ | & | \\ (ClSiCH_2SiCl)_x \\ | & | \\ H & H \end{matrix} \xrightarrow[\text{toluene refl.} \atop -2\,NaCl]{2\,Na} \begin{matrix} Me & Me & Me \\ | & | & | \\ -(Si)_{1-x}-(SiCH_2Si)_x- \\ | & | & | \\ Me & H & H \end{matrix}$$

Polysilacarbosilane (PSCS)

The formula given for PSCS does not indicate its structure (alternate sequence or block copolymer). Some important features have to be pointed out : i) the involved disilylmethane, previously described[14], was prepared by condensation of $MeSiHCl_2$ on CH_2Br_2 in the presence of magnesium. By adding a catalytic amount of zinc, the Grignard condensation takes place with CH_2Cl_2[15] ; ii) in contrast with Yajima's PDMS, the resulting PSCS was soluble ; iii) the conversion of PSCS into PCS occurs at 450°C (instead of 550°C for PDMS at 1 atm.) and the obtained PCS would be less branched ; iv) the introduction of $MeHSiCH_2$-$SiHMe$ units was desirable since they prefigure the expected final sequence.

Ceramic yields given by these PCS are very close to those of Yajima's ones ($\approx$ 60 %), for spinnable precursors.

Electrochemical Synthesis of Polysilanes and Polycarbosilanes.

Polysilanes. Problems due to the handling of massive amounts of (alkali) metals to create silicon-silicon linkages encouraged several authors to involve electrochemistry for the same purpose[17]. However extrapolation of common electrochemical techniques to an industrial scale raises a number of questions. Taking into account these observations, we have proposed a technique previously developed by Périchon et al[17]. This technique is based on the use of a single compartment cell with a sacrificial anode, here an aluminum bar. Electrolysis was carried out at constant current density which only requires a cheap stabilized supply, with stirring. The electrochemical oxidation of the anode forms aluminum trichloride and this allows to strongly decrease the amount of the supporting electrolyte (0.02 mol l^{-1} instead of 0.2-0.8 mol l^{-1}). Moreover it was possible to perform electrochemical reactions under anhydrous conditions, by the addition of trimethylchlorosilane in small amounts to the medium (solvent and supporting electrolyte): the reaction affords $(Me_3Si)_2O$ which is not electroactive and HCl, eliminated by pre-electrolysis. Under these conditions, with a mixture of THF-HMPA or THF-TDA-1[19] as the solvent and LiCl or Et_4NBF_4 as the supporting electrolyte, $Me_3SiSiMe_3$ (74 % from Me_3SiCl), $PhMe_2Si$-$SiMe_3$ (91 % from $PhMe_2SiCl$ and Me_3SiCl), or $Me_3SiSiMe_2$-$SiMe_3$ (60 % from Me_2SiCl_2 and Me_3SiCl) were synthesized.

From Me_2SiCl_2, PDMS identical to that obtained with sodium, was isolated in 47 % yield (not optimized) :

$$Me_2SiCl_2 \xrightarrow{2.2} PDMS\ (47\%)$$

Anode : Al ; solvent : THF-HMPA : 80-20 (vol.) ; supporting electrolyte : Et_4NBF_4.

This preliminary result suggests that electrosynthesis can constitute an alternative route for the preparation of PDMS.

Direct Synthesis of Polycarbosilanes[20]. Regarding the perspectives offered by electrosynthesis for the creation of the Si-C bond either from common electrochemistry[21] or using the Périchon technique[22], a direct synthesis of polycarbosilanes from CH_2Cl_2 and Me_2SiCl_2 for instance, seemed very attractive. Unfortunately, under our reaction conditions the reduction potentials of these compounds are very close and we could not isolate $(Me_2SiCH_2)_n$ in satisfactory yields. In contrast, electrochemical reduction of $ClMe_2SiCH_2Cl$ afforded $ClMe_2SiCH_2SiMe_2CH_2Cl$[20], a key intermediate in the synthesis of $(Me_2SiCH_2)_2$[23], our starting material for the preparation of polysilapropylene (see below).

$$ClMe_2SiCH_2Cl \xrightarrow{1.1 \text{ (theor.)}} ClMe_2SiCH_2SiMe_2CH_2Cl \ (43\%)$$

Anode : Al ; solvent : THF-TDA-1 : 80-20 (vol.) supporting electrolyte : Et_4NBF_4.

Under appropriate conditions $(Me_2SiCH_2)_2$ was directly formed (34 %)[20].

These preliminary results which are not yet optimized suggest that electrosynthesis of $ClMe_2SiCH_2Cl$ could be an attractive route to polycarbosilanes. To reinforce this view, bis(dimethylchlorosilyl)methane was obtained according to[20] :

$$ClCH_2SiMe_2Cl + Me_2SiCl_2 \xrightarrow{2.2 \text{ F.mol}^{-1}} ClMe_2SiCH_2SiMe_2Cl$$

(60% after distn.)

We have not yet investigated the preparation of materials from precursors derived from polycarbosilanes prepared by electrosynthesis.

<u>Preparation of a Polysilapropylene Model</u>. Yajima's approach to PCS precursors can be summarized as follows[3] :

$$Me_2SiCl_2 \xrightarrow[(Na)]{\text{polycondensation}} (Me_2Si)_n \xrightarrow{\text{thermolysis}} \text{"-(MeHSiCH}_2)_n\text{-"}$$

$$\qquad\qquad\qquad\qquad\quad \text{PDMS} \qquad\qquad\qquad\qquad \text{PCS}$$

The structure of this PCS has been not entirely determined and its formula is different from the linear representation given here which does not take into account the strong deficiency of Si-H bonds and the highly crosslinked character of the polymer[25]. For studying the conversion :

$$\text{organosilicon polymer} \longrightarrow \text{inorganic material}$$

in its fundamental aspects, it was necessary to have a well defined preceramic polymer. To allow a possible comparison with Yajima's PCSs, we have chosen the polysilapropylene synthesized according to the following scheme[24]:

$$(Me_2SiCH_2)_2 \xrightarrow{\text{Pt cat.}[26]} -(Me_2SiCH_2)_n- \xrightarrow[-Me_4Si]{Me_3SiCl \text{ (excess)} - AlCl_3 \text{ cat.}}$$

$$-(MeClSiCH_2)_n \cdot - \xrightarrow{LiAlH_4} -(MeHSiCH_2)_n \cdot -$$

polysilapropylene

In fact, the splitting of Si-CH$_3$ bonds was accompanied by a cleavage of some Si-CH$_2$ bonds :

$$-Me_2SiCH_2- \xrightarrow{Me_3SiCl \text{ }(AlCl_3)} -SiMe_2Cl + Me_3SiCH_2-$$

$$Me_3SiCH_2- \xrightarrow{Me_3SiCl \text{ }(AlCl_3)} -SiMe_2Cl + Me_4Si$$

Therefore the chlorinated polymer consisted of a polymeric sequence $-(MeClSiCH_2)-_n$. and defined end chains $-SiMe_2Cl$. Consequently, the polysilapropylene resulting from reduction with LiAlH$_4$ possesses the following structure :

$$HMe_2Si(CH_2SiMeH)_n \cdot CH_2SiMe_2H$$

The cleavage of some Si-CH$_2$ bonds leads to a decrease in M$_n$ from ≈ 250,000 to ≈ 2,300 and the obtained polysilapropylene and the spinnable Yajima's PCS exhibit comparable Mn values. However, the obtained polysilapropylene is a mobile oil whereas Yajima's PCSs is a solid (softening temp. : 220-300°C). This is due to the linear character of the former while the latter is highly branched. To confirm this linear structure and for more detailed physicochemical studies, five short chain oligomers were synthesized[24].

$$XMe_2Si(CH_2SiXMe)_mMe \quad X = Cl, H ; m = 1,2,3,4,5.$$

Configurational effects were observed by ^{29}Si, ^{13}C and ^{1}H NMR and the end chains well characterized.

On the other hand, comparative thermolysis of PDMS and polysilapropylene led to the following conclusions :

Me_2SiH_2 is the main gas first formed when PDMS is thermolyzed ; ii) hydrogen formation is essentially connected with the presence of Si-H bonds ($(Me_2SiCH_2)_n$ is stable even at 550° under inert atm.) ; iii) the formation of Me_3SiH is connected with the branching process ; iv) the polysilapropylene model is decomposed faster than PDMS under the same conditions and this confirms the impossibility to form $-(HMeSiCH_2)-_n$ upon thermolysis of PDMS : v) thermolysis involves, at least partially, homolytic cleavages ; vi) comparison of thermolyses giving Mark I or Mark III from PDMS[27] confirms that the cleavage of Si-Si bonds is not complete in the case of Mark III (Me_2SiH_2 is less abundant), but Mark III is much more branched than Mark I (Me_3SiH formation is in the range of ten times more for Mark III than for Mark I).

So, polysilapropylene, which is not a convenient precursor of ceramics when used without reticulation, constitutes a good model for basic studies. Moreover, investigation in this field allowed us to synthesize new functional polycarbosilanes[24] :

$-(XMeSiCH_2)_n-$ X = D, NHSi≡, NHMe, NMe_2

By warming, N-H bonds were converted into N-Si ones by transamination. Thus, new precursors of SiCN-based materials are available from this approach and the ceramic yield could be correlated to the nature of cross-linkages.

Precursors of SiCN-Based Materials

In our approach towards SiC precursors, we have postulated that the structure of the preceramic polymer might prefigure that of the final material formed upon pyrolysis at high temperature (over 1000°C). For the reasons given at the beginning of this report and in order to delay the crystallization of SiC in the material, we desired to synthesize SiCN precursors. In agreement with our postulate, we have chosen to introduce both Si-C and Si-N linkages in the main chain of the preceramic polymer : this strategy differs for the one involving polysilazanes[6-9] and from that of Corriu et al.[28,29] involving vinyldichlorosilane. Indeed in the last case, hydrosilylation leads to SiCCSi sequences different from SiCSiC desired in our strategy. An approach based on co-condensation of Me_2SiCl_2 and $(HMeClSi)_2NR$ (R = H, alkyl ...) was successfully used. These disilazanes, previously unknown, were easily prepared[30]. For instance $(HMeClSi)_2NH$ was synthesized as follows :

2 $MeSiHCl_2$ + $(Me_3Si)_2NH$ ⟶ $(HMeClSi)_2NH$ + 2 Me_3SiCl

(95%)

In the case of the N-methylated corresponding derivative, the reaction was favoured by addition of a catalytic amount of Bu_4NF.

SiCN precursors were obtained according to the scheme[31] :

$$(Me_2SiCl_2)_{1-x} + [(HMeClSi)_2NH]_x \xrightarrow[\substack{\text{xylene refl.} \\ -2\,NaCl}]{2\,Na}$$

$$-(Me_2Si)_{1-x}-[HMeSi-N-SiMeH]_x-$$

polysilasilazane (PSSZ)

$$PSSZ \xrightarrow{\text{thermolysis}} \text{polycarbosilazane (PCSZ)}$$

The formula given for our PSSZ is not representative of its structure (block copolymer or alternate sequences) but it summarizes important features : i) the nitrogen ratio in the PSSZ corresponds approximatively to that of the formula ; ii) Si-Si and Si-N linkages have been characterized by IR and multinuclear NMR ; iii) N-H bonds were converted into the corresponding N-Si bonds essentially by condensation with silylanions involving the formation of a branched copolymer. From a number of observations, especially the formed gases and low-boiling products, as well as the behaviour of some models, our results are consistent with the following mechanisms :

$$\equiv SiCl \xrightarrow{Na} \equiv Si^{\ominus}\ ^{\oplus}Na$$

$$\equiv Si^{\ominus}\ ^{\oplus}Na \xrightarrow{\equiv SiNHSi\equiv} \equiv SiH + \equiv SiN^{\ominus}Si\equiv\ ^{\oplus}Na$$

$$\xrightarrow{\equiv SiCl} \equiv SiSi\equiv + NaCl$$

$$\xrightarrow{\equiv SiH} \equiv SiSi\equiv + NaH$$

$$\equiv SiN^{\ominus}Si\equiv\ ^{\oplus}Na \xrightarrow{\equiv SiCl} \equiv Si\diagdown_{\substack{N \\ | \\ Si\equiv}}\diagup Si\equiv + NaCl$$

$$\equiv SiNHSi\equiv + NaH \longrightarrow \equiv SiN^{\ominus}Si\equiv\ ^{\oplus}Na + H_2$$

We also suggest the formation of sila-imines[7,32] but we have not any evidence for a mechanism involving these intermediates.

$$ClSiN^{\ominus}Si\equiv {}^{\oplus}Na \xrightarrow{-NaCl} \left[\begin{array}{c} \diagdown \\ Si=N \\ \diagup \end{array} \diagup Si\equiv \right]^{32} \begin{array}{l} \nearrow \text{insertion in NH bonds}^{33} \\ \\ \searrow \text{dimerization}^{34} \end{array}$$

PSSZ are not convenient SiCN precursors : like PDMS or PSCS, they need thermolysis to afford (spinnable) PCSZ with a good ceramic yield ($\approx$ 60 %). From the pyrolysis of these PCSZ[35] one can assume that this strategy, comparable to that reported by Dow Corning Corp.[10], allows to prepare SiCN-based materials containing variable amounts of nitrogen (between 0 and 22 % (at.) fixed by the value of x). As we expected, the presence of nitrogen delays crystallization of the material (studies concerning the quantification of this phenomenon are in progress) but, at very high temp., nitrogen was eliminated : so when x = 0.5, pyrolysis at 1600°C under argon affords silicon carbide containing free carbon.

The versatility of this approach offers attractive perspectives for the comprehension of the elaboration of SiCN- based materials especially for determining the role of the structure and the composition of the precursor on the structure and the properties of the final inorganic material.

Acknowledgments

The authors would like to thank E. Bacqué, P. Pons, C. Richard (Theses), O. Babot, P. Lapouyade, the members of the "G.S. Céramiques à base SiCN"[35] and of the CESAMO (Univ. Bordeaux I). They are indebted to CNRS, DRET, Région Aquitaine, Rhône-Poulenc Minérale Fine and Société Européenne de Propulsion for their financial and logistical assistance.

REFERENCES*

*Because of the number of reports only some important reviews are mentioned in refs. 1-11. In particular reports concerning the synthesis of polysilanes (R. West, J.F. Harrod, K. Matyjaszewski, J. Corey, etc.,) are not mentioned here but most of them are summarized in ACS Symp. Ser., 1988, <u>360</u>. Moreover an excellent overview has been published by R.H. Baney and G. Chandra (Dow Corning Corp.) in this area ('Encyclopedia of Polymer Science and Engineering', Wiley, NY, 1988, <u>13</u>, 312.

1. W. Verbeek and G. Winter, Ger. Demand 2, 236, 078 (1974).
2. G. Winter and W. Verbeek, US Pat. 3, 892, 53 (1975).

3. See, for instance, S. Yajima, **Am. Ceram. Soc. Bull.**, 1983, <u>62</u>, 893 ; from the report of Y. Hasegawa, **J. Mater. Sci.**, 1989, <u>24</u>, 1177 it is possible to go further back and find all the reports from this group.
4. See, for instance, T. Yamamura, T. Hurushima, M. Kimoto, T. Ishikawa, M. Shibuya and T. Iwai, 'High Tech Ceramics', Elsevier, 1987, p. 737 and ref. therein.
5. Among the most remarkable Reviews see, for instance, C.L. Schilling, Jr., J.P. Wesson and T.C. Williams, **J. Polymer Sci. Polymer Symp.**, 1983, <u>70</u>, 121 ; id., **Am. Ceram. Soc. Bull.**, 1983, <u>62</u>, 912 and refs. therein ; R.H. Baney, J.H. Gaul and T.K. Hilty, **Organometallics** 1983, <u>2</u>, 859 and refs.. therein ; D. Seyferth, ACS Symp. Ser., 1988, <u>360</u>, 21 and refs. therein.
6. G. Simon and A.R. Bunsell, **J. Mater.Sci. Lett.**, 1983, 80 ; T. Mah, N.L. Hecht, D.E. McCullum, J.R. Hoenigman, H.M. Kim, A.P. Katz and K.A. Lipsitt, **J. Mater. Sci.**, 1984, <u>19</u>, 1191.
7. D. Seyferth, G.H. Wiseman, J.M. Schwark, Y.-F. Yu and C.R. Poutasse, ACS Symp. Ser., 1988, <u>360</u>, 143 and refs. therein.
8. R. Laine, Y.D. Blum, D. Tse and R. Glaser, ACS Symp. Ser., 1988, <u>360</u>, 124 and refs. therein.
9. J.-J. Lebrun and H. Porte (Rhône-Poulenc), French Pat. 2, 577, 933 (1985) ; 2, 579, 602 (1985) ; 2, 581, 386 (1985) ; 2, 581, 391 (1985) ; 2, 584, 079 (1985) ; 2, 584, 080 (1985) ; 2, 590, 580 (1985) ; 2, 590, 584 (1985).
10. G.E. Le Grow, T.F. Lim, J. Lipowitz, R.S. Reaoch, 'Better Ceramics Through Chemistry II', Mater.Res. Soc., Pittsburgh, 1986, p. 533. ; J. Lipowitz, J.A. Rabe and T.M. Carr, ACS Symp. Ser., 1988, <u>360</u>, 156 and refs. therein.
11. See, for instance, S. Yajima, Y. Hasegawa, K. Okamura and T. Matsuzawa, **Nature**, 1978, <u>273</u>, 525.
12. See, for instance, refs. 3, 11 and T. Iwai, T. Kawahito and M. Tokuse, Eur. Pat., 51, 855 (1980) ; M. Tokushu and K. Zairyo, Jap. Pat., 110, 733 (1981).
13. J.-P. Pillot, E. Bacqué, J. Dunoguès and P. Olry, French Pat., 2, 399, 371 (1986).
14. D.J. Cooke, N.C. Lloyd and W.J. Owen, **J. Organometal. Chem.**, 1970, <u>22</u>, 55.
15. J.-P. Pillot, C. Biran, E. Bacqué, P. Lapouyade, J. Dunoguès and P. Olry, French Pat., 2, 599, 369 (1986).
16. C. Biran, M. Bordeau, P. Pons, M.-P. Léger and J. Dunoguès, **J. Organometal. Chem.**, 1990, <u>382</u>, C 17.
17. E. Hengge and G. Litscher, **Angew. Chem. Int. Ed. Engl.**, 1976, <u>15</u>, 370 ; R.J.P. Corriu, G. Dabosi and M. Martineau, **J. Chem. Soc. Chem. Commun.**, 1979, 457 ; id., **J. Organometal. Chem.**, 1980, <u>186</u>, 19 ; id., ibid., 1981, <u>222</u>, 195 ; E. Hengge and H. Firgo, **J. Organometal. Chem.**, 1981, <u>212</u>, 155.
18. S. Sibille, E. D'Incan, L. Leport and M. Périchon, **Tetrahedron Lett.**, 1986, <u>27</u>, 3129 and refs. therein.

19. TDA-1 : tris(3,6-dioxaheptyl)-amine. Cf. G. Soula, *J. Org. Chem.*, 1985, **50**, 3717.
20. M. Bordeau, C. Biran, P. Pons, M.-P. Léger and J. Dunoguès, *J. Organometal. Chem.*, 1990, **382**, C21.
21. See, for pioneering work, T. Shono, Y. Matsumara, S. Katoh and N. Kise, *Chem. Lett.*, 1985, 463 ; J. Yoshida, K. Muraki, H. Funakashi and N. Kawabata, *J. Organometal. Chem.*, 1985, **284**, C 33 ; id., *J. Org. Chem.*, 1986, **51**, 3996.
22. P. Pons, C. Biran, M. Bordeau, J. Dunoguès, S. Sibille and J. Périchon, *J. Organometal. Chem.*, 1987, **321**, C27 ; P. Pons, C. Biran, M. Bordeau and J. Dunoguès, *J. Organometal. Chem.*, 1988, **358**, 31.
23. G. Greber and G. Degler, *Makromol. Chem.*, 1962, **52**, 174 ; W.A. Kriner, *J. Org. Chem.*, 1964, **29**, 1601.
24. E. Bacqué, J.-P. Pillot, M. Birot and J. Dunoguès, *Macromol.*, 1988, **21**, 30 and 34.
25. Y. Hasegawa and K. Okamura, *J. Mater. Sci.*, 1986, **21**, 321.
26. D.R. Weyenberg and L.E. Nelson, *J. Org. Chem.*, 1965, **30**, 2618.
27. M. Birot, E. Bacqué, J.-P. Pillot and J. Dunoguès, *J. Organometal. Chem.*, 1987, **319**, C 41.
28. N.S. Choong Kwet Yive, Thesis, Université Montpellier II, 1990.
29. B. Boury, R.J.P. Corriu and L. Carpenter, *Angew. Chem. Int. Ed. Engl.*, in press.
30. E. Bacqué, J.-P. Pillot, J. Dunoguès, C. Biran and P. Olry, French Pat., 2, 599, 037 (1986).
31. E. Bacqué, J.-P. Pillot, J. Dunoguès and P. Olry, Eur. Pat., 88, 401, 401 (1988).
32. U. Klingebiel, D. Bentmann and A. Meller, *J.Organometal.Chem.*, 1978, **144**, 381.
33. D.R. Parker and L.H. Sommer, *J.Organometal.Chem.*, 1976, **110**, C1.
34. W. Fink, *Chem.Ber.*, 1963, **96**, 1071.
35. Pyrolysis of precursors was studied by R. Naslain et al. (LCTS, Univ. of Bordeaux I, Fr.) assisted in the characterization by Drs. A. Oberlin, M. Monthioux (Pau, Fr.), F. Taulelle (Paris VI, Fr.), J. Dixmier (Meudon, Fr.) and their coworkers. All belong to the "Groupement Scientifique Céramiques à base SiCN" founded in 1988 by the CNRS.

Heterochain Polysilahydrocarbons Containing Silicon–Silicon Bonds in the Main Chain

L. E. Gusel'nikov and Yu. P. Polyakov
TOPCHIEV INSTITUTE OF PETROCHEMICAL SYNTHESIS, USSR ACADEMY OF SCIENCES, MOSCOW, USSR

1 INTRODUCTION

Since the late 1960s much attention has been paid to the chemistry of such low coordinated compounds of silicon as multiple $p_\pi-p_\pi$ bonded intermediates (e.g. silaalkenes)[1] and silylenes.[2] The interest is due not only to their novel bonding structure and reactivity, but also to possibilities of their use (like silacyclobutanes) as bricks to build up oxygen-free organosilicon polymeric materials (polysilahydrocarbons, PSHC).

Depending on the structure of the main chain one may divide PSHC into the four types:

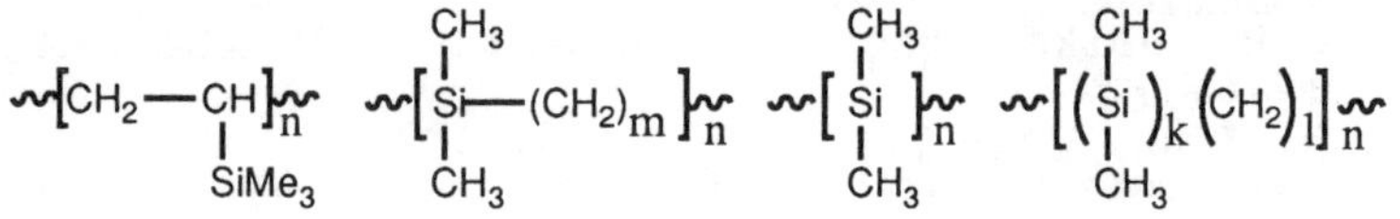

$$\sim\!\!\left[CH_2\!-\!\underset{SiMe_3}{\overset{}{C}H}\right]_n\!\!\sim \quad \sim\!\!\left[\underset{CH_3}{\overset{CH_3}{\underset{|}{Si}}}\!-\!(CH_2)_m\right]_n\!\!\sim \quad \sim\!\!\left[\underset{CH_3}{\overset{CH_3}{\underset{|}{Si}}}\right]_n\!\!\sim \quad \sim\!\!\left[\left(\underset{CH_3}{\overset{CH_3}{\underset{|}{Si}}}\right)_k\!(CH_2)_l\right]_n\!\!\sim$$

$$\text{I} \qquad\qquad \text{II} \qquad\qquad \text{III} \qquad\qquad \text{IV}$$

The first type (I) involves carbochain polymers, such as polyvinyltrimethylsilane (PVTMS).[3] Films made of PVTMS possess good gas separation properties to be used as membranes. Heterochain polymers belong to PSHC of the second type (II). They are built up of single silicon atoms bonded with bridges of carbons. A large number of such polymers were prepared by ring-opening polymerization of both mono- and 1,3-disila-cyclobutanes.[4] As films they may be obtained via gas phase silaalkene polymerization.[5-7] Polysilylenes (polysilanes), which are homochain polymers built up of silicon atoms, present the third type (III) of PSCH. They are extremely useful for production of ceramics, high strength silicon carbide fibres, photoresists, etc.[8-10] The fourth type (IV) of PSHC is a miscellany of types II and III. It involves PSHC whose main chains consist of alternating polysilylene groupings bonded with carbon bridges.

That means the main chains of such polymers definitely have both Si-Si and Si-C bonds but also may have C-C bonds. In this report we discuss preparation and properties of PSHC of the fourth type; especially of polyethylenedisilenes – new polymers derived from spontaneous polymerization of 1,2-disilacyclobutanes.

2 PREPARATION

Synthesis and Polymerization of 1,2-Disilacyclobutanes

Although ring-opening polymerization is known to be a conventional way to produce PSHC (II), its application to synthesis of PSHC (IV) was not visible because of the absence of polymerizable 1,2-disilacyclobutanes. The only exception was 1,1,2,2-tetrafluoro-1,2-disilacyclobutane **1a** (formed in the reaction of difluorosilylene with ethylene) which was stable at 77K but polymerizes upon warming up to room temperature.[11] Other fluoro-1,2-disilacyclobutanes **1b** and **1c** seem to be stable at room temperature but also polymerized, whereas nothing is known about polymerization of highly substituted molecules **1d-1f, 2a-2d, 3a-3d**. The behaviour of **4** which dimerises at room temperature is rather specific.

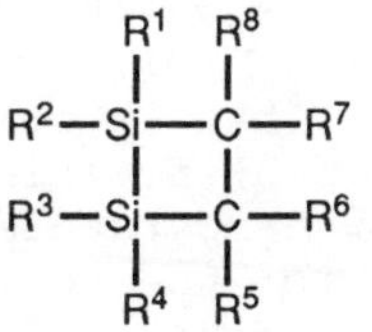

	$R^1 - R^4$	R^5, R^6	R^7, R^8	Ref.
1a:	F	H	H	11
1b:	F	H	H, Me	12
1c:	F	Me	Me	13
1d:	Me$_3$Si	Me$_3$C	Me$_3$SiO	14
1e:	Me$_3$Si	Ph	Me$_3$SiO	15
1f:	Me	Me	Me	16

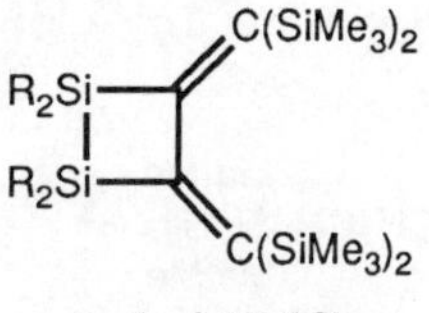

2a,3a: R=Ph
2b,3b R=*o*-tolyl
2c,3c R=*m*-tolyl
2d,3d: R=*p*-tolyl

2 (ref 17,18)

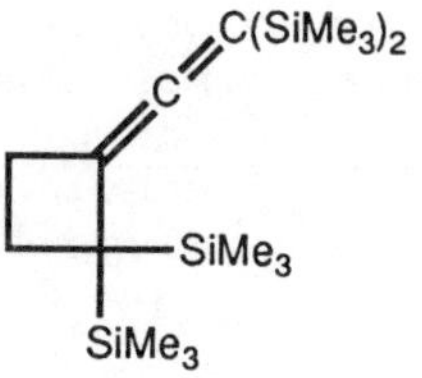

3 (ref 17,18) 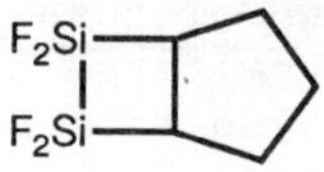   **4** (ref 19)

For a long time 1,1,2,2-tetramethyl-1,2-disilacyclo-butane **1h** was not available. After several failed attempts to obtain it by both Hg sensitized photolysis of 1,2-bis-(dimethylsilyl)ethane and Wurtz reaction of 1,2-bis(dimethylchlorosilyl)ethane we were successful when we turned to the alkali metal vapors dehalogenation of 1,2-bis(dialkylchlorosilyl)alkanes.[20,21]

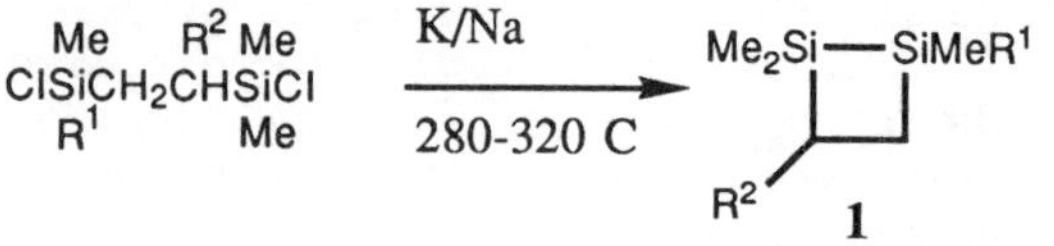

$$\text{ClSiCH}_2\text{CHSiCl} \text{ (Me, R}^2\text{Me, R}^1\text{, Me)} \xrightarrow[280-320\ C]{K/Na} \text{Me}_2\text{Si—SiMeR}^1 \text{ with R}^2 \quad \mathbf{1}$$

1g, R^1=Me, R^2=H
1h, R^1=R^2=Me
1i, R^1=nBu, R^2=H

The technique we used was similar to that earlier described.[21] It provided: (1) continuous evaporation of the sample, (2) interaction of the latter with alkali metal vapors in a flow reactor, (3) freezing of the reaction product vapors in the low temperature (77K) trap. On the whole it made allowances for the necessity of quick removal and stabilization of the reaction products. The reaction was carried out as follows: vapors of 1,2-bis(dimethylchlorosilyl)ethane were passed at low pressure over sodium/potassium alloy at 280-320°C. The products were collected in a trap cooled with liquid nitrogen. At room temperature they represented a clear, colourless, volatile liquid having the smell of geranium. The liquid allows trap to trap vacuum distillation. On standing in the absence of air it shortly becomes cloudy, changes to a white residue, and finally turns within a few minutes into a white solid representing high crystalline polymer of **1g** - polyethylenetetramethyldisilene (PETMDS):

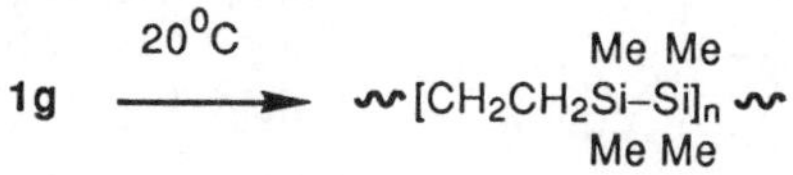

$$\mathbf{1g} \xrightarrow{20°C} \sim[\text{CH}_2\text{CH}_2\text{Si–Si}]_n \sim \text{ (Me Me / Me Me)}$$

To avoid spontaneous polymerization taking place with noticeable liberation of heat **1g** had to be kept in the sealed tube at low temperature (dry ice or liquid nitrogen). Therefore all spectroscopic characteristics of **1g** were determined at low temperatures.

Table 1 NMR Data of **1g**[a], δ, ppm

	^1H[b]	^{13}C[c]	^{29}Si
4 (CH_3)	-0.23	-2.95	-
2 (CH_2)	0.63	8.70	-
2 (Si)	-	-	7.8

[a] without solution, -50°C,   [b] external standard $CHCl_3$,
[c] external standard CD_3OD.

In Table 1 NMR-parameters are given for the monomer sample at -50°C.[22] For obtaining IR-spectra, the sample and argon matrix were deposited under vacuum from the gas phase on to the target at 10K. It shows the absorption bands characterizing 1,1,2,2-tetramethyl-1,2-disilacyclobutane: 654, 765, 795, 810, 837, 905, 927, 985, 1059 and 1135 cm^{-1}. Polymerization of **1g** is accompanied by a variation in the general nature of the spectrum over the range 700-850 cm^{-1} with simultaneous disappearance of the 654, 905, 927 and 985 cm^{-1} bands. The 1136 cm^{-1} band shifts down to 1127 cm^{-1} and its intensity increases.

Addition reactions of **1g** with halides (Cl_2, Br_2) and hydrogen halides (HCl, HBr) provide supplementary evidence for the structure of this unusually reactive silacycle.[22] They occur with Si-Si bond cleavage even more readily than polymerization.

Wurtz Reaction

There are few reports concerning the preparation of PSHC (IV).[23-25] High molecular solids (M up to 36000) were obtained upon dehalogenation of both 1,4-bis(methylethylchlorosilyl)- and 1,4-bis(methylphenylchlorosilyl)benzenes with sodium dispersion in toluene. Low molecular PSHC (IV) were also prepared by Wurtz reaction from 1,4-bis(dimethylchlorosilyl)benzene[23,24] or bis(dimethylchlorosilyl)methane.[25]

We failed to prepare high molecular PETMDS by conventional Wurtz reaction of 1,2-bis(dimethylchlorosilyl)ethane with sodium suspension in hydrocarbon solvents. Only oligomers melting at 40°C were obtained. High molecular PETMDS was synthesized using a modified technique.[26,27]

$$\begin{array}{c}\text{Me} \quad \text{Me} \\ \text{ClSiCH}_2\text{CH}_2\text{SiCl} \\ \text{Me} \quad \text{Me}\end{array} \xrightarrow{\text{Na}} \sim\!\![\text{CH}_2\text{CH}_2\overset{\text{Me}}{\underset{\text{Me}}{\text{Si}}}\!-\!\overset{\text{Me}}{\underset{\text{Me}}{\text{Si}}}]_n\!\!\sim$$

The same technique used for the mixture of 1,2-bis(dimethylchlorosilyl)ethane and dimethyldichlorosilane as starting materials resulted in PSHC (IV) containing groupings of three, four, etc. silicon atoms in the main chain. Depending on the $(ClSiMe_2CH_2)_2/Me_2SiCl_2$ ratio either melting and soluble polymers or nonmelting and nonsoluble solids were obtained.

$$k \begin{array}{c} Me \\ ClSiCH_2CH_2SiCl \\ Me \end{array} \begin{array}{c} Me \\ \\ Me \end{array} + l\ Me_2SiCl_2 \xrightarrow{Na} \sim\!\![CH_2CH_2Si\text{-}Si\text{-}(Si)_m]_n\!\!\sim \begin{array}{c} Me\ Me\ Me \\ \\ Me\ Me\ Me \end{array}$$

$$l/k = m$$

Similarly high molecular polymethylenetetramethyldisilene (PMTMDS) – a polymer in which the main chain consists of alternating methylene and disilene groups – was obtained.[26]

$$\begin{array}{c} Me\ \ Me \\ ClSiCH_2SiCl \\ Me\ \ Me \end{array} \xrightarrow{Na} \begin{array}{c} MeMe \\ \uparrow\!\!\!\!\!\uparrow CH_2Si\text{-}Si]_n\!\!\sim \\ MeMe \end{array}$$

Partial Rearrangement of Polysilanes

Thermal rearrangement of polydimethylsilylene is an important stage of its conversion into ceramic materials.[8-10]

$$\begin{array}{c} Me\ Me \\ \sim\!\!Si\!\!-\!\!Si\!\!\sim \\ Me\ Me \end{array} \xrightarrow{\Delta} \begin{array}{c} Me\ \ H \\ \sim\!\!Si\text{-}CH_2\text{-}Si\!\!\sim \\ Me\ \ Me \end{array}$$

Unless the reaction is accomplished to produce PSHC (II) intermediates PSHC (IV) are present. Based on this idea we were interested in laser decomposition of polydimethylsilylenes to produce PSHC (IV).[28] Experiments were carried out in Dr. J. Pola's group (Prague) with a glass optical cell having NaCl windows and a tunable continuous-wave CO_2-laser.

The most intriguing result was almost complete evaporation of starting polymer followed by deposition of white solid films throughout the inside of the cell. In the IR-spectrum of the deposit is present an absorption at $2100\,cm^{-1}$ indicating Si-H bonds. Therefore rearrangement of PSHC (III) into PSHC (II) via intermediate formation of PSHC (IV) also occurs upon laser pyrolysis. Formation of solid films during this process may be of interest for their further conversion into ceramics.

3 CHARACTERIZATION

Polyethylenetetramethyldisilene

As follows from the chemical formula PETMDS may be described as regular co-polymer of ethylene and tetramethyldisilene (Me$_2$Si=SiMe$_2$). Obtained by polymerization of **1g**, it is a white solid which can be easily machined and moulded. By its appearance the polymer bears a resemblence to high-density polyethylene. It melts at 106-108°C but is not soluble in hexane, benzene, toluene, carbon tetrachloride, dichloroethane, <u>etc</u>. at room temperature. Above 60°C it can be gradually dissolved in the above mentioned solvents. However, it reprecipitates as the temperature is reduced to 40°C.

IR and Raman Spectra.[22,29] Infrared spectra of PETMDS and the model compound - 2,2,5,5,6,6,9,9-octamethyl-2,5,6,9-tetrasiladecane:

$$Me_3SiCH_2CH_2Me_2SiSiMe_2CH_2CH_2SiMe_3$$

5

are almost the same. Both spectra contain absorption bands at 1050 and 1127 cm^{-1} characterizing a dimethylene bridge between two silicon atoms. Absorptions due to Si-Si bonds are not pronounced in IR spectra even when no symmetry restriction is involved, but we expected their manifestation in the Raman spectra as for polydimethylsilylene[30] (strong band at 373 cm^{-1}). However, for PETMDS we found several weak absorptions in the region 300-500 cm^{-1} of which the strongest one was at 460 cm^{-1}. Having obtained this rather unexpected result we turned to model compound **5** for which the X-ray structure was obtained.

Again, we found three absorptions in the region 300-500 cm^{-1} due to Si-Si stretching. Assuming the existence of different conformers at room temperature we obtained the Raman spectrum at 77K. As a result only the band at 460 cm^{-1} was present in the region 300-500 cm^{-1}. Therefore the band at 460 nm^{-1} was assigned to ν_{Si-Si} vibrations.

NMR Spectra. Table 2 contains NMR-parameters of PETMDS which are in good agreement with the structure of the main chain built up by alternative dimethylene and methyl substituted disilene units, the value of ^{29}Si chemical shift indicating silicon-silicon bond. The measured half-width of a signal in the ^{29}Si spectrum of the solid sample suggests high regularity of the polymer structure.

Intrinsic Viscosity/Molecular Mass Relationship[32]

The technique of thermofield fractioning and the ebulliometric method were used to get and describe fractions of PETMDS. The following type of Kune-Mark-Howink equation was evaluated:

$$[\eta] = 1.56 \times 10^{-3} M^{0.57}$$

Table 2   NMR Data of PETMDS, δ (ppm)

	$^1H^a$	$^{13}C^b$	$^{29}Si^c$
4(CH$_3$)	0.275	-0.45	-
2(CH$_2$)	0.875	13.84	-
2(Si)	-	-	-15.5^d

^{a}solution in C$_6$H$_6$ (Tesla BS-467), ^{b}early stage of polymerization at -30 °C (HX-90 Bruker), ^{c}solid sample (CXF-200), ^{d}half-width 20±2Δ Hz.

Molecules of PETMDS have the effect of steric hindrance to rotation around the main chain bond which is similar to that of polyethylene molecules. However, the flexibility of PETMDS molecules is lower (26,5Å) than that of polyethylene molecules (19,4Å).

X-ray, Differential Scanning Calorimetry (DSC) and Electron Microscopy Data.[33-35] PETMDS is a high crystalline polymer. According to X-ray, thermomechanical and DSC data the first order transition was found at 35-53°C, this being diffusable on the temperature scale and has a small heat.

In order to obtain information for suggesting the most likely molecular and crystal structure of OETDMS an X-ray study of the structural precursor 5 was carried out. The molecule **5** is centrosymmetric, the CSiCCSiSiCCSiC chain having an almost planar *all-trans* conformation. In the crystal of **5** one can distinguish molecular layers parallel to the *a b* plane; the long axes of the molecules form an angle of 24.4° with the layer plane; intermolecular contacts in layers are shorter, on the average, than interlayer contacts.

According to electron microscopy data PETMDS forms planar type crystalline structures with sizes at least several tens of nanometers.

Polymethylenetetramethyldisilene (PMTMDS)

This amorphous-crystalline solid which can be described as a regular co-polymer of methylene with tetramethyldisilene has m.p. 70°C. It dissolves in a number of organic solvents at room temperature. According to X-ray, thermomechanical and differential scanning calorimetry data the first order transition occurs at 35°C.[34] At this moment crystalline reflections disappear. However, isotropization of the sample is accomplished at 70°C. In the region 35-70°C the polymer stays in a mesomorphic state. By analogy with polydi-n-hexylsilylene,[36,37] one may suggest two stage melting which involves conformational distortion of macromolecules followed by loss of the positional order.

Films made of PMTMDS are characterized by fairly high parameters of gas permeability. Having the same density as polyethylene it has much higher permeability for oxygen, nitrogen and carbon dioxide. The parameters given in Table 3 together with films noticeable elasticity allow us to consider them as possible membrane materials.[38]

Table 3   Permeability Parameters (P)

Polymer	Density	$P \cdot 10^{10}$, cc cm. cm^{-2} s^{-1} cm^{-1} Hg			
	g cc^{-1}	H_2	O_2	N_2	CO_2
PMTMDS	0.968	88	13	4.9	31
Polyethylene	0.973	1.4	0.4	0.14	1.7

Polyethylenepolysilylenes (PEPS)

PEPS of general formula:

$$\sim[CH_2CH_2\underset{Me}{\overset{Me}{Si}}-\underset{Me}{\overset{Me}{Si}}-(\underset{Me}{\overset{Me}{Si}})_m]_n\sim$$

when m = 0.33, 0.66 and 1.5 were white solids melting at 106-108°C and dissolving on heating in aromatic hydrocarbons, whereas the sample with m = 4 was non-melting and non-soluble.

There is a little difference in IR spectra of PETMDS and PEPS due to intensities of the bands. In Raman spectra of PEPS together with absorptions 460 cm^{-1} (ν_{Si-Si}) and 530 cm^{-1} (ν_{Si-C}) which are characteristic of PETMDS new bands at 373 and 662 cm^{-1} due to polysilylene groupings were also observed. The greater "m", the higher the intensities of the bands.

Dependence of the UV maximum on the length of the polysilylene chain is well known.[8] Likewise a bathochromic shift from 316 nm to 339 nm was observed for PEPS on going from m = 0.95 to m = 4.5.[29]

4 CHEMICAL PROPERTIES

Thermal Decomposition

According to DTA/TG data PETMDS decomposition starts in the air at 240°C. In vacuum it is stable up to 350°C. Thermal decomposition of PETMDS was studied also using combined TG/GC/MS system.[39] The main volatile decomposition product appeared to be **1g**, but others were cyclooligomers and 1,1,2,2,3,3,-hexamethyl-1,2,3-trisilacyclopentane **6**

$$\text{PETMDS} \xrightarrow{\Delta} (CH_2CH_2SiMe_2SiMe_2)_k + \underset{\mathbf{6}}{Me_2Si\overset{\frown}{\underset{SiMe_2}{}}SiMe_2}$$

$$k = 1,2,3...$$

Identification of **1g** was made also by its ability for spontaneous polymerization. Some of the volatile thermal decomposition products specially prepared in a flow vacuum system were polymerized on warming from 77K to room temperature. In terms of a diradical mechanism the above-mentioned volatile products are formed via homolytic cleavage of the weaker Si-Si bonds, followed by intra-molecular combination of biradicals. The mechanism for formation of compound **6** probably is related to homolytic cleavage of both Si-Si and Si-C bonds. As $Me_2Si-SiMe_2$ groupings in polysilylenes rearrange into Me_2SiCH_2SiHMe we were interested whether or not such rearrangement occurs for PSHC (IV). Thermal treatment of PETMDS in an autoclave at 350° for two hours resulted in a non-volatile product which was soluble in hydrocarbons and its IR spectrum showed prominent Si-H absorption.

CO_2-Laser Pyrolysis[40]

We were interested in CO_2-laser pyrolysis (LP) of PSHC (IV). After ten seconds some transfer of material from the irradiated spot was observed accompanied by formation of small amount of gaseous products. Finally, most if not all samples were evaporated to form continuous polymer films on the wall. Like polysilylene films those obtained from PMTMDS contained Si-H bonds (absorption at 2100 cm^{-1}). This is not the case in laser pyrolysis of PETMDS. No absorption due to Si-H bond was observed in the IR spectrum of films obtained. Therefore for some reason rearrangement does not occur. Presumably the difference is due to the nature of monomer formed upon decomposition, namely 1,3-disilacyclobutanes (LP of PMTMDS) and **1g** (LP of PETMDS).

$$\text{\textasciitilde}[\underset{\underset{Me}{|}}{\overset{\overset{Me}{|}}{Si}}CH_2\underset{\underset{Me}{|}}{\overset{\overset{Me}{|}}{Si}}]_n\text{\textasciitilde} \longrightarrow Me_2Si\overset{\displaystyle CH_2}{\underset{\displaystyle \diagup\ \ \diagdown}{-\!-\!-}}SiHMe$$

$$\swarrow \qquad \searrow$$

$$Me_2Si\sqcap + \sqcap \qquad MeHSi\sqcap$$
$$\sqcup\!-\!SiH_2 \qquad\qquad \sqcup\!-\!SiHMe$$

$$\downarrow \qquad\qquad\qquad \downarrow$$

$$\text{\textasciitilde}[\underset{\underset{Me}{|}}{\overset{\overset{Me}{|}}{Si}}CH_2\underset{\underset{H}{|}}{\overset{\overset{H}{|}}{Si}}CH_2]_n\text{\textasciitilde} \qquad \text{\textasciitilde}[\underset{\underset{H}{|}}{\overset{\overset{Me}{|}}{Si}}CH_2\underset{\underset{H}{|}}{\overset{\overset{Me}{|}}{Si}}CH_2]_n\text{\textasciitilde}$$

$$\text{\textasciitilde}[\underset{\underset{Me}{|}}{\overset{\overset{Me}{|}}{Si}}CH_2CH_2\underset{\underset{Me}{|}}{\overset{\overset{Me}{|}}{Si}}]_n\text{\textasciitilde} \quad \overset{\Delta}{\rightleftharpoons} \quad Me_2Si\!-\!SiMe_2$$

The first is known to polymerize on heating,[1] whereas the second undergoes spontaneous polymerization.[22] Experiments with added butadiene show no reactive intermediates like silylenes or disilenes are involved.

Oxidative Cleavage of Si-Si bonds

PSHC (IV) do not change upon heating in water, concentrated hydrochloric and sulphuric acids or sodium (potassium) hydroxides. However, in solutions or dispersed in carbon tetrachloride they readily react with halogens. This results in a cleavage of Si-Si bonds which in excess of chlorine produces bis(dimethylchlorosilyl)methane or 1,2-bis(dimethylchlorosilyl)ethane. In the case of PEPS dimethyldichlorosilane is also formed . Reactions with bromine causes the loss of colour of the starting solution. PETMDS dispersed in benzene as well as compound **5** in benzene solution react with peroxybenzoic acid at 25°C resulting in the corresponding siloxanes. Depending on the amount of peroxybenzoic acid taken the product contains either siloxane only or both siloxane and disilane bonds. The ability to oxidatively cleave PETMDS Si-Si bonds was also used for modification of silicates.[41]

5 CONCLUSIONS

Heterochain polysilahydrocarbons containing silicon-silicon bonds in the main chain do not simply combine properties of PSHC (II) and polysilylenes. They may be interesting for some applications (*e.g.* membrane materials, liquid crystals, photoresists, silicon carbide ceramics, etc.). Some of them are easily available. Thus PETMDS may be obtained by Wurtz reaction from commercial 1,2-bis(dimethylchlorosilyl)ethane. Further investigations of these polymers which can be also called heterochain polysilylenes are in progress.

REFERENCES AND FOOTNOTES

1. L.E. Gusel'nikov and V.G. Avakyan, *Sov. Sci. Rev. B, Chem.*, 1989, **13**, 39.
2. P.P. Gaspar, 'Silylenes', (M. Jones, Jr. and R.A. Moss, Eds.), Reactive Intermediates, v.2, Wiley New York, 1985, p.333.
3. N.S. Nametkin and S.G. Durgar'yan, *Plast. Massy*, 1980, 13.
4. N.S. Nametkin and V.M. Vdovin, *Izv. Acad. Nauk SSSR, Ser. Khim.*, 1974, 1163.
5. E.A. Volnina, PhD Thesis, Topchiev Institute of Petrochemical Synthesis, Moscow, 1975.
6. J. Pola, V. Chvalovsky, E.A. Volnina, and L.E. Gusel'nikov, *J. Organomet. Chem.*, 1988, **341**, C13.
7. M. Sedlackova, J. Pola, E.A. Volnina, and L.E. Gusel'nikov, *J. Anal. Appl. Pyrolysis*, 1989, **14**, 345.
8. R. West, *J. Organomet. Chem.*, 1986, **300**, 327.
9. S. Yajima. J. Hayashi, and M. Omori, *Chem. Lett.*, 1975, 931.
10. Y. Hasegawa, M. Iimura and S. Yajima, *J. Mat. Sci.*, 1980, **15**, 720.
11. J.C. Thompson, P.J. Timms, and J.J. Margrave, *J. Chem. Soc., Chem. Comm.*, 1966, 566.
12. C-C Shiau, T-L Hwang and C-S Liu, *J. Organomet. Chem.*, 1981, **214**, 31.
13. D. Seyferth and D.P. Duncan, *J. Amer. Chem. Soc.*, 1978, **100**, 7734.
14. A.G. Brook, J.W. Harris, J. Lennon and M. El Sheikh, *J. Amer. Chem. Soc.*, 1979, **100**, 83.
15. A.G. Brook and J.W. Harris, *J. Amer. Chem. Soc.*, 1976, **98**, 3381.
16. D. Seyferth and S.C. Vick, *J. Organomet. Chem.*, 1977, **125**, C11.
17. M. Ishikawa, T. Fuchikami and M. Kumada, *J. Amer. Chem. Soc.*, 1979, **101**, 13.
18. M. Ishikawa, K. Nishimura, H. Ochiai and M. Kumada, *J. Organomet. Chem.*, 1982, **236**, 7.
19. Y-M. Pai, C.-K. Chen, and C.-S. Liu, *J. Organomet. Chem.*, 1982, **226**, 21.
20. L.E. Gusel'nikov, Yu. P. Polyakov, E.A. Volnina, and N.S. Nametkin, USSR Patent 810701, 1981; USA Patent 4328350.
21. L.E. Gusel'nikov and Yu. P. Polyakov, *Sov. Sci. Rev. B, Chem.*, 1989, **13**, 3.
22. L.E. Gusel'nikov, Yu. P. Polyakov, and N.S. Nametkin, *Dokl. Akad. Nauk SSSR*, 1980, **235**, 1133.
23. H.-Z. Ni, M. Ishikawa, K. Nate, K. Matsuzaki and M. Kumada, Seventh International Symposium on Organosilicon Chemistry, Kyoto, Japan, 1984, Abstracts of Papers, p. M02.
24. Dow Corning Limited, Brit. Pat. 667435, **1952**; *Chem. Abstr.*, 1952, **46**, 6429.
25. H.A. Clark, USA Pat. 2563004, 1951; *Chem. Abstr.*, 1952, **46**, 5846.
26. E.A. Chernyshev, S.A. Bashkirova, P.M. Matveichev, L.E. Gusel'nikov, and Yu.P. Polyakov - to be published.
27. Our attempt to obtain high molecular PETMDS via polyaddition reaction of 1,1,2,2-tetramethyldisilane to 1,2-divinyl-1,1,2,2-

tetramethyldisilane with H_2PtCl_6 catalyst failed. Only low molecular oily product was formed.

28. J. Pola, J. Vitec, Yu. P. Polyakov, L.E. Gusel'nikov, P.M. Matveychev, S.A. Bashkirova, J. Tlaskal, and R. Mayer, International COnference on Laser Induced Chemistry, Abstracts of Papers, September 18-22, 1989, Bechyne, Czechoslovakia, p. 108.
29. L.A. Leites, L.E. Gusel'nikov, Yu. P. Polyakov, S.A. Bashkirova, and P.M. Matveychev - to be published.
30. L.A. Leites, S.S. Bukalov, V.V. Dement'ev, T.S. Yadritseva, M.K. Mochov, and T.M. Frunze, *Izv. Akad. Nauk SSSR, Ser. Khim.*, 1989, 2869.
31. Yu. E. Ovchinnikov, V.E. Schklover, Yu. T. Struchkov, Yu. P. Polyakov, and L.E. Gusel'nikov, *Macromolecul. Chem.*, 1986, **187**, 2011.
32. I.I. Tverdokhlebova, O.I. Sutkevich, I.A. Ronova, L.E. Gusel'nikov, and Yu. P. Polyakov, *Vysokomolek. Soed., A.*, 1988, **30**, 1070.
33. Yu. E. Ocvhinnikov, V.E. Schklover, Yu. T. Struchkov, Yu. P. Polyakov, and L.E. Gusel'nikov, *Acta Crystallogr. Sect. C.*, 1985, **C41**, 1055.
34. V.M. Polikarpov, E.M. Antipov, E.V. Matukhina, Yu. P. Polyakov, L.E. Gusel'nikov, S.A. Bashkirova and P.M. Matveychev - to be published.
35. Yu. E. Ovchinnikov, V.E. Shklover, Yu. T. Struchkov, L.E. Gusel'nikov, Yu.P. Polyakov, E.A. Volnina, I.A. Litvinov, C.D. Artamonova, and Yu.H. Dyshlevski, 15th Conference on Macromolecular Chemistry, Abstracts of Papers, 1985, Alma-Ata, USSR, p.54 (in Russ.).
36. P. Weber, D. Guillon, S. Houlios, and R.D. Miller, *J. Phys. France*, 1989, **50**, 793.
37. A.J. Lovinger, F.C. Schilling, F.A. Bovey and J.M. Zeigler, *Macromolecules.*, 1986, **19**, 2660.
38. V.V. Teplyakov, L.E. Gusel'nikov, Yu.P. Polyakov, S.A. Bashkirova, and P.M. Matveychev - to be published.
39. M. Blazso, T. Szekely, T. Till, G. Varhegyi, L.E. Gusel'nikov, and Yu.P. Polyakov - unpublished results.
40. J. Pola, J. Vitec, Yu.P. Polyakov, L.E. Gusel'nikov, P.M. Matveychev, S.A. Bashkirova, J. Tlaskal, and R. Mayer - in press.
41. T.A. Kochina, H.E. Glushkova, Yu.P. Polyakov and L.E. Gusel'nikov, 7th Conference on Chemistry and Practical Application of Organosilicon Compounds, Abstracts of Papers, February, 1989, Leningrad, USSR, p.106 (in Russ.).

Small Polysilylalkane Molecules: Feed Stock Gases for Chemical Vapour Deposition of Silicon and Silicon Alloys

H. Schmidbaur,* R. Hager, and J. Zech
ANORGANISCH-CHEMISCHES INSTITUT, TECHNISCHE UNIVERSITÄT MÜNCHEN, LICHTENBERGSTRASSE 4, D-8046 GARCHING, GERMANY

1 INTRODUCTION

Most modern photovoltaic devices for solar energy conversion are based on either polycrystalline or amorphous silicon. While the crystalline material prevails for larger scale, high efficiency installations, the amorphous material (a-Si) has distinct advantages for small scale and low price systems.

Current a-Si production is largely based on plasma enhanced chemical vapour deposition using SiH_4 as the feed stock gas, with either phosphine or diborane admixed for doping. The thin-film devices thus obtained contain very significant amounts of hydrogen (a-Si:H). The characteristic semiconductor energy band-gap ΔE_g of this material can be altered by introducing carbon or germanium as alloy constituents, and standard technology is therefore using CH_4 or GeH_4 as sources for C and Ge in the PE-CVD process. The properties of the films generated by these techniques are not fully satisfactory, however, and alternative sources for C (and Ge) may offer distinct advantages, if the microstructure of the films can be adjusted through selection of the appropriate stoichiometry and constitution.

Volatile polysilylmethanes (Scheme 1) and their homologues have been the main target molecules in pertinent studies in this laboratory, as reviewed in this presentation [1,2], and elsewhere [3,4]. Efficient syntheses for these compounds based on cheap starting materials, and results on the physical properties of some thin-film products are now available.

C/Si-Isomers or C/Si-homologues of the polysilylmethanes (Scheme 2) may have advantages for certain purposes, especially regarding refractory or insulating carbon-rich coatings, and representative examples have also been included in the studies.

2 RESULTS

Polysilylmethanes

Members of the polysilylmethane series have the general formula $C(SiH_3)_n H_{4-n}$ with the silicon contents increasing as n increases from 1 to 4:

> Scheme 1 The four polysilylmethanes
>
> CH_3SiH_3 $CH_2(SiH_3)_2$ $CH(SiH_3)_3$ $C(SiH_3)_4$

<u>Methylsilane</u> CH_3SiH_3, the n = 1 species, is a standard commercial chemical with attractive properties as a CVD feed-stock material. With a convenient boiling point (-53°C) it is easily separated and purified and obtained as a non-toxic and not spontaneously inflammable gas, which can be stored in common pressure vessels without decomposition. It showed very limited advantages over the traditional methane/silane mixture, however, when applied for the deposition of amorphous silicon carbide films. Spectroscopic analyses and effusion measurements of the products generated in glow discharge experiments (with or without silane and hydrogen gas admixtures) suggested that the carbon contents of the films is largely associated with methyl or methylene moieties. These structural units lead to severe distortions and voids in the silicon network, and are thus detractive for the electrical properties of the material [5]. Electrical conductivity in particuar is strongly reduced.

<u>Disilylmethane</u> $CH_2(SiH_3)_2$, the n = 2 species, has been known in the literature for some time and several preparative routes have been tested for its synthesis. Virtually all of these used bis(trichlorosilyl)methane as the precursor. This hexachloro-derivative is most readily available from a Direct Synthesis employing silicon/copper alloy and polyhalomethanes [6], but also in a Benkeser reaction starting from silicochloroform, polyhalomethanes and a tertiary amine [7]. Substitution of chlorine by hydrogen in $CH_2(SiCl_3)_2$ is accomplished by reaction with excess complex alkali hydrides, mostly $LiAlH_4$, and leads directly to disilylmethane. The compound is also accessible from silylpotassium and dihalomethanes, but this method is only feasible on the laboratory scale [8].

A new route to $CH_2(SiH_3)_2$ starts from commercially available trichlorophenylsilane or dichlorodiphenylsilane, which are easily converted into (di)phenylsilane (with complex alkali hydrides), and finally into monochloro- or monobromo-phenylsilane by $AlCl_3$-catalyzed reaction with HCl or treatment with anhydrous HBr at low temperature, respectively. All steps afford good yields of the intermediates, which are easily purified. The remaining phenyl groups can be removed by HBr-cleavage to give the dibromide $CH_2(SiH_2Br)_2$ in quantitative yield [9]:

$$PhSiCl_3 \xrightarrow{MH} PhSiH_3 \xrightarrow{HX} PhSiH_2X$$
$$Ph_2SiCl_2 \xrightarrow{MH} Ph_2SiH_2 \xrightarrow{HX} PhSiH_2X$$

$$2\ PhSiH_2X + CH_2X_2 + 2\ Mg \rightarrow CH_2(SiH_2Ph)_2 + 2\ MgX_2$$

$$CH_2(SiH_2Ph)_2 + 2\ HBr \longrightarrow CH_2(SiH_2Br)_2 + 2\ PhH$$

$$CH_2(SiH_2Br)_2 + 2\ MH \longrightarrow CH_2(SiH_3)_2 + 2\ MBr$$

(X = Cl, Br; M = metal; Ph = phenyl.)

Like methylsilane, disilylmethane is not spontaneously inflammable gas(b. p. 15°C), which is stable at room temperature and can be stored indefinitely in common low-pressure cylinders or glass ampoules. Distillation and fractional condensation afford very high purity samples. A complete set of analytical, spectroscopic and structural data is available [9,10].

Amorphous hydrogenated silicon/carbon alloy films obtained in glow-discharge experiments using disilylmethane and its admixtures with silane and/or hydrogen as feed-stock gases showed greatly improved physical characteristics. Owing to changes in the basic structure of the films (largely a reduced number of methyl and methylene units as indicated by thermal effusion experiments and vibrational spectroscopy), a significant increase in the band-gap and a higher photoconductivity are achieved at lower carbon contents [11].

<u>Trisilylmethane</u> $CH(SiH_3)_3$, the n = 3 species, has been prepared only recently. Early attempts of synthesis had failed or given extremely low yields owing to an unexpected cleavage of Si-C bonds in the final Cl^-/H^- substitution step [12]. The nona<u>chloro</u> precursor $CH(SiCl_3)_3$, although available in low yields through the methods also used for $CH_2(SiCl_3)_2$ (above), <u>i.e.</u> mainly the Direct Synthesis or the Benkeser reaction, is thus not an appropriate starting material in this case.

Again the new route via a triphenylated precursor provided a more useful alternative. The Merker-Scott reaction of chlorophenylsilane with bromoform and magnesium in tetrahydrofuran gives acceptable yields of tris-(phenylsilyl)methane, which is then readily converted into the tribromide and finally into trisilylmethane following the procedures already described for disilylmethane (above).

$$3\ PhSiH_2X + CHX_3 + 3\ Mg \rightarrow CH(SiH_2Ph)_3 + 3\ MgX_2$$

$$CH(SiH_2Ph)_3 \longrightarrow CH(SiH_2X)_3 \longrightarrow CH(SiH_3)_3$$

Trisilylmethane is a colourless liquid (b.p. 61°C), not spontaneously inflammable in air, which can be stored at room temperature for a prolonged time without decomposition. Physical, analytical and spectroscopic data of high purity samples have been determined including the results of electron diffraction and photoelectron spectroscopy studies. The crystal structure of the triphenylated precursor $CH(SiH_2Ph)_3$ has been determined by X-ray diffraction [13].

PE-CVD experiments carried out in a similar way as described for disilylmethane afforded amorphous hydrogenated silicon carbon alloy films and further improved photoelectrical quality. The superior properties appear to be associated again with structure modifications originating from a decreasing number of C-H bonds, as induced by the stoichiometry and structure of the precursor molecule.

<u>Tetrasilylmethane</u> $C(SiH_3)_4$, the n = 4 species, although provisionally included in patents [14], has not been reported previously. Its synthesis has now also been accomplished via the halo(phenyl)silane route according to the following sequence:

$$PhSiH_2X \rangle C(SiH_2Ph)_4 \rangle C(SiH_2Br)_4 \rangle C(SiH_4)_4$$

However, several complications arose in these investigations, which rendered the initially proposed procedure unsatisfactory: a) The yields of the Merker-Scott reaction are low. b) Owing to the high boiling point of the tetraphenyl compound (and the by-products), purification has to rely on crystallisation of this intermediate. More often than not, however, the crystallisation of $C(SiH_2Ph)_4$ is unsuccessful, or at least unpredictable. c) The yields of the tetrabromo intermediate are surprisingly low, and it was found that this compound is inherently unstable in the absence of a solvent, and undergoes disproportionation (HBr redistribution) and finally decomposition even below room temperature. d) The final Cl^-/H^- substitution is accompanied by significant Si-C cleavage.

The seemingly trivial step from phenyl to <u>p</u>-tolyl solved the first two of these problems [15]: The yield of the Merker-Scott reaction is slightly improved with a tolyl precursor, and $(\underline{p}\text{-}MeC_6H_4SiH_2)_4C$ turned out to crystallize much better (m.p. 69°C from ethanol) than the tetraphenyl homologue (m.p. 47°C). Regarding problems c) and d), it is now recommended to avoid removal of solvent from $C(SiH_2Br)_4$ and to work at low temperature, or to turn to $C(SiH_2Cl)_4$ as an alternative.

This new compound is available from $(PhSiH_2)_4C$ or
$(p-MeC_6H_4SiH_2)_4C$ in the $AlCl_3$-catalyzed cleavage using
HCl in a polar solvent, and shows greatly improved stability (^{1}H NMR data: δ SiH_2 4.80 ppm, $^1J(SiH)$ = 255 Hz,
in benzene-d_6). Trichloro-p-tolylsilane is commercially
available and can be converted into p-tolylsilane
through Cl^-/H^- substitution and into chloro-p-tolylsilane with $AlCl_3/HCl$ in diethylether. Finally,
improvements in the yield of the halide/hydride
substitution [problem d)] are gained through the use of
a phase transfer catalyst [16].

Tetrasilylmethane (the Si/C inverse of tetramethylsilane) is a colourless liquid (b.p. 86.5°C), which is
not spontaneously inflammable in air and can be stored
indefinitely at room temperature. Its physical, analytical, and spectroscopic properties have been compiled,
including electron diffraction data. The crystal structure of $(PhSiH_2)_4C$ has been determined. PE-CVD experiments are in progress [17].

(Methyl/silyl)methanes

The poly(silyl)methanes are easily recognized as a
special set of species taken from a larger series of
compounds where mixing of hydrogen, methyl and silyl
substituents at a central carbon atom is considered
(Scheme 2):

Scheme 2 The fifteen methyl/silyl-methanes

CH_4 H_3CCH_3 $H_2C(CH_3)_2$ $HC(CH_3)_3$ $C(CH_3)_4$
  H_3CSiH_3 $H_2C(CH_3)SiH_3$ $HC(CH_3)_2SiH_3$ $C(CH_3)_3SiH_3$
    $H_2C(SiH_3)_2$ $HC(CH_3)(SiH_3)_2$ $C(CH_3)_2(SiH_3)_2$
      $HC(SiH_3)_3$ $C(CH_3)(SiH_3)_3$
        $C(SiH_3)_4$

The first line of the matrix represents the poly(methyl)methanes, and the second line contains the four
alkylsilanes $MeSiH_3$, $EtSiH_3$, $i-ProSiH_3$, and $t-ButSiH_3$,
while the diagonal lists the poly(silyl)methanes. The
remaining "off-diagonal elements", which – surprisingly
– had not been reported, have now also been considered
as candidates for CVD experiments. Their more carbon-rich stoichiometry suggests an approach to silicon carbide type phases, however, with very different physical
properties.

1,1,1-Tri(silyl)ethane $CH_3C(SiH_3)_3$ is available via
the halo(phenyl)silane route from 1,1,1-trichloroethane.
The synthesis suffers from low yields of the step towards the $CH_3C(SiH_2Ph)_3$ precursor (m.p. 80°C), but all
other parts of the preparation proceed as usual [18]. The
compound is a colourless, distillable liquid (b.p. 81°C).

1,1-Disilyl-ethane $CH_3CH(SiH_3)_2$ is best prepared through catalytic hydrosilylation of trichlorovinylsilane with silicochloroform followed by Cl^-/H^- substitution of the $CH_3CH(SiCl_3)_2$ precursor [19]. The product is a colourless, distillable liquid (b.p. 41°C). Thermodynamically it is less stable than its isomer $H_3SiCH_2CH_2SiH_3$.

2,2-Di(silyl)propane $(CH_3)_2C(SiH_3)_2$ could be obtained through similar synthetic procedures, but purification has not yet been fully satisfactory [20]. As soon as this problem is solved, the complete family of compounds contained in Scheme 1 is available for CVD test experiments.

Isomeric compounds

For disilylmethane only one isomer can be written. This compound, **methyldisilane** $CH_3SiH_2SiH_3$, has appeared in the literature several times, but it could never be isolated in significant amounts. In new attempts the result has not been any more encouraging. Neither the halo(phenyl)silane route starting from CH_3SiH_2Ph via CH_3SiH_2Br and its coupling with H_3SiK, nor the methylation of a halodisilane proved more useful than the earlier approaches. The product is a colourless, distillable liquid (b.p. 16°C), which ignites in air [21].

1,2-Disilylethane $H_3SiCH_2CH_2SiH_3$ presents no synthetic difficulties. The hexachloro-derivative, also available through $HSiCl_3$ addition to trichlorovinylsilane (above), is readily converted into the hexahydride. Like the disilylmethane isomer, but unlike methyldisilane *e.g.*, the compound is not spontaneously inflammable in air [22].

3 DISCUSSION

In the systematic studies presented here a large family of small, simple, ternary compounds Si/C/H of high volatility has been completed, which hold great promise as starting materials for chemical vapour deposition experiments aiming at the preparation of amorphous or polycrystalline thin films containing the three elements in different stoichiometries and with a variety of structural characteristics.

These compounds offer a number of distinct advantages over standard mixtures of silane, methane, and hydrogen. The species are all accessible starting from cheap and commercially available materials and through procedures which are easily carried out in standard equipment. Being not spontaneously igniting in air, they are easy and safe to handle. Their high thermal stability allows indefinite storage at room temperature in ordinary glass or steel containers.

The only chemical bonds allowed in the system are Si-C, Si-H, and C-H. With the Si-H bond as the weakest among these, and with few or no C-H bonds present in the feed-stock gas (as e.g in tetrasilylmethane), the formation of a silicon-silicon / silicon-carbon lattice or network should be favoured in the thermally, photochemically or plasma-induced decomposition. Pertinent experiments carried out to date seem to confirm this reasoning. Further work is under way.

ACKNOWLEDGEMENT

This work has been supported by the Council of the European Community, by Deutsche Forschungsgemeinschaft, by Fonds der Chemischen Industrie, and by Wacker Chemie GmbH. (through the donation of chemicals). The authors are also indebted to the colleagues who have carried out the initial studies for helpful discussion (J. Ebenhöch, C. Dörzbach-Lange, and J. Rott). G. Müller, O. Steigelmann and J. Riede are thanked for establishing X-ray data sets, D. W. H. Rankin and H. E. Robertson of the Edinburgh electron diffraction unit for determining two crucial structures, and H. Bock (Universität Frankfurt) for measuring photoelectron spectra. G. Winterling and M. Gorn (MBB, Phototronics) are thanked for many helpful discussions.

REFERENCES

1. H. Schmidbaur, J. Ebenhöch, Z. Naturforsch. B 41 (1986) 1527.
2. H. Schmidbaur, C. Dörzbach, Z. Naturforsch. B 42 (1987) 1088.
3. B. F. Fieselmann, C. R. Dickson, J. Organometal. Chem. 363 (1989) 1.
4. B. Goldstein, C. R. Dickson, I. H. Campbell, P. M. Fauchet, Int. Photovolt. Conf., Florence 1988. B. F. Fieselmann, M. Milligan, A. Wilczynski, J. Pickens, C. R. Dickson, IEEE Photovoltaic Spec. Conf., New York 1987, Abstr. 19, p. 1510.
5. A. Bubenzer, M. Gorn, N. Kniffler, K. Thalheimer, G. Winterling in R. Urban (Ed.): Status Report 1987 - Photovoltaic, PBE-KFA Jülich, 1988, p. 349.
6. R. Müller, G. Seitz, Chem. Ber. 91 (1958) 22.
7. R. A. Benkeser, J. M. Gaul, W. E. Smith, J. Am. Chem. Soc. 91 (1969) 3666; R. A. Benkeser, W. A. Smith, ibid. 90 (1968) 5307; R. L. Merker, M. J. Scott, J. Org. Chem. 29 (1964) 953; G. Fritz, Angew. Chem. 79 (1967) 657.
8. E. Amberger, H. D. Boeters, Chem. Ber. 97 (1964) 1999; J. A. Morrison, J. M. Bellama, J. Organometal. Chem. 92 (1975) 163.

9. R. Hager, O. Steigelmann, G. Müller, H. Schmidbaur, *Chem. Ber.* **122** (1989) 2115.
10. R. Hager, PhD Thesis, Techn. Univ. Munich 1990.
11. W. Beyer, R. Hager, H. Schmidbaur, G. Winterling, *Appl. Phys. Lett.* **54** (1989) 1666.
12. J. Ebenhöch, R. Hager, J. Zech, H. Schmidbaur, unpublished work 1986-90.
13. Ref. 9 and 10; O. Steigelmann, Diploma Thesis, Techn. Univ. Munich 1988.
14. C. R. Dickson, Eur. Pat. Appl. 0.233.613 (1987); Solarex Corp.
15. J. Zech, H. Schmidbaur, unpublished results 1989/-90.
16. V. N. Gevorgyan, L. M. Ignatovich, E. Lukevics, *J. Organmometal. Chem.* **284** (1985) C 31.
17. H. Bock *et al.*, unpublished results 1989/90.
18. J. Zech, H. Schmidbaur, unpublished results 1990.
19. H. Schmidbaur, R. Hager, *Z. Naturforsch.* B **43** (1988) 571.
20. J. Zech, H. Schmidbaur, submitted (1990).
21. J. Zech, Diploma Thesis, Techn. Univ. Munich 1988.
22. Ref. 2; see also H. Schmidbaur, J. Rott, *Z. Naturforsch.* B. in press.

Synthesis and Properties of Some Polysilaethers – Polyoxymultisilylenes

J. Chojnowski,* J. Kurjata, S. Rubinsztajn, and M. Ścibiorek

CENTRE OF MOLECULAR AND MACROMOLECULAR STUDIES OF THE POLISH ACADEMY OF SCIENCES, SIENKIEWICZA 112, 90–363 ŁÓDŹ, POLAND

1 INTRODUCTION

Among polymers bearing silicon and oxygen in the main polymer chain the most important are polysiloxanes. It is generally accepted to classify as polysiloxanes all polymers having the -SiOSi- fragment in the backbone. However, classical siloxane polymers are only those with alternating arrangement of silicon and oxygen atoms. They contain both the -SiOSi- and -OSiO- groupings and may be considered as silicon analogues of polyacetals. There are, however, several classes of organosilicon polymers of regular structure with silicon and oxygen atoms in the backbone, which have no silaacetal linkage. They can be derived from polyethers by replacement of all carbon atoms or a part of them in the main polymer chain by silicon atoms. Being silicon analogues of polyethers they can be referred to as polysilaethers. Several classes of linear polymers of this type could be distinguished:

    1. Polysilaethers with siloxane linkage
       a) polyoxymultisilylenes - $\{O(Si)_m\}_n$
       b) polycarbosiloxanes - $\{OSi(C)_m Si\}_n$

    2. Polysilaethers with mixed -SiOC- ether linkage

    3. Polysilaethers with classical -COC- ether linkage and silicon non-bonded to oxygen atom.

Polymers of these classes differ to a considerable extent in their reactivity and physical behavior. Although there has been considerable interest in studies of polycarbosiloxanes and polysilaethers bearing the SiOC linkages, very little is known of polysilaethers having exclusively silicon atoms besides oxygen in the backbone[1]. Recently we have synthesized four polymers of this type which are fully methylated analogues of common polyethers: polydisilaoxirane m=2 (permethyl-polyoxydisilylene), polytrisilaoxetane m=3 (permethyl-polyoxytrisilylene), polytetrasilatetrahydrofuran m=4 (permethyl-polyoxytetrasilylene) and polyhexasilaoxepane m=6 (permethyl-polyoxyhexasilylene).

$$\{(SiMe_2)_mO\}_n \qquad (1)$$

We also synthesized some copolymers of polysiloxanes containing corresponding polysilylene sequencies.

Polyoxymultisilylenes may constitute interesting new materials because they combine the structural features of two most common classes of organosilicon polymers, i.e., polysiloxanes and polysilylenes. In particular permethyl polymers of this type may be of interest. Polydimethylsiloxane having an unusually flexible chain is an elastomer of a very low Tg and a low temperature of crystallization, perfectly soluble in many solvents. In contrast, polydimethylsilylene is a highly crystalline, insoluble and unformable material. Polymers of ordered structure comprising siloxane groupings separating polysilylene sequences of equal sizes are crystalline, but soluble and meltable. They preserve some interesting properties of polysilylenes.

2 SYNTHESIS

Oxidation of Polysilylenes

The first approach to synthesis of polysilaethers explored the oxidation of polysilylenes[2]. This reaction was believed to produce a new material containing siloxane and silicon-silicon linkages, although silaether polymers of a regular chain structure were not likely to be formed by this route. It was shown that oxidation of polycyclohexylmethylsilylene with m-chloroperbenzoic acid leads to the insertion of oxygen to the Si-Si bond, changing strongly the properties of the polymer[2]. The distribution of the oxygen atom along the chain is far from statistical as the formation of the SiOSi grouping makes the neighboring Si-Si bond more susceptible to the oxygen attack leading to a copolymer block like structure. The oxidation of cyclic polysilylenes is important as it may give an easy access to some cyclic polysilaethers - monomers for the ring-opening polymerization. Particularly easily oxidized are cyclic polysilylenes with strained four--membered rings[3].

Hydrolytic Polycondensation of α,ω-Dichlorooligosilanes

The hydrolytic polycondensation is a method of synthesis of both cyclic and linear polysilylene ethers. Basic monomers used here are α,ω-dichloropermethyloligosilylenes, which may be synthesized by a two step route comprising synthesis of permethylcyclohexasilylene by the Würtz method followed by chlorination with chlorine[4]. The dimer 1,2-dichlorotetramethyldisilylene may also be afforded by the Würtz synthesis of hexamethyldisilylene followed by the Kumada chlorination process[5].

Hydrolytic polycondensation of the silyl chloride ended oligomers may be carried out in a solvent using 1/2 equivalent of water in the presence of stoichiometric amounts of Et_3N as HCl-acceptor and catalytic amounts of 4-N,N-dimethylaminopyridine (DMAP) or N-methylimidazole. Water is slowly introduced to the polycondensation system, thus securing the coexistance of the SiOH and SiCl functional groups during the polycondensation process. The reaction presumably occurs according to the scheme of the heterofunctional polycondensation (eq.2), as the high effectiveness of this catalytic system in the promotion of the condensation involving the SiOH and SiCl groups has recently been demonstrated[6].

$$n\ Cl(Me_2Si)_mCl + n\ H_2O \xrightarrow[-2n[Et_3N \cdot HCl]]{2n\ Et_3N + DMAP\ (in\ dioxane)} \xrightarrow{H_2O}$$

$$HO[(Me_2Si)_mO]_nH + cyclic\ oligomers \qquad (2)$$

The yield of the linear polymer amounts to 70%. The number average molecular weight of the polymer is within the range of $0.5 \times 10^4 - 3 \times 10^4$.

The heterofunctional polycondensation has been also successfully used for synthesis of copolymers having polysiloxane and polysilylene sequencies, e.g., see eq.3. A ^{29}Si NMR study confirmed the regular structure of the polysilaethers and the preservation of the polysilylene sequencies in the copolymers.

$$n\ HO[Si(Me)_2O]_3H + n\ Cl[Si(Me)_2]_3Cl \longrightarrow \{[Si(Me)_2O]_3[Si(Me)_2]_3\}_n \qquad (3)$$

A modified hydrolytic polycondensation method has been explored to convert α,ω-dichlorooligodimethylsilylenes into some cyclic silaethers used as monomers in the ring opening polymerization described in the next subsection.

Ring Opening Polymerization

Thermodynamics. The ring-chain equilibrium was studied for the permethylpolyoxydisilylene system. It is difficult to attain full equilibrium conditions for this system because of some undesired processes accompanying cationic polymerization of the cyclic silaethers. On the other hand a low chemoselectivity in the anionic polymerization of these monomers makes this reaction useless for studies of the thermodynamics of the specific ring opening polymerization involving the exclusive SiOSi bond cleavage. The equilibration was obtained to a reasonable degree by the cationic polymerization of octamethyl--1,4-dioxatetrasila-cyclohexane in methylene chloride solution initiated with trifluoromethanesulfonic acid. Approximate contributions to the equilibration mixture from the cyclics of $(Me_2SiMe_2SiO)_n$ series counting on the bulk polymerization system were estimated to be 0,8,6,3, 0.8,0.2 w % for n=1-6, respectively. The total content of cyclics up to 18 membered ring was about 18%, being thus higher than that in the bulk polydimethylsiloxane equilibrium mixture containing about 12% of rings of the corresponding size, i.e. $(Me_2SiO)_n$ n=3-9[7]. None the less thermodynamics is not a serious obstacle for the synthesis of this polyoxymultisilane by the ring opening polymerization.

The cyclic dimer (a fully methylated silicon analogue of dioxane), used here as the monomer, presumably has some small strain in the ring which may result from a smaller SiOSi bond angle as compared with unstrained siloxanes[8]. This cyclic compound readily undergoes the oxygen atom insertion to one of its Si-Si bonds[9] which also could be explained by a ring strain release.

Some initiators of the anionic polymerization of the cyclic silaethers produce active propagating centers which are very reactive towards splitting of both the Si-O and Si-Si bonds. Application of such an initiator makes possible the non-selective equilibration of the system involving cleavage of all types of bonds in ring and chain skeleton. The chromatogram of the cyclic oligomers in such equilibrium mixture is presented in Fig.1.

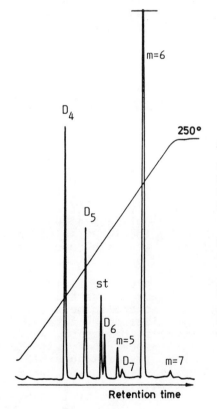

Figure 1 Gas liquid chromatogram of the cyclic oligomers in the equilibration mixture obtained by the nonselective ring opening polymerization of $[(Me)_2Si(Me)_2SiO]_2$, 2.2 mol·dm^{-3} in THF, 25°C, catalyst Me_3SiOK + HMPT, involving cleavage of both the Si-O and Si-Si bonds. Oligomers of the $(Me_2SiO)_n$ series are denoted by D_n n=4-7 while those of the $(Me_2Si)_m$ series by numbers corresponding to the m value, st is standard.

The result is surprising as the reaction eventually leads to almost exclusive formation of cyclic oligomers of the polydimethylsiloxane series and the polydimethylsilylene series. No cyclics of polyoxytetramethyldisilene series are observed in the equilibration mixture. The linear polymer in the equilibrium mixture thus must have the structure of block copolymer of PDMS and polydimethylsilylene. This observation proves that oxymultisilylene arrangement of the polymer chain is thermodynamically unstable. The reason is the increase of the average energy of the Si-Si bond with increasing the length of the polysilylene sequence due to the ability of the $\sigma-\sigma^*$ conjugation of the polysilylene[10].

Cationic Ring-Opening Polymerization. The ability of cyclic oxymultisilylene monomers to polymerize under the action of cationic initiators has not been obvious in the light of known reactivity of polysilylene sequencies towards electrophilic reagents[11]. Prior to our work[1] only oligomers have been obtained by this route[12]. However, the octamethyltetrasila-1,4-dioxane ring is opened readily with a strong protonic acid like CF_3SO_3H. The polymerization goes smoothly giving about 60-70% of the polymer when the reaction is carried out in bulk or in a concentrated solution, and quenched at 80-90% of the monomer conversion. The polymerization of octamethyltetrasila-THF (eq.4) goes even faster and leads to the full monomer conversion. Evidently there is a considerable strain in the five-membered oxatetrasilapentane ring.

$$\begin{array}{c}\text{Me} \\ \text{Me}-\text{Si}-\text{Si}-\text{Me} \\ \text{Me}-\text{Si} \quad \text{Si}-\text{Me} \\ \text{Me} \quad \text{O} \quad \text{Me}\end{array} \xrightarrow{CF_3SO_3H} \left[\begin{array}{c}\text{Me Me Me Me} \\ | \; | \; | \; | \\ \text{Si-Si-Si-Si-O} \\ | \; | \; | \; | \\ \text{Me Me Me Me}\end{array}\right]_n \quad (4)$$

In contrast the polymerization of permethylhexasilaoxepane proceeds very slowly even with a strong protonic acid indicating that the elongation of the polysilylene sequence in the monomer makes more difficult the cationic polymerization. The regular oxymultisilylene structure of polymers obtained by the cationic polymerization was confirmed by ^{29}Si NMR spectra and by the identity of the NMR, UV and IR spectra of these polymers with respect to those obtained by the hydrolytic polycondensation.

The polymerization is accompanied by the formation of higher cyclic oligomers. In the case of the polymerization of the permethyl-tetrasiladioxane, oligomers having even numbers of the repeating oxydisilylene units are formed simultaneously with the monomer conversion, whereas those with odd numbers of the units appear in a late stage of the process, (Fig.2). Thus the polymerization exhibits similar behavior to that of the polymerization of hexamethylcyclotrisiloxane[13]. This observation eliminates thus the possibility of the formation of the even cyclics by back biting depolymerization. They presumably arise as a result of the reaction between end groups as proved for the polydimethylsiloxanes. The mechanism of the polymerization is supposed to be analogous to that postulated for the cationic polymerization of cyclic siloxanes[14], Scheme 1.

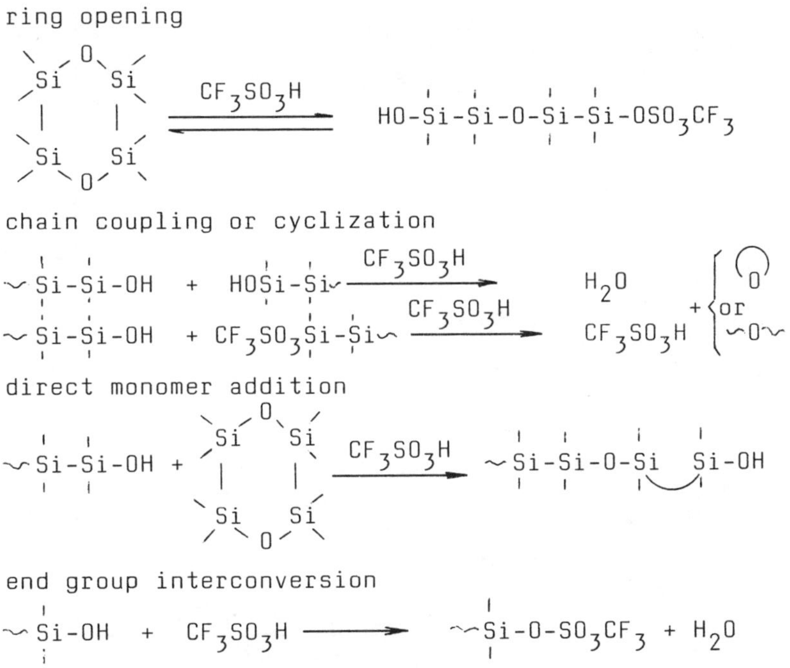

ring opening

$$\text{cyclic}(Si-Si-O-Si-Si-O) \xrightleftharpoons{CF_3SO_3H} HO-Si-Si-O-Si-Si-OSO_3CF_3$$

chain coupling or cyclization

$$\sim Si-Si-OH + HOSi-Si\sim \xrightarrow{CF_3SO_3H} H_2O + \begin{cases} \text{cyclic} \\ \text{or} \\ \sim O \sim \end{cases}$$
$$\sim Si-Si-OH + CF_3SO_3Si-Si\sim \xrightarrow{CF_3SO_3H} CF_3SO_3H$$

direct monomer addition

$$\sim Si-Si-OH + \text{cyclic}(Si-Si-O-Si-Si-O) \xrightarrow{CF_3SO_3H} \sim Si-Si-O-Si\quad Si-OH$$

end group interconversion

$$\sim Si-OH + CF_3SO_3H \longrightarrow \sim Si-O-SO_3CF_3 + H_2O$$

Scheme 1

There is, however, a substantial difference in polymerizations of hexamethylcyclotrisiloxane and cyclic oxymultisilylenes as the latter slows down as it proceeds. An interaction of polysilylene with the initiator leading to its deactivation is responsible for this phenomenon. This interaction may have also a detrimental influence on the polymer formed. It may lead to decrease of the molecular weight, to the formation of reactive side groups being potential branching and cross-linking points and finally it may introduce some other modification to the polymer structure spoiling its regularity. Particularly dangerous is a possible process of cleavage of alkyl groups by acid, as similar reactions have been reported[15].

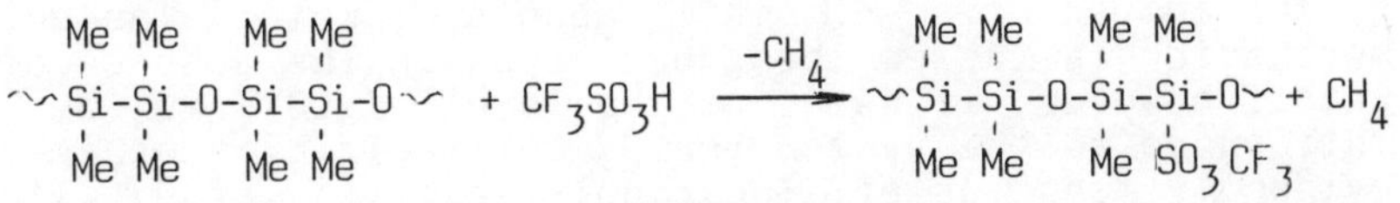

$$\underset{\underset{Me\ Me}{|\ \ |}}{\sim Si-Si-O-Si-Si-O\sim} + CF_3SO_3H \xrightarrow{-CH_4} \underset{\underset{Me\ Me\quad Me\ SO_3CF_3}{|\ \ |\quad\ \ \ \ |\ \ \ \ \ \ |}}{\sim Si-Si-O-Si-Si-O\sim} + CH_4$$

In order to reduce the detrimental effect of the side reactions, the polymerization was quenched at about

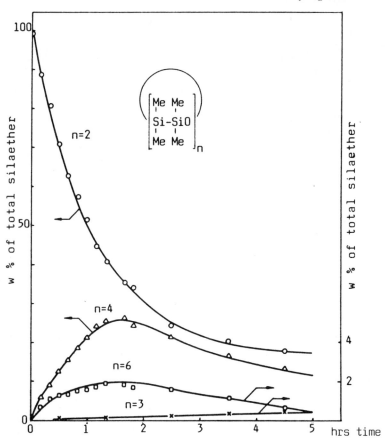

Figure 2 The oligomer formation during the monomer conversion in the polymerization of permethyltetrasila-1,4-dioxane. 57% in CH_2Cl_2, 25°C, initiator $[CF_3SO_3H]_0 = 7 \times 10^{-4}$ mol·dm^{-3}, $[H_2O]_0 = 4.4 \times 10^{-3}$ mol·dm^{-3}.

85% of the monomer conversion, the polymer was carefully washed with water, dried and treated with the mixture of Me$_3$SiCl, Et$_3$N and DMAP 1:1:0.1 to silylate all OH groups including also those possibly formed pendant to the chain. The polymer treated this way was not only highly soluble but it showed unimodal molecular weight distribution function (Fig.3), which proved that branching was of a minor importance. In contrast to the polymerization of hexamethylcyclotrisiloxane (D$_3$) the initial rate of the polymerization of the tetrasiladioxane is little affected by the introduction of small amounts of water. The polymerization is faster than that of D$_3$ in the absence of water; however, if water at the level of 10^{-2} mol dm^{-3} is introduced to the system the D$_3$ polymerization proceeds evidently much faster. The copolymerization of D$_3$ with the tetrasiladioxane occurs smoothly making possible synthesis of polydimethylsiloxane containing a determined number of the Si-Si bonds.

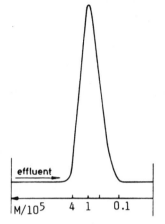

Figure 3 The gel permeation chromatogram of permethyl-polyoxydisilylene, synthesized by ring opening polymerization, 50°C, initiated with CF_3SO_3H in bulk 4.4×10^{-3} mol dm^{-3}. Cyclic oligomers were removed by precipitation of the polymer from its CH_2Cl_2 solution with methanol. $M_w/M_n=1.8$. GPC calibrated with polystyrene.

Anionic Ring-Opening Polymerization. Cyclic permethyl oxymultisilylenes undergo anionic ring opening polymerization initiated with a strong base[16]; however the polymer formed has not the regular oxymultisilylene structure. The reason is that silanolate active propagating centres open both the siloxane bond as well as the silicon to silicon bond. The method of the generation of the silyl anion by the action of trialkylsilanolate on hexaalkyl-disilane (eq.6) has been proposed earlier[17].

$$Me_3SiSiMe_3 + Me_3SiOMt^+ \xrightarrow{HMPT/THF} Me_3Si^-Mt^+ + Me_3SiOSiMe_3 \quad (6)$$

$$Mt^+ = Na^+, K^+, Cs^+$$

The polymerization may thus occur according to the Scheme 2.

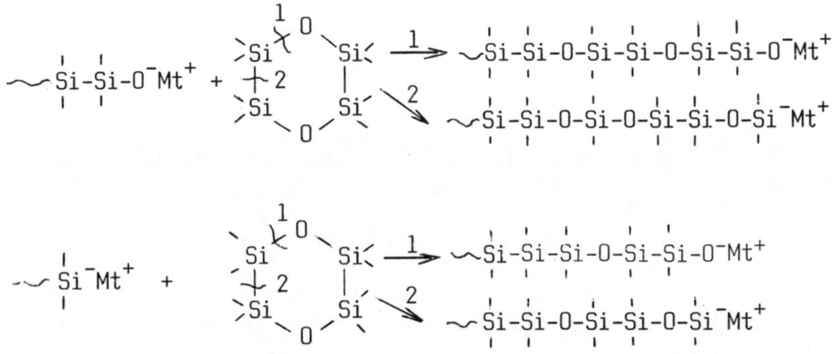

Scheme 2

Longer polysilylene sequences appear as the reaction proceeds. The silyl anion having the negative charge on the silicon atom bonded to oxygen is presumably very unstable and we suppose that it is also converted intramolecularly to a more stable silanolate or silylanion with a longer polysilylene sequence.

The relative reactivity in opening of the siloxane and disilylene grouping differs with a structure of the active propagating centre. If KOH is used as the initiator the preference for the opening of the siloxane bond is rather high and the polymer contains mostly disilylene sequencies, Fig.4.

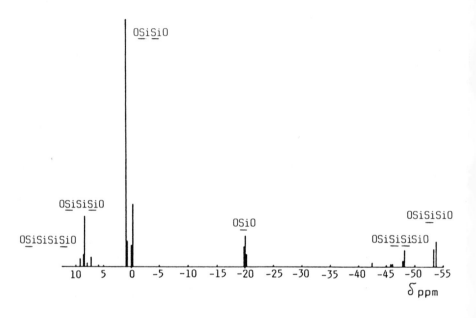

Figure 4 ^{29}Si NMR spectrum of the polysilaether synthesized by the ring opening polymerization of octamethyl--1,4-dioxatetrasilacyclohexane quenched at 93.5% of the monomer conversion in the presence of $[KOH]_{\bar{o}}=5\times10^{-3}$·mol·dm^{-3}, temp.110°C.

The formation of the silylanion active centre makes possible the synthesis of block copolymers with vinyl and diene monomers able to undergo anionic polymerization. For example, if to the living polymer obtained from a cyclic silaether, well-dried styrene is introduced then the yellow color of the silyl anion is turned into the red color of polystyryl anion, and fast styrene polymerization takes place.

$$\sim O(\overset{|}{\underset{|}{Si}})^-_n Mt^+ + CH_2=CH\text{-}Ph \longrightarrow \sim O(\overset{|}{\underset{|}{Si}})_n CH_2\text{-}CH^-\text{-}Ph\ Mt^+ \quad (7)$$

3 PROPERTIES

<u>Morphological Structure of the Polymers</u>   In contrast to polydimethylsiloxane which is known to be completely amorphous at room temperature permethylpolyoxymultisilylenes are highly crystalline, waxy tough-elastomeric materials. The melting point of the crystalline phase is dependent on the number of silicon atoms in the polysilylene sequence being for m=2,3,4,6, 45°C, 60°C, 145°C and 140°C respectively. The polymers are very soluble in many organic solvents including ethers, hydrocarbons and chlorinated hydrocarbons.

Wide-angle X-Ray scattering studies showed a high contribution from the order structure area reaching about 70% for the oxydisilylene polymer[18]. The nature of the crystalline phase is different for oxydisilylene and oxytetrasilylene materials. The latter shows small crystallites while the former has well shaped and relatively large spherulites.

<u>Spectral Properties</u>. Permethylpolyoxymultisilylenes show absorption in UV region dependent upon the length of the polysilylene sequence. As it could be expected there is a tendency to shift the absorption towards a longer wavelength with the elongation of the silylene sequence (Table 1). Since the polymers containing the Si-Si bonds are known for their photodegradability, these polymers are potential materials in lithographic technique.

<u>Thermal Properties</u>. Thermogravimetric investigations[18] showed that peroxymultisilylenes are thermally much less stable than polydimethylsiloxanes. Onset temperatures for polymers $-[(SiMe_2)_mO]_n$ m=2,3 and 4 were usually within the range 230-270°C, when the temperature was increased at a rate 20°C/min. A slow decomposition of the polymers was, however, observed in some cases even at temperatures below 200°C. The order of thermal stability of the polymers seemed to be following m=4 < 3 < 2 < 6. The shape of the TGA curves disclosed that the decomposition of these polymers is a very complex process depending also on the sample shape, its purity, way of its preparation and on the atmosphere of the TGA experiment. Polymers m=2,3 and 4 show relatively regular TGA curves in the range up to about 350-400°C. Mass spectroscopic studies showed that the main decomposition process is the depolymerization which is also accompanied by some restructurization process in polymers leading to cross-linking and retarding the depolymerization. The competition of these restructurization processes and depolymerization has a great impact on the decomposition course at higher temperatures and on the residue which in temp.900°C was in the range of 10-45 w %. A considerable mass loss is observed in the range 400-600°C. The TGA curve is irregular in this range. IR studies confirmed that extensive cleavage of organic groups occurs in this temperature region.

Table 1   Some spectral properties of the polymers $\text{-}(\text{Me}_2\text{Si})_m\text{O}\text{-}_n$

Polymer m	UV maxima[a) value/nm	^{29}Si NMR[b) value/ppm		
2	201	0.8		
3	224	8.3	-53.5	
4	242	8.9	-48.1	
6	264	9.1	-39.9	-45.6

a) in n-heptane, b) in CCl_4

REFERENCES

1. J. Chojnowski, J. Kurjata and S. Rubinsztajn, Makromol.Chem.Rapid Commun., 1988, 15, 469.
2. P. Tréfonas III and R. West, J.Polym.Sci., Polym. Lett.Ed., 1985, 23, 469.
3. M. Ishikawa and M. Kumada, J.Organomet.Chem., 1972, 42, 325.
4. W. Wojnowski, C.J. Hurt and R. West, J.Organomet. Chem., 1977, 124, 271.
5. M. Ishikawa, M. Kumada, H. Sakurai, J.Organomet. Chem., 1970, 23, 63.
6. S. Rubinsztajn, M. Cypryk and J. Chojnowski, J. Organomet.Chem., 1989, 367, 27.
7. P.V. Wright and M.S. Beevers, "Cyclic Polymers" ed. J.A. Semlyen, Elsevier Applied Science Publishers, London 1986, page 85.
8. T. Takano, N. Kasai and M. Kakudo, Bull.Chem.Soc.Jpn., 1963, 36, 585.
9. G.A. Razuvaev, T.N. Brevnova and V.V. Semenov, J. Organomet.Chem., 1984, 271, 261.
10. R.D. Miller and J. Michl. Chem.Rev., 1989, 89, 1359.
11. R. West "The Chemistry of Organic Silicon Compounds" eds. S. Patai and Z. Rappoport, Wiley and Sons, Chichester, 1989, Part 2, page 1207.
12. M. Kumada, M. Ishikawa and B. Murai, Kogyo Kagaku Zasschi, 1963, 66, 637.
13. J. Chojnowski, M. Ścibiorek and J. Kowalski, Makromol. Chem., 1977, 178, 1351.
14. J. Chojnowski "Silicon Chemistry" eds. E.R. Corey, J.Y. Corey and P.P. Gaspar, Ellis-Horwood, New York 1988, page 297.
15. Y.L. Chen and K. Matyjaszewski, 8th International Symposium on Organosilicon Chemistry, St.Louis, 1987, Abstract, page 85.
16. J. Chojnowski and J. Kurjata in preparation.
17. M. Mazurek and J. Chojnowski, 6th International Symposium on Organosilicon Chemistry, Budapest 1981, Abstracts page 148.
18. J. Chojnowski, J. Kurjata, S. Rubinsztajn, M. Ścibiorek and M. Zeldin in preparation.

Recycling – Magic or Tragic?

A. Stroh
DIVISION S, WACKER-CHEMIE GMBH, PRINZREGENTENSTRAßE 22, D-8000 MÜNCHEN 22, GERMANY

In 1968 the famous MIT submitted its report entitled "The limits to growth" to The Club of Rome, which had commissioned it.

Scientists, managers in industry and politicians of the highest calibre from all around the world established that the resources on our planet are finite. From this they concluded that every time capital stock and population grow, the world's equilibrium is upset and the existence of mankind and even the whole planet is put in jeopardy. They wrote of an urgent need to attain a state of global equilibrium as quickly as possible because time was pressing. I quote: "All input and output rates - births, deaths, investment, and depreciation - are kept to a minimum." End of quotation.

In order to attain this equilibrium, and especially to maintain it "technological advance would be both necessary and welcome". They wrote further that - quote - "A few obvious examples of the kinds of practical discoveries that would enhance the workings of a steady-state society include:

- new methods of waste collection, to decrease pollution and make discarded material available for recycling;

- more efficient techniques of recycling, to reduce rates of resource depletion; ..."

This list of "practical discoveries" extended from the harnessing of solar energy, through natural pest control to simultaneous advances in medicine to decrease the death rate and in contraceptives to reduce the birth rate accordingly.

But at the top of the list they placed recycling methods. Recycling appeared on the scene as a magic charm for saving humanity and was snatched up greedily by politicians and the media who broadcast it further with their own ideas.

Meanwhile, scientists and industrialists attempted to view the new method from a more realistic perspective.

Discussions took place but proved fruitless. The word recycling and the method behind it - although not new by any means, as I will show you later - remained and nowadays are firmly entrenched in many of our modern thought processes. Nevertheless, an emotive label is attached to the word recycling time and time again. And it often happens that, when recycling is hailed as the new magic cure-all, pseudo-realists pop up and point out its tragic components.

Back then, there was no mention of silicones as a class of products worth recycling. After all, they were made from sand and relatively minor amounts of organic substances. The high amounts of energy consumed - and therefore also lost - in converting the sand into silicon were regarded as being negligible because production at that time was low compared with that of other plastics.

From this point of view, therefore, silicones are plastics which do not urgently need recycling because they do not suffer from a lack of raw materials.

And yet, ever since the first Rochow syntheses were implemented in the late forties and early fifties, a considerable amount of recycling had already taken place - and not because of public pressure.

Three figures make this quite clear. In 1950, approximately half of the silanes used in the Rochow synthesis were discarded. In 1968, *i. e.* when The Club of Rome report appeared, this figure had fallen to 25 %. Nowadays it is less than 5 %!

The fact that the by-products could not be re-used was not a problem in terms of recycling because so little was being manufactured at that time. Economically there was no problem either because the sales prices were still extremely high and they more than covered costs.

The aim of the Rochow synthesis was the preparation of dimethyldichlorosilane. The first step taken by the silicone chemists was to find a suitable application for the other silanes that were also formed during the synthesis. Methyltrichlorosilane, which was formed initially in quantities of up to 25 % was used to make silicone resins, which at that time astonished everybody with their resistance to high temperatures and their hydrophobic nature. They were first used in building electric motors and in protecting masonry.

Methyl hydrogen dichlorosilane, which was formed in quantities of up to 15 %, considerably broadened the range of possible syntheses in silicone chemistry by virtue of the hydrogen atom attached to the silicon atom.

Besides these attempts to re-use or recirculate the by-products instead of destroying them, work was commenced on optimizing the synthesis process itself. The aim here was to allow only as many by-products to be formed as could be re-used for known applications.

But optimization also meant recycling the unusable by-products and unconsumed reagents. With this in mind, processes were developed for separating low-boiling and high-boiling mixtures of silanes, cleaving the products and converting them to useful silanes.

Unreacted methyl chloride is nowadays reintroduced into the circuit by means of special processes.
Consumed contact mass is no longer discarded but instead is regenerated and used again.
A process has even been developed for re-using unreacted silicon dust, instead of dumping it as was done in the past.

The most important recycling scheme to have been developed since the early days of the Rochow synthesis is that conceived for chlorine. The empirical formula for the manufacture of silicone clearly shows that chlorine mainly acts as a carrier to bring the methyl group to the silicon atom. All attempts so far to attach the methyl group without using chlorine have failed to produce a solution that can be used on an industrial scale. So in the early days, they aimed to reintroduce the chlorine into the circuit and not to discard it. Nowadays two processes are employed for recovering the chlorine. One involves hydrolysis followed by preparation of methyl chloride and the other, namely methanolysis, combines both of these steps. Through these two processes, approximately 70 % of the chlorine is recycled.

But processes also exist for recovering the chlorine from the other methyl chlorosilanes, *i. e.* the monochloro and trichloro products, and introducing it into the process for making methyl chloride. Of course, this is only done when these silanes are not going to be processed to chlorinated sales products.

One of the most important of these processes consists in oxidizing the unused chlorosilanes to highly dispersed silicic acid, *i. e.* extremely fine silicon dioxide, which finds its main application in the manufacturing of silicone elastomers. With the aid of this technique, the theoretical limits of chlorine recycling can almost be reached - *i.e.* a product with high value derived from a product which was in the past discarded.

Most of the silane-producing companies are sending their silanes not used, sometimes over hundreds of miles to a producer of highly dispersed silicic acid. This is both dangerous and uneconomic and it consequently interrupts the recycling circuit.

All of these techniques - and I have only mentioned the most important - have enabled us to increase the yield of the Rochow synthesis to 95 %, *i. e.*, only 5 % of the silanes generated now have to be disposed of. In absolute terms, approximately twice as much waste is generated now as at the time of The Club of Rome's report 22 years ago even though we are now producing 700 000 tons p. a. compared with the 50 000 tons and wastage rates of up to 30 % back in 1968. In other words, production has increased by a factor of 15 whereas wastage has only doubled. This is surely a spendid recycling achievement by the silicone industry.

Furthermore, it is worth comparing what becomes of the 35 000 tons of waste from silane production with that of other plastics. The techniques employed today, such as incineration, effluent treatment _etc._ generate 25 000 tons of silicon dioxide, _i. e._ sand, and 10 000 tons of sodium chloride, _i. e._ common salt, which, if the techniques I have already mentioned are consequently applied, can be reduced further. Lastly, 20 000 tons of carbon dioxide are also produced. This amount is equivalent to 0.003 thousandths of annual total emissions or 0.08 thousandths of the quantity of CO_2 given off during combustion of fossil fuels and burning of rain forests for cultivation purposes. This comparison with the figures published by the Association of the German Chemicals Industry (VCI) clearly illustrates how successful our endeavours in recycling have been and also demonstrates that silicones are plastics that hardly cause any pollution.

In contrast to the general recommendations issued by The Club of Rome on the subject of recycling, a much more specific and sophisticated approach is taken nowadays. For example, the Association of the German Chemicals Industry requires from the German Chemical Industry

1. to design its processes in such a way as to minimize formation of by-products, and

2. to consider how new products under development can be disposed of after use.

We in the silicone industry have gone a long way towards fulfilling the first of these requirements, as the examples I have quoted have shown. But as we gain more and more scientific knowledge, we should be able to increase our results in this special field to a maximum in the future.

The second requirement is much more difficult to satisfy. Silicones are used in tiny amounts for countless applications that are hard to keep track of. But here too there is scope for measures that are economically and ecologically sound.

For one thing, methods will have to be developed of returning products that can be collected in sufficient quantities to silicone producers who can then convert them to re-usable products. A number of processes already exist for this and are in widespread use.

For those silicone products which cannot be collected in sufficient quantities we will have to devise ways of making them degrade automatically after use. This subject is highly problematical as, for example, phosphate-free detergents have shown: and yet phosphates themselves are degradable. So the problem will have to be tackled not only from the point of view of what is feasible, which in itself is very difficult, but also from the point of view of the effect that the degradation products have on other systems. We chemists are all called upon to propose solutions to this problem!

Let me summarize what I have been talking about today. Major steps forward have been taken in the field of recycling silicone products. The latest developments that occurred in science were implemented immediately by the silicone industry. By virtue of the basic siloxane structure, we have our hands on a plastic that will be a true plastic of the future. We silicone chemists and silicone users will therefore have to take pains to adapt scientific progress to capitalize on the initial successes we have made in recycling. For us, recycling is neither tragic nor magic. It was, is and shall remain a routine matter that is not only the ethical but also the economic responsibility of every scientist.

REFERENCES

Dennis Meadows, Donella Meadows, Erich Zahn, Peter Milling
'The Limits to Growth', Universe Books, New York, 1973.

Verband der Chemischen Industrie e.V.; Chemie und Umwelt, Luft, Dezember 1989.

Section II: Mechanistic Organosilicon Chemistry
              a) Gas Phase and Photochemical Reactions

Mechanism of Pyrolysis of Some Polysilanes

Iain M.T. Davidson,[a] Terry Simpson,[a] and
Richard G. Taylor[b]
[a]DEPARTMENT OF CHEMISTRY, THE UNIVERSITY, LEICESTER LE1 7RH, UK
[b]DOW CORNING SUSSEX RESEARCH, SCHOOL OF MOLECULAR SCIENCES, UNIVERSITY OF SUSSEX, BRIGHTON BN1 9QJ, UK

Polysilanes are of current importance as photoresists[1,2] and as precursors to new polycarbosilane polymers. Both photochemical and thermal processes are believed to be involved in the laser-induced ablation of these materials.[2] Accordingly, we are undertaking separate studies of the thermal and photochemical decomposition of model polysilanes. Most of our work so far has been on the kinetics and mechanism of thermal decomposition of permethylated trisilane and tetrasilanes in the gas phase, using our stirred-flow (SFR) technique with analysis by gas chromatography[3] or gc/mass spectrometry;[4] but we have also initiated some work on the mechanism of photolysis of similar polysilanes in solution, in co-operation with Professor Josef Michl at the University of Texas at Austin.

Our gas-phase work has been concerned with three simple models for polysilanes, viz. octamethyltrisilane [Me_8Si_3] and the two decamethyltetrasilanes, linear [n-$Me_{10}Si_4$] and branched [$(Me_3Si)_3SiMe$, or iso-$Me_{10}Si_4$]. A preliminary account of our results for Me_8Si_3 has been published.[5] Each polysilane was pyrolysed between 560° C and 650° C in SFR apparatus, using gc/ms for product identification[4] and gc for kinetic measurements.[3] The major silicon-containing product was invariably Me_3SiH, with some other methylmonosilanes, but the main products with more than one silicon per molecule were **cyclic** carbosilanes. Product composition was complex, especially in the pyrolysis of the tetrasilanes, with numerous isomeric products. Some were identified by comparison with authentic samples, but most structural assignments were based on analogy with similar compounds; it is not difficult to deduce from mass spectra whether or not cyclic carbosilanes have Si-H bonds or Me_3Si side-chains, for instance. In these pyrolyses, isomeric products could arise from the existence of isomeric intermediates, as in Schemes III and V below, and also from radical-induced isomerisation of cyclic carbosilanes formed initially, e.g.[6]

$$\underset{Me_2Si-SiMe_2}{\overset{SiMe_2}{\diagup\diagdown}} + R\cdot \longrightarrow \underset{Me_2Si-SiCH_2}{\overset{SiMe_2}{\diagup\diagdown}}\underset{Me}{\overset{\cdot}{}} \longrightarrow \underset{Me_2Si\diagdown\diagup SiMe}{\overset{SiMe_2}{\diagup\diagdown}}\cdot \longrightarrow \underset{Me_2Si\diagdown\diagup SiMe}{\overset{SiMe_2}{\diagup\diagdown}}_H$$

The product composition was pressure-dependent, carbosilanes of higher molecular weight being formed at higher pressure; pressure-dependence of product composition is also a feature of the pyrolysis of hexamethyldisilane,[7] but in that case the major product at higher pressure is a linear carbosilane isomeric with Me_6Si_2; corresponding **linear** carbosilanes in the pyrolyses described here were minor or negligible products.

Trapping and kinetic experiments were undertaken to identify reactive intermediates, and to assess their relative importance. Pyrolysis of all three polysilanes in the presence of toluene gave ethylbenzene, benzyltrimethylsilane and dibenzyl, with substantial suppression of the formation of cyclic carbosilanes, indicating the presence of $Me_3Si\cdot$ and $Me\cdot$ radicals. Pyrolysis in the presence of butadiene or 1,3-dimethylbutadiene also suppressed formation of cyclic carbosilanes; new products were cyclic silacycloalkenes resulting from trapping of silylenes, silenes or disilenes. The major intermediate trapped in this way was invariably $:SiMe_2$, but small quantities of $Me_2Si=CH_2$ and $Me_2Si=SiMe_2$ were also trapped. In the case of the tetrasilanes, there was also evidence for $:Si(Me)SiMe_3$.

Kinetics of formation of the principal trapped products in the pyrolysis of the two tetrasilanes are shown in Table I.

TABLE I

KINETIC RESULTS IN THE PRESENCE OF TRAPPING AGENT

		n-Me$_{10}$Si$_4$		iso-Me$_{10}$Si$_4$	
		log A	E/kJ	log A	E/kJ
+ PhCH$_3$ →	PhCH$_2$CH$_3$	16.3±0.4	317±6	15.8±0.7	307±12
→	PhCH$_2$SiMe$_3$	15.8±0.1	294±3	16.6±0.6	303±10
→	CH$_4$	17.4±0.6	322±10	17.9±0.4	326±7
+ (isobutene) →	(silacyclopentene w/ SiMe$_2$)	16.0±1.2	305±1.2	16.4±0.8	292±13

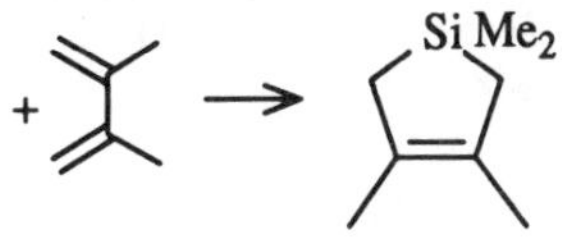

In each pyrolysis the Arrhenius parameters for the formation of the trapped product from Me$_3$Si·, Me·, and :SiMe$_2$ were the same within experimental error. The Arrhenius parameters are close to those to be expected for formation of Me$_3$Si· radicals by Si-Si bond homolysis,[7] but not for direct elimination of :SiMe$_2$.[10] These kinetic results indicate that the primary process in these pyrolyses is Si-Si bond homolysis to form Me$_3$Si· radicals, with :SiMe$_2$ and Me· resulting from secondary decomposition of Me$_3$Si· radicals, thus: Me$_3$Si· ⟶ Me$_2$Si: + Me· . Radical mechanisms therefore predominate in the Schemes proposed below for the formation of cyclic carbosilanes.

Kinetic measurements in the absence of trapping agents were less informative; for instance, Arrhenius plots for the formation of Me$_3$SiH curved sharply upwards at higher temperature, indicating a rapidly-increasing contribution from secondary reactions. In these circumstances, it would be rash to make any further deductions from the Arrhenius parameters in Table I.

Pyrolysis of Me$_8$Si$_3$ had previously been studied by Barton and his co-workers, using flash vacuum pyrolysis (FVP) at low pressure;[8] under these conditions there was only one cyclic carbosilane product, 1,1,3-trimethyl-1,3-

-disilacyclobutane (I). This was also a major product in our experiments at higher pressure, but under our conditions 1,1,2,2,4,4-hexamethyl-1,2,4--trisilacyclopentane (II) was also prominent.[5] Formation of I is readily explained by Scheme I, based on reactions suggested by Barton.[8,9]

SCHEME I

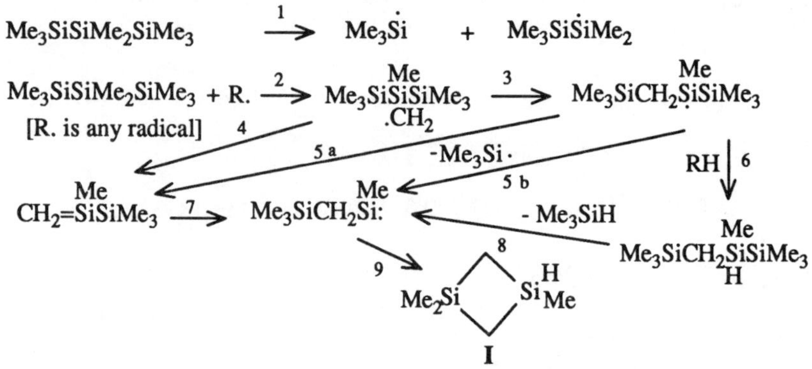

$$Me_3SiSiMe_2SiMe_3 \xrightarrow{1} Me_3\dot{S}i + Me_3SiSiMe_2$$

$$Me_3SiSiMe_2SiMe_3 + R. \xrightarrow{2} Me_3Si\dot{S}iSiMe_3 \xrightarrow{3} Me_3SiCH_2\overset{Me}{\underset{.CH_2}{SiSiMe_3}}$$

[R. is any radical]

$$CH_2=\underset{Me}{SiSiMe_3} \xrightarrow{7} Me_3SiCH_2\overset{Me}{Si:} \xleftarrow{5b} -Me_3SiH$$

$$\xrightarrow{RH} \overset{Me}{\underset{H}{Me_3SiCH_2SiSiMe_3}}$$

(structure I: four-membered ring with Me_2Si and $\overset{H}{\underset{Me}{Si}}$)

Quite different chemistry must be involved in the formation of **II**, a product with no Si-H bonds. That fact and the fact that formation of **II** was favoured at higher pressure, implying that bimolecular reactions were important, led us to the unusual suggestions in Scheme II.[5]

SCHEME II

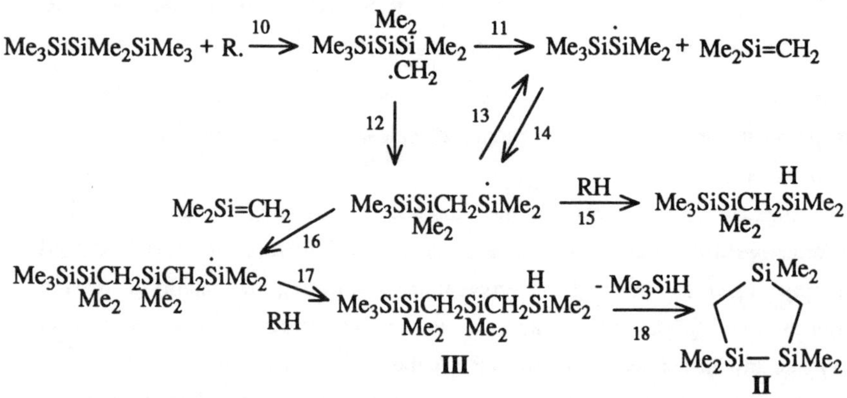

$$Me_3SiSiMe_2SiMe_3 + R. \xrightarrow{10} Me_3SiSi\overset{Me_2}{\underset{.CH_2}{Si}} Me_2 \xrightarrow{11} Me_3Si\dot{S}iMe_2 + Me_2Si=CH_2$$

$$Me_2Si=CH_2 \quad Me_3SiSiCH_2\dot{S}iMe_2 \xrightarrow[15]{RH} Me_3SiSiCH_2\overset{H}{\underset{Me_2}{SiMe_2}}$$

$$Me_3SiSiCH_2SiCH_2\dot{S}iMe_2 \xrightarrow{17} Me_3SiSiCH_2SiCH_2\overset{H}{\underset{Me_2}{SiMe_2}} \xrightarrow[18]{-Me_3SiH} \overset{SiMe_2}{\underset{Me_2Si-SiMe_2}{}}$$

III → II

We suggested that compound **III** is formed, then ring-closes with elimination of Me_3SiH, a somewhat analogous process to the formation of silylenes from hydrido-disilanes. We confirmed the feasibility of this suggestion by synthesising **III** and showing that **II** was a major product of its pyrolysis; we also showed that the components of Me_3SiH had to be attached to silicon; an isomer of **III** with the Me_3Si group attached to carbon did not undergo this reaction.[5] We believe that reaction 18 may be an example of a reaction of rather general importance; further possible examples are discussed below.

Product composition in the pyrolysis of the tetrasilanes was more complex, but the prominent products in the pyrolysis of iso-$Me_{10}Si_4$ could be accounted for by Schemes III and IV, analogous to Schemes I and II respectively.

SCHEME III

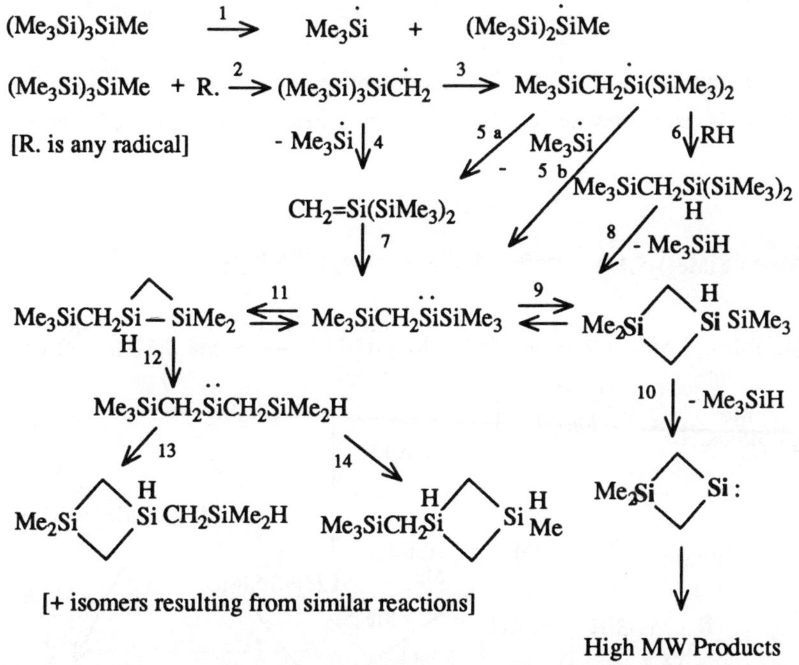

$(Me_3Si)_3SiMe \xrightarrow{1} Me_3Si\cdot + (Me_3Si)_2\dot{S}iMe$

$(Me_3Si)_3SiMe + R\cdot \xrightarrow{2} (Me_3Si)_3SiCH_2\cdot \xrightarrow{3} Me_3SiCH_2\dot{S}i(SiMe_3)_2$

[R. is any radical] $-Me_3Si\cdot \downarrow 4$ $\overset{5a}{\swarrow} Me_3Si\cdot$ $\overset{6}{\downarrow} RH$

$CH_2=Si(SiMe_3)_2$ $\overset{5b}{\searrow} Me_3SiCH_2Si(SiMe_3)_2$
                                        $ H$
$\downarrow 7$ $\overset{8}{\swarrow} -Me_3SiH$

$Me_3SiCH_2\overset{\frown}{Si-SiMe_2} \underset{}{\overset{11}{\rightleftarrows}} Me_3SiCH_2\ddot{S}iSiMe_3 \overset{9}{\rightleftarrows} Me_2\dot{S}i\overset{\frown}{}\overset{H}{Si}SiMe_3$
H
$_{12}\downarrow$ $10 \downarrow -Me_3SiH$

$Me_3SiCH_2SiCH_2SiMe_2H$

$\swarrow 13 \downarrow 14 Me_2\dot{S}i\overset{\frown}{}Si:$

$Me_2Si\overset{\frown}{}\overset{H}{Si}CH_2SiMe_2H Me_3SiCH_2\overset{\frown}{\underset{}{Si}}\overset{H}{}\overset{H}{Si}Me$

[+ isomers resulting from similar reactions] $\downarrow$

 High MW Products

SCHEME IV

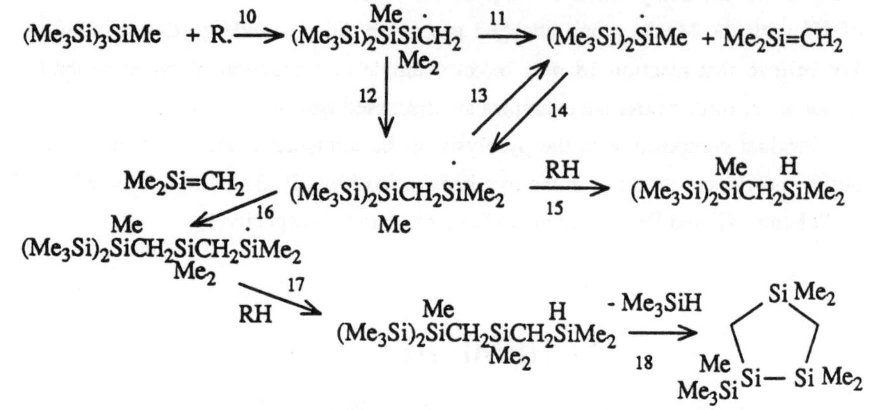

$$(Me_3Si)_3SiMe + R· \xrightarrow{10} (Me_3Si)_2\overset{Me}{\underset{Me_2}{Si}}SiCH_2 \xrightarrow{11} (Me_3Si)_2\dot{S}iMe + Me_2Si=CH_2$$

$$\downarrow 12 \quad \overset{13}{\swarrow} \overset{}{\nwarrow} 14$$

$$Me_2Si=CH_2 \overset{16}{\swarrow} \quad (Me_3Si)_2\underset{Me}{SiCH_2SiMe_2} \xrightarrow{RH}{15} (Me_3Si)_2\underset{}{SiCH_2\overset{Me\ H}{SiMe_2}}$$

$$\underset{Me}{Me}$$
$$(Me_3Si)_2SiCH_2\dot{S}iCH_2SiMe_2$$
$$\underset{Me_2}{\quad}$$

$$\overset{17}{\underset{RH}{\searrow}} \quad (Me_3Si)_2\underset{Me_2}{SiCH_2\overset{Me\ H}{SiCH_2SiMe_2}} \overset{-Me_3SiH}{\underset{18}{\longrightarrow}} \quad \overset{Si\ Me_2}{\underset{Me_3Si}{\overset{}{\diagup}}\underset{Si-\ Si\ Me_2}{\diagdown}}$$

In the pyrolysis of n-Me$_{10}$Si$_4$, prominent products under "low pressure" conditions could likewise be accounted for by Scheme V, analogous to Scheme I.

SCHEME V

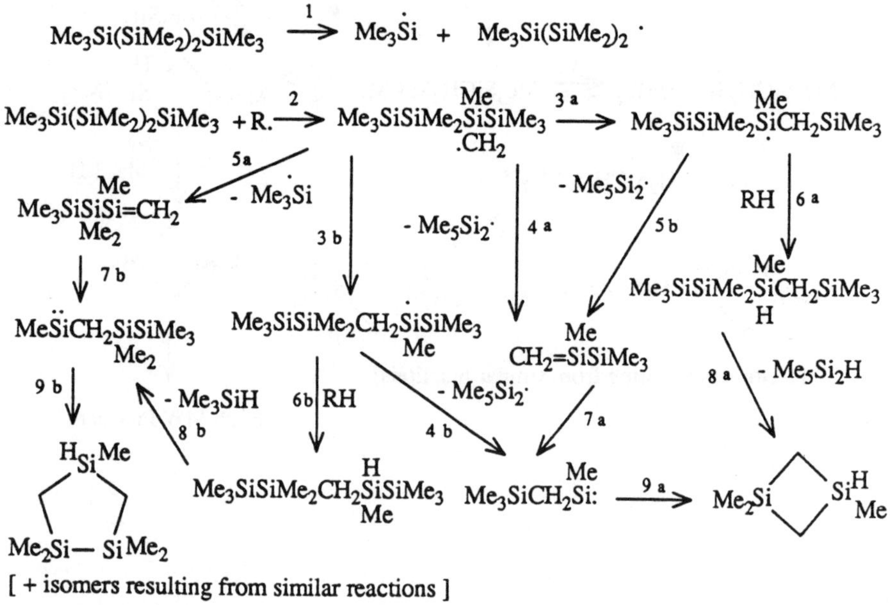

$$Me_3Si(SiMe_2)_2SiMe_3 \xrightarrow{1} Me_3\dot{S}i + Me_3Si(SiMe_2)_2 ·$$

$$Me_3Si(SiMe_2)_2SiMe_3 + R· \xrightarrow{2} Me_3SiSiMe_2\overset{Me}{\underset{·CH_2}{Si}}SiMe_3 \xrightarrow{3a} Me_3SiSiMe_2\overset{Me}{\underset{}{Si}}CH_2\dot{S}iMe_3$$

$$\overset{5a}{\swarrow} \quad -Me_3\dot{S}i \qquad \qquad -Me_5Si_2· \quad \nearrow \qquad RH \Big| 6a$$

$$Me_3Si\dot{S}iSi=CH_2 \qquad \quad 3b\Big| \quad -Me_5Si_2· \Big| 4a \quad 5b \qquad \qquad Me$$
$$\underset{Me_2}{Me}\qquad \qquad \qquad \qquad \qquad \qquad \qquad Me_3SiSiMe_2\overset{Me}{\underset{H}{Si}}CH_2\dot{S}iMe_3$$

$$\downarrow 7b$$
$$Me\ddot{S}iCH_2SiSiMe_3 \quad Me_3SiSiMe_2CH_2\dot{S}iSiMe_3$$
$$\underset{Me_2}{\quad} \qquad \qquad \underset{Me}{\quad} \quad \underset{CH_2=\dot{S}iMe_3}{Me} \qquad 8a \quad -Me_5Si_2H$$

$$9b\Big| \quad \overset{-Me_3SiH}{\nwarrow} \quad 6b\Big|RH \qquad -Me_5Si_2· \qquad \qquad \downarrow$$
$$\quad \qquad 8b \qquad \qquad \qquad 4b \qquad 7a$$

$$\overset{H_{Si}Me}{\diagup \diagdown} \qquad \quad Me_3SiSiMe_2CH_2\overset{H}{\underset{Me}{Si}}SiMe_3 \quad Me_3SiCH_2\ddot{S}i: \xrightarrow{9a} Me_2Si\overset{}{\underset{}{\diagup \diagdown}}\overset{Si}{\underset{Me}{\diagdown H}}$$
$$Me_2Si-SiMe_2$$

[+ isomers resulting from similar reactions]

However, the parallels with the pyrolysis of Me$_8$Si$_3$ now break down. Scheme VI, analogous to Scheme II, fails to predict the remaining main products in the pyrolysis of n-Me$_{10}$Si$_4$.

SCHEME VI

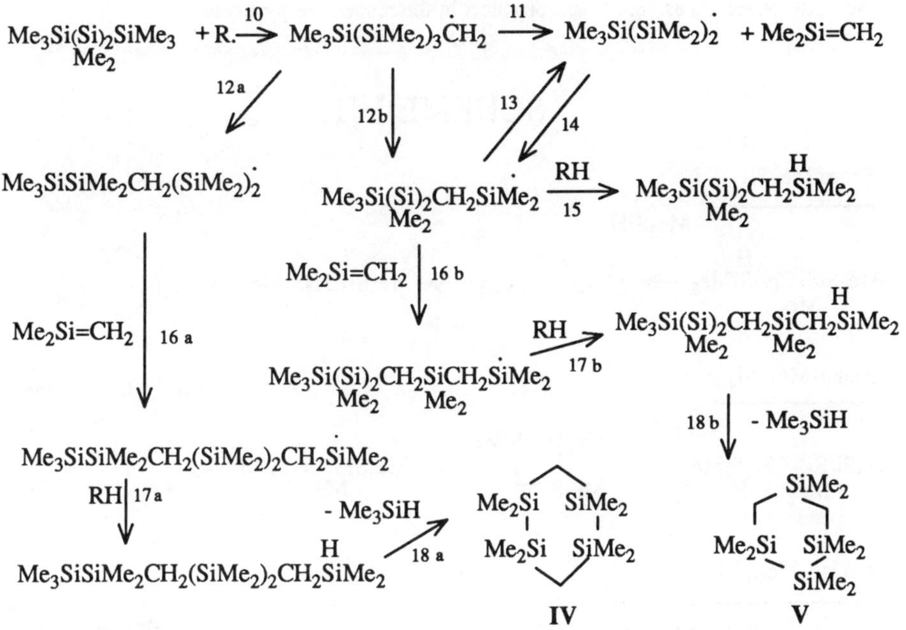

$$Me_3Si(SiMe_2)_2SiMe_3 + R\cdot \xrightarrow{10} Me_3Si(SiMe_2)_3\dot{C}H_2 \xrightarrow{11} Me_3Si(SiMe_2)_2\cdot + Me_2Si=CH_2$$

12a ↙   12b ↓   13 ↗   14 ↗

$Me_3SiSiMe_2CH_2(SiMe_2)_2\cdot$

$Me_3Si(SiMe_2)_2CH_2SiMe_2\cdot \xrightarrow[15]{RH} Me_3Si(SiMe_2)_2CH_2SiMe_2H$

16a ↓ $Me_2Si=CH_2$ 16b ↓ $Me_2Si=CH_2$

$Me_2Si=CH_2$

$Me_3Si(SiMe_2)_2CH_2SiCH_2SiMe_2\cdot \xrightarrow[17b]{RH} Me_3Si(SiMe_2)_2CH_2SiCH_2SiMe_2H$
  (Me_2)   (Me_2)   (Me_2)   (Me_2)

18b ↓ $-Me_3SiH$

$Me_3SiSiMe_2CH_2(SiMe_2)_2CH_2SiMe_2\cdot$

RH ↓ 17a

$Me_3SiSiMe_2CH_2(SiMe_2)_2CH_2SiMe_2H$

$-Me_3SiH$, 18a →

IV: cyclic $Me_2Si-SiMe_2 / Me_2Si-SiMe_2$ (6-membered ring with Me_2Si–$SiMe_2$ bridges)

V: cyclic structure with $SiMe_2$ / Me_2Si–$SiMe_2$ / $SiMe_2$

The products exemplified by **IV** and **V** in Scheme VI were indeed observed, but were minor. Substantially more prominent was a group of products, especially two with MW 256 and 242 (products **IV** and **V** have MW 260). These products have not yet been identified, but appear to be cyclic carbosilanes with more than one ring and/or with C-C π-bonds. They were also observed in all of the other pyrolyses, but in those cases they were less prominent than the products of Schemes I - IV. They may simply result from secondary decomposition of the main products, but in that case we have yet to explain why the reactions in Scheme VI are less efficient than those in Schemes II and IV.

Hence, although our work on these pyrolyses is incomplete, there is considerable consistency in the pyrolysis mechanisms, with some substantiation of the suggestion that formation of cyclic carbosilanes by elimination of Me_3SiH from suitable precursors is quite a general reaction. Indeed, that type of reaction might account for the striking difference noted above between the pyrolyses reported here and the pyrolysis of Me_6Si_2, where a linear carbosilane isomer was a prominent product.[7] Corresponding isomers of Me_8Si_3 and $Me_{10}Si_4$ were not observed. However, they might have been formed and could then have eliminated

Me$_3$SiH as shown in Scheme VII, to give cyclic carbosilanes which were all actually observed as significant products in the respective pyrolyses.

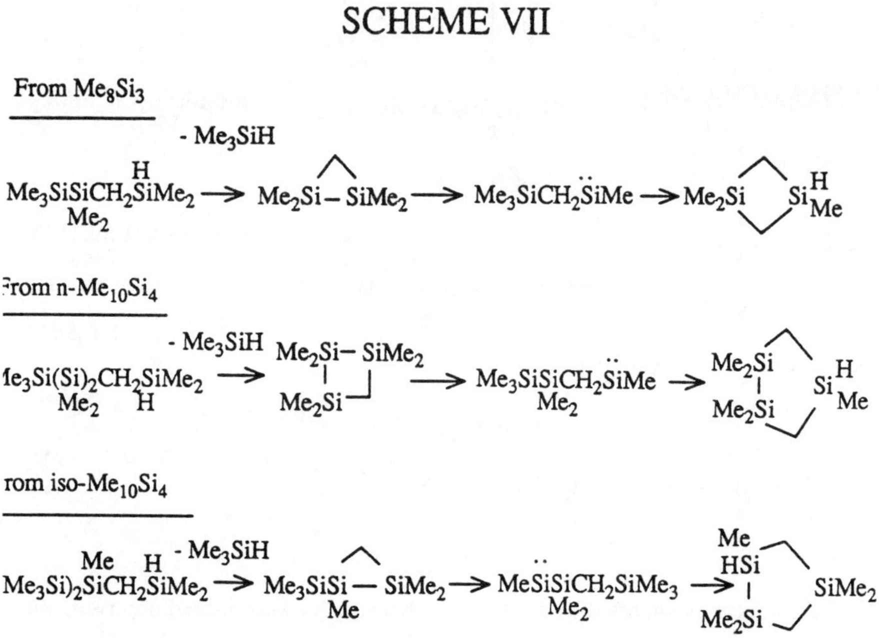

SCHEME VII

From Me$_8$Si$_3$

$$\text{Me}_3\text{SiSiCH}_2\overset{H}{\underset{Me_2}{\text{SiMe}_2}} \xrightarrow{-\text{Me}_3\text{SiH}} \text{Me}_2\text{Si}-\overset{\triangle}{\text{SiMe}_2} \rightarrow \text{Me}_3\text{SiCH}_2\overset{..}{\text{SiMe}} \rightarrow \text{Me}_2\text{Si}\overset{\triangle}{\underset{\diamondsuit}{}}\text{Si}\overset{H}{\underset{Me}{}}$$

From n-Me$_{10}$Si$_4$

$$\text{Me}_3\text{Si(Si)}_2\text{CH}_2\text{SiMe}_2 \xrightarrow{-\text{Me}_3\text{SiH}} \begin{array}{c}\text{Me}_2\text{Si}-\text{SiMe}_2\\||\\\text{Me}_2\text{Si}\end{array} \rightarrow \text{Me}_3\text{SiSiCH}_2\overset{..}{\underset{Me_2}{\text{SiMe}}} \rightarrow \begin{array}{c}\text{Me}_2\text{Si}\\|\\\text{Me}_2\text{Si}\end{array}\overset{\triangle}{}\text{Si}\overset{H}{\underset{Me}{}}$$

From iso-Me$_{10}$Si$_4$

$$(\text{Me}_3\text{Si})_2\overset{Me}{\underset{}{\text{SiCH}}}_2\overset{H}{\text{SiMe}_2} \xrightarrow{-\text{Me}_3\text{SiH}} \text{Me}_3\text{SiSi}\underset{Me}{-}\overset{\triangle}{\text{SiMe}_2} \rightarrow \text{MeSiSiCH}_2\underset{Me_2}{\overset{..}{\text{SiMe}_3}} \rightarrow \overset{Me}{\underset{Me_2\text{Si}}{\overset{HSi}{|}}}\overset{\triangle}{}\text{SiMe}_2$$

We have embarked on photolysis studies to complement the pyrolysis work with the help and encouragement of Professor Josef Michl of the University of Texas at Austin. Initial experiments were carried out there during a visit by one of us (TS).

Photolyses were carried out at 254 nm in n-hexane solution, with analysis by gc and gc/ms. Silylene intermediates were trapped by insertion into Et$_3$SiH, which also trapped radicals by hydrogen-abstraction; primary and secondary products were distinguished from each other by checking the linearity of product-time plots. Primary processes identified in this way in the photolysis of Me$_{14}$Si$_6$ are in Scheme VIII.

SCHEME VIII

PRIMARY PROCESSES IN THE PHOTOLYSIS OF $Me_{14}Si_6$

Main Process:

$(Me_3Si)_2\overset{Me}{\underset{Me}{Si}}Si(SiMe_3)_2 \xrightarrow{h\nu} (Me_3Si)_3SiMe + Me\ddot{S}iSiMe_3$ 87 %

Minor Processes:

$(Me_3Si)_2\overset{Me}{\underset{Me}{Si}}Si(SiMe_3)_2 \xrightarrow{h\nu} (Me_3Si)_2\overset{Me}{\underset{Me_2}{Si}}SiSiMe_3 + Me_2Si:$ 3 %

$(Me_3Si)_2\overset{Me}{\underset{Me}{Si}}Si(SiMe_3)_2 \xrightarrow{h\nu} (Me_3Si)_2SiMe_2 + (Me_3Si)_2Si:$ 3 %

$(Me_3Si)_2\overset{Me}{\underset{Me}{Si}}Si(SiMe_3)_2 \xrightarrow{h\nu} (Me_3Si)_2\dot{S}iMe + (Me_3Si)_2\dot{S}iMe$ 4 %

$(Me_3Si)_2\overset{Me}{\underset{Me}{Si}}Si(SiMe_3)_2 \xrightarrow{h\nu} (Me_3Si)_2\overset{Me\,\cdot}{\underset{Me}{Si}}SiSiMe_3 + Me_3Si\cdot$ 2 %

Quantum yields have not yet been measured for other photolyses, to quantify the relative importance of primary pathways, but the photolysis of iso-$Me_{10}Si_4$ and n-$Me_{10}Si_4$ each followed a similar course, as shown in Schemes IX and X.

SCHEME IX

PRIMARY PROCESSES IN THE PHOTOLYSIS OF iso - $Me_{10}Si_4$

Main Process:

$(Me_3Si)_3SiMe \xrightarrow{h\nu} Me_3SiSiMe_3 + Me_3Si\ddot{S}iMe$

Minor Processes:

$(Me_3Si)_3SiMe \xrightarrow{h\nu} (Me_3Si)_2SiMe_2 + Me_2Si:$

$(Me_3Si)_3SiMe \xrightarrow{h\nu} (Me_3Si)_2\dot{S}iMe + Me_3Si\cdot$

SCHEME X

PRIMARY PROCESSES IN THE PHOTOLYSIS OF n - $Me_{10}Si_4$

Main Process:

$Me_3Si(Si)_2SiMe_3$ $\xrightarrow{h\nu}$ $Me_3SiSiSiMe_3$ + $Me_2Si:$
$\quad\quad Me_2$ $\quad\quad\quad\quad\quad\quad\quad\quad Me_2$

Minor Processes:

$Me_3Si(Si)_2SiMe$ $\xrightarrow{h\nu}$ $Me_3SiSiMe_2\dot{S}iMe_2$ + $Me_3\dot{S}i$
$\quad\quad Me_2$

$\xrightarrow{h\nu}$ $Me_3Si\dot{S}iMe_2$ + $Me_3Si\dot{S}iMe_2$

In view of the importance of side-chain reactions in the ablation of polysilanes with higher alkyl groups attached to silicon,[2] our studies should be extended to cover model compounds other than permethylated polysilanes. As the first step in this programme we have started some work on $Me_3SiSiEt_2SiMe_3$, with the results shown in Scheme XI. Although the side-chain remained intact in this simple case, two silylene-forming pathways were observed.

SCHEME XI

PRIMARY PROCESSES IN THE PHOTOLYSIS OF $Et_2Me_6Si_3$

Main Process:

$(Me_3Si)_2SiEt_2$ $\xrightarrow{h\nu}$ $Me_3SiSiMe_3$ + $Et_2Si:$

Minor Processes:

$(Me_3Si)_2SiEt_2$ $\xrightarrow{h\nu}$ $Me_3SiSiEt_2^{Me}$ + $Me_2Si:$

$(Me_3Si)_2SiEt_2$ $\xrightarrow{h\nu}$ $Me_3Si\dot{S}iEt_2$ + $Me_3\dot{S}i$

It is clear from these experiments that silylene formation is the predominant process in the photolysis of these simple polysilanes, with radical formation as a very minor process. It is noteworthy that formation of various silylenes occurs, as it does in the laser ablation of polysilanes of high MW.[2] On the other hand, thermal decomposition proceeds predominantly by radical mechanisms to form cyclic carbosilanes, with cyclisation by elimination of Me$_3$SiH from silicon centres as a general reaction of some importance. These distinctive features could be exploited to direct the course of the decomposition of polysilanes in different applications of these interesting compounds.

Acknowledgements

We are most grateful to Dow Corning Ltd. for support of this work, help and encouragement from Dr. Peter Lo being particularly appreciated. We are also grateful to the SERC for support through Research Grants, the CASE Award Scheme, and the IT Initiative of the Chemistry Committee. We are indebted to Prof. Josef Michl for his enthusiastic co-operation and support, and to Dr. Allan McKinley of the University of Texas at Austin for help with the photolysis experiments. We are also grateful to Geraint Morgan, Paul Gibbs, and David Walmsley for help with the pyrolysis experiments.

References:

1   R.D. Miller in "Silicon Chemistry" (Proceedings of the Eighth International Symposium on Organosilicon Chemistry, St. Louis, Missouri, U.S.A., 1987), p. 377.
2   T.F. Magnera, V. Balaji, J. Michl and R.D. Miller in "Silicon Chemistry" (Proceedings of the Eighth International Symposium on Organosilicon Chemistry, St. Louis, Missouri, U.S.A., 1987), p. 491.
3   A.C. Baldwin, I.M.T. Davidson and A.V. Howard, J. Chem. Soc., Faraday Trans. 1, 71, (1975), 972
4   I.M.T. Davidson, G. Eaton and K.J. Hughes, J. Organometal. Chem., 347, (1988), 17.
5   B.N. Bortolin, I.M.T. Davidson, D. Lancaster, T. Simpson and D.A. Wild, Organometallics, 9, (1990), 281.
6   I.M.T. Davidson, G. Fritz, F.T. Lawrence and A.E. Matern, Organometallics, 1, (1982), 1453.
7   I.M.T. Davidson, C. Eaborn, and J.M. Simmie, J. Chem. Soc., Faraday Trans. 1, 70, (1974), 249. I.M.T. Davidson and A.V. Howard, J. Chem. Soc., Faraday Trans. 1, 71, (1975), 69. I.M.T. Davidson, P. Potzinger and B. Reimann, Berichte Bunsenges Phys. Chem., 86, (1982), 13.
8   T.J. Barton, S.A. Burns and G.T. Burns, Organometallics, 1, (1982), 210.
9   T.J. Barton and S.A. Jacobi, J. Am. Chem. Soc., 102, (1980), 7979.
10 I.M.T. Davidson, K.J. Hughes and S. Ijadi-Maghsoodi, Organometallics, 6, (1987), 639.

Recent Results in the Chemistry of Silylenes and Germylenes

P.P. Gaspar, K.L. Bobbitt, M.E. Lee, D. Lei, V.M. Maloney, D.H. Pae, and M. Xiao
DEPARTMENT OF CHEMISTRY, WASHINGTON UNIVERSITY, SAINT LOUIS, MO 63130, USA

1 INTRODUCTION

The three topics to be discussed in this chapter are related by our interest in mechanistic differences between reactive intermediates based on first-row elements and their heavier analogs. These differences are clues that may lead to the refinement of the ideas of organic chemistry to the point where they can explain and predict the making and breaking of covalent bonds anywhere in the periodic table.

2 GENERATION OF STERICALLY CROWDED SILYLENES

A major difference between silylenes and carbenes is the tendency for silylenes to have singlet electronic ground states, while many carbenes have triplet ground states. One approach to a triplet ground state silylene is to attach such bulky substituents that the bond angle is increased substantially from the 92° found for SiH_2. <u>Ab Initio</u> calculations by Gordon predict that the potential energy curves for the lowest singlet and triplet states of SiH_2 cross at <u>ca.</u> 129°.[1] In the limit of linearity, the nonbonding orbitals should be degenerate, and Hund's rule dictates a triplet ground state.

Diadamantylsilylene Ad_2Si was chosen for study, and a partially optimized MINDO/3 calculation predicted a 124° C-Si-C bond angle. Our first attempts to generate Ad_2Si employed trisilane precursors expected to extrude the desired silylene upon photolysis:

$$(RMe_2Si)_2SiAd_2 \xrightarrow{h\nu} (RSiMe_2)_2 + :SiAd_2 \quad (1)$$

R = Me, Ph

Both these 2,2-diadamantyltrisilanes were synthesized from the corresponding trisilanes $(RSiMe_2)_2SiH_2$ by a little-known, novel 'adamantylation' reaction,[2] the hydrosilylation of 1,3-dehydroadamantane.[2] The photolysis of

$(Me_3Si)_2SiAd_2$ was slow and inconclusive, while the photoproducts from $(PhSiMe_2)_2SiAd_2$ suggested the occurence of radical reactions. Irradiation in the presence of 2,3-dimethylbutadiene gave no 1-silacyclopent-3-ene, expected from a singlet silylene.

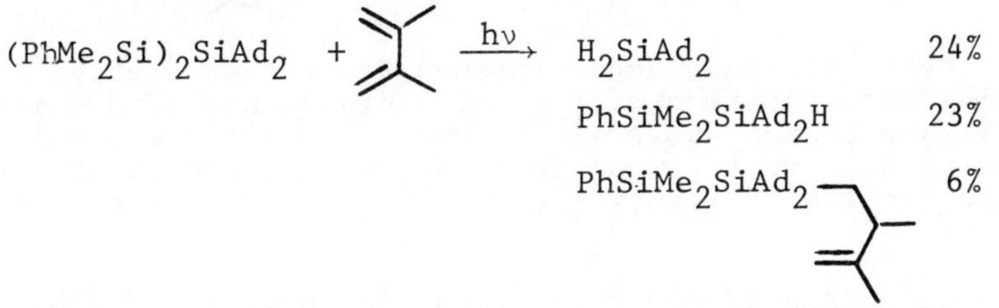

$(PhMe_2Si)_2SiAd_2$ + [2,3-dimethylbutadiene] $\xrightarrow{h\nu}$

H_2SiAd_2 24%   (2)

$PhSiMe_2SiAd_2H$ 23%

$PhSiMe_2SiAd_2$—[alkenyl]   6%

While it might be argued that the formation of H_2SiAd_2 (a possible product from triplet Ad_2Si) and $PhSiMe_2SiAd_2H$ in the presence of an excellent radical scavenger speaks for molecular elimination, the diene adduct is surely from a radical reaction.

The decomposition of 1,1-diadamantylsiliranes was therefore chosen to provide Ad_2Si, with the precedent of Seyferth's discovery that thermolysis of hexamethylsilirane yielded Me_2Si,[3] and Boudjouk's finding that a 1,1-di(tert-butyl)silirane is a photochemical as well as a thermal precursor of $(t-Bu)_2Si$.[4] To prepare the required precursors and authentic products of Ad_2Si reactions, we also employed another Boudjouk discovery, Li-induced dehalogenation of a dihalosilane assisted by ultrasonic irradiation. A species is produced that mimics a silylene in its addition and insertion reactions. Treatment of Ad_2SiI_2 (synthesized by adamantylation of H_2SiI_2) with Li in the presence of various substrates led to the following products:

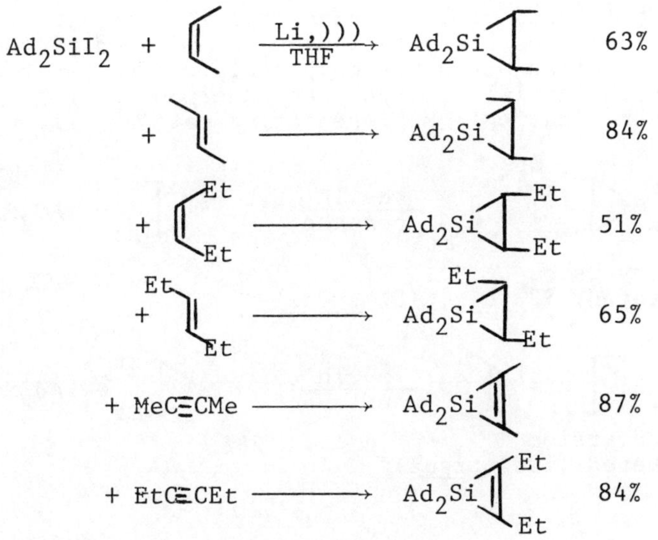

Ad_2SiI_2 + [cyclopentene] $\xrightarrow{Li,)))\atop THF}$ Ad_2Si[cyclopentane]   63%   (3)

+ [alkene] $\longrightarrow$ Ad_2Si[ring]   84%   (4)

+ [Et,Et-alkene] $\longrightarrow$ Ad_2Si[Et,Et ring]   51%   (5)

+ [Et,Et-alkene] $\longrightarrow$ Ad_2Si[Et,Et ring]   65%   (6)

+ $MeC{\equiv}CMe$ $\longrightarrow$ Ad_2Si[ring]   87%   (7)

+ $EtC{\equiv}CEt$ $\longrightarrow$ Ad_2Si[Et,Et ring]   84%   (8)

$$Ad_2SiI_2 + HSiEt_3 \xrightarrow{Li,)))} Ad_2HSiSiEt_3 \quad 90\% \quad (9)$$

$$+ \quad \diagup\!\!\!=\!\!\!\diagdown \quad \longrightarrow \quad Ad_2Si\!\!\diagdown\!\!\bigtriangleup \quad 80\% \quad (10)$$

These siliranes and silirenes are rather stable. Thermal decomposition of trans-1,1-diadamantyl-2,3-dimethylsilirane begins above 110°C, and it is not air-sensitive. Its crystal structure has been determined, and the key bond angles are 116.8° (exocyclic C-Si-C) and 49.5° (endocyclic C-Si-C).

Ad_2Si is extruded from 1,1-diadamantyl-2,3-dialkyl-siliranes upon pyrolysis or photolysis. Pyrolysis of the 2,3-dimethylsiliranes at 140° in the presence of 3-hexenes leads to stereospecific extrusion of Ad_2Si and addition to the olefin. Recovered initial silirane had not undergone cis,trans-isomerization, nor had the olefin substrate. Within experimental error (2%) no geometric isomers of the silirane products were detected. Representative data are given below:

$$Ad_2Si\underset{Et}{\overset{Et}{\sqsubset}} \quad + \quad \underset{Et}{\overset{Et}{\diagdown\!=\!\diagup}} \quad \xrightarrow[C_6D_6]{\Delta, 140°C, 1h} \quad Ad_2Si\underset{Et}{\overset{Et}{\sqsubset}} \quad + \quad \diagup\!=\!\diagdown \quad (11)$$
$$ 17\%$$

$$Ad_2Si\underset{}{\overset{Et}{\sqsupset}} \quad + \quad \underset{Et}{\overset{Et}{\diagup\!=\!\diagdown}} \quad \xrightarrow[C_6D_6]{\Delta, 140°C, 1h} \quad Ad_2Si\underset{Et}{\overset{Et}{\sqsupset}} \quad + \quad \underset{Et}{\overset{Et}{\diagup\!=\!\diagdown}} \quad (12)$$
$$ 100\%$$

Photolysis of 1,1-diadamantylsiliranes in the presence of olefins presents a more complicated picture. Transfer of Ad_2Si is not completely stereospecific, and recovered initial silirane has undergone partial geometric isomerization. Representative data are given below:

$$Ad_2Si\underset{}{\overset{Et}{\sqsubset}} \quad + \quad \underset{Et}{\overset{Et}{\diagup\!=\!\diagdown}} \quad \xrightarrow[pentane]{h\nu, 1h} \quad Ad_2Si\underset{Et}{\overset{Et}{\sqsubset}} \quad + \quad Ad_2Si\underset{Et}{\overset{Et}{\sqsupset}} \quad (13)$$

76% conversion $\qquad\qquad\qquad\qquad\qquad$ 7% $\qquad$ 65%
(recovered: 82% cis, 18% trans)

$$Ad_2Si\underset{}{\overset{Et}{\sqsupset}} \quad + \quad \underset{Et}{\overset{Et}{\diagdown\!=\!\diagup}} \quad \xrightarrow[pentane]{h\nu, 1h} \quad Ad_2Si\underset{Et}{\overset{Et}{\sqsubset}} \quad + \quad Ad_2Si\underset{Et}{\overset{Et}{\sqsupset}} \quad (14)$$

43% conversion $\qquad\qquad\qquad\qquad$ 26% $\qquad$ 5%
(recovered: 100% trans)

In all cases studied, photochemical extrusion of Ad_2Si from 1,1-diadamantyl-2,3-dimethyl- (and diethyl-) silirane, and addition to a 2-butene or a 3-hexene, led to a predominant product stereoisomer with the configuration of the olefin substrate, the retention varying from 75 to 97%. These results differ dramatically from those found by Ando for photochemically generated dimesityl- and bis(triisopropylphenyl)silylene, which seemed to show a high preference for the formation of the cis-isomer of the 2-butene adducts.[5]

In our experiments, recovered initial siliranes underwent some cis,trans-photoisomerization, 3% to 18% for the 2,3-dimethyl compounds and 18% to 42% for the 2,3-diethylsiliranes. In each case, retention of configuration was higher in the Ad_2Si adduct than in its recovered precursor. This observation supports the view that photoextrusion of Ad_2Si, like the thermal process, is stereospecific, as is the addition of the photogenerated silylene. The photoisomerization appears to be an independent process, possibly involving a diradical intermediate. Both the Ad_2Si extrusion and addition are likely to be concerted processes. Thus we suggest the following reaction scheme:

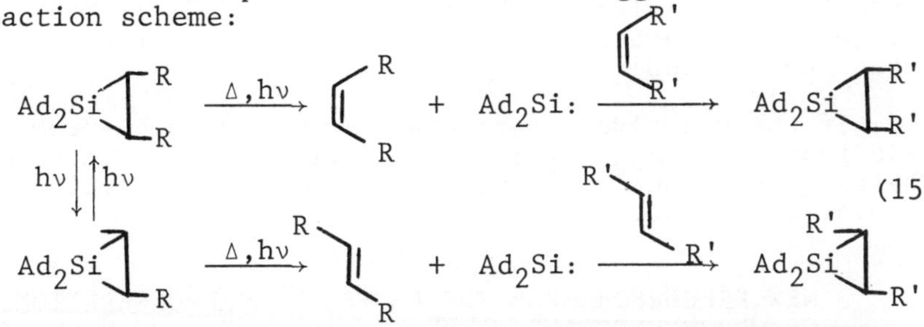

$$Ad_2Si \begin{bmatrix} R \\ R \end{bmatrix} \xrightarrow{\Delta, h\nu} \begin{bmatrix} R \\ R \end{bmatrix} + Ad_2Si: \xrightarrow{R'} Ad_2Si \begin{bmatrix} R' \\ R' \end{bmatrix}$$

$h\nu \updownarrow h\nu$ (15)

$$Ad_2Si \begin{bmatrix} R \\ R \end{bmatrix} \xrightarrow{\Delta, h\nu} \begin{bmatrix} R \\ R \end{bmatrix} + Ad_2Si: \xrightarrow{R'} Ad_2Si \begin{bmatrix} R' \\ R' \end{bmatrix}$$

That free Ad_2Si is an intermediate formed in the thermal and photochemical reactions of the diadamantylsiliranes studied seems assured by the lack of a dependence of the reaction rate on the nature of the substrate, even when the substrate is one that is unlikely to react directly with a silirane, such as Et_3SiH. Concerted, stereospecific addition suggests that Ad_2Si undergoes reaction in a singlet electronic state.

What is the nature of the intermediate in the Li-induced dehalogenation of Ad_2SiI_2? Competition between an alkyne and a silane allowed comparison of the selectivity of the reactive intermediates formed from photolysis and pyrolysis of a 1,1-diadamantylsilirane with those formed upon Li-induced dehalogenation of Ad_2SiI_2:

$$Ad_2SiI_2 + HSiEt_3 + EtC{\equiv}CEt \xrightarrow[\text{THF} \atop 0.5h]{Li,)))} Ad_2HSiSiEt_3 + Ad_2Si\begin{bmatrix} Et \\ Et \end{bmatrix}$$

67% -

(+17% Ad_2SiH_2)

(16)

$$Ad_2Si{\sqcup} + HSiEt_3 + EtC{\equiv}CEt \xrightarrow[\text{pentane}]{h\nu, 1h} Ad_2HSiSiEt_3 + Ad_2Si{\sqcup}\!\!{}^{Et}_{Et}$$
$$\phantom{Ad_2Si{\sqcup} + HSiEt_3 + EtC{\equiv}CEt \xrightarrow[\text{pentane}]{h\nu, 1h}} 17\% 80\% \qquad (17)$$

$$Ad_2Si{\sqcup} + HSiEt_3 + EtC{\equiv}CEt \xrightarrow[\text{pentane}]{\Delta, 180°C} Ad_2HSiSiEt_3 + Ad_2Si{\sqcup}\!\!{}^{Et}_{Et}$$
$$\phantom{Ad_2Si{\sqcup} + HSiEt_3 + EtC{\equiv}CEt \xrightarrow[\text{pentane}]{\Delta, 180°C}} - 54\% \qquad (18)$$

In the room-temperature, metal-induced reaction, Si-H insertion predominates, while pi-addition is the dominant process in both the room-temperature photolysis and the high-temperature pyrolysis experiments. There is a selectivity difference, possibly due to the temperature dependence of the rate constants, between the photolysis and the pyrolysis experiments, although both are believed to involve free Ad_2Si as the reactive intermediate. However, the selectivity inversion, relative to both pyrolysis and photolysis, observed in the Li-promoted deiodination, suggests that a silylenoid (here Ad_2SiLi or $Ad_2Si[THF]$) is the reactive species in the metal-induced reaction, as suggested by Boudjouk.[4]

If the reactive intermediate formed upon Li-promoted deiodination of Ad_2SiI_2 is not a free silylene, it nevertheless mimics one closely, even undergoing insertion into an Si-H bond.

3 NEW PRECURSORS FOR THE PHOTOCHEMICAL GENERATION OF GERMYLENES AND GERMENES, AND THE MECHANISM OF ADDITION OF GERMYLENES TO 1,3-DIENES

It has been observed in Neumann's laboratory,[6] and in ours,[7] that germylenes, unlike silylenes, undergo stereospecific addition to Z- and E-1,3-dienes. Neumann has suggested that this is due to a concerted 1,4-addition mechanism,[8] quite different from the stepwise addition of silylenes that we have championed.[9] Evidence has recently been presented for vinylsilirane intermediates even in the addition of Me_2Si to unsubstituted butadiene,[10] and Conlin has isolated sterically shielded vinylsiliranes and studied their conversion to 1-silacyclopent-3-enes.[11]

The discovery of a new photochemical route to the generation of germylenes has made it possible for us to make kinetic measurements that are relevant to, one might almost say germane to, this mechanistic question. We were scrutinizing the products from the photogeneration of diphenylgermylene by the elimination of hexamethyldisilane from $(Me_3Si)_2GePh_2$,[12] the elimination of a disilane upon irradiation of a disilylgermane[13] paralleling the well-known extrusion of silylenes from chains of three or more silicon atoms.[14]

$$(Me_3Si)_2GeRR' \xrightarrow{h\nu} Me_6Si_2 + :GeRR' \qquad (19)$$

Irradiation of $(Me_3Si)_2GePh_2$ in the presence of 2,3-dimethylbutadiene yielded products that made it clear that two different germylenes were formed. The unexpected elimination of $PhSiMe_3$ evidently competes with the well-known elimination of Me_6Si_2. Similar results were obtained from a monophenyl(disilyl)germane:

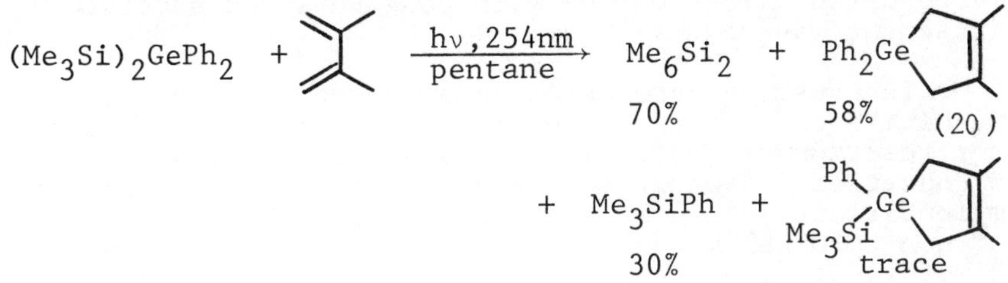

$$(Me_3Si)_2GePh_2 + \text{\small diene} \xrightarrow[\text{pentane}]{h\nu, 254nm} Me_6Si_2 + Ph_2Ge\text{-cycle} \qquad (20)$$
$$70\% \qquad 58\%$$

$$+ Me_3SiPh + \underset{Me_3Si}{Ph}Ge\text{-cycle}$$
$$30\% \qquad \text{trace}$$

$$(Me_3Si)_2GeMePh + \text{\small diene} \xrightarrow[\text{pentane}]{h\nu, 254nm} Me_6Si_2 + \underset{Me}{Ph}Ge\text{-cycle} \qquad (21)$$
$$60\% \qquad 55\%$$

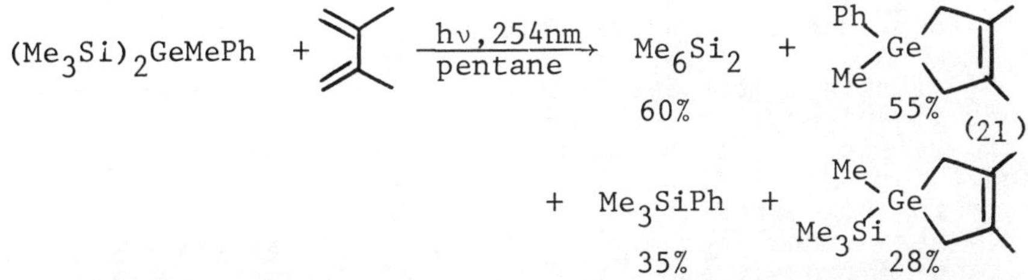

$$+ Me_3SiPh + \underset{Me_3Si}{Me}Ge\text{-cycle}$$
$$35\% \qquad 28\%$$

We wondered if elimination of Me_3SiPh would become the dominant process when a phenyl(monosilyl)germane was subjected to ultraviolet irradiation. We predicted that, for the first time, a chain of only two higher group XIV elements could be made to efficiently extrude one of them in the form of a divalent radical. This proved to be the case with $PhGeMe_2SiMe_3$, but another unexpected reaction accompanied the extrusion of Me_2Ge. Formal 1,3-migration of the trimethylsilyl group to the aromatic ring, generating a germene, is evident from the trapping products:

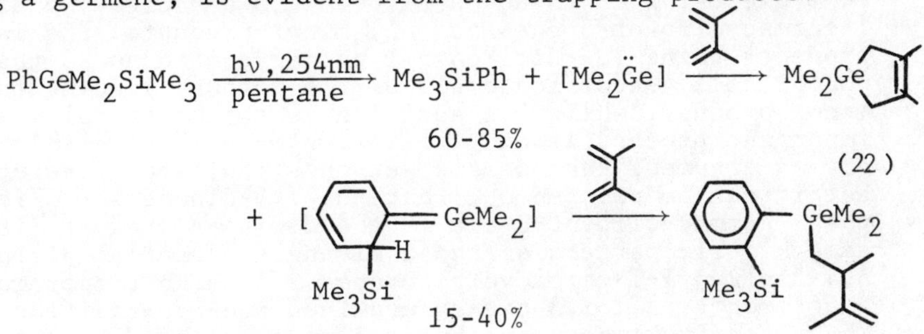

$$PhGeMe_2SiMe_3 \xrightarrow[\text{pentane}]{h\nu, 254nm} Me_3SiPh + [Me_2\ddot{G}e] \xrightarrow{\text{diene}} Me_2Ge\text{-cycle} \qquad (22)$$
$$60-85\%$$

$$+ [\text{cyclohexadienyl-}GeMe_2\text{, }Me_3Si,H] \xrightarrow{\text{diene}} \text{aryl-}GeMe_2\text{-adduct, }Me_3Si$$
$$15-40\%$$

PhGeMe$_2$SiMe$_3$ has proven to be a useful precursor of Me$_2$Ge for time-resolved kinetic spectroscopy. Laser-flash photolysis in cyclohexane at 266nm (Nd-YAG) or 248nm (KrF) leads to a transient absorption at λ_{max} = 425nm, in good agreement with that reported by other workers for Me$_2$Ge using various precursors in both matrix isolation and kinetic spectroscopy experiments.[15] A series of rate measurements for the disappearance of this transient in the presence of various substrates in cyclohexane solution at room temperature was carried out. The results are presented in Table I. Where comparison with the few previously reported values of the rate constants was possible, agreement was good.[15]

Interesting conclusions concerning the addition mechanism can be drawn from the rate constants observed for reaction of Me$_2$Ge with the eight 1,3-dienes studied. A concerted 1,4-addition mechanism requires that the germylene reacts with the s-cis-conformer. Yet the presence of non-terminal methyl substituents in isoprene and 2,3-dimethylbutadiene leads to only a modest increase of less than 50% in the rate constant for reaction with Me$_2$Ge, while greatly increasing the equilibrium mole-fraction of the s-cis-conformer. The small rate constant increase could be due to an electronic rather than a steric effect.

Terminal methyl-substituents cause a decrease in k$_2$, but cis-piperylene reacts more rapidly than does trans-piperylene, although the trans-compound has the higher equilibrium mole-fraction of the s-cis-conformer necessary for concerted 1,4-addition. The reactivity of the 2,4-hexadienes decreases in the order: cis,cis- > cis,-trans- > trans,trans-, while the mole fraction of the s-cis-conformer increases in this order. Thus the steric effect of terminal methyl substituents on 1,3-dienes seems to be due to the substitution pattern for the individual olefinic units, cis- reacting more rapidly than trans-propenyl units, but both reacting less rapidly than unsubstituted vinyl.

These kinetic results speak for a 1,2-addition as the dominant primary process in the reaction of Me$_2$Ge with 1,3-dienes. It can be argued, of course, that since some dienes, such as unsubstituted 1,3-butadiene, yield 1-germacyclopent-3-enes as only minor products, the major products being 3,4-dialkenyl-1-germacyclopentanes, that concerted 1,4-addition could be the pathway leading to the minor product, while 1,2-addition is the kinetically more important process leading to the major product. This is not an argument that we are yet in a position to refute, but it can be pointed out that the 1,3-dienes for which the 1-germacyclopent-3-ene is the dominant product fit the same kinetic pattern as those giving substantial yields of 3,4-dialkenyl-1-germacyclopentanes. In high-temperature experiments, all 1,3-dienes examined undergo addition by Me$_2$Ge to give 1-germacyclopent-3-enes as the dominant products.[16] We have also found that at room temperature,

Table I Second-order rate constants for the disappearance of the 425nm transient absorption in the 266nm photolysis of PhGeMe$_2$SiMe$_3$ in cyclohexane.

Substrate	k_2 (M^{-1}s^{-1})	Concentration range (M)
butadiene	1.24 ± 0.07 x 10^7	0.01 - 0.07
isoprene	1.63 ± 0.13 x 10^7	0.01 - 0.07
2,3-dimethylbutadiene	1.71 ± 0.38 x 10^7	0.0088 - 0.062
cis-piperylene	9.19 ± 0.41 x 10^6	0.0202 - 0.202
trans-piperylene	2.88 ± 0.24 x 10^6	0.0138 - 0.104
cis,cis-2,4-hexadiene	1.22 ± 0.09 x 10^6	0.035 - 0.210
cis,trans-2,4-hexadiene	3.46 ± 0.48 x 10^5	0.07 - 0.35
trans,trans-2,4-hexadiene	<5 x 10^5	0.07
1-butene	<10^5	>0.45
1-hexene	<10^4	6.40
1-hexyne	3.00 ± 0.23 x 10^4	0.86 - 5.22
1-propyne	<10^5	>0.45
dimethylsulfide	<10^7	0.00555
tetrahydrofuran	<10^4	neat
ethanol	1.86 ± 0.12 x 10^4	1.71 - 12
carbon tetrachloride	3.15 ± 0.27 x 10^8	0.0005 - 0.005
triethylsilane	<10^4	1.51
oxygen	2 x 10^9	0.001

silylenes, like germylenes, can give substantial yields of 3,4-dialkenyl-1-silacyclopentanes, refuting the notion that the formation of 1:2 germylene-diene adducts signals a different addition mechanism from that of silylenes.[17]

We favor the mechanism shown below for addition of germylenes to 1,3-dienes. A vinylgermirane plays a key role, analagous to the addition of silylenes to 1,3-dienes. In the germylene case, we believe that ring-cleavage to a diradical that can react with a second diene molecule competes with sigmatropic rearrangement to a germacyclopent-3-ene, a process that can be stereospecific. No open chain adducts or germacyclopent-2-enes are formed, common in the case of silylene-diene additions,[10] because the carbon-carbon bond of the germirane ring is much stronger than the germanium-carbon bonds.

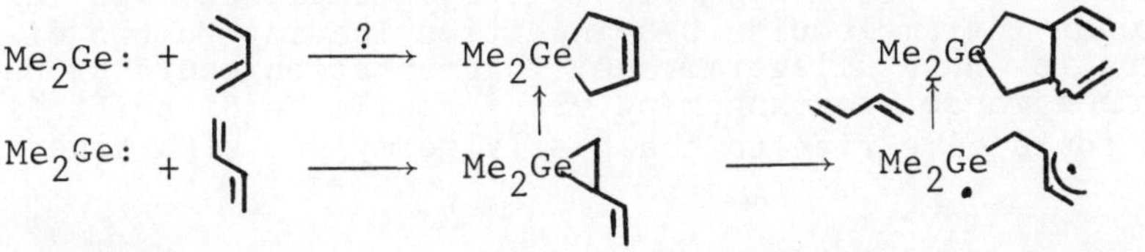

Me$_2$Ge: + ⟶ Me$_2$Ge ⟶ Me$_2$Ge (23)

Me$_2$Ge: + ⟶ Me$_2$Ge ⟶ Me$_2$Ge (24)

4 GAS-PHASE REARRANGEMENTS OF SILYLGERMYLENES

The seminal work of Wulff, Goure and Barton on the rearrangements of methyl(trimethylsilyl)silylene $Me_3Si\text{-}\ddot{S}i\text{-}Me$ led to the recognition that intramolecular C-H insertion by an α-silylsilylene could give rise to a disilirane, while methyl-migration could lead to a disilene.[18] A third process, hydrogen migration to the divalent silicon atom, would produce a silene.[19,20] In the scheme below, one asks whether analogous processes occur for methyl(trimethylsilyl)germylene, accessible by pyrolysis of methylbis(trimethylsilyl)germane.

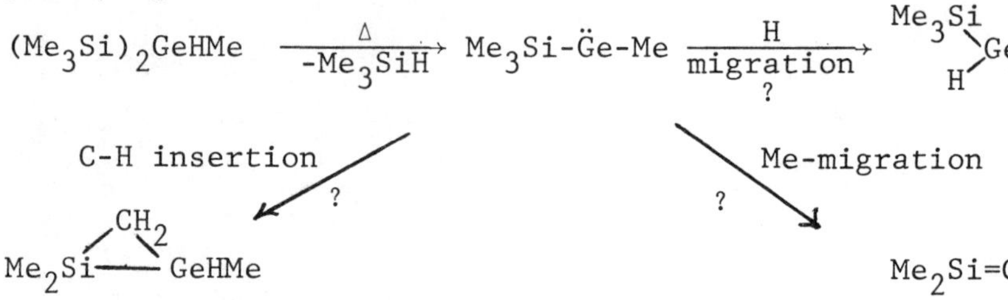

$$(Me_3Si)_2GeHMe \xrightarrow[-Me_3SiH]{\Delta} Me_3Si\text{-}\ddot{G}e\text{-}Me \xrightarrow[?]{H \text{ migration}} \underset{H}{\overset{Me_3Si}{>}}Ge=CH_2$$

C-H insertion ? / \ Me-migration (25)

$Me_2Si\overset{CH_2}{\underline{\quad}}GeHMe \qquad\qquad Me_2Si=GeMe_2$

The required germylene precursor was synthesized by a three-step sequence:

$$Me_2GeCl_2 + 2\ Me_3SiCl \xrightarrow[THF]{Li} (Me_3Si)_2GeMe_2 \quad 46\% \quad (26)$$

$$(Me_3Si)_2GeMe_2 \xrightarrow[MeNO_2]{SnCl_4} (Me_3Si)_2GeClMe \quad 95\% \quad (27)$$

$$(Me_3Si)_2GeClMe \xrightarrow[THF]{LAH} (Me_3Si)_2GeHMe \quad 95\% \quad (28)$$

When $(Me_3Si)_2GeHMe$ was subjected to gas-phase flow-pyrolysis in the presence of 2,3-dimethylbutadiene, ten products were obtained. Nine of these can be rationalized on the basis of the reaction scheme shown below, whose key step is the intramolecular C-H insertion by the divalent germanium center of $Me_3Si\text{-}Ge\text{-}Me$, in competition with its addition. The product ratios suggest that fragmentation of the silagermirane intermediate is a more facile process than its ring-opening to β-silylgermylenes. Fragmentation to both GeHMe and to Me_2Si seems to take place, while ring-opening is more selective and gives no sign of producing a β-germylsilylene.

There is a fly in the ointment, the tenth product, shown in reaction 36 below, that forces us to consider another mechanism, one capable of yielding $Me_3SiCH_2\text{-}Ge\text{-}Me$, a germylene with one extra carbon atom! The reaction scheme given in equations 37 and 38 begins with the dimerization of $Me_3Si\text{-}Ge\text{-}Me$ to a digermene whose competetive Me_3Si- and Me-shifts could form isomeric α-germylgermylenes. If these underwent intramolecular C-H insertion leading to two digermiranes and a silagermirane, fragmentation could produce a germene whose rearrangement via a facile Me_3Si-shift finally could give rise to the β-silylgermylene with five carbon atoms.

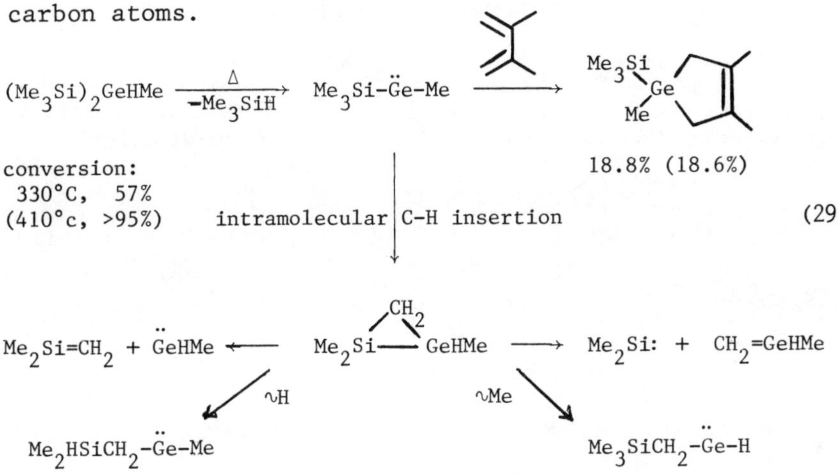

$(Me_3Si)_2GeHMe \xrightarrow[-Me_3SiH]{\Delta} Me_3Si-\ddot{G}e-Me \longrightarrow$ [Me₃Si,Me-Ge cyclopentene product]

conversion:
330°C, 57%
(410°C, >95%)

18.8% (18.6%)

intramolecular C-H insertion (29)

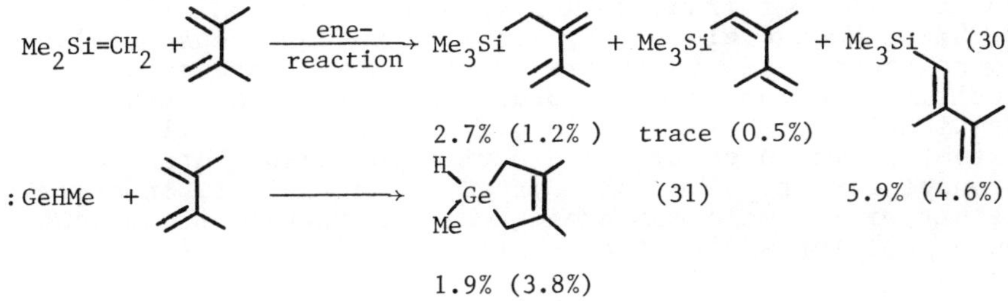

$Me_2Si=CH_2 + \ddot{G}eHMe \longleftarrow Me_2Si\overset{CH_2}{-}GeHMe \longrightarrow Me_2Si: + CH_2=GeHMe$

↙ ∿H   ∿Me ↘

$Me_2HSiCH_2-\ddot{G}e-Me$ $Me_3SiCH_2-\ddot{G}e-H$

$Me_2Si=CH_2 + $ [diene] $\xrightarrow[\text{reaction}]{\text{ene-}} Me_3Si$[product] $+ Me_3Si$[product] $+ Me_3Si$[product] (30)

2.7% (1.2%) trace (0.5%)   5.9% (4.6%)

:GeHMe + [diene] ⟶ [H,Ge,Me cyclopentene] (31)

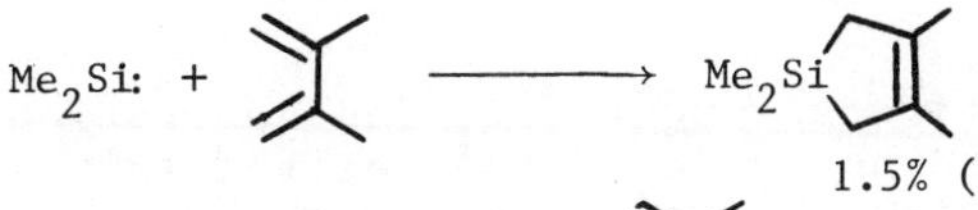

1.9% (3.8%)

$Me_2Si: + $ [diene] ⟶ Me_2Si[cyclopentene] (32)

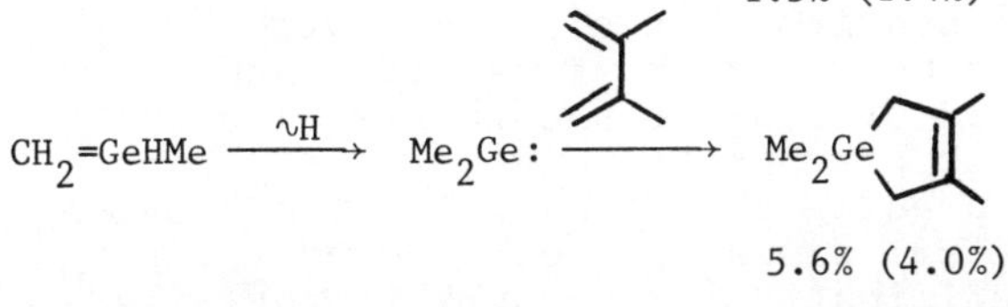

1.5% (1.4%)

$CH_2=GeHMe \xrightarrow{\sim H} Me_2Ge: \longrightarrow Me_2Ge$[cyclopentene] (33)

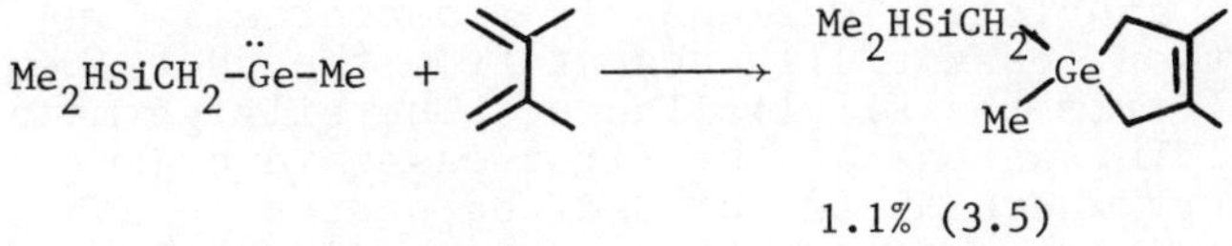

5.6% (4.0%)

$Me_2HSiCH_2-\ddot{G}e-Me + $ [diene] ⟶ [Me₂HSiCH₂,Me-Ge cyclopentene] (34)

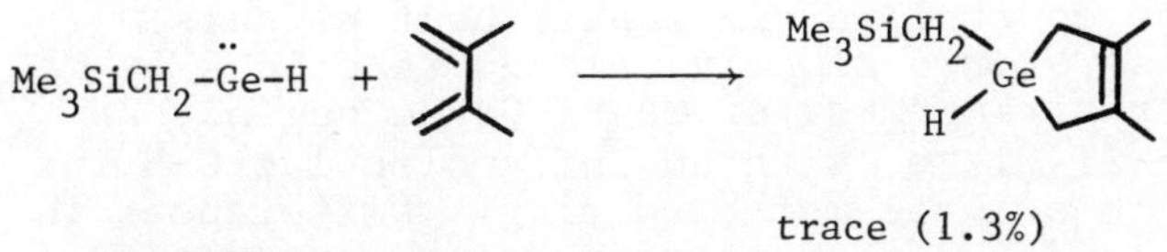

1.1% (3.5)

$Me_3SiCH_2-\ddot{G}e-H + $ [diene] ⟶ [Me₃SiCH₂,H-Ge cyclopentene] (35)

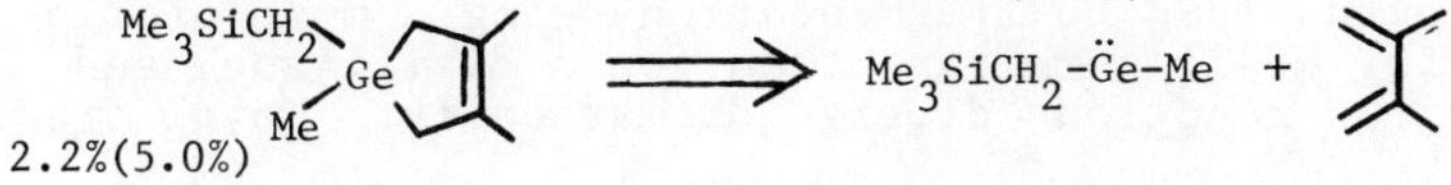

trace (1.3%)

[Me₃SiCH₂,Me-Ge cyclopentene] $\Longrightarrow Me_3SiCH_2-\ddot{G}e-Me + $ [diene] (36)

2.2%(5.0%)

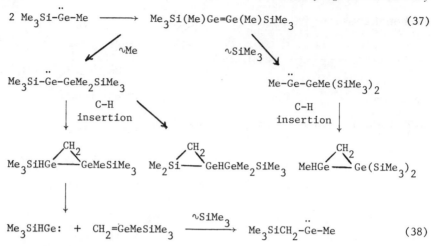

Perversely, other fragmentations of the digermiranes and the silagermirane formed upon dimerization of Me$_3$Si-Ge-Me can give rise, via a series of steps too long to be shown here, to several of the other species whose addition and ene-reaction products were found: Me$_2$Si, Me$_3$SiCH$_2$-Ge-H, Me$_2$Si=CH$_2$, Me$_2$Ge and MeHGe. Until the extent to which germylene dimerization contributes to product formation in this system is determined, an uncertainty will remain concerning which germylene it is that undergoes intramolecular C-H insertion.

5 CONCLUSION

What have these three projects taught us? The generation of diadamantylsilylene did not give us the desired triplet silylene, but it did confirm that siliranes with bulky substituents at silicon are the most easily synthesized and most efficient precursors for the thermal and photochemical generation of sterically congested silylenes. One can confidently predict that they will soon provide ground state triplet silylenes. In our second investigation we found that elimination of phenyltrimethylsilane from phenyl(trimethylsilyl)germanes is a potentially instructive reaction, essentially absent from the photochemistry of phenyl(trimethylsilyl)silanes. The silylgermane system is unfolding as one of the first cases in organosilicon chemistry where different excited states responsible for competing reaction pathways can be identified, and their ratios controlled via substituent effects. In our third glimpse at two closely related reactions it was found that the rearrangement of Me$_3$Si-Ge-Me begins, as does that of Me$_3$Si-Si-Me, with an intramolecular C-H insertion leading to a three-membered ring. Differences in bond strengths lead to fragmentation being a more prominent fate for a silagermirane than for a disilirane, and hence the two reactions diverge dramatically. Novel steps in the reaction sequence proposed for Me$_3$Si-Ge-Me include rearrangements of germenes to germylenes, these being apparently more facile than the corresponding rearrangements of silenes to silylenes.

Acknowledgments. We are grateful to the United States National Science Foundation for financial support under Grant No. CHE-88-02176. Tom Barton, Rob Conlin, Iain Davidson and Mait Jones provided helpful suggestions and good cheer.

REFERENCES

1. M.S. Gordon, Chem.Phys.Lett., 1985, 114, 348.
2. B.I. No, V.V. Son, T.V. Belyakova, V.P. Ushchenko and N.I. Kulikova, Zhur.Obshch.Khim., 1982, 52, 2138; T. Sasaki, K. Shimizu and M. Ohno, Synth.Commun., 1984 14, 853.
3. D. Seyferth and D.C. Annarelli, J.Am.Chem.Soc., 1975, 97, 7162; J.Organometal.Chem., 1976, 117, C51.
4. P. Boudjouk, U. Samaraweera, R. Sooriyakumaran, J. Chrusciel and K.B. Anderson, Angew.Chem.Int.Ed.Engl., 1988, 28, 1355.
5. W. Ando, M. Fujita, H. Yoshida and A. Sekiguchi, J.Am. Chem.Soc., 1988, 110, 3310.
6. M. Schriewer and W.P. Neumann, Angew.Chem.Int.Ed. Engl., 1981, 20, 1019.
7. E.C.-L. Ma, K. Kobayashi, M. Barzilai and P.P. Gaspar, J.Organometal.Chem., 1982, 224, C13.
8. W.P. Neumann, E. Michels and J. Köcher, Tetrahedron Lett., 1987, 28, 3783.
9. D. Lei, R.-J. Hwang and P.P. Gaspar, J.Organometal. Chem., 1984, 271, 1.
10. D. Lei and P.P. Gaspar, Res.Chem.Int., 1989, 12, 103.
11. S. Zhang and R.T. Conlin, private communication.
12. S. Konieczny, S.J. Jacobs, J.K. Braddock Wilking and P.P. Gaspar, J.Organometal.Chem., 1988, 341, C17. The rate constants for reactions of Ph$_2$Ge and Ph$_2$Si reported there are under reinvestigation.
13. S. Collins, S. Murakami, J.T. Snow and S. Masamune, Tetrahedron Lett., 1985, 26, 1281; W. Ando, T. Tsumuraya and A. Sekiguchi, Chem.Lett., 1987, 317.
14. M. Ishikawa and M. Kumada, Chem.Commun., 1970, 612 and 1971, 489.
15. H. Sakurai, K. Sakamoto, M. Kira, Chem.Lett., 1984, 1379; W. Ando, T. Tsumuraya, A. Sekiguchi, ibid., 1987, 317; S. Tomoda, M. Shimoda, Y. Takeuchi, Y. Kajii, K. Obi, I. Tanaka, K. Honda, Chem.Commun., 1988, 910; M. Wakasa, I, Yoneda, K. Mochida, J.Organometal.Chem., 1989, 366, Cl; J. Barrau, D.L. Bean, K.M. Walsh, R. West, J. Michl, Organometallics, 1989, 8, 2606.
16. D. Lei and P.P. Gaspar, manuscript in preparation.
17. K.L. Bobbitt and P.P. Gaspar, ibid.
18. W.D. Wulff, W.F. Goure, T.J. Barton, J.Am.Chem.Soc., 1978, 100, 6236.
19. R.T. Conlin, P.P. Gaspar, J.Am.Chem.Soc., 1976, 98, 878.
20. I.M.T. Davidson, R.J. Scampton, J.Organometal.Chem., 1984, 271, 249

Direct Kinetic Studies of Silicon Hydride Radicals in the Gas Phase

Joseph M. Jasinski
IBM RESEARCH DIVISION, T.J. WATSON RESEARCH CENTER, PO BOX 218, YORKTOWN HEIGHTS, NY 10598, USA

1 INTRODUCTION

The mono-silicon hydride radicals, SiH, SiH_2, and SiH_3, are among the simplest and most fundamental reactive intermediates in both organo- and inorganic silicon chemistry. Silylene, SiH_2, is an important intermediate in the thermal decomposition of silanes, while silylidyne, SiH, and silyl, SiH_3 can be generated in photochemical and glow discharge processes involving silanes. Silyl may also be generated by atom abstraction reactions from silanes.

Over the past five years, significant experimental and theoretical progress has been made in understanding the structure, spectroscopy, reactivity and thermochemistry of these species. In particular, direct, time-resolved spectroscopic techniques have now been developed for use in kinetic studies of *all three* of these radicals in the gas phase. This paper describes our kinetic studies of SiH, SiH_2 and SiH_3 using laser flash photolysis - time resolved laser spectroscopic techniques, and summarizes the new photochemical, kinetic and thermochemical information which has been obtained. The paper is intended as a summary and guide to the recent literature rather than an exhaustive discourse on the chemistry of SiH, SiH_2 and SiH_3.

2 GENERATION AND DETECTION OF RADICALS

Figure 1 shows a generalized schematic of the type of apparatus used to generate and detect SiH, SiH_2 and SiH_3 in our laser flash photolysis - laser spectroscopy experiments. Appropriate precursor molecules, dilute in a buffer gas such as helium, are flash photolyzed in a gas cell under slowly flowing conditions by the UV output of an excimer laser operating at either 193 or 248 nm. Radicals generated by this 10-20 ns pulse of light are detected in a time resolved manner by an appropriate probe laser, which may be pulsed or continuous, operating in the IR, visible or near UV. The detection scheme is either direct absorption spectroscopy or laser induced fluorescence (LIF) spectroscopy, depending on the species being studied. Both techniques provide signals which are proportional to the concentration of thermalized ground electronic state radicals as a function of time after the photolysis pulse. The generation and detection of each SiH_x radical is discussed briefly below, and full experimental details are given in the appropriate references.

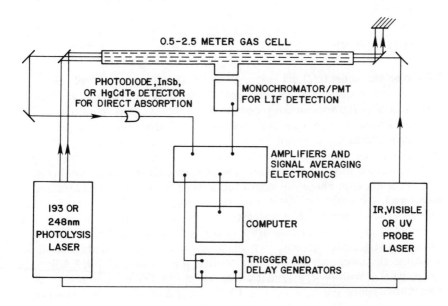

Figure 1 Schematic of laser flash photolysis – laser spectroscopy kinetic apparatus.

Kinetic studies are performed by monitoring the loss of a specific radical as a function of added reaction partner. Rate constants have been measured at room temperature, unless otherwise noted, and the values frequently depend on the total pressure in the gas cell, since many of the reactions of SiH, SiH_2 and SiH_3 are addition or insertion reactions which exhibit three-body kinetics. Most of our measurements have been at 5 or 10 Torr total pressure in helium buffer gas. Some reactions of silylene have been studied over a wider pressure range, typically 1-100 Torr.

The kinetic techniques we employ operate on timescales of 10 ns to 10 ms, depending on the detection scheme and type of kinetics (radical-molecule or radical-radical) under study. For typical radical-molecule reactions, this means that we can easily measure rate constants which range from gas kinetic ($\geq 10^{-10}$ cm^3 molecule^{-1} s^{-1}) to about 10^{-4} gas kinetic. Processes slower than this cannot generally be distinguished from reaction with impurities, etc. For radical-radical reactions, the dynamic range of the experiment is reduced from gas kinetic to about 10^{-2} gas kinetic by limitations on the concentrations of radicals which can typically be generated and the severity of loss mechanisms other than the reaction under study.

3 SILYLIDYNE

Silylidyne is generated by excimer laser flash photolysis of either phenylsilane or disilane at 193 nm, and detected by pulsed laser induced fluorescence (LIF) spectroscopy employing various vibronic transitions of the well known A ← X electronic transition.[1,2] Formation of SiH in this manner is a complicated photochemical process which is not well studied. Both ground (X state) electronic state and excited (A state) electronic state SiH are formed in the dissociation. The X state concentration, monitored by LIF, is linear in photolysis laser energy while the A state concentration, monitored by emission spectroscopy, is approximately quadratic in photolysis laser energy. This suggests that the X state SiH is formed in a one photon process, while formation of the A state from either disilane or phenylsilane requires at least two photons. This is consistent with the available thermochemistry, and our results[1] for the case of phenylsilane are consistent with observations by Nemoto et al.[2] These authors also report formation of the A state of SiH by multiphoton photodissociation of phenylsilane at 248 nm. No measurements of the quantum yields for production of either the X or A states have been made. Chu et al.[3] have suggested that the total quantum yield for SiH formation in the 193 nm photolysis of disilane must be somewhat less than 0.1, but the argument is rather indirect.

Absolute rate constants for the reaction of SiH with H_2, D_2 and SiH_4 have been reported by Begemann et al.[1] Rate constants for the reaction of SiH with NO, O_2, SiH_4 and phenylsilane have been reported by Nemoto et al.[2] Selected values for these rate constants are given in Table 1. Values are at room temperature, unless otherwise noted. With the exception of H_2 and D_2, SiH reacts remarkably rapidly, at or near the gas kinetic rate ($k \geq 10^{-10}$ cm^3 molecule^{-1} s^{-1}), with all of these partners. The reaction with H_2 and D_2, however, is too slow to be studied reliably ($k \leq 10^{-14}$ cm^3 molecule^{-1} s^{-1}), suggesting that a significant potential barrier exists. The anticipated reaction pathway for SiH with H_2, D_2 and silane is insertion to form the respective silyl or disilyl radicals. Evidence for this process in the reaction of SiH with SiH_4 is that the absolute rate constant appears to be pressure dependent.[1,2,4,5] Detailed studies of this pressure dependence remain to be carried out. Further evidence[1] for the existence of a several Kcal/mole barrier to insertion of SiH into molecular hydrogen is that the loss of SiH in the v = 1 vibrational level is at least 10^3 times faster than loss of SiH in v = 0. The temperature dependence and Arrhenius parameters for this reaction remain to be determined.

Table 1 Absolute Rate Constants for SiH Reactions

Reactant	Rate Constant (cm^3 $molecule^{-1}$ s^{-1})	Conditions	Reference
H_2	$\leq (1.2\pm0.2)\times10^{-14}$	5 Torr He	(1)
D_2	$\leq (1.8\pm0.2)\times10^{-14}$	5 Torr He	(1)
NO	$(2.5\pm0.3)\times10^{-10}$	2 Torr Ar	(2)
O_2	$(1.7\pm0.2)\times10^{-10}$	2 Torr Ar	(2)
$C_6H_5SiH_3$	$\sim3\times10^{-10}$	2 Torr Ar	(2)
SiH_4	$(4.3\pm0.3)\times10^{-10}$	5 Torr He	(1)
	$(2.7\pm0.5)\times10^{-10}$	2 Torr He	(5)
	$(2.8\pm0.6)\times10^{-10}$	2 Torr Ar	(2)
	$(3.3\pm0.5)\times10^{-12}$	4-50 mTorr, 500 K	(4)
H_2	$(1.6\pm0.1)\times10^{-11}$	SiH v = 1, 5 Torr He	(1)

4 SILYLENE

Silylene is generated by photodissociation of phenylsilane or disilane at 193 nm or by photodissociation of iodosilane at 248 nm. It is detected either by pulsed LIF or by direct laser absorption spectroscopy employing the A ← X electronic transition. The quantum yield for silylene formation in the photolysis of either phenylsilane or disilane is estimated[3,6] to be small, ≤ 0.1.

Direct kinetic studies of silylene reactions have been the subject of a substantial amount of work[7-14] beginning in 1985. Absolute rate constants for the reaction of silylene with H_2, D_2, a variety of diatomics such as NO, O_2, Cl_2, HCl, CO, Si-H bonds of silanes, alkanes, olefins, acetylenes and dienes have been measured, by three independent research groups. Selected values from these studies are given in Table 2. Values are at room temperature, unless otherwise noted. When comparisons are possible, rate constants for the same reaction measured by different groups generally agree, or differ by at worst a factor of 2-3. Although such discrepancies are outside the quoted error limits for any individual study and remain to be reconciled, all direct studies to date agree that silylene is significantly more reactive than previously thought.

The key findings of the direct kinetic studies of silylene are that it inserts into Si-H bonds and adds to olefins, dienes and acetylene at or near the gas kinetic rate and exhibits a range or reactivity with simple diatomics. It is unreactive with C-H bonds of alkanes. The kinetics of the insertion reaction of silylene with hydrogen, silane and disilane, combined with the kinetics of the reverse reactions, dissociation of silane, disilane and trisilane, have been used to derive[11] a heat of formation of ≃65 Kcal/mole, in excellent agreement with other recent experiments and theoretical results.[15] Preliminary results on the temperature dependence for silylene insertions have been reported by the Reading group.[13]

5 SILYL

Silyl is generated by flash photolysis of iodosilane at 248 nm or bromosilane at 193 nm. Small amounts of silyl are also generated in the photolysis of disilane at 193 nm. No silyl is observed in the photolysis of phenylsilane at 193 nm. Quantum yields are estimated to be 0.2, 0.1 and 0.08 for the cases of iodosilane, bromosilane and disilane, respectively.[16,17] It is clear that other radical species are also produced in all of these processes, and this complicates the use of any of these direct photochemical sources of silyl in kinetic studies. Silyl is also generated by rapid H atom abstraction from silane by chlorine atoms. The atomic chlorine can be formed by photodissociation of carbon tetrachloride or HCl at 193 nm, or by infrared multiphoton dissociation of chlorofluorocarbons. Photolysis of molecular chlorine at 351 nm in the presence of silane would be a very desirable source of silyl. This process is not useful, however, due to spontaneous reaction of chlorine and silane in the absence of photolysis.[16] Thus far, optical detection of silyl in kinetic studies has been by infrared diode laser absorption spectroscopy,[16,18,19] employing various vibration-rotation transitions of the v_2 umbrella mode. Silyl has also been detected by multiphoton ionization spectroscopy,[20] photoionization mass spectrometry[21] and laser magnetic resonance spectroscopy.[22]

Absolute rate constants[16,18,19,21-23] for the reaction of silyl with S_2Cl_2, NOCl, N_2O, NO_2, O_2 and NO, ethylene, propylene and propyne have been reported, as have values for the silyl recombination rate constant. Selected values are summarized in Table 3. Values are at room temperature, unless otherwise noted. Silyl is unreactive with N_2O. It is also markedly sluggish in its addition to carbon multiple bonds, suggesting that this reaction has a significant activation barrier. Silyl reacts much more readily with S_2Cl_2 and NOCl, open shell oxidizing species, and itself. Two groups have reported rate constants for the recombination of silyl radicals.[16,18] The rate constants for recombination are near gas kinetic, but require further refinement. The major difficulty in measuring this rate constant is that, unlike the case of radical-molecule reactions which can be studied under pseudo-first order conditions, the recombination kinetics are bimolecular and the absolute number density of silyl must be known accurately. Preliminary investigations of the reaction of silyl with SiD_4 and Si_2H_6 show no observable reaction.[17]

Table 2 Absolute Rate Constants for SiH_2 Reactions

Reactant	Rate Constant (cm^3 molecule^{-1} s^{-1})	Conditions	Reference
H_2	$(1.0\pm0.4)\times10^{-13}$	1.8 Torr He	(7)
	$(2.6\pm0.6)\times10^{-13}$	2 Torr He	(11)
D_2	$(2.1\pm0.4)\times10^{-12}$	5 Torr He	(14)
	$(1.9\pm0.2)\times10^{-12}$	5 Torr SF_6, 268-330K	(13)

Table 2 (continued)

SiH_4	$(6.7\pm0.7)\times10^{-11}$	1 Torr He	(11)
	$(1.1\pm0.2)\times10^{-10}$	1 Torr He	(7)
	1.3×10^{-10}	1 Torr He	(12)
CH_3SiH_3	$(3.7\pm0.2)\times10^{-10}$	5 Torr SF_6	(12)
$(CH_3)_2SiH_3$	$(3.3\pm0.3)\times10^{-10}$	5 Torr SF_6	(12)
$(CH_3)_3SiH$	$(2.5\pm0.1)\times10^{-10}$	5 Torr SF_6	(12)
Si_2H_6	$(5.7\pm0.2)\times10^{-10}$	1 Torr He	(7)
	$(1.5\pm0.2)\times10^{-10}$	1 Torr He	(11)
	$(2.8\pm0.3)\times10^{-10}$	5 Torr He	(11)
	$(4.6\pm0.7)\times10^{-10}$	5 Torr Ar	(12)
CH_4	$\leq (2.5\pm0.5)\times10^{-14}$	5 Torr He	(10)
C_2H_6	$\leq (1.2\pm0.5)\times10^{-14}$	5 Torr He	(10)
C_2H_4	$(2.6\pm0.3)\times10^{-11}$	1 Torr He	(10)
	$(9.7\pm1.2)\times10^{-11}$	1 Torr He	(7)
C_3H_6 (propylene)	$(1.2\pm0.1)\times10^{-10}$	5 Torr He	(10)
	2.4×10^{-10}	–	(12)
C_4H_6 (butadiene)	$(1.9\pm0.2)\times10^{-10}$	5 Torr He	(10)
C_2H_2	$(9.8\pm1.2)\times10^{-11}$	5 Torr He	(10)
HCl	$(6.8\pm1.0)\times10^{-12}$	5 Torr He	(9)
Cl_2	$(1.4\pm0.2)\times10^{-10}$	5 Torr He	(9)
NO	$(1.7\pm0.2)\times10^{-11}$	5 Torr He	(9)
O_2	$(7.7\pm1.0)\times10^{-12}$	5 Torr He	(9)
CO	$< 10^{-13}$	5 Torr He	(9)
N_2	$< 10^{-13}$	5 Torr He	(9)

Table 3 Absolute Rate Constants for SiH_3 Reactions

Reactant	Rate Constant (cm^3 molecule^{-1} s^{-1})	Conditions	Reference
SiH_3	$\leq (6.1\pm3.5)\times10^{-11}$	9.5 Torr He	(16)
	$(1.5\pm0.6)\times10^{-10}$	0.9 Torr H_2	(18)
O_2	$(9.7\pm1.0)\times10^{-12}$	1-27 Torr Ar	(23)
	$(1.3\pm0.4)\times10^{-11}$	1-6 Torr He	(21)
	$(1.3\pm0.3)\times10^{-11}$	1-10 Torr N_2	(19)
NO	$(2.5\pm0.5)\times10^{-12}$	9.5 Torr He	(16)
	$(8.2\pm0.9)\times10^{-30}$ a	3-11 Torr N_2	(19)
N_2O	$< 5\times10^{-15}$	500 K	(21)
NO_2	$(5.1\pm0.9)\times10^{-11}$	3-10 Torr N_2	(19)
C_2H_4	$\leq (3\pm3)\times10^{-15}$	9.5 Torr He	(16)
C_3H_6 (propylene)	$\leq (1.5\pm0.5)\times10^{-14}$	9.5 Torr He	(16)
C_3H_4 (propyne)	$\leq (1.8\pm0.4)\times10^{-14}$	9.5 Torr He	(16)
NOCl	$(1.3\pm0.3)\times10^{-11}$	5 Torr Ar	(23)
S_2Cl_2	$(2.4\pm0.5)\times10^{-11}$	4-7 Torr He, 326K	(22)

a Rate constant is linear in total pressure. Units are cm^6 molecule^{-2}s^{-1}

6 CONCLUSIONS

Direct spectroscopic kinetic techniques are now available to study the chemistry of SiH, SiH_2 and SiH_3. A wealth of new information has already been obtained in the past few years, and the potential now exists for a very complete understanding of the reactivity of these fundamental silicon centered radicals.

REFERENCES

1. M. H. Begemann, R. W. Dreyfus and J. M. Jasinski, Chem. Phys. Lett., 1989, 155, 351.
2. M. Nemoto, A. Suzuki, H. Nakamura, K. Shibuya and K. Obi, Chem. Phys. Lett., 1989, 162, 467.
3. J. O. Chu, M. H. Begemann, J. S. McKillop and J. M. Jasinski, Chem. Phys. Lett., 1989, 155, 576.
4. J. P. M. Schmitt, P. Gressier, M. Krishnan, G. DeRosny and J. Perrin, Chem. Phys., 1984, 84, 281.
5. J. M. Jasinski, Mat. Res. Soc. Symp. Proc., 1990, 165, in press.
6. J. E. Baggott, H. M. Frey, P. D. Lightfoot, and R. Walsh, Chem. Phys. Lett., 1986, 125, 22.
7. G. Inoue and M. Suzuki, Chem. Phys. Lett., 1985, 122, 361.
8. J. M. Jasinski, J. Phys. Chem., 1986, 90, 555.
9. J. O. Chu, D. B. Beach, R. D. Estes and J. M. Jasinski, Chem. Phys. Lett., 1988, 143, 135.
10. J. O. Chu, D. B. Beach and J. M. Jasinski, J. Phys. Chem., 1987, 91, 5340.
11. J. M. Jasinski and J. O. Chu, J. Chem. Phys., 1988, 88, 27.
12. J. E. Baggott, H. M. Frey, P. D. Lightfoot, R. Walsh and I. M. Watts, J. Chem. Soc. Faraday Trans., 1990, 86, 27.
13. J. E. Baggott, H. M. Frey, K. D. King, P. D. Lightfoot, R. Walsh and I. M. Watts, J. Phys. Chem., 1988, 92, 4025.
14. J. M. Jasinski and J. O. Chu, in "Silicon Chemistry", J. Y. Corey, E. R. Corey and P. P. Gaspar, Eds., Ellis Horwood, Chichester, 1988, Chapter 40.
15. P. Ho and C. F. Melius, J. Phys. Chem., 1990, 94, 5120, and references therein.
16. S. K. Loh, D. B. Beach and J. M. Jasinski, Chem. Phys. Lett., 1990, 169, 55.
17. S. K. Loh and J. M. Jasinski, unpublished.

18. N. Itabashi, K. Kato, N. Nishiwaki, T. Goto, C. Yamada and E. Hirota, Jpn. J. Appl. Phys., 1989, 28, L325.
19. K. Sugawara, T. Nakanaga, H. Takeo and C. Matsumura, Chem. Phys. Lett., 1989, 157, 309.
20. R. D. Johnson, B. P. Tsai and J. W. Hudgens, J. Chem. Phys., 1989, 91, 3340.
21. I. R. Slagle, J. R. Bernhardt and D. Gutman, Chem. Phys. Lett., 1988, 149, 180.
22. L. N. Krasnoperov, E. N. Chesnokov and V. N. Panfilov, Chem. Phys., 1984, 89, 297.
23. S. A. Chasovnikov and L. N. Krasnoperov, Khim. Fiz., 1987, 6, 956.

Silylene and Disilene Addition Reactions

Manfred Weidenbruch
FACHBEREICH CHEMIE DER UNIVERSITÄT OLDENBURG,
POSTFACH 25 03, D-2900 OLDENBURG, FRG

Introduction

The synthesis of stable disilenes[1], compounds containing silicon-silicon double bonds, as well as the discovery of stable silenes[2,3] have been milestones in modern chemistry. The formation of stable disilenes is accomplished by photolysis of open-chain[4] or cyclic[5] trisilanes bearing at least one aryl group per silicon atom.

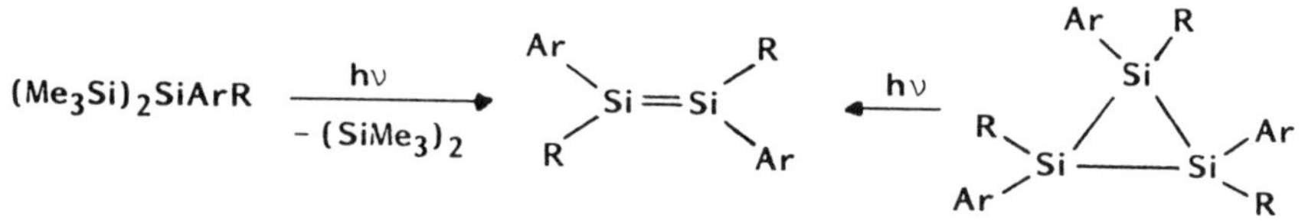

Stable tetraalkyldisilenes are not accessible by these routes. Photolysis of hexa-tert-butylcyclotrisilane (1)[6], for example, produces the marginally stable tetra-tert-butyldisilene (2) along with di-tert-butylsilylene (3) both of which can be trapped by several addition reactions to unsaturated compounds.

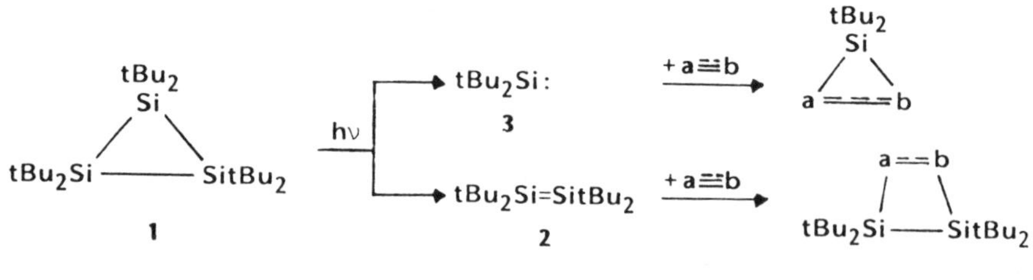

The formation of two reactive intermediates on photolysis of 1 has the advantage of giving at least one isolable addition product in most cases. On the other hand, in some reactions of 1 it cannot clearly be decided whether the isolated product results from an addition of 2 or of 3. Despite this uncertainty photolysis of 1 in the presence of unsaturated compounds or of molecules containing carbene-like carbon atoms has led to several novel ring systems which are documented below.

[2+1]- and [2+2]-Cycloaddition Reactions

The compounds 2 and 3 can be simultaneously or individually trapped by means of a variety of addition reactions. For example, 3 undergoes [2+1]-addition to the triple bond of nitriles to give silaazirines, which, presumably owing to the high ring strain, undergo spontaneous σ-dimerization to give 1,4- (4) or 1,3-diaza-2,5-disila-1(6),3-cyclohexadienes (5)[7].

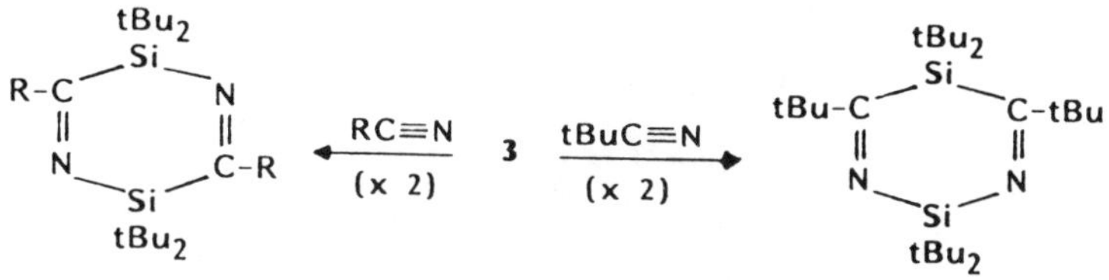

Formation of the silaazirine intermediates could be confirmed indirectly by the analogous reaction of 3 with phosphaalkynes, from which stable phosphasilirenes (6) are isolable[8]. The molecules 6 were the first three-membered rings containing a PC double bond. Meanwhile, analogous compounds containing carbon[9], germanium[10] or phosphorus[11] instead of silicon have been synthesized.

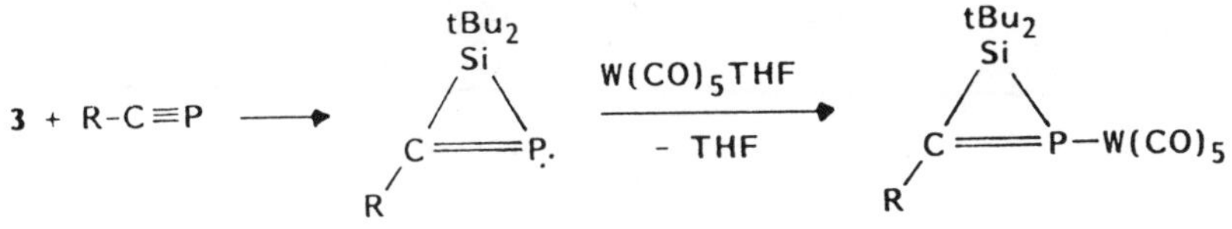

The structure of the three-membered ring compound 7 has been confirmed by X-ray crystallography of its end-on (P-) coordinated complex with tungsten pentacarbonyl (figure 1). The PC double bond length in 7 amounts to 1.686 Å and is comparable to open-chain phosphaalkenes.

It appears surprising that photolysis of 1 in the presence of phenyl isothiocyanate does not give rise to a product of [2+1]-cycloaddition of 3 to the CN double bond but rather to its isomer 8. The silathiirane 8 is assumed to undergo spontaneous σ-dimerization to give the isolated six-membered ring 9[12]. Recently, it has been shown

$$3 + PhN=C=S \longrightarrow \left(\begin{array}{c} tBu_2 \\ Si \\ C\!\!-\!\!S \\ PhN \\ 8 \end{array} \right) \xrightarrow{\times 2} \begin{array}{c} tBu_2 \\ Si \\ S \\ N=C \\ Ph \\ tBu_2 \end{array} \begin{array}{c} Ph \\ C=N \\ S \\ 9 \end{array}$$

that the reaction of decamethylsilicocene with phenyl isothiocyanate presumably is initiated by a similar [2+1]-cycloaddition. By a sequence of elimination and addition reactions the primarily formed silathiirane is converted to a dithiasiletane[13].

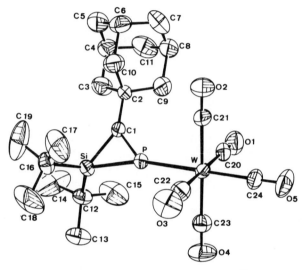

Figure 1. Crystal structure of **7** (H-atoms omitted)[8]. Selected bond lengths (Å) and angles (°) (standard deviations): P-C(1) 1.686(6), P-Si 2.196(2), C(1)-Si 1.842(6), P-W 2.475(2); Si-P-C(1) 54.8(2), P-C(1)-Si 76.9(3), C(1)-Si-P 48.4(2).

In contrast to the [2+1]-cycloadditions of the silylene **3**, which are extremely rare, [2+2]-cycloadditions of **2** as well as of stable disilenes[1] to numerous multiply-bonded compounds have thoroughly been investigated. Tetramesityldisilene or the nearly-stable **2** undergo [2+2]-cycloadditions with unsymmetrically substituted alkynes[14,15] as well as with carbon-carbon double bonds.

$$\begin{array}{c} \diagdown \diagup \\ -C-C- \\ \diagup \diagdown \\ tBu_2Si \!-\!\!-\! SitBu_2 \end{array} \xleftarrow{>C=C<} 2 \xrightarrow{RC\equiv CH} \begin{array}{c} R \qquad H \\ \diagdown \diagup \\ C=C \\ \diagup \diagdown \\ tBu_2Si \!-\!\!-\! SitBu_2 \end{array}$$

Silylene and Disilene Addition Reactions

Whereas unhindered disilenes have long been known to be excellent Diels-Alder reagents[16], [2+4]-cycloadditions between stable or nearly-stable disilenes have not been observed as yet[1]. For example, 2 reacts with 1,3-cyclohexadiene[17] or with 1,3-diphenylisobenzofuran [18] to give the [2+2]- (11) rather than the [2+4]-cycloadduct (10).

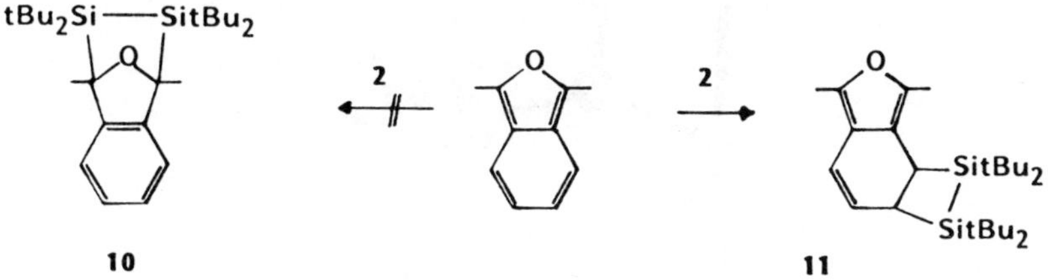

From a thermodynamic point of view, formation of the aromatic compound 10 should be expected rather than formation of isomer 11.

Disilene 2 adds also to ketones to give 1,2,3-oxadisiletanes (12). The adducts 12, formed from the enolizable ketones acetone and acetophenone, undergo further rearrangement reactions to give the disilanylenolethers (13)[14].

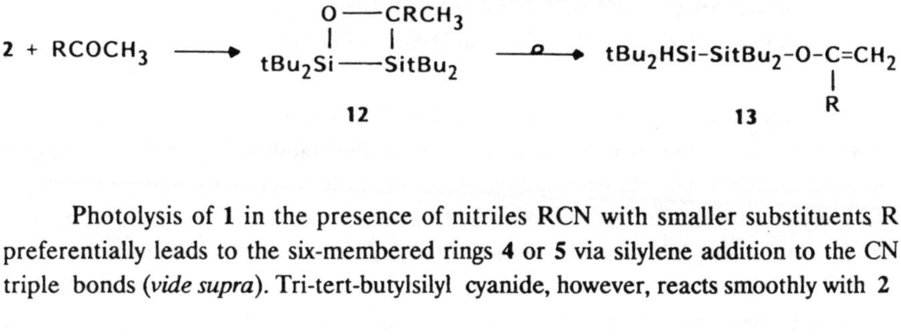

Photolysis of 1 in the presence of nitriles RCN with smaller substituents R preferentially leads to the six-membered rings 4 or 5 via silylene addition to the CN triple bonds (*vide supra*). Tri-tert-butylsilyl cyanide, however, reacts smoothly with 2

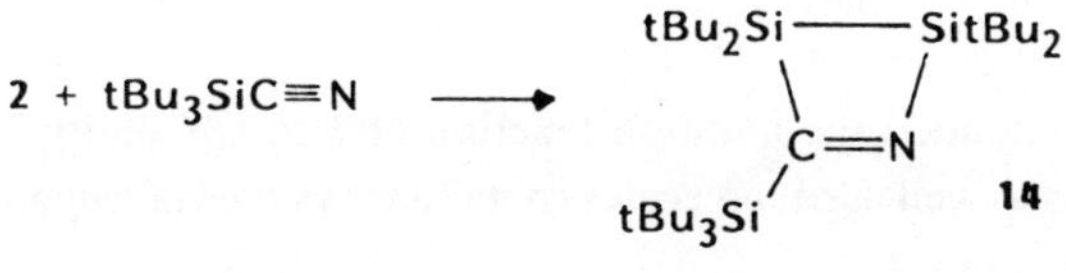

to give the 2,3-disila-1-azetine 14. Apparently the bulky tri-tert-butylsilyl group prevents addition of 3 to the CN triple bond and thus allows the [2+2]-cycloaddition to give 14. The structure of the ring 14 has been unambiguously confirmed by X-ray crystallography (figure 2)[19].

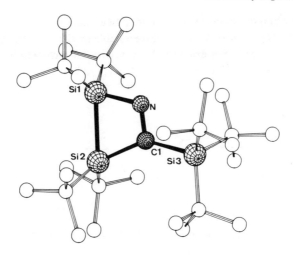

Figure 2. Crystal structure of **14** (H-atoms omitted). Selected bond lengths (Å) and angles (°) (Standard deviations): Si(1)-Si(2) 2.435(1), Si(1)-N 1.798(2), C(1)-N 1.304(3), C(1)-Si(2) 1.996(3), C(1)-Si(3) 1.954(3); Si(1)-N-C(1) 109.7(2), N-C(1)-Si(2) 104.8(2), C(1)-Si(2)-Si(1) 69.5(1), N-Si(2)-Si(2) 75.8(1)

Interestingly, alkenes do not undergo [2+2]-cycloadditions with nitriles[20]. Despite bulky ligands, which are neccessary for the stabilization of the SiSi double bond, stable or nearly-stable disilenes are often more reactive than alkenes bearing sterically less demanding substituents, and, in some cases undergo reactions atypical for CC double bonds[1]. Another example for the unusual reactivity of **2** is reflected in its addition reaction to a racemic 3-thiazoline derivative which proceeds diastereoselectively to yield the 2,3-disila-1-azetidine **15**. The structure of **15** has been confirmed by NMR data and by an X-ray structure analysis[21].

Most recently, we have extended the addition reaction of **2** to CN multiple bonds of heterocyclic compounds as well as of molecules containing systems of cumulated double bonds. Disilene **2** reacts smoothly with isoquinoline to give regioselectively the 2,3-disila-1-azetidine derivative **16**. In contrast to the above discussed 2,3-disila-1-azetidine **15**, compound **16** reveals an almost planar environment at the ni-

trogen atom. As a consequence of the bulky substituents at the silicon atoms, the Si-N-C and Si-C-C angles are considerably widened[22].

In the presence of N,N'-dicyclohexylcarbodiimide photolysis of **1** seems to follow an analogous route. Initially, we assumed that the [2+2]-cycloaddition product of **2** with one of the NC double bond (**17**) was formed. The X-ray structure analysis of the pale pink crystals, however, demonstrated clearly that the rearranged 2,4-disila-1-azacyclobutanimine (**18**) has been isolated[22]. The formation of **18** can be reasonably

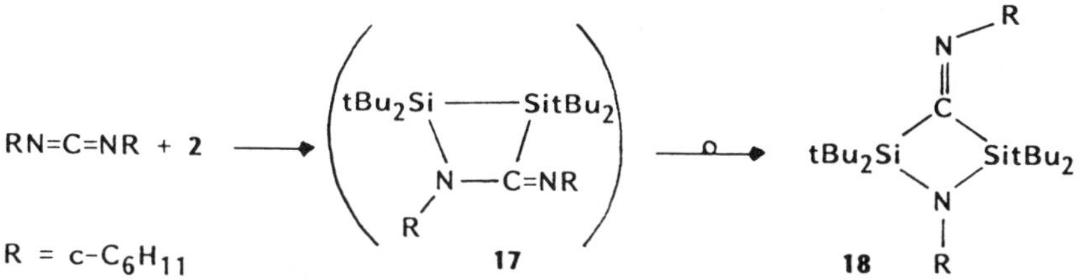

explained by the assumption of a 2,3-disila-1-azacyclobutanimine intermediate which undergoes a rearrangement (**17** → **18**) with formation of the disilazane moiety. Similar rearrangements have been observed in the reactions of tetramesityldisilene with nitrobenzene[23] or of **2** with an isocyanate[19] (see below).

[3+1]- and [3+2]-Cycloaddition Reactions

In contrast to the above mentioned [2+1]-cycloadditions of **3** or [2+2]-cycloaddition reactions of **2**, examples for [3+1]- and [3+2]-cycloadditions of these molecules and related compounds are exceedingly rare. Recently, *West* et al.[24] have shown by X-ray structure analyses of two hindered silyl azides that both NN bond lengths in silyl azides are nearly equal. Therefore, molecules of this type may be considered as systems of cumulated double bonds.

Photolysis of **1** in the presence of tri-tert-butylsilyl azide gives two compounds, both of which have been previously obtained by independent routes[25]. The 1,7-diaza-2,6-disilaheptane **20** is the dominating product of this reaction. The probably primary

$$\text{t-Bu}_3\text{SiN}_3 + 3 \longrightarrow \left(\text{tBu}_3\text{Si-N} \underset{\underset{\text{tBu}_2}{\text{Si}}}{\overset{\text{N}}{\diagup\diagdown}} \text{N} \right) \xrightarrow{-\text{N}_2} \underset{\mathbf{19}}{\text{tBu}_3\text{Si-N=SitBu}_2}$$

$$\xrightarrow{\text{i-C}_4\text{H}_8} \underset{\mathbf{20}}{\text{H}_2\text{C=C(CH}_2\text{SitBu}_2\text{NHSitBu}_3)_2}$$

intermediate, the silanimine **19**, is also isolated in minor amounts. The formation of **19** and **20** is most easily explained when it is assumed that the [3+1]-cycloaddition of **3** to the silyl azide followed by loss of nitrogen gives **19** which by an ene reaction with photolytically formed isobutylene produces the final product **20**[26]. A related reaction between dimesitylsilylene, generated photochemically from 2,2-dimesitylhexamethyltrisilane[27], and the crowded silyl azide gives rise to the silabenzocyclobutene derivative **22** whose structure has been confirmed by the X-ray structure analysis[26].

$$\text{tBu}_3\text{SiN}_3 + \text{Mes}_2\text{Si:} \xrightarrow{-\text{N}_2} \underset{\mathbf{21}}{\text{t-Bu}_3\text{Si-N=SiMes}_2} \longrightarrow$$

22 tBu₃Si-NH-Si—[benzene ring with CH₃, CH₃, Mes substituents]

It is believed that the reaction sequence initially gives the silanimine **21** which rearranges by a [1,5]-sigmatropic shift of hydrogen from one of the *ortho*-methyl groups to the nitrogen atom, followed by a conrotatory ring closure. The thermal rearrangement of tetramesityldisilene to a silabenzocyclobutene derivative[1,28] proceeds similarly.

Only a few examples of [2+3]-cycloaddition reactions of disilenes are known. The reaction of tetrakis(2,6-dimethylphenyl)disilene with diazomethane may proceed via a [2+3]-cycloaddition intermediate to give disilacyclopropane[19] finally. Tetramesityldisilene reacts with aryl azides to give triazadihydrosiloles along with disilaaziridines[1,30].

Photolysis of **1** in the presence of tri-tert-butylsilyl isocyanate may also be initiated by a [2+3]-addition of **2** to the oxygen and nitrogen atoms of the isocyanato group. By migration of the bulky tri-tert-butylsilyl group from the N atom to the C atom and by cleavage of the CO bond and insertion of the oxygen between two Si atoms finally the five-membered ring **23** is isolated whose structure has been unequivocally confirmed by the X-ray structure analysis.

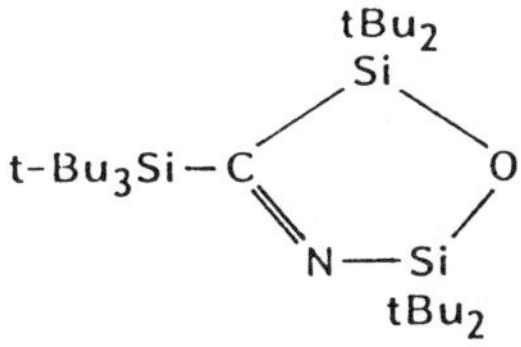

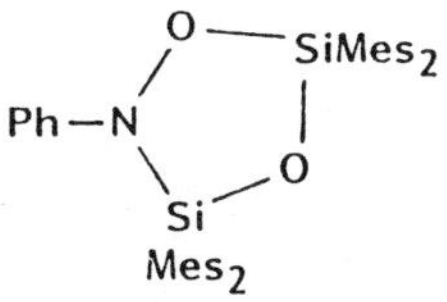

Simultaneously, *West* et al.[27] found that the reaction of tetramesityldisilene with nitrobenzene is also initiated by a [2+3]-cycloaddition of the SiSi double bond to the oxygen atoms of the nitro group. Subsequently, as in the case of **23**, a rearrangement with formation of **24** finally takes place.

[4+1]- and [4+2]-Cycloaddition Reactions

Photolysis of **1** in the presence of 2,2'-bipyridyl proceeds by [4+1]-cycloaddition of **3** to the nitrogen atoms and by [2+2]-cycloaddition of **2** to the CC double bond in the 3,4-position giving the twofold addition product **25**[18].

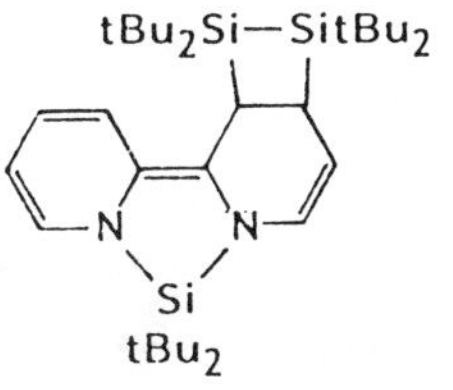

a: R = t-Bu, R' = H
b: R = t-Bu, R' = CH_3
c: R = Mes, R' = CH_3

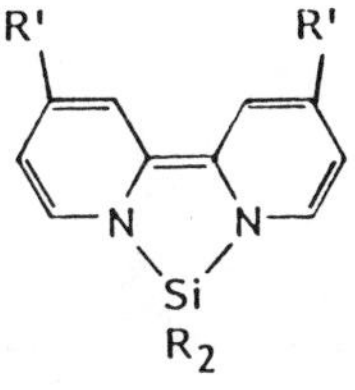

When 2,2'-bipyridyl or 4,4'-dimethyl-2,2'-bipyridyl are allowed to react with 1,1-di-tert-butyl-*trans*-2,3-dimethylsilirane[31] or with hexamethyl-2,2-dimesityltrisilane solely the [4+1]-cycloaddition products **26** are obtained in which the heteroaromatic nitrogen compounds have also been converted into conjugated systems of double bonds[32]. The rings **26** as well as ring **25** are dark violet coloured and show longest wave-length absorptions at ca. 560 nm.

Photolysis of **1** in the presence of N,N'-dicyclohexyl-1,4-diazabutadiene, which possesses the same C_2N_2-framework as 2,2'-bipyridyl, gives not the expected five-membered ring but leads to addition of **2** to the nitrogen atoms in a [2+4]-fashion. The formation of the six-membered ring **27** (figure 3) can be explained by a hetero Diels-Alder reaction between **2** and the 1,4-diazabutadiene. Obviously, an alternative

R = c-C_6H_{11}

27

28

mechanistic possibility is that the ring **27** is formed in a stepwise manner. At present we have no evidence for or against the intermediacy of a [1+4]-cycloaddition product like **28**[33].

However, compound **28** can easily be isolated from the reaction of the dilithiated 1,4-diazabutandiene with di-tert-butyldichlorosilane[33].

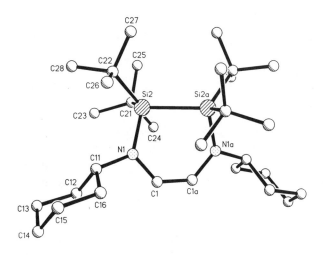

Figure 3. Crystal structure of **27** (H-atoms omitted). Selected bond lengths (Å) and angles (°) (Standard deviations): Si(2-Si(2a) 2.469(1), Si(2)-N(1) 1.753(2), N(1)-C(1) 1.408(2), Si(2)-C 1.951; N(1)-Si(2)-Si(2a) 98.4(1), C(1)-N(1)-Si(2) 120.9(1), N(1)-C(1)-C(1a) 131.8(1)

Addition Reactions

In contrast to the numerous cycloaddition reactions of **2** or **3** to multiply bonded compounds cophotolyses of **1** and several aryl isonitriles seem to follow an alternative route. In the presence of phenyl isocyanide[34] or mesityl isocyanide dark-red crystals of the corresponding 2,4-disilacyclobutane-1,3-diimines are isolated. Also in the presence of 2,6-diisopropylphenyl isocyanide violet crystals of the four-membered heterocycle 29 are obtained. In this case minor amounts of the isomeric ring 30 are also isolated as green crystals[35]. The structures of both ring systems have been confirmed by X-ray crystallography[34,35].

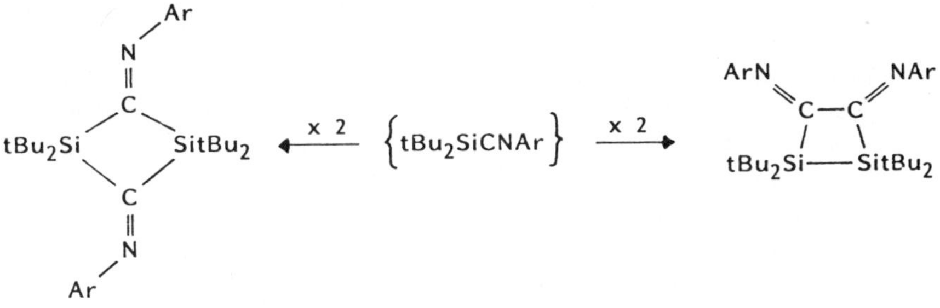

$$29 \quad\quad\quad 31 \quad\quad\quad 30$$

Since both, 29 and 30, are formed simultaneously in the photolysis reaction, it is possible that they arise from a common intermediate such as a silaketenimine or a Lewis-acid base adduct of typ 31. Head-to-tail dimerization then should give the four-membered ring 29 whereas head-to-head dimerization should yield the isomer 30. To verify that the isolated rings result from dimerizations of the suggested silaketenimine of type 31 and not from a twofold addition of the aryl isocyanides to the disilene 2 we allowed 1 to react with 2,4,6-tri-tert-butylphenyl isocyanide, whose bulky *ortho*-tert-butyl groups should disfavour both cyclodimerization and insertion into the SiSi bond of a possible disilacylopropanimine[36] intermediate.

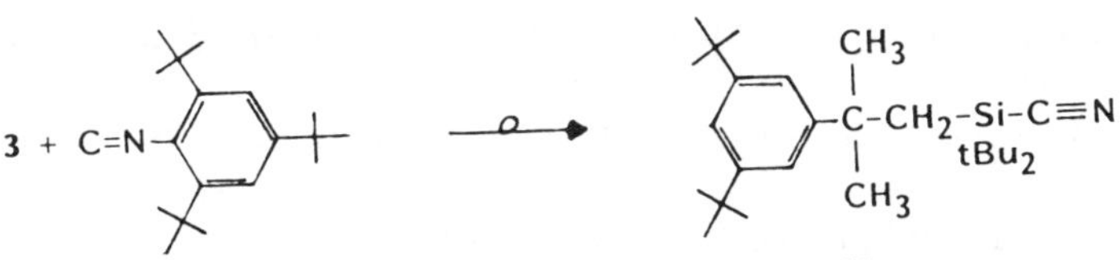

$$32$$

In fact, a monomeric compound consisting of 3 and the aryl isocyanide in a ratio of 1:1 was isolated. The spectroscopic data indicated however, that the expected intermediate of type 31 had rearranged to the trialkylsilyl cyanide 32[34]. These results seem to confirm the suggested formation of the rings 29 and 30. However, it must be pointed out that other mechanisms can not definitely be ruled out.

Acknowledgements. Financial support from the Deutsche Forschungsgemeinschaft, the Volkswagen-Stiftung, and the Fonds der Chemischen Industrie is gratefully acknowledged. I would also like to acknowledge the enthusiasm and experimental skill of my co-workers whose names are given in the references. Thanks are also due to the groups of Professor von Schnering (Stuttgart) and Professor Pohl (Oldenburg) for the crystal structure determinations. The author is particularly grateful to Professor R. West for information furnished in advance of publication and for helpful advice.

REFERENCES
1. Reviews: R. West, *Angew. Chem.* 1987, *99*, 1231; *Angew. Chem. Int. Ed. Engl.* 1987, *26*, 1201; G. Raabe and J. Michl in *The Chemistry of Organic Silicon Compounds* (Edited by S. Patai and Z. Rappoport), Part 2, Wiley, Chichester 1989, p. 1015.
2. A.G. Brook, F. Abdesaken, B. Gutekunst, G. Gutekunst and R.K. Kallury, *J. Chem. Soc., Chem. Commun.* 1981, 191.
3. N. Wiberg, G. Wagner and G. Müller, *Angew. Chem.* 1985, *97*, 220; *Angew. Chem. Int. Ed. Engl.* 1985, *24*, 229.
4. R. West, M.J. Fink and J. Michl, *Science (Washington)* 1981, *214*, 1343.
5. S. Masamune, Y. Hanzawa, S. Murakami, T. Bally and J.F. Blount, *J. Am. Chem. Soc.* 1982, *104*, 1150.
6. A. Schäfer, M. Weidenbruch, K. Peters and H.G. von Schnering, *Angew. Chem.* 1984, *96*, 311; *Angew. Chem. Int. Ed. Engl.* 1984, *23*, 302.
7. M. Weidenbruch, A. Schäfer, K. Peters and H.G. von Schnering, *J. Organomet. Chem.* 1986, *314*, 25.
8. A. Schäfer, M. Weidenbruch, W. Saak and S. Pohl, *Angew. Chem.* 1987, *99*, 806; *Angew. Chem. Int. Ed. Engl.* 1987, *26*, 776.
9. O. Wagner, G. Maas and M. Regitz, *Angew. Chem.* 1987, *99*, 1328; *Angew. Chem. Int. Ed. Engl.* 1987, *26*, 1257.
10. A.H. Cowley, S.W. Hall, C.M. Nunn and J.M. Power, *J. Chem. Soc., Chem. Commun.* 1988, 753.
11. E. Niecke, R. Streubel, M. Nieger and D. Stalke, *Angew. Chem.* 1989, *101*, 1708; *Angew. Chem. Int. Ed. Engl.* 1989, *28*, 1673.
12. M. Weidenbruch and B. Brand-Roth, unpublished results.
13. P. Jutzi and A. Möhrke, *Angew. Chem.* 1989, *101*, 769; *Angew. Chem. Int. Ed. Engl.* 1989, *28*, 762.
14. A. Schäfer, M. Weidenbruch and S. Pohl, *J. Organomet. Chem.* 1985, *282*, 305.
15. D.J. De Young, M.J. Fink, J. Michl and R. West, *Main Group Met. Chem* 1987, *1*, 19.
16. L.E. Gusel'nikov and N.S. Nametkin, *Chem. Rev.* 1979, *79*, 529.
17. M. Weidenbruch, A. Schäfer and K.-L. Thom, *Z. Naturforsch., Part B*, 1983, *38*, 1695.
18. M. Weidenbruch, A. Schäfer and H. Marsmann, *J. Organomet. Chem.* 1988, *354*, C12.
19. M. Weidenbruch, B. Flintjer, S. Pohl and W. Saak, *Angew. Chem.* 1989, *101*, 89; *Angew. Chem. Int. Ed. Engl.* 1989, *28*, 95.

20. See, for example: G. Tennant in *Comprehensive Organic Chemistry, Vol. 2* (Edited by D. Barton, V.D. Ollis and I.O. Sutherland), Pergamon Press, Oxford 1979, p. 385.
21. M. Weidenbruch, B. Flintjer, S. Pohl, W. Saak and J. Martens, *J. Organomet. Chem.* 1988, *338*, C1.
22. M. Weidenbruch, A. Lesch, K. Peters and H.G. von Schnering, *Chem. Ber.*, 1990, *123*, in press.
23. G.R. Gillette, J. Maxda and R. West, *Angew. Chem.* 1989, *101*, 90; *Angew. Chem. Int. Ed. Engl.* 1989, *28*, 54.
24. S.S. Zigler, K.J. Haller, R. West and M.S. Gordon, *Organometallics* 1989, *8*, 1656.
25. N. Wiberg and K. Schurz, *Chem. Ber.* 1988, *121*, 581.
26. M. Weidenbruch, B. Brand-Roth, S. Pohl and W. Saak, *J. Organomet. Chem.* 1989, *379*, 217.
27. M.J. Fink, M.J. Michalczyk, K.J. Haller, R. West and J. Michl, *Organometallics* 1984, *3*, 793.
28. M.J. Fink, D.J. De Young, R. West and J. Michl, *J. Am. Chem. Soc.* 1983, *105*, 1070.
29. S. Masamune, S. Murakami and H. Tobita, *J. Am. Chem. Soc.* 1987, *105*, 7776.
30. G.R. Gillette and R. West, submitted for publication.
31. P. Boudjouk, U. Samaraweera, R. Sooriyakumaran, J. Chrisciel and K.R. Anderson, *Angew. Chem.* 1988, *100*, 1406; *Angew. Chem. Int. Ed. Engl.* 1988, *27*, 1355.
32. M. Weidenbruch, A. Lesch and H. Marsmann, *J. Organomet. Chem.* 1990, *385*, C47; M. Weidenbruch and A. Lesch, unpublished results.
33. M. Weidenbruch, A. Lesch, K. Peters and H.G. von Schnering, unpublished results.
34. M. Weidenbruch, B. Brand-Roth, S. Pohl and W. Saak, *Angew. Chem.* 1990, *102*, 93; *Angew. Chem. Int. Ed. Engl.* 1990, *29*, 90.
35. M. Weidenbruch, B. Brand-Roth, S. Pohl and W. Saak, *Polyhedron*, submitted.
36. H.B. Yokelson, A.J. Millevolte, K.J. Haller and R. West, *J. Chem. Soc., Chem. Commun.* 1987, 1605.

Reactive Intermediates and Mechanistic Aspects of Organosilicon Photoreactions

M. Kira,* T. Taki, T. Maruyama, and H. Sakurai
DEPARTMENT OF CHEMISTRY, FACULTY OF SCIENCE, TOHOKU UNIVERSITY, AOBA-KU SENDAI 980, JAPAN

1 INTRODUCTION

Photochemistry of organosilicon compounds[1,2] has been the subject of extensive studies for many years, since σ-π hybrid systems such as benzylsilanes, vinyl-, alkynyl- and arylpolysilanes as well as peralkylpolysilanes are responsible for generation of interesting reactive intermediates, e.g. silyl radicals and diradicals, silylenes, silenes, disilenes. Relatively few efforts have been put into the elucidation of structure and reactivity of the electronically excited states of the organosilicon compounds.[3] Organosilicon photochemistry should be carried a stage further so as to delineate the mechanistic details of photo-reactions in terms of the surface pathways which are followed starting from initial exited states, while the studies of the designed organosilicon reactive intermediates generated by using the authentic photochemical processes are still the major subjects.

We wish herein to report our recent studies of the organosilicon photochemistry. They includes mechanistic details of the photoisomerization of allylsilanes[4a] and silyl enol ethers,[4b] photochemical generation of elaborated silenes and silylenes,[5] and the investigation of the structure and reaction mechanisms of these reactive intermediates.[5] Unless otherwise stated, a 125 W low pressure mercury arc lamp was used as a light source.

2 PHOTOCHEMICAL 1,3-SILYL MIGRATION IN ALLYLSILANES

Among a large number of facile 1,3-silyl migration reported, the migration between carbon atoms in an allylsilane system is of particular interest from mechanistic view points. Whereas many 1,3-silyl migrations are attributed to intramolecular nucleophilic substitutions proceeding with retention of configuration at silicon,[6] the carbon to carbon migration has been fully characterized as a symmetry-allowed [1,3] sigmatropic rearrangement accompanied by the inversion of configuration at silicon and by rather high activation energy of ca. 50 kcal/mol. The transition state has been represented by the structure **A** in eq 1.[7]

$$R\underset{R}{\diagdown}\diagup\diagdown Si\underset{b}{\overset{a}{\diagdown}}c \;\rightleftarrows\; \left[\begin{array}{c} \text{A} \end{array} \right] \;\rightleftarrows\; c\underset{b}{\overset{a}{\diagdown}}Si\diagdown\underset{R\ R}{\diagup}\diagdown \quad (1)$$

Photochemical versions of the 1,3-silyl migration have been reported but mostly without the mechanistic details.[8-10] We found concerted 1,3-silyl migration in allylsilanes with aromatic substituents at the silicon atom, in contrast to the related photo-rearrangement of benzylsilanes occurring with the homolytic cleavage of the benzylic carbon-silicon bond.[8b]

Typically, irradiation of ca. 0.05 M hexane solution of prenyldimethylphenylsilane (**1a**) for 1 h gave a mixture of **1a** and the isomer, (1,1-dimethyl-2-propenyl)dimethylphenylsilane (**2a**) in 29 and 52% yields, respectively, together with 5% of phenyldimethylsilane (**3**).[11] In contrast to the reported thermal reaction,[7] the present photo-migration occurs reversibly giving a sterically more crowded allylsilane as the major product at the photo-stationary state.

$$\underset{R}{Me}\diagdown\diagup\diagdown SiMe_2Ar \;\underset{}{\overset{h\nu}{\rightleftarrows}}\; \underset{Me\ \ R}{\diagup\diagdown}SiMe_2Ar \quad (2)$$

1a, R= Me, Ar= Ph   **2a**, R= Me, Ar= Ph
1b, R= Me, Ar= Naphthyl   **2b**, R= Me, Ar= Naphthyl
1c, R= H, Ar= Ph   **2c**, R= H, Ar= Ph

A similar but much slower photo-isomerization was observed for propargyldimethylphenylsilane. The 1,3-silyl migration across a linear framework gave allenyldimethylphenylsilane in 28% after irradiation for 18 h with the recovery of the propargylsilane (25%). Prenyltrimethylsilane was inert under the similar photochemical conditions.

Intramolecularity of the reaction was confirmed by a crossover experiment. Thus, when a 1:1 mixture of ethylmethylphenylprenylsilane (**4**) and **1c** in hexane was photolyzed for 4.7 h, (1,1-dimethyl-2-propenyl)ethylmethylphenylsilane (54%) and **2c** (54%) were obtained together with the recovery of **4** (29%) and **1c** (27%). No cross products were detected in the reaction mixture.

A prenylsilane having 1-pyrenyl group as an aromatic group, prenyldimethyl-(1-pyrenyl)silane, was inert, while the 1,3-silyl migration of prenyldimethyl-(1-naphthyl)silane (**1b**) occurred under the similar conditions. Comparing the singlet energies among benzene, naphthalene and pyrene, which values are 110, 92 and 77 kcal/mol, respectively,[12] the result may suggest that the photochemical migration in a prenylarylsilane needs the energy more than

80 kcal/mole of the excited singlet state, which would be an aromatic π,π^* state. Triplet sensitized excitation of a prenylarylsilane results in no 1,3-silyl migration: **1c** was intact when irradiated in the presence of benzophenone with longer wavelength light than 350 nm. The quantum yield for the isomerization from **1a** to **2a** was determined to be 0.49 in pentane at 0 °C.

Irradiation of optical active methyl-(1-naphthyl)phenylprenylsilane (**5**, $[\alpha]_D$ -3.6, c 3.6, cyclohexane) in hexane gave 2-methyl-3-buten-2-ylsilane **6**, which showed $[\alpha]_D$ -3.8 (c 3.2, cyclohexane) after purified with GPC. Although absolute configuration as well as optical purities of both the starting **5** and produced **6** were not determined, the stereochemical consequence of the photochemical migration was derived from the result of the further thermolysis of the isolated **6** ($[\alpha]_D$ -3.8) at 590 °C. The thermal isomerization of **6** gave **5** having $[\alpha]_D$ -3.5 (c 2.3, cyclohexane) after purified, which value was essentially the same as that for the starting **5**, being indicative of the overall retention of configuration at silicon during the reaction sequence from **5** to **6** to **5** (eq 3). Repeated runs gave similar results. Since the thermal 1,3-silyl migration has been confirmed to occur with complete inversion of configuration at the silicon atom,[7] the photochemical migration must also take place with inversion of configuration!

$$\underset{\textbf{5, }[\alpha]_D \text{ -3.6}}{\overset{\text{Me}}{\underset{\text{Me}}{\diagup}}\!\!\!\diagdown\!\!\text{Si}^*\text{R}_3} \xrightarrow{h\nu} \underset{\textbf{6, }[\alpha]_D \text{ -3.8}}{\overset{}{\diagup\!\!\!\diagdown}\!\!\overset{\text{Si}^*\text{R}_3}{\underset{\text{Me Me}}{\big|}}} \xrightarrow{\Delta} \underset{\textbf{5, }[\alpha]_D \text{ -3.5}}{\overset{\text{Me}}{\underset{\text{Me}}{\diagup}}\!\!\!\diagdown\!\!\text{Si}^*\text{R}_3} \quad (3)$$

$$\text{Si}^*\text{R}_3 = \text{SiMePhNaph-1}$$

The results imply that the photochemical 1,3-silyl migration of allylsilanes follows suprafacial [1,3]-shift with inversion at silicon, in apparent disagreement with the prediction by the Woodward-Hoffmann rules.[13] The migration would proceed via the similar transition structure to **A** in eq 1 which is assumed as the transition structure for the thermal migration.

The isomer ratio between **1a** and **2a** at the photostationary state in pentane changed dramatically depending on temperatures. The ratio, **2a/1a** changed from 2.06 to 6.22 between 0 and -90 °C. Excellent linear correlation between ln(**2a/1a**) and 1/T was observed with the correlation coefficient of 0.999_6. Since the ratio of the absorption coefficients at 254 nm between **1a** and **1b** does not depend on the temperatures, the results suggest that there exists the significant barriers on the excited-state surfaces along the reaction pathway. The origin of the temperature dependence of the photostationary state would be ascribed to the difference of the energy barriers between the forward- and backward-reactions. Both the ratios and slopes increased with increasing electron donating ability of the substituent on the phenyl ring. Good linear relationship was found between the slopes and Brown-Okamoto's σ^+. It is noteworthy that irradiation of prenyldimethyl-p-tolylsilane (**7**) gave a mixture of (1,1-dimethyl-2-propenyl)dimethyl-p-tolylsilane (**8**) and **7** with the ratio **8/7** of 95.5/4.5 at the photostationary state at - 50 °C in pentane. The present photomigration is promising as a method for obtaining synthetically useful 1,1-dimethyl-2-propenylsilanes.

Interestingly, the photostationary states depended significantly also on the solvents: thus, for pentane, a 5:1 mixture of pentane/THF, and THF as a solvent, the slopes/K for plots of ln(**2a/1a**) vs. 1/T were 628, 291, and 0, respectively. These results may suggest that rather polar transition structure is involved for the photochemical 1,3-silyl migration, whereas further works should be required to reach the origin of the interesting dependence of the photostationary state on temperature, solvent and substituent.

3 PHOTOCHEMICAL 1,3-SILYL MIGRATION IN SILYL ENOL ETHERS

Although a number of 1,3-silyl migrations between carbon and oxygen occurring under simple thermal and catalyzed conditions have been reported,[14] no analogous migration under irradiation has been known until now. Irradiation of β-ketosilanes was reported to give only fragmentation products.[15] Whereas very recently, formal 1,3-silyl migration from carbon to oxygen of 1-silyl 1,2-diones has been reported, the actual mechanism is still open.[10f]

We found a novel irreversible 1,3-silyl migration from oxygen to carbon.[4b] Thus, silyl enol ethers afforded the corresponding β-ketosilanes. The present photoisomerization would be interesting not only from mechanistic view points but also important because it constitutes a new method for synthetically useful β-ketosilanes[16] from silyl enol ethers.

The rearrangement was proved by a cross-over experiment to occur intramolecularly. Thus, irradiation of a mixture of 2-(dimethylphenylsiloxy)propene (**9a**, 0.05 M) and 1,1,3,3,3-pentadeuterio-2-(ethylmethylphenylsiloxy)propene (**9c**, 0.05 M) in benzene under the similar conditions for 12 h, isolation of the isomerized products by preparative GPC, and then analysis by ^{1}H NMR spectroscopy revealed nonexistence of the cross products.

$$\underset{\mathbf{9}}{\underset{R^2}{\overset{R^3}{\underset{\diagup}{R^1}}}\!\!\!\!=\!\!\!\!\underset{}{\overset{}{\diagdown}}\text{OSiMeR'Ph}} \xrightarrow{h\nu\ (254\text{nm})} \underset{\underset{R^1\ R^2}{}}{R^3\!\!-\!\!\overset{O}{\overset{\|}{C}}\!\!-\!\!\text{SiMeR'Ph}} \quad (4)$$

10, 40-60% yield

a, $R^1=R^2=$ H, $R^3=$ CH$_3$, R'= Me
b, $R^1=$ CH$_3$, $R^2=$ H, $R^3=$ Et, R'= Me
c, $R^1=R^2=$ D, $R^3=$ CD$_3$, R' = Et

The 1,3-silyl migration was not found during the photolysis of 2-(dimethyl-1-naphthylsiloxy)propene. The higher bond energy of Si-O bonds than that of Si-C bonds would be reflected in the reactivity difference between allylsilanes and silyl enol ethers.

Irreversibility of the photo-migration would be attributed to the existence of the low-lying $n\pi^*$ singlet and triplet states in β-ketosilanes. The latter excited states may not contribute to the sigmatropic rearrangement.

4 STEREOCHEMISTRY AND MECHANISM FOR THE ADDITION OF ALCOHOLS TO A PHOTOCHEMICALLY GENERATED CYCLIC SILENE

Although silenes[17] are known to be trapped efficiently by alcohols giving adducts, the stereochemistry and mechanism for the reaction are not yet known with certainty. Nonstereospecific addition of methanol to isolable silenes has been observed by Brook et al.,[17a,18] while a silene reported by Jones et al. adds to methanol stereospecifically.[19] Wiberg has proposed a two-step mechanism involving the first formation of a silene-alcohol complex followed by proton migration to the carbon of the silene.[20] The mechanism may be compatible with the results by Brook et al., if rotation around the Si-C(silene) bond occurs faster than the proton migration.

In disagreement with the simple two-step or a concerted four-centered mechanism, we found that alcohols added nonstereospecifically even to the cyclic silene where the bond rotation is prohibited.[5] The stereochemical outcome for the addition of various alcohols to the silene depends remarkably on the concentration and on the acidity of the alcohols.[21] We propose a more elaborate mechanism for the reaction where the intramolecular proton transfer in the first generated silene-alcohol complex competes with the intermolecular proton transfer from an extra alcohol in the complex. The stereochemical complexity is fully explained by this mechanism.

Typically, irradiation of a cyclic divinyldisilane (11, 0.01 M) in methanol gave a 27/73 mixture of the cis (13, R = Me) and trans adduct (14, R = Me) (eq 5) in a total yield of 71 % after 59 % consumption of 11.[22] The ratio of 13/14 increased in the following

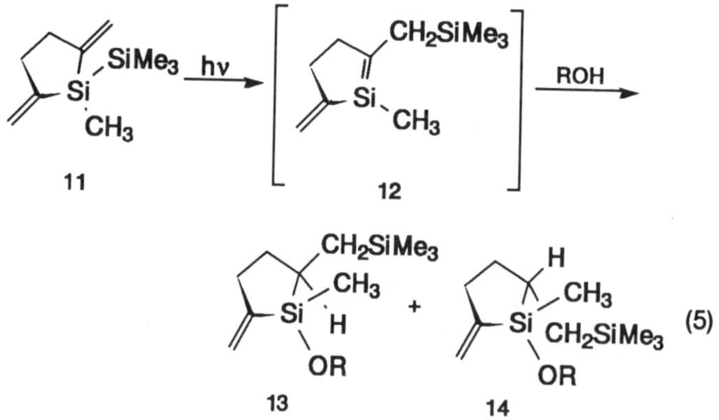

a, R= CH$_3$; b, R= CH$_2$CH$_2$CH$_3$; c, R= CH(CH$_3$)$_2$; d, R= C(CH$_3$)$_3$.

order of the used alcohols: MeOH < n-PrOH < i-PrOH << t-BuOH. Tert-Butyl alcohol gave only a <u>cis</u> adduct. No interconversion between **13** and **14** was observed under irradiation. Further interesting findings are the dependence of the initial product ratio **13/14** on the concentration of alcohol. Thus, plots of **13/14** <u>vs.</u> the inverse of alcohol concentration gave straight lines, when **11** was irradiated for 2 - 5 min in the presence of various amounts of alcohols in acetonitrile. The values of the slope are 4.6, 9.2, and 32 for MeOH, n-PrOH, and i-PrOH, respectively. Virtually, infinitely large slope was estimated for t-BuOH. The intercept fell always near zero, while it depended on the slope.

On the basis of these results, we propose a more elaborate mechanism for the addition of alcohol to a silene as shown in Scheme I. Thus, after the first formation of an alcohol-silene complex **15** as suggested by Wiberg,[20] the intramolecular proton migration in **15** (the first-order rate constant, k_1) competes with the intermolecular proton transfer from an extra alcohol to **15** (the second-order rate constant, k_2), which two processes give the <u>cis</u> and <u>trans</u> isomers, **13** and **14**, respectively.[23] The mechanism is fully compatible with the observed linear relationship between the product ratio **13/14** and 1/[ROH], since the initial product ratio should be represented by the following equation.

$$d[13]/d[14] = (k_1/k_2)/[ROH] \qquad (6)$$

The slope which means the relative rate constant (k_1/k_2) would thus reflect the relative ease between the intra- and intermolecular proton transfer. According to the Brønsted catalysis

Scheme I

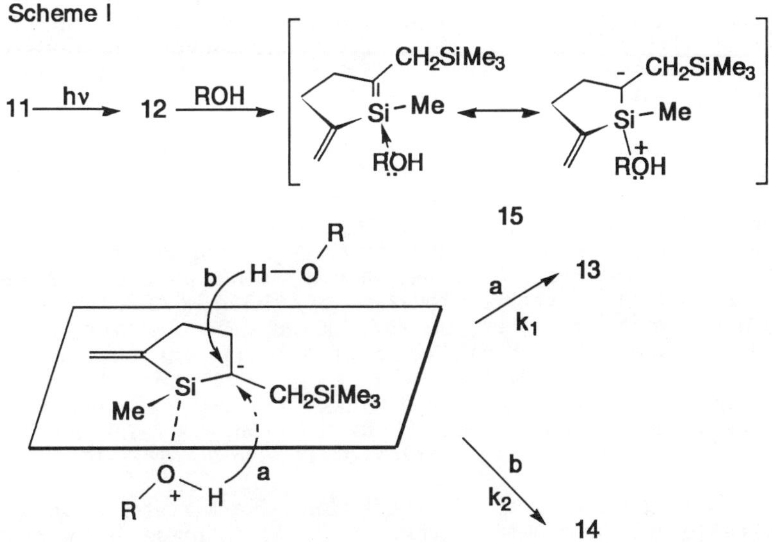

a, Intramolecular H^+ transfer
b, Intermolecular H^+ transfer

law, k_2 and k_1 are expected to increase with increasing acidity of ROH and the protonated, respectively. The pK_a values of alcohols decrease in the following order: MeOH> n-PrOH> i-PrOH> t-BuOH.[24] The inverse order is known for the protonated alcohols, RO^+H_2: t-BuO$^+H_2$> i-Pr$^+OH_2$> n-PrO$^+H_2$> MeO$^+H_2$.[25] The more acidic the alcohol is, the less acidic the corresponding protonated alcohol. Thus k_1/k_2 should increase in the following order: MeOH< n-PrOH< i-PrOH< t-BuOH. The observed dependence of the slope on the kinds of alcohol is in good agreement with the above prediction.

5 THE FIRST MATRIX-ISOLATION AND ELECTRONIC STRUCTURE OF A PHOTOCHEMICALLY GENERATED DIVINYLSILYLENE

Finally structure of a divinylsilylene generated by the photolysis of the corresponding trisilane is discussed. Much attention has been currently focused on the electronic structure of silylenes. Drahnak, Michl, and West have first reported that the irradiation of dodecamethylcyclohexasilane affords a characteristic absorption maximum at 450 nm in an argon or a 3-methylpentane (3-MP) matrix at 77 K.[26] Recent laser flash photolysis[27] as well as the theoretical[28,29] studies have confirmed that the 450 nm absorption is due to the ground-state singlet dimethylsilylene and assigned to the n(Si) - 3p(Si) transition. The absorption maximum is shifted to 350 nm in a 2-methyltetrahydrofuran (2-MeTHF), the formation of a silylene-ether complex being indicated.[30] Although the remarkable substituent effects on the n(Si) - 3p(Si) absorption band maxima are known, the origin has been discussed only on the basis of the insufficient experimental and theoretical results.[28]

When 2,5-bis(methylene)-1,1-bis(trimethylsilyl)-1-silacyclopentane (**16**) was photolyzed in a 3-MP matrix at 77 K with a 125-W low pressure Hg arc lamp, a purple color was produced, being indicative of the formation of the corresponding cyclic divinylsilylene **17**. The absorption spectrum showed a broad band at 505 nm and a relatively intense band at 290 nm (Table 1).

Upon annealing of the matrix both the two bands disappeared, whereas the expected formation of a disilene via the dimerization of **17** was not observed; **17** may react more favorably with the double bonds in the precursor **16**. Convincing evidence for the formation of a cyclic divinylsilylene, **17** is provided by the following experimental results. Thus, when a similar photolysis of **16** was carried out in a matrix composed of 95:5 3-MP/2-MeTHF at 77 K, initially two bands at 290 nm and 505 nm were observed, but these bands disappeared upon annealing of a matrix for several seconds to give rise to a new band with λ_{max} = 390 nm. After further warming, this band disappeared too. The 390 nm band was also observed by the irradiation of **16** in a pure 2-MeTHF matrix, and therefore assigned to the complex of **17** with 2-MeTHF, **18**.

During the course of irradiation of a mixture of **16** and t-butyl alcohol in a 3-MP matrix at 77 K, followed by warming to melt, a purple color was produced and then disappeared. After the procedure was repeated several times, GC-MS analysis of the solution showed the expected production of a t-butyl alcohol adduct (**19**) of the silylene. Irradiation of a mixture of **16** and an excess

Table 1. Theoretical and experimental transition energies of divinylsilylene **17**

Transition Energies/eV		Assignment
Obsd[a]	Calcd[b]	
2.46 (1.0)	2.459 (0.00031)	n(Si) − 3p(Si)
4.28 (7.4)	3.935 (0.132)	π − 3p(Si)

a. Relative absorption coefficients are shown in parentheses. b. Oscillator strengths are shown in parentheses.

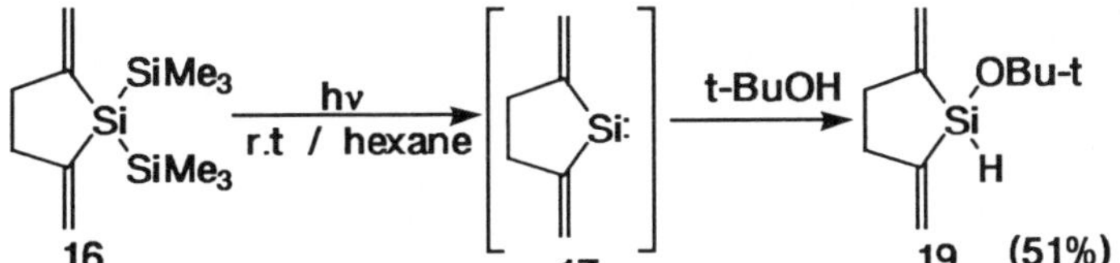

$$\mathbf{16} \xrightarrow[\text{r.t / hexane}]{h\nu} [\mathbf{17}] \xrightarrow{\text{t-BuOH}} \mathbf{19}\ (51\%) \qquad (7)$$

t-butyl alcohol in hexane at ambient temperature afforded **19** in the isolated yield of 51 %.

The UV absorption band with λ_{max} = 505 nm, which corresponds to the 450 nm band of dimethylsilylene, is easily assigned to the electronic transition from n(Si) to 3p(Si). The band maximum for **17** is shifted ca. 50 nm longer wavelength than that for dimethylsilylene, in contrast to the ab initio theoretical prediction by Apeloig et al.,[28] who expected the blue shift by introducing vinyl substituents, which act as π donors, to silylene. The theory predicts also that the absorption maximum of silylene is shifted to the longer wavelength as the apex angle increases. The ground state structure of the cyclic divinylsilylene **17** fully optimized with the 3-21G basis set showed, however that the apex angle was 87.4°, which is even smaller than the angle for dimethylsilylene (98 − 99°).[29,31] Their study may be inconclusive, since it used small basis set for the geometrical optimization of the ground-state singlet and did not include the effects of electronic correlation on the excitation energy.

To get insight into the electronic structure of vinyl-substituted silylenes, we have calculated the electronic transition for **17** with the 3-21 G optimized geometry, by using the semi-empirical CNDO/S theory which involves the electronic correlation effects with use of configuration interaction (CI) method.[32] Both the theoretical and experimental transition energies of **17** and their assignment are shown in Table 1. Good agreement was observed between the calculated and theoretical values, indicating that the

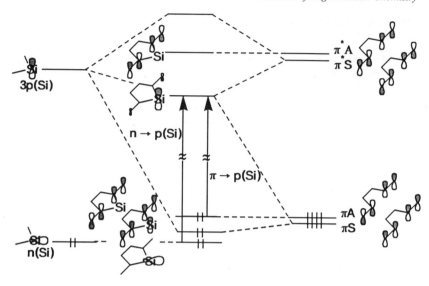

Figure 1. Schematic orbital interaction diagram among two vinyl π and a 3p(Si) orbitals.

505 nm band is correctly assigned to the n(Si) - 3p(Si) transition. Additionally, the relatively intense band with λ_{max} = 290 nm is assigned to the π - 3p(Si) transition. The assignment is compatible to the blue shift of the both two bands in the silylene-ether complex, because the 3p(Si) level should be lifted by the partial bonding with the oxygen base. However, the CNDO/S as well as the previous theoretical calculation could not predict the blue shift of the 450 nm band of dimethylsilylene by vinyl substitution: the CNDO/S calculation predicted the λ_{max} for the <u>ab initio</u> optimized structure of dimethylsilylene as 551 nm. This may be caused by the inadequate empirical parameters adopted by the silicon atom for the CNDO/S calculations.

Qualitative MO consideration suggests that the red shift of the n(Si) - 3p(Si) transition caused by the vinyl substitution may be due to the lowering of the 3p(Si) orbital level. The 3p(Si) orbital can interact with the two orbitals made by the symmetric combination of the two vinyl π and π^* orbitals (π_S and π^*_S), respectively. If the intrinsic level of 3p(Si) lie higher than the middle of the π_S and π^*_S levels, the 3p(Si) level would be lowered by the more significant interaction with the antibonding π^*_S than the bonding π_S orbitals.

REFERENCES

1. For comprehensive reviews, see (a) A. G. Brook, 'The Chemistry of Organosilicon Compounds', Part 2, Ed. S. Patai and Z.

Rappoport, John Wiley & Sons, 1989, Chapter 15, p. 965. (b) J. C. Dalton, 'Organic Photochemistry', Ed. A. Padwa, Marcel Dekker, New York, 1985, Vol. 7, p. 149.

2. (a) H. Sakurai, J. Organomet. Chem., 1980, 200, 261. (d) M. Ishikawa and M. Kumada, Adv. Organomet. Chem., 1981, 19, 51. (b) G. Raabe and J. Michl, Chem. Rev., 1985, 85, 419. (c) R. D. Miller and J. Michl, ibid., 1989, 89, 1359.

3. Unusual exited-state structure of aryldisilanes and phenylethynyldisilanes have been investigated extensively: (a) H. Sakurai, H. Sugiyama, and M. Kira, J. Phys. Chem., 1990, 94, 1837. (b) K. A. Horn, R. B. Grossman, J. R. G. Thorne, and A. A. Whitenack, J. Am. Chem. Soc., 1989, 111, 4809. (c) H. Shizuka, Y. Sato, Y. Ueki, M. Ishikawa, M. Kumada, J. Chem. Soc., Faraday Trans. 1, 1984, 80, 341. See also references cited therein.

4. (a) M. Kira, T. Taki, and H. Sakurai, J. Org. Chem., 1989, 54, 5647. (b) idem, to be published.

5. M. Kira, T. Maruyama, and H. Sakurai, to be published.

6. For a review, see: A. G. Brook and A. R. Bassindale, 'Rearrangements in Ground and Excited States', Ed. P. de Mayo, Academic Press, New York, N. Y., 1980, Vol. 2, Essay 9,

7. (a) H. Kwart and J. Slutsky, J. Am. Chem. Soc., 1972, 94, 2515. (b) J. Slutsky and H. Kwart, Ibid., 1973, 95, 8678.

8. (a) M. Ishikawa, K. Nakagawa, M. Ishiguro, F. Ohi, and M. Kumada, J. Organomet. Chem., 1978, 152, 155. (b) M. Kira, H. Yoshida, and H. Sakurai, J. Am. Chem. Soc., 1985, 107, 7767.

9. Photochemical 1,3-germyl migration in cinnamylgermanes was reported without mechanistic details: M. Kobayashi and M. Kobayashi, Chem. Lett., 1986, 385. In contrast to our system, the germyl migration should involve direct photoexcitation of the cinnamyl π system.

10. A number of examples for formal photochemical 1,3-silyl migration have been reported. Si--O: (a) A. G. Brook, J. W. Haris, J. Lennon, and M. E. Sheikh, J. Am. Chem. Soc., 1979, 109, 83. (b) K. M. Baines and A. G. Brook, Organometallics, 1987, 6, 692. Si--C: (c) M. Ishikawa, T. Fuchikami, T. Sugaya, and M. Kumada, J. Am. Chem. Soc., 1975, 97, 5923. (d) M. Ishikawa, T. Fuchigami, and M. Kumada, J. Organomet. Chem., 1978, 149, 37. (e) H. Sakurai, Y. Kamiyama, and Y. Nakadaira, J. Am. Chem. Soc., 1976, 98, 7424. C--O: (f) B. B. Wright, J. Am. Chem. Soc., 1988, 110, 4456.

11. The hydrosilane 3 would be formed by simple δ-elimination. Intermediacy of free silyl radicals is incompatible with the results of a crossover experiment. Thus, irradiation of a mixture of $(CH_3)_2C=CHCH_2SiMeEtPh$ and $(CD_3)_2C=CHCH_2SiMe_2Ph$ in hexane gave only $HSiMeEtPh$ and $DSiMe_2Ph$ as hydrosilanes; neither $DSiMeEtPh$ nor $HSiMe_2Ph$ was detected.

12. S. L. Murov, 'Handbook of Photochemistry,' Marcel Dekker, New York, N. Y., 1973.

13. R. B. Woodward and R. Hoffmann, Angew. Chem. Int. Ed. Engl., 1969, 8, 781.

14. (a) I. F. Lutsenko, Yu. I. Baukov, G. S. Burlachenko, and B. N. Khasapov, J. Organomet. Chem., 1966, 5, 20. (b) I. F. Lutsenko, Yu. I. Baukov, A. S. Kostyuk, N. I. Sovel'eva, and V. K. Krysina, ibid., 1969, 17, 241. (c) O. V. Litvinova, Yu. I. Baukov, and I. F. Lutsenko, Dokl. Akad. Nauk SSSR, 1967, 173, 578. (d) A. G. Brook, D. M. MacRae, and W. W. Limburg, J. Am. Chem. Soc., 1967, 89, 5493. (e) A. G. Brook, Acc. Chem. Res., 1974, 7, 77.

15. H. G. Kuivila and P. L. Maxfield, J. Organomet. Chem., 1967, 10, 41.
16. (a) P. F. Hudrlik and D. Peterson, Tetrahedron Lett., 1972, 1785. (b) P. F. Hudrlik, A. M. Hudrlik, R. N. Misra, D. Peterson, G. P. Withers, and A. K. Kulkarni, J. Org. Chem., 1980, 45, 4444. (c) K. Utimoto, M. Obayashi, and H. Nozaki, J. Org. Chem., 1976, 41, 2940.
17. For recent reviews, see: (a) A. G. Brook and K. M. Baines, Adv. Organomet. Chem., 1986, 25, 1. (b) G. Raabe and J. Michl, Chem. Rev., 1985, 85, 419.
18. A. G. Brook, K. D. Safa, P. D. Lickiss, and K. M. Baines, J. Am. Chem. Soc., 1985, 107, 4339.
19. P. R. Jones, T. F. Bates, J. Am. Chem. Soc., 1987, 109, 913.
20. (a) N. Wiberg, G. Wagner, G. Muller, and J. Riede, J. Organomet. Chem., 1984, 271, 381. (b) N. Wiberg, J. Organomet. Chem., 1984, 273, 141.
21. Recently, Fink et al. reported nonstereospecific addition of alcohols to a silacyclobutadiene in a glass matrix. The results have been explained however by the photoisomerization of an initially formed diastereomer. (a) M. J. Fink, D. B. Puranik, and M. P. Johnson, J. Am. Chem. Soc., 1988, 110, 1315. (b) D. B. Puranik and M. J. Fink, J. Am. Chem. Soc., 1989, 111, 5951.
22. Vinyldisilanes have been established to give efficiently the corresponding silenes via 1,3-silyl migration under irradiation. (a) H. Sakurai, Y. Kamiyama, and Y. Nakadaira, J. Am. Chem. Soc., 1976, 98, 7424. (b) M. Ishikawa, T. Fuchikami, and M. Kumada, J. Organomet. Chem., 1978, 149, 37. (c) R. T. Conlin and K. L. Bobbit, Organometallics, 1987, 6, 1406.
23. If the intermolecular proton transfer occurs at the same side of the complexed alcohol with the rate constant of k_2', the intercept will correspond to k_2'/k_2. Apparently meaningless values of the intercept suggest that the intermolecular syn addition can be neglected.
24. F. G. Bordwell, Acc. Chem. Res., 1988, 21, 456.
25. E. M. Arnett, Prog. Phys. Org. Chem., 1963, 1, 223.
26. T. J. Drahnak, J. Michl, and R. West, J. Am. Chem. Soc., 1979, 101, 5427.
27. (a) G. Levin, P. K. Das, and C. L. Lee, Organometallics, 1988, 7, 1231. (b) G. Levin, P. K. Das, C. Bilgrien, and C. L. Lee, ibid., 1989, 8, 1206.
28. Y. Apeloig and M. Karni, J. Chem. Soc., Chem. Commun., 1985, 1048.
29. R. S. Grev and H. F. Schaefer, III, J. Am. Chem. Soc., 1986, 108, 5804.
30. (a) G. R. Gillett, G. H. Noren, and R. West, Organometallics, 1987, 6, 2617. (b) idem, ibid., 1988, 8, 487. (c) W. Ando, A. Sekiguchi, K. Hagiwara, A. Sakakibara, and H. Yoshida, Organometallics, 1988, 7, 558.
31. M. S. Gordon and M. W. Schmidt, Chem. Phys. Lett., 1986, 132, 294.
32. J. D. Bene and H. H. Jaffe, J. Chem. Phys., 1968, 48, 1807.

Structure and Reactivity of 7-Sila- and 7-Germanorbornadienes as Precursors of Silylenes and Germylenes

O. M. Nefedov and M.P. Egorov
THE INSTITUTE OF ORGANIC CHEMISTRY USSR, ACADEMY OF SCIENCES, LENINSKY PROSP. 47, MOSCOW, USSR

1 INTRODUCTION

Silylenes and germylenes have become a very important class of reactive intermediates both from theoretical and synthetic points of view. The chemistry of these species is developing rapidly. Its fundamental problems are the search for suitable precursors and methods for the generation of R_2E species; the study of both the mechanism of their formation and their further reactions; and the direct spectroscopic detection of these short lived intermediates.

In fact not so many types of convenient precursors of silylenes and germylenes are known, and 7-sila- and 7-germanorbornadienes are playing an important role among them. The first representatives of heterocycles of this class were synthesized and proposed as thermal sources of R_2E (E=Si, Ge) species in the earlier 1960s[1,2]. The different derivatives of 7-sila- and 7-germanorbornadienes have been synthesized later[3-9]. However, the systematic investigation of these compounds and their application as silylene and germylene precursors started only recently[10-17].

We report here the study of the mechanism of photochemical generation of dimethylsilylene and dimethylgermylene from the 7-heteronorbornadienes of the same structure 7,7-dimethyl-1,4,5,6-tetraphenyl-2,3-benzo-7-sila(germa)-norbornadienes (**1** and **2**).

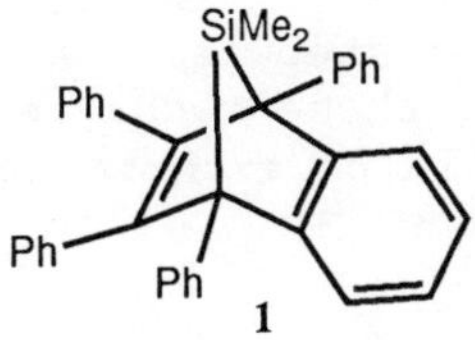

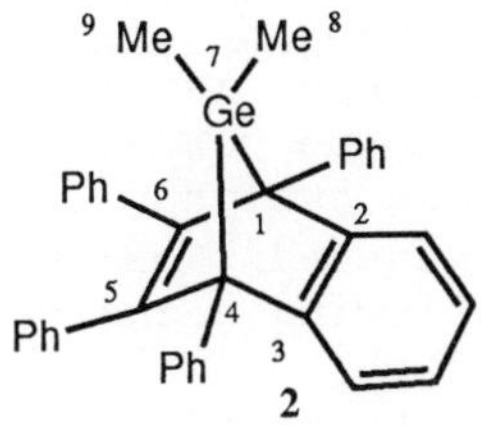

The first observation of Me$_2$Ge produced by the photolysis of **2** in the liquid phase, its complexation and reactivity will be discussed in detail. Finally we shall describe a new, catalytic method of R$_2$E generation. However, first of all we briefly discuss the structural features of 7-germanorbornadiene **2**.

2 MOLECULAR STRUCTURE OF **2**

The X-ray analysis has been done both for **1** and **2**. Their molecular structures are very similar. Therefore we shall discuss here only the structural parameters of **2**.

The exocyclic Ge-C(9) (1.948(5) Å) and Ge-C(8) (1.938(4) Å) bonds are shorter and the endocyclic Ge-C(1) (2.022(4)Å) and Ge-C(4) (2.024(4) Å) bonds are longer than those in Me$_4$Ge (1.98 Å). The valence angle C(8)-Ge-C(9) is 112.4(2)°. A structural feature of **2** is the small value of the endocyclic C(1)-Ge-C(4) angle (78.5(2)°). This value is smaller than that in **1** and the other 7-silanorbornadienes (81.7-82.6°)[18] and considerably smaller than that in norbornadiene (94°).[19]

The C(2), C(3), C(5), C(6), atoms are situated in one plane (±0.001 Å) (plane A). The planes of phenyl substituents at C(1), C(6), C(5), C(4) carbon atoms are perpendicular to plane A (80.6, 88.8, 84.6, 91.6° respectively). Such orientation of phenyl rings excludes the possibility of their participation in the delocalization of electron density of a radical formed in the course of the homolysis of an endocyclic Ge-C bond - the key step of 7-heteronorbornadiene decomposition. This should result[7,8] in the increased stability of a 7-heteronorbornadiene. Indeed in **1**, that decomposes at ~300°, all phenyl rings are also perpendicular to the plane A. Apparently the high angle strain caused by a small endocyclic bond angle at the bridge Ge atom is an important factor in decreasing the thermal stability of **2** (decomp. at 70-80°) in comparison with the silicon analogue **1**.

3 THE MECHANISM OF PHOTOCHEMICAL DECOMPOSITION OF 7-SILA(GERMA)NORBORNADIENES

The formation of silylenes and germylenes from 7-heteronorbornadienes may occur by two pathways, including symmetry allowed concerted (path A) or stepwise (path B) cleavage of endocyclic E-C bonds. Both these mechanisms have been discussed,[5-9,20,21] but there is no experimental evidence allowing a choice between them.

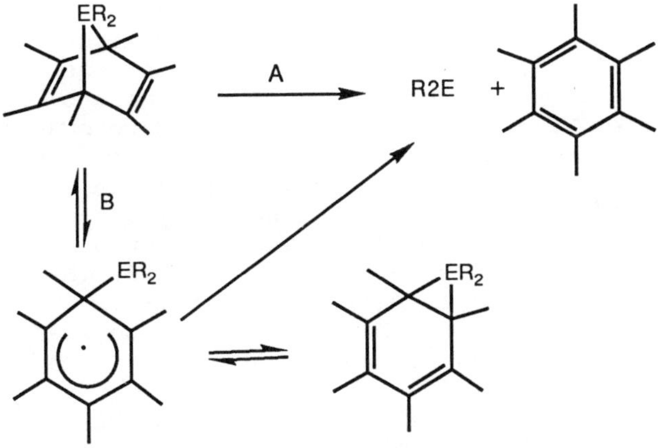

7-Sila- and 7-Germanorbornadienes as Precursors

We have studied the photolysis of **1** and **2** by the CIDNP ^{1}H NMR technique. In the course of photolysis of **1** and **2** we observed CIDNP effects both for the starting compounds **1(2)** (emission of MeE-group signals of **1** and **2**) and for their decomposition product, 1,2,3,4-tetraphenylnaphthalene (TPN) (enhanced absorption of phenyl proton signals). The same pattern of **1(2)** polarization has been found when **1(2)** were photolyzed in the presence of Me$_2$E-trapping agents (for example, Me$_3$SnCl). These results show that the polarized **1**

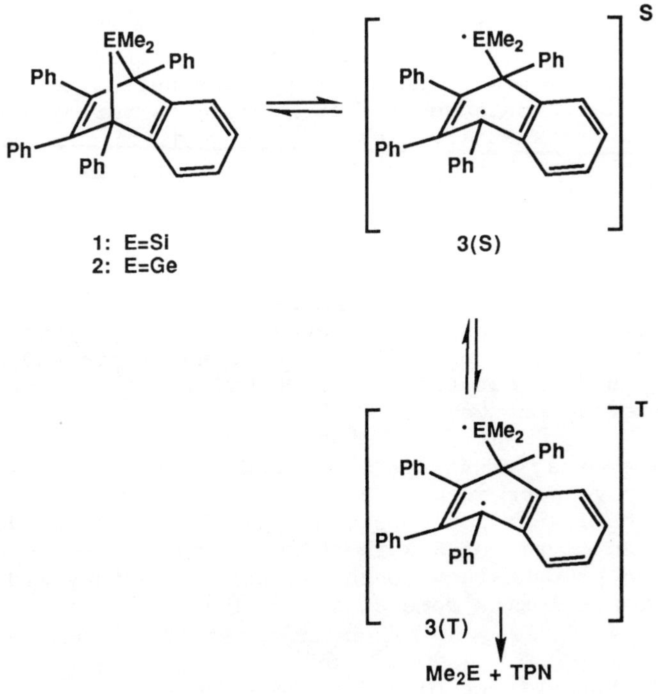

SCHEME 1

2 and TPN should have a paramagnetic precursor, namely, a biradical **3**. CIDNP effects of opposite signs for the recombination (**1,2**) and decomposition (TPN) products of biradical **3** testify in favor of an S-T mechanism of CIDNP effects. The CIDNP effects of **1,2** and TPN were analysed according to Kaptein's rules. The analysis shows that the singlet biradical (**3**)(**S**) is formed from the S_1 state of **1,2**. The singlet biradical **3** recombines giving the polarized starting norbornadienes **1,2**. The triplet biradical **3** obtained as a result of S-T evolution of (**3**)(**S**) irreversibly decomposes with the formation of Me_2E (E=Si,Ge) and a polarized TPN. Thus, the mechanism of photochemical decomposition could be presented as shown in Scheme 1.:

4 ELECTRONIC ABSORPTION SPECTRA AND REACTIVITY OF Me_2Ge AND ITS COMPLEX, $Me_2Ge-PPh_3$.

The investigation of the spectral properties and reactivity of short lived silylenes has become a subject of growing interest.[10,11,20-24] In contrast to silylenes the same aspect of the chemistry of germylenes has begun to be explored only very recently.[16,17,25-31] To obtain the transient absorption spectra of the simplest dialkylgermylene, dimethylgermylene, we have studied the intermediates arising from flash photolysis of **2**.

Two transient absorptions at 380 and 460 nm appear upon the flash photolysis of **2** in degassed heptane solution (20°, pulse width 5 μs). These absorptions belong to different intermediates because the rate constants for disappearance of these transient species, k_{dis} have different values; k_{dis} = 3500 s^{-1} for the 380 nm transient **4** and k_{dis} = 500 s^{-1} for the 460 nm transient **5**. They appear at the same time immediately after the UV-pulse, *i.e.* they are formed by parallel paths. Trapping experiments were carried out to investigate the nature of the transients. Thus, the addition of Me_3SnCl, an effective germylene scavenger, to the solution of **2** progressively enhanced the decay of the 380 nm species, but does not affect the decay of the 460 nm species **5**, *i.e.* **5** does not react with Me_3SnCl. The intermediate **5** is therefore not Me_2Ge.

The decay of **4** also increased in the presence of other Me_2Ge-scavengers such as MeOH, CCl_4, O_2, styrene, 2,2,6,6-tetramethyl-4-thiacycloheptyne (**6**). In preparative steady state photolysis of **2** with these scavengers the expected trapping products have been isolated in good yields. These experiments let us come to the conclusion that the intermediate **4** with λ_{max} = 380 nm is dimethylgermylene.[16]

In the absence of any Me_2Ge scavengers the decay of the Me_2Ge occurred with the pseudo-first-order rate constant linearly depending on the concentration of **2**, *i.e.* the reaction of Me₂Ge with **2** takes place. Indeed, the insertion reaction of Me₂Ge into the endocyclic Ge-C(sp³) bond of **2** upon melting of **2** was reported by Neumann and co-workers.[32]

From the flash photolysis experiments it follows that the concentration of TPN, a stable product of photolysis of **2** does not change in the course of the dark reactions that follow. Thus the intermediate **5** appearing in a parallel way with Me₂Ge does not produce TPN under the decomposition, i.e. [Me₂Ge] = [TPN]. Based on this experimental fact the extinction coefficient of Me₂Ge has been obtained, ε_{380} = 1.3×10^3 M⁻¹ cm⁻¹. This value is close to that of the long wavelength absorption maximum (414 nm) of the stable dialkylgermylene, [(Me₃Si)₂CH]₂Ge (ε_{414} = 970 M⁻¹ cm⁻¹)[33] and of the short-lived dimesitylsilylene (ε_{580} = 1.96×10^3 M⁻¹cm⁻¹).[24]

The quantum yield of Me₂Ge is decreased when the flash photolysis of **2** is carried out in the presence of triplet state quenching agents, such as oxygen, and nitroxyl radicals. This shows that the dimethylgermylene is formed from the triplet photoexcited state of **2**.

The lifetime of the **2*(T)** obtained from the quenching experiments with different quenching agents is 2-5 μs.

The singlet character of Me₂Ge generated from **2** in the flash photolysis experiments follows from the reaction with oxygen. The reaction is fast (k_{O_2} = 2.7×10^7 M⁻¹s⁻¹) but in

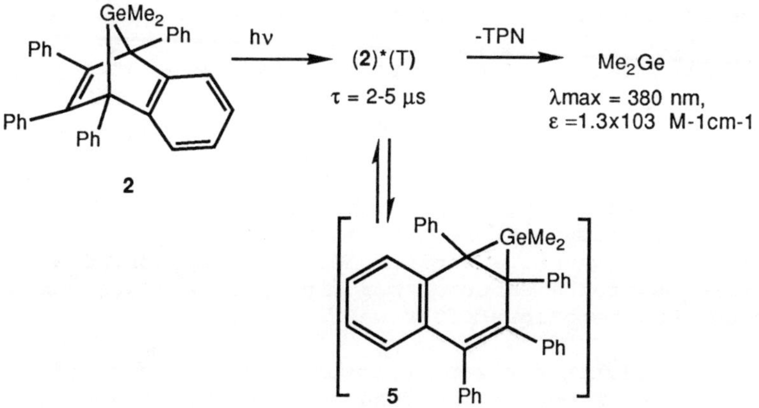

fact is substantially slower(~100 times) than the diffusion-controlled rate constant expected for the triplet Me₂Ge species. For example, for the triplet diphenylcarbene the rate constant for the reaction of triplet diphenylcarbene with oxygen is 5.0×10^9 M⁻¹s⁻¹.[34] The singlet state of dimethylgermylene thermally or photochemically generated from **2** was also confirmed by CIDNP experiments.[12]

What is the identity of the 460 nm transient? This is not a product of Me$_2$Ge **4** dimerization, the tetramethyldigermene, because a dimethylgermylene and an intermediate **5** appear by a parallel but not a successive way. If **5** is a radical, one could expect the reaction of **5** with Me$_3$SnCl and CCl$_4$ - a radical scavenger, but this is not the case. One of the most likely structures which can be suggested for the 460 nm transient **5** is a germabenzonorcaradiene. This intermediate containing a strained germacyclopropane ring forms as a result of photorearrangement of **2**. The possibility of isomerizations of this type has been discussed for 7-silanorbornadiene.[1,4,20]

The absorption maximum of Me$_2$Ge (380 nm) obtained from the flash photolysis experiment in the liquid phase differs from that of 420-440 nm assigned to the Me$_2$Ge in hydrocarbon matrices.[25-27] In fact this blue shift between the low temperature matrix and the room temperature solution spectrum is of similar magnitude to that observed for diphenylgermylene[30] and silylenes.[10,11,22,23]

We have also studied the photolysis of **2** in hydrocarbon matrices. Indeed the UV irradiation (high pressure Hg lamp, 1000 W) of **2** in 3-methylpentane at 77 K produced a yellow glass with an absorption band at 420 nm. The colour does not disappear at 77 K for a long time. Therefore the colour does not belong to a triplet excited state of Me$_2$Ge.

The ESR study of the yellow glass shows the presence of radical species with a concentration of several per cent. The colour and the ESR signal disappear upon annealing of the matrix. Preparative prolonged (9-12 hrs) photolysis of **2** ([**2**] = 0.1 M) in a toluene matrix (77 K) both in the absence and in the presence of Me$_2$Ge scavengers (Me$_3$SnCl, thiacyclo-heptyne **6**) results in the appearance of the 420 nm species. However, a noticeable decomposition of **2** or the formation of the expected trapping products was not detected by the NMR technique after warming-up.

Photolytic decomposition of **2** without formation of coloured species occurs when the matrix temperature is increased (~140 K). Dimethylgermylene generated under these conditions could be trapped by thiacycloheptyne **6**. The expected addition product - germacyclopropene **11** was isolated in a good yield (80-90%).

The results obtained from the matrix photolysis of **2** could be explained in a framework of the proposed two-step mechanism by the formation of biradical **3** which is stable at 77 K and has the absorption maximum at 420 nm. One can suggest that at 77 K the rate constant for recombination of **3** is higher than that for homolysis of a second Ge-C bond, and thus **3** completely transforms into the starting **2** upon matrix annealing.

It is possible that at higher temperatures (e.g. 140 K), the ratio between these rate constants is opposite to that which resulted in the photolysis of **2**.

Complexation of Me$_2$Ge with PPh$_3$

In a singlet state, silylenes and germylenes have a vacant p-orbital, providing a possibility to form complexes with n-donor agents such as amines, phosphines, ethers, etc. Indeed, stable complexes of dihalogermylenes with Lewis bases have been obtained and their structure determined by X-ray analysis.[35] The formation of complexes of diorganosilylenes (germylenes) with Lewis bases have been also discussed, but in fact there has been no direct evidence of their existence for a long time. Only recently the formation of Lewis base adducts of diorganosilylenes both in the liquid phase[23] and in matrices[36] was reported. Complexation of diorganogermylenes with n-donor agents was also detected in a low temperature matrix.[27]

Here we shall discuss the results concerning the first detection of a Lewis base adduct of Me$_2$Ge with PPh$_3$.[17]

Flash photolysis of **2** (10^{-4} M) with PPh$_3$ (6×10^{-5} M) (heptane 20°) gives rise to a transient species **7** with absorption maximum at 370 nm. The absorption maximum shifts to the blue compared to that of Me$_2$Ge generated under the same conditions but without PPh$_3$.

Both the trapping reactions and kinetic data (second-order kinetics for the decay of **7**, the increase of lifetime and decreasing of reactivity of **7** compared to those of Me$_2$Ge) show that the 370 nm species is the complex Me$_2$Ge·PPh$_3$. The molar extinction coefficient of the complex **7** was obtained as $\varepsilon_{370} = 7.5 \times 10^3$ M^{-1}cm^{-1}. The ε_{370} value of Me$_2$Ge·PPh$_3$ is ~6 times as much as that for Me$_2$Ge.

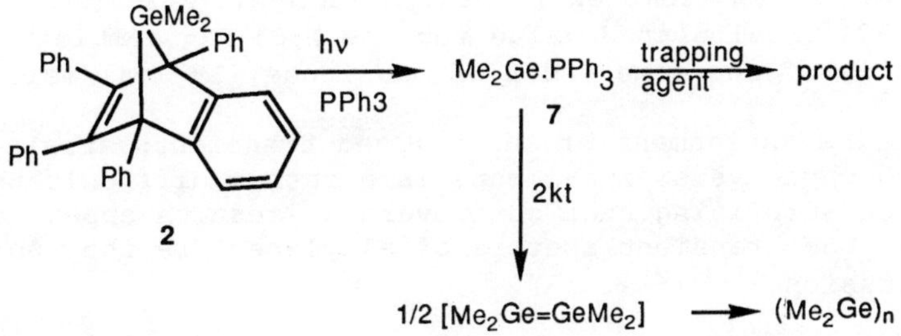

The complexation of Me_2Ge with PPh_3 results in the decrease of its reactivity. Thus in contrast to the "free" dimethylgermylene which reacts with **2**, the reaction between **2** and $Me_2Ge \cdot PPh_3$ is not observed. The decay of **7** occurs under second-order kinetics suggesting the formation of tetramethyldigermene. The dimerization rate constant, $2k_t$ is $1.5 \cdot 10^9$ $M^{-1}s^{-1}$. This value is smaller than the dimerization rate constant of a "free" dimesitylsilylene ($2k_t = 1.7 \cdot 10^{10}$ $M^{-1}s^{-1}$)[24], which is close to the encounter control limit in cyclohexane. The decreasing of Me_2Ge reactivity upon complexation results also in the absence of a reaction between $Me_2Ge \cdot PPh_3$ and O_2.

However, when active Me_2Ge-scavengers are used, reaction between $Me_2Ge \cdot PPh_3$ and a trapping agent occurs. The second-order rate constants for the insertion and addition reactions of Me_2Ge and of $Me_2Ge \cdot PPh_3$ are presented in the Table. The k_r values for $Me_2Ge \cdot PPh_3$ are smaller than those for a "free" Me_2Ge. This is a result of both a steric factor and the filling of a germanium empty p-orbital which undoubtedly plays an important role in germylene reactivity.

The question of $Me_2Ge \cdot PPh_3$ thermodynamic stability is of interest. To answer the question we have estimated the value of the equilibrium constant, $K = \dfrac{k_1}{k_{-1}}$, where k_1 is the rate constant for complex formation and k_{-1} is the rate constant for decomposition of the complex.

The observed second-order kinetics for $Me_2Ge \cdot PPh_3$ decay shows that the rate of dimerization is higher than the rate of decomposition of $Me_2Ge \cdot PPh_3$, i.e. $k_{-1} << k_t [Me_2Ge \cdot PPh_3] = 1.6 \times 10^3 s^{-1}$. The bimolecular rate constant for the reaction of Me_2Ge with PPh_3 is obtained as 6.5×10^8 $M^{-1} s^{-1}$. (The value is lower than that for the reaction of Me_2Si with THF, $1.3 \times 10^{10} M^{-1}$.[23] Thus the equilibrium constant for the complexation between Me_2Ge and PPh_3, $K > 4 \times 10^5$ M^{-1} (heptane, 20°). For comparison, the equilibrium constant for complexation of $GeCl_2$ with PPh_3 in nBu_2O has been determined earlier as 2.02×10^3 M^{-1}.[37] Thermodynamic parameters for $GeCl_2 \cdot PPh_3$ are: $\Delta H = 10.74$ kcal/mol, $\Delta S_{296} = 25$ eu.[37] If one can suppppose that the $\Delta S_{GeCl_2 \cdot PPh_3} \approx \Delta S_{Me_2Ge \cdot PPh_3}$, then the ΔH for $Me_2Ge \cdot PPh_3$ complex formation is $\Delta H > 14$ kcal/mol. The HF/6-31*G calculated value for the $H_2Si \cdot PH_3$ complex formation is of the same order of magnitude, 18 kcal/mol.[38]

The assignment of short lived transients arising in flash photolysis experiments is a rather difficult task. It is not surprising that controversial results appear often. Thus, the transient spectra of silylenes are the subject of discussion.[11,22,23]

Table. Second-Order Rate Constants for the Reaction of Me_2Ge and $Me_2Ge \cdot PPh_3$ with Trapping Agents (heptane, 20°)

Reaction	k_r^a, $M^{-1}s^{-1}$	
	Me_2Ge	$Me_2Ge \cdot PPh_3$
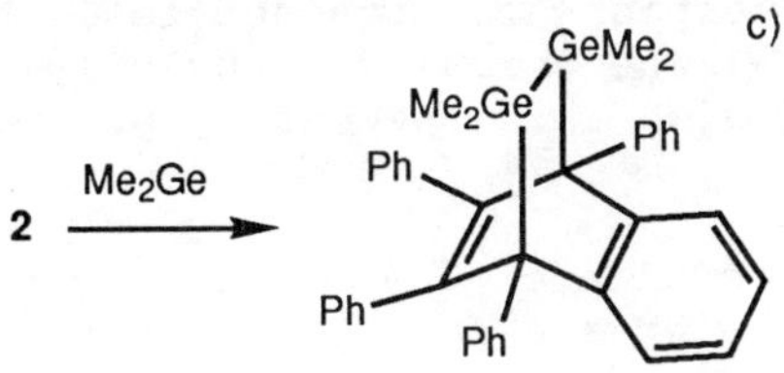 2 Me_2Ge → (tetraphenyl benzogermanorbornadiene-type adduct) c)	1.27×10^7	no reaction
$Me_3SnCl \xrightarrow{Me_2Ge} Me_3SnGeMe_2Cl$	3.5×10^8	1.0×10^8
$MeOH \xrightarrow{Me_2Ge} MeGe(H)OMe$	$3.0 \times 10^{7\,b}$	-
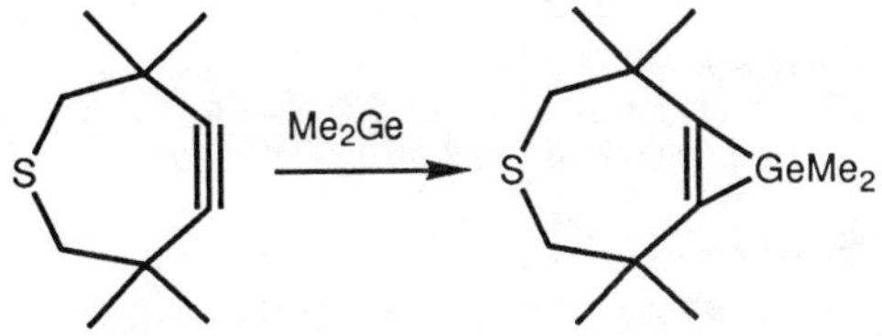 (dimethyl-thiacycloheptyne) $\xrightarrow{Me_2Ge}$ (germylene-bridged adduct)	$5.0 \times 10^{8'}$	7.0×10^7
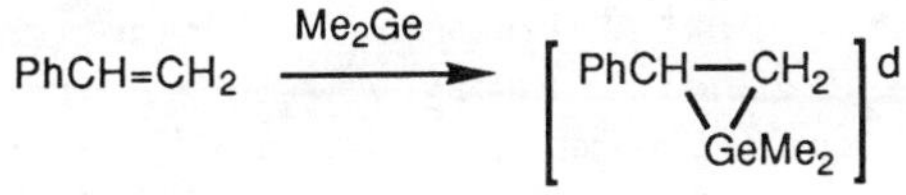 $PhCH=CH_2 \xrightarrow{Me_2Ge} [PhCH-CH_2 \backslash / GeMe_2]^d$	7.0×10^7	-
$CCl_4 \xrightarrow{Me_2Ge} [Me_2GeCl + CCl_3]^e$	1.2×10^7	3.0×10^6
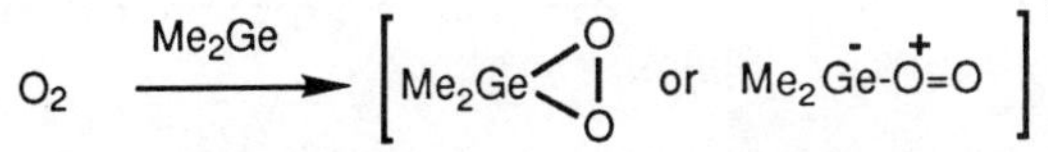 $O_2 \xrightarrow{Me_2Ge} [Me_2Ge\langle{}^O_O \text{ or } Me_2\overset{-}{Ge}\text{-}\overset{+}{O}\text{=}O]^f$	2.0×10^7	no reaction

a) ± 20%; b) extrapolated to zero MeOH concentration; isolated products are: c) $(Me_2Ge)_n$; d) 1,1-dimethyl-2,3-diphenyl-1-germacyclopentane; e) Me_2GeCl_2 ; f) $(Me_2GeO)_n$

Recently the 420-440 nm transients generated by laser photolysis of substituted dimethylgermanes $Me_2GeR'R"$, where R'=R"=SePh,[28] GeMe_2Ph;[29] R'=Ph, R"=GeMe_2Ph[31] have been assigned to Me_2Ge. It is difficult to compare these results with ours because there were no data on the reactivity of the former species. However, it should be pointed out that the photochemistry of the precursors is rather complex and along with 420-440 nm species other transients assigned to radical species have been observed.[28,29,31]

5 CATALYTIC DECOMPOSITION OF 2

Most methods which are used now to generate R_2E (E=Si,Ge) species are based on thermal and photochemical reactions of the precursors. We shall discuss here a novel, catalytic method of R_2E generation. In fact, the first step to this approach was reported by Seyferth and co-workers.[39] who studied the Pd-catalyzed reactions of strained small heterocycles - silacyclopropenes. The formal dimethylsilylene extrusion followed by trapping of this species by unsaturated compounds has been found.

We studied the decomposition of **1** and **2** in the presence of Pd-containing compounds. When a catalytic amount (1-5 mol. %) of PdL_2X_2 is added to a benzene solution of **2**, the decomposition of **2** with the formation of TPN and oligomers $(Me_2Ge)_n$ takes place. The reaction proceeds at a lower temperature than that of decomposition of **2** without a catalyst (70-75°C). Thus, the half-life time (t1/2) of **2** at 50° (benzene, [**2**] = 7×10^{-5} M) in the presence of 3-5 mol.% of $PdCl_2(PPh_3)_2$ (**8**) is 1.5 h. Under similar conditions, but without a catalyst, the t1/2 of **2** is 9 h. The catalytic activity of other catalysts decreases in the order: $(PPh_3)_2PdCl_2 \sim (PhCN)_2PdCl_2 > (PPh_3)_4Pd$: $NiCl_2(PPh_3)_2$ does not catalyze the decomposition of **2** at all.

The possible mechanism of palladium-catalyzed decomposition of **2** is of interest. The first point one should consider is the oxidation state of active catalyst. It has been shown recently that a Pd(II) compound is reduced to a Pd(0) species in the reactions of $PdCl_2L_2$ with silirene,[39] 1,2-disilacyclobutane[40] and digermacyclopropane[41] derivatives. We suggest that the low valent $(PPh_3)_2Pd(0)$ species is the active catalyst in the decomposition of **2**. Indeed, it was confirmed by the experiment in which **2** was treated with a stoichiometric amount of **8**. The reaction produced TPN, Pd-metal and Me_2GeCl_2.

$$\mathbf{2(1)} + PdL_2Cl_2 \longrightarrow Me_2ECl_2 + TPN + [PdL_2] \rightarrow [Pd]_m$$

Then the coordinatively unsaturated $L_2Pd(0)$ species inserts into an endocyclic Ge-C(sp^3) bond of **2** with the formation of unstable metallocycle **9**. Decomposition of **9** yields TPN and a dimethylgermylene complex, Me$_2$Ge-PdL$_2$ **10** which oligomerizes in the absence of trapping agents.

Dimethylgermylene incorporated into a complex with PdL$_2$ could be trapped by a suitable agent, such as thiacycloheptyne **6**. The addition product, the germirene **11** was obtained in quantitative yield. In contrast to 2,7-silanorbornadiene, **1** does not decompose in the presence of a catalytic amount of **8** (benzene, 160°C, 1h.). Only a stoichiometric reaction of **1** with **8** resulted in the reduction of Pd(II) to Pd(0) with the formation of Pd-metal, TPN and Me$_2$SiCl$_2$ observed. Apparently in this case the insertion of L$_2$Pd(0) species into an endocyclic Si-C(sp^3) bond of **1** does not occur. It may be a result of less strain in molecule **1** in comparison with **2**.

Palladium complexes are able to cause a catalytic decomposition of the other strained germanium containing heterocycles, e.g. the germirene **11**. Germirene **11** possesses a high thermal stability both in solid form (it is stable in vacuum at 120-150°C) and in solution (it withstands prolonged boiling in benzene).[14,15] However, when a catalytic amount (2-5 mol. %) of **8** is present, germirene **11** decomposes in benzene at as low a temperature as 40°C. The reaction terminates within 5-6 h. and leads to the formation of a product of Me$_2$Ge insertion into the Ge-C(sp^2) bond of the initial **11** - 1,2-digermacyclobutene-3, **12** (~60% yield), together with oligomers (Me$_2$Ge)$_n$ and thiacycloheptyne **6**.

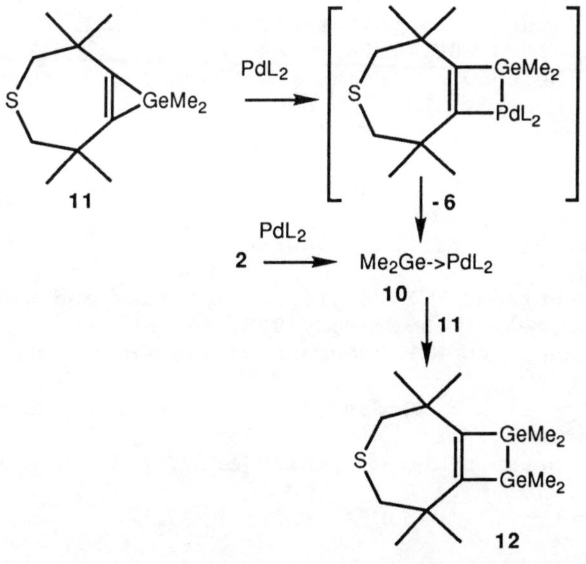

A reasonable mechanism for this reaction includes the intermediate formation of the palladiumgermacyclobutene **12**. It should be pointed out that a nickelsilacyclobutene derivative was obtained by the stoichiometric reaction of a silirene with $Ni(PEt_3)_4$.[42]

Complexation of Me_2Ge with ML_n considerably changes the reactivity of Me_2Ge in insertion reactions. For example, in the absence of a catalyst the yield of the insertion product of Me_2Ge, thermally or photochemically generated from **2**, into $Ge-C(sp^2)$ bond of **11** does not exceed 5-10%. Under a catalytic decomposition of **2** in the presence of **8** the yield of the insertion product - the 1,2-digermacyclobutene **12** is close to quantitative. Similarly the yield of $PhCH_2GeMe_2Br$, the insertion product of Me_2Ge into the C-Br bond of $PhCH_2Br$ increases from 4 to 34% when the reaction of **2** with $PhCH_2Br$ proceeds in the presence of a catalytic amount of $Pd(PPh_3)_4$ or $PdCl_2(PPh_3)_2$.

In conclusion it seems likely that a catalytic method for generation of carbene analogues could be useful both from mechanistic and synthetic points of view. Investigations in this direction are in progress in our group.

This study was carried out in cooperation with the Institute of Chemical Physics (Moscow), the Institute of Organoelement Compounds (Moscow), the Institute of Kinetics and Combustion (Novosibirsk) of the USSR Academy of Sciences and Hamburg University (F.R.G.).

REFERENCES

1. H. Gilman, S.G. Cottis and W.H. Atwell, *J. Am. Chem. Soc.*, 1964, **86**, 1596, 5584.
2. O.M. Nefedov, T. Szekey, G. Garzo, S.P. Kolesnikov, M.N. Manakov and V.I. Schiryaev in: *Internat. Symposium on Organosilicon Chem.*, *Sci. Commun. Prague.* 1965, p.65.
3. H. Sakurai, H. Sakaba and Y. Nakadaira, *J. Am. Chem. Soc.*, 1982, **104**, 6156.
4. H. Sakurai, K. Oharu and Y. Nakadaira, *Chem. Lett.*, 1986, 1797.
5. T.J. Barton, W.F. Goure, J.L. Witiak and W.D. Wulff, *J. Organomet. Chem.*, 1982, **225**, 87.
6. B. Mayer and W.P. Neumann, *Tetrahedron Lett.*, 1980, **21**, 4887.
7. H. Appler L.W. Gross, B. Mayer and W.P. Neumann, *J. Organomet. Chem.*, 1985, **291**, 9.
8. W.P. Neumann and M. Schriewer, *Tetrahedron Lett.*, 1980, **21**, 3273.
9. M. Schriewer and W.P. Neumann, *J. Am. Chem. Soc.*, 1983, **105**, 897.
10. P.P. Gaspar in *Reactive Intermediates*; M. Jones Ed., Wiley, New York, 1978, Vol. 1, p.229; 1981, vol. 2, p.335; 1985, vol. 3, p.333.

11. J.A. Hawari and D. Griller, *Organometallics*, 1984, **3**, 1123.
12. J. Köcher, M. Lehnig and W.P. Neumann, *Organometallics*, 1988, **7**, 1201; ibid., 1984, **3**, 937.
13. G. Billeb, H. Brauer, S. Maslov and W.P. Neumann, *J. Organomet. Chem.*, 1989, **373**, 11.
14. A. Krebs and J. Berndt, *Tetrahedron Lett.*, 1983, **24**, 4083; M.P. Egorov, S.P. Kolesnikov, Yu.T. Struchkov, M. Yu. Antipin, S.V. Sereda and O.M. Nefedov, *J. Organomet. Chem.*, 1985, **290**, C27.
15. O.M. Nefedov, M.P. Egorov and S.P. Kolesnikov, *Sov. Sci. Rev. B. Chem.*, 1988, **12**, 53.
16. M.P. Egorov, A.S. Dvornikov, S.P. Kolesnikov, V.A. Kuzmin, and O.M. Nefedov, *Izv. Akad. Nauk SSSR, Ser. Khim.*, 1987, 1200.
17. S.P. Kolesnikov, M.P. Egorov, A.S. Dvornokov, V.A. Kuzmin and O.M. Nefedov, *Izv. Akad. Nauk SSSR, Ser. Khim.*, 1988, 2654; *Metalloorganicheskaya Khim.*, 1989, **2**, 799.
18. H. Preut, B. Mayer and W.P. Neumann, *Acta. Cryst.*, 1983, **C39**, 1118.
19. A. Yokozeki and A. Kuchitsu, *Bull. Chem. Soc. Jpn.*, 1971, **44**, 2356.
20. R. Balasubrumanian and H.V. Georg. *J. Organomet. Chem.*, 1975, **85**, 123.
21. A. Shusterman, B.E. Landrum and R.L. Millér, *Organometallics*, 1989, **8**, 1851.
22. P.P. Gaspar, D. Holten, S. Konieczny and J.Y. Corey, *Acc. Chem. Res.*, 1987, **20**, 1329.
23. G.L. Levin, P.K. Das, C. Bilgrien and C.L. Lee, *Organometallics*, 1984, **8**, 1206.
24. R.T. Conlin, J.C. Netto-Ferrein, S. Zhang and J.C. Scaiano, *Organometallics*, 1990, **9**, 1332.
25. H. Sakurai, K. Sakamoto and M. Kira, *Chem. Lett.*, 1984, 1379.
26. W. Ando, T. Tsumuraya and A. Sekiguchi, *Chem. Lett.*, 1987, 1317.
27. W. Ando, H. Itoh and T. Tsumuraya, *Organometallics*, 1989, **8**, 2759; ibid., 1988, **7**, 1880.
28. S. Tomoda, M. Shimoda, Y. Takeuchi, Y. Kajii, K. Obi, I. Tanaka and K. Honda, *J. Chem. Soc., Chem. Commun.*, 1988, 910.
29. M. Wakasa, I. Yoneda and K. Mochida, *J. Organomet. Chem.*, 1989, **366**, C1.
30. S. Konieczny, S.J. Jakobs, J.K.B. Wilking and P.P. Gaspar, *J. Organomet. Chem.*, 1988, **341**, C17.
31. J. Mochida, M. Wakasa, Y. Nakadaira, Y. Sakaguchi and H. Hayashi, *Organometallics*, 1988, **7**, 1869.
32. P. Bleckmann, R. Minkwitz, W.P. Neumann, M. Schriewer, M. Thibud and B. Watta, *Tetrahedron Lett.*, 1984, **25**, 2467.
33. P.J. Davidson, H.D. Harris and M.F. Lappert, *J. Chem. Soc., Dalton Trans.*, 1976, 2268.

34. J.C. Scaiano, W.G. McGimpsey and H.L. Casal, *J. Org. Chem.*, 1989, **54**, 1612.
35. O.M. Nefedov, S.P. Kolesnikov and A.I. Ioffe, *J. Organomet. Chem. Libr.* **5**, 1977, 181.
36. G.R. Gillette, G.H. Noren and R. West, *Organometallics*, 1987, **6**, 2617; ibid., 1989, **8**, 487.
37. S.P. Kolesnikov, I.S. Rogozhin, A.Ya. Schteinschneider and O.M. Nefedov, *Izv. Akad. Nauk SSSR, Ser. Khim.*, 1980, 799.
38. K. Raghavachari, J. Chanderasekhar, M.S. Gordon and J. Dykema, *J. Am. Chem. Soc.*, 1984, **106**, 5853.
39. D. Seyferth, M.L. Shannon, S.C. Vick and T.F.O. Lim, *Organometallics*, 1985, **4**, 57.
40. D. Seyferth, E.W. Goldman and J. Escudie, *J. Organomet. Chem.*, 1984, **271**, 337.
41. T. Tsumuraya and W. Ando, *Organometallics*, 1989, **8**, 2286.
42. J. Ohshita, Y. Isomura and M. Ishikawa, *Organometallics*, 1989, **8**, 2050; *J. Am. Chem. Soc.*, 1986, **108**, 7417.

Reactions of Silane with Some Halogen and Oxygen-containing Organic Compounds Induced by Infrared Laser Radiation

J. Pola
INSTITUTE OF CHEMICAL PROCESS FUNDAMENTALS,
CZECHOSLOVAK ACADEMY OF SCIENCES, 165 02 PRAGUE

1 INTRODUCTION

Infrared laser induced bimolecular reactions have attracted much less attention compared to their unimolecular counterparts, and the studies of laser induced reactions of silane with organic compounds are restricted to only those of silane with hexafluorobenzene[1], methane[2] and ethene[2]. Related thermally or photochemically induced reactions of silane with olefins and halogenoolefins were studied almost 40 years previously and it is known[3-6] that they consist of radical steps and yield products of hydrosilylation and alkylchlorosilanes.

In this paper we wish to present CO_2 laser induced reactions of silane with some perhaloethenes that can be initiated by irradiation tuned to either perhaloethene or silane and show that these reactions yield gaseous products and that they differ depending upon the irradiating wavelength and total pressure. Reactions of silane with olefins bearing various oxygenated functional groups will be reported next and shown to result in the formation of solid materials the structure of which consists of different siloxane units. The feasibility of these reactions is revealed to depend on the structure of the olefin. The latter reactions contribute to the new technique for the laser-induced chemical vapor deposition of novel organosilicon materials and they will be, in this regard, compared to CO_2 laser photosensitized decomposition of 1,1-dimethyl-1-silacyclobutane in the presence of the oxygenated olefins or to CO_2 laser photosensitized decomposition of 1-methyl-1-silacyclobutane and 1-methyl-1-vinyl-1-silacyclobutane. Finally, CO_2 laser induced reactions of silane with methyl trifluoroacetate, trifluoroacetic acid and hexafluoroacetone will be introduced as a potential approach for the gas-phase preparation of solid organosilicon or carbon layers.

2 TEA CO_2 LASER INDUCED REACTIONS OF SILANE WITH PERHALOETHENES

The laser absorption data of the perhaloethene/silane mixtures were obtained by using unfocused radiation and they nearly parallel the infrared absorption curves (Fig. 1). Their pattern is not changed whether they are

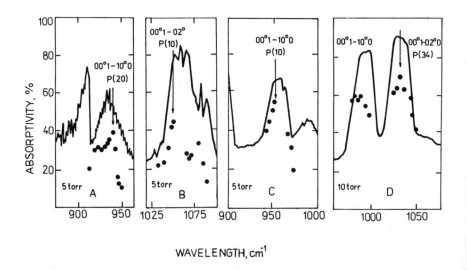

Figure 1 The i.r. spectrum (solid curve) and laser absorption (full circles) of SiH_4 (A), $ClFC:CF_2$ (B), $ClFC:CFCl$ (C) and $CF_2:CCl_2$ (D) in the region of CO_2 laser emission

obtained for pure components or mixtures of perhaloethene/silane. For the irradiation of the mixtures most strongly absorbed lines, i.e. the P(20) at 944.2 cm^{-1} absorbed in silane, the P(10) at 1055.6 cm^{-1} absorbed in chlorotrifluoroethene, the P(10) at 952.9 cm^{-1} for 1,2--dichlorodifluoroethene, and the P(34) line at 1033.5 cm^{-1} absorbed in 1,1-dichlorodifluoroethene were used. The photochemistry can be observed only with focused radiation and it exhibits different features when performed at medium (0.5-2.5 kPa) or low (less than 0.1 kPa) pressures.

Reactions at Medium Pressures

Irradiation at the i.r. transitions of SiH_4 or perhaloethenes leads, regardless of the wavelength, to the formation of ethyne, trifluorosilane, tetrafluorosilane, hydrogen chloride and minor amounts of fluorinated ethenes. The reactions are accompanied with visible luminescence. The absence of compounds obtained by the decomposition of silane (higher silanes, or deposited silicon) or of haloethenes (cyclodimers or cyclotrimers) is in line with the presumption that the reactions are not controlled by the decomposition of either or both primary species. The reaction progress depends on the relative amounts of primary species, which is shown on Figs 2 and 3. Interestingly, the reaction of $CFCl:CF_2$ is most favo-

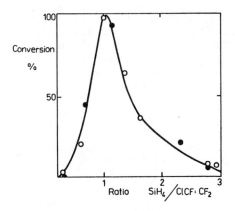 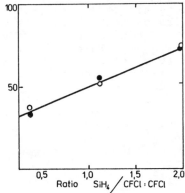

Figure 2   Reaction progress in $SiH_4/ClFC:CF_2$ mixtures at different initial ratios of parent species after 50 pulses of laser radiation at 944.2 cm^{-1} (O) and 1055 cm^{-1} (●). Total pressure 1.5 kPa, energy in pulse 0.5 J

Figure 3   Reaction progress in $SiH_4/CFCl:CFCl$ mixtures at different initial ratios of parent species after 50 pulses of laser radiation at 944.2 cm^{-1} (O) and 952.9 cm^{-1} (●). Total pressure 1.5 kPa, energy in pulse 0.8 J

red at $SiH_4/ClFC:CF_2$ ratio around 1, while the reaction of $ClFC:CFCl$ is enhanced with a higher proportion of SiH_4. The explanation certainly involves several factors, one of them possibly being the difference in the positions of the absorption bands of SiH_4 and perhaloethene, which can imply that more efficient intermolecular V-V energy transfer occurs between SiH_4 and $ClFC:CFCl$. The importance of equal amounts of SiH_4 and $ClFC:CF_2$ lends support to the operation of a common collisional reaction channel opened by the absorption in one particular species and initiated by an encounter of sufficiently energized SiH_4 and $ClFC:CF_2$ molecules.

Reactions at Low Pressures

The experiments performed with different SiH_4/ClFC:CF_2 mixtures show[7] that the reactions can be induced only by radiation at 1055.6 cm^{-1} and that the absorption in silane is ineffective. 1055.6 cm^{-1} radiation induces reactions that result in the formation of C_2F_4 and minor amounts of $C_2F_2H_2$, C_2F_3H, C_3F_4, C_2F_2ClH, Si_2H_6, C_3F_6, C_4F_4, $C_2F_2Cl_2$, C_3F_4ClH and $C_3F_4Cl_2$ compounds. These products are consistent with the chemistry initiated with infrared multiphoton dissociation (IRMPD) of $ClFC:CF_2$ into ClFC: and $F_2C:$, reduction of ClFC: by silane, addition of primary (ClFC: , $F_2C:$) and secondary (HFC:) carbenes to $ClFC:CF_2$ and also with carbenes recombinations. The proposed reaction of ClFC: with SiH_4 can take place as a sequence of the insertion of ClFC: into the Si-H bond and of a decomposition of such a product facilitated due to some extent of vibrational energy in attacking carbene:

$$SiH_4 + ClFC: \longrightarrow \left[H_3Si-CFCl-H \xrightarrow{-HCl} \begin{matrix} H_3Si-CF: \\ \uparrow \downarrow \\ H_2Si=CFH \end{matrix} \right] \longrightarrow H_2Si: + HFC:$$

$$SiH_4 \xrightarrow{SiH_2} Si_2H_6$$

This mechanism is in line with the occurrence of disilane among the products, since insertion of silylene into silane can provide reasonable explanation of disilane formation.

In the case of SiH_4/ClFC:CFCl mixture, the single irradiating wavelengths tuned to each compound are very near and the appropriate i.r. absorption bands weakly overlap, which means that a selective absorption of the radiation is less probable than with SiH_4/ClFC:CF_2 mixture. Different main products are, however, formed depending whether 944.2 or 952.9 cm^{-1} wavelength is used:

$$SiH_4 + ClFC:CFCl \begin{cases} \xrightarrow{944.2 \text{ cm}^{-1}} C_2F_2H_2, CFCl:CFH, Si_2H_6, C_2H_2, C_2F_4 \text{ (traces)} \\ \xrightarrow{952.9 \text{ cm}^{-1}} C_2F_4 \text{ (major product)}, C_2H_2, CFH:CFH, CFH:CFCl, Si_2H_6 \end{cases}$$

With 944.2 cm^{-1}, the formation of $C_2F_2H_2$, C_2F_2ClH and

Si_2H_6 is supportive of the reduction of transient ClFC: to HFC: and of the recombination of HFC: and ClFC: species. Much less silane is consumed in reactions occurring with 952.9 cm^{-1} radiation and consequently C_2F_4 is the main product. C_2F_4 can be formed by the 1,2-migration of halogen in transient CFCl:CF· radical

$$CFCl:CFCl \underset{Cl}{\overset{-Cl}{\rightleftarrows}} FC:CFCl \overset{F\ shift}{\longrightarrow} F_2C:CCl \overset{Cl^{\bullet}}{\longrightarrow} F_2C:CCl_2$$

and by reactions of carbenes originating from 1,1-dichlorodifluoroethene

$$2\ F_2C: \longrightarrow C_2F_4$$

$$Cl_2C: \overset{SiH_4}{\longrightarrow} H_2C:\ \ or\ \ ClHC: \underset{HCl\ elimination}{\overset{recombination}{\longrightarrow}} C_2H_2$$

These results show that (1) the laser irradiation of the silane/perhaloethene mixtures induces the IRMPD of the perhaloethenes and (2) that silane reacts with carbenes, despite that it can undergo the process of IRMPD itself.

3 TEA CO_2 LASER INDUCED REACTIONS OF SILANE WITH OXYGENATED OLEFINS

The irradiation of silane in the presence of methyl acrylate (MA), acrolein (AC), allyl methyl ether (AME), allyl acetate (AA), methyl vinyl ether (MVE) and vinyl acetate (VA) with total pressure 2-3.5 kPa results in the depletion of both parent compounds, the formation of various gaseous compounds and gas-phase solid deposition on the reactor surface[8] (Table 1).
The P(20) line of the 10.6 μm band of CO_2 laser coincides with an i.r. transition in the ν_4 mode of silane and induces $SiH_4 \longrightarrow SiH_2 + H_2$ dissociation, as it has been reported previously[9,10]. The olefins absorb the radiation much less than SiH_4. Silylene is known[11] to add to C=C double bonds to form three-membered rings which can react to either ethylsilylene or vinylsilane. We postulate the initial formation of compounds with oxygen not directly bonded to silicon but susceptible to O→Si coordination bond which can[12] induce further bond fissions as e.g. below.

$$CH_2=C(CH_3)CO_2CH_3 + :SiH_2 \longrightarrow \left[\begin{array}{c} SiH_2\diagdown O-CH_3 \\ CH_2-C-C \\ |\ \ \ \ \diagup O \\ CH_3 \end{array} \right] \longrightarrow$$

$$\longrightarrow H_2Si(OCH_3)C(CH_3)=CH_2\ +\ CO$$

The compounds formed can undergo numerous scrambling re-

Table 1 Laser Induced Reaction between Silane and Oxygenated Olefin[a]

Olefin	Gaseous products	ESCA analysis of deposit	A_{SiO}[b]/number of pulses
MA	$CH_4, C_2H_2, C_2H_4,$ $CH_2CO, CH_2:CHCHO$	SiC_2O_2	0.13/200
AC	$C_2H_4, C_2H_2, CH_2CO,$ C_3H_6	SiC_2O_2	0.09/500
AME	$CH_4, C_2H_2, CH_2O,$ C_3H_6, CH_2CO	$SiCO_{1.5}$ $Si^{o}(18\%)$	0.05/2500
AA	$CH_4, C_2H_2, CH_2CO,$ $C_2H_4, C_3H_6, C_4H_8,$ $CH_3CHO, CH_2:CHCHO,$ $CH_2:CHCH_2OCH_3$	$SiC_{1.3}O_{1.6}$	0.06/3000
MVE	$CH_4, C_2H_2, C_2H_4,$ CH_3CHO, CH_2CO	$SiCO_{1.4}$ $Si^{o}(28\%)$	0.01/500
VA	$C_2H_2, C_2H_4, CH_2CO,$ $CH_3CHO, CH_3C(O)CH_3$	$SiC_{1.3}O_{1.8}$	0.05/1800

[a] Focused radiation, energy in pulse $\sim$ 2 J, SiH_4/olefin ratio $\sim$ 2.

[b] Absorptivity of ν_{Si-O}^{as} band of solid deposited on reactor window after 0.1 kPa depletion of SiH_4.

actions[13] and intermolecular condensations which eventually lead to siloxanes. The other primary pathway can, however include a specific interaction of silylene with the oxygen moiety[14].

An additional channel that may, at least partly, take place is the SiH_4-photosensitized decomposition of the oxygenated olefin. This pathway seems, however, less probable as it follows from the estimation of log k vs 1/T plots for thermal decomposition of the oxygenated olefins and the comparison to the same plot known for the thermal homogeneous decomposition of SiH_4. The IRMPD of silane is known[10] to be disfavored with low partial pressure of silane since SiH_4 can act only as energy conveying agent. This is also obvious in our experiments with mixtures of low SiH_4 content, where the laser photosensitized decomposition of only olefin can occur.

The i.r. spectra of the solid deposit (Fig. 4) corroborate the view that final product has a siloxane structure. The Si 2p core level binding energies (102.7 eV) of the deposit as well as the sum of the Si 2p bin-

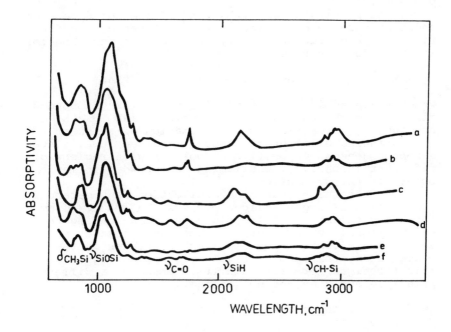

Figure 4 IR spectra of the deposit from reaction of silane with MA(a), AC(b), AME(c), AA(d), MVE(e) and VA(f)

ding energy and Si KLL kinetic energy (1711.8 eV) fit well into the values reported for methylsilicone resins.

Silylene is thought to be an important intermediate in the chemical vapor deposition of silicon and hydrogenated amorphous silicon formed through the insertion of SiH_2 into the Si-H bonds. Trapping rate constants[15,16] of silylene reactions with silane and some unsaturated hydrocarbons show SiH_2 as a highly reactive and undiscriminating species. The absence of disilane or trisilane and the lack of the i.r. absorption of the deposit at 2100 and 880 cm^{-1} after the irradiation of SiH_4-MMA mixture[17], a characteristic of hydrogenated films, were taken by us as support for the statement that SiH_2 reaction with MMA is strongly favored over the insertion of SiH_2 into the Si-H bond of silane[17]. Absorption of the deposit in the region of the stretching Si-H vibration (Fig. 4) does not allow us to draw similar conclusions with other mixtures. It is possible that SiH_2 reacts simultaneously with both oxygenated olefin and silane and that disilane formed by the second reaction undergoes dissociation and is thus removed from the observed gaseous products.

Absorptivity of the ν_{SiO}^{as} band of the solid deposits under comparable conditions has been taken as a criterion of the olefin capability to form siloxane. The following order was revealed

$$MA > AC \sim AME > AA > MVE \sim VA$$

which seems to imply that olefins having oxygen bonded directly to the unsaturated carbon bond (as $CH_2=CH-OCH_3$) are the least reactive in the formation of siloxane.

Scanning electron microscopy revealed in all cases a nonhomogeneous deposition of polymer with regions of different morphology, confirming that powder formation occurs in the gas phase.

This technique can find application as a method for preparation of organosilicon powders. A similar technique not using silane but silacyclobutanes and leading to novel organosilicon polymers is based on the laser photosensitized decomposition of silacyclobutanes in the presence of oxygenated olefins. We wish to mention also this technique here although it is based on a different principle and does not use silane.

CW CO_2 Laser Induced Reactions of Silaethenes with Oxygenated Olefins

Conventional thermal decomposition (CTD) of silacyclobutanes is known[18] to yield reactive silaethenes that, in the absence of traps, undergo a multitude of reactions such as isomerization into silylenes, insertion into Si-H bonds, cyclodimerization and polymerization. It was therefore surprising to reveal that silaethenes generated during laser photosensitized (SF_6) decomposition (LPD) of 1-methyl-1-silacyclobutane[19], 4-silaspiro(3.4)octane[20] and 1-methyl-1-vinyl-1-silacyclobutane[21] undergo only polymerization. This laser induced chemical vapor deposition is a promising technique for preparation of new polymers and their deposition on thermally unstable surfaces.

$$\diagdown Si \diagdown\!\!\!\square \xrightarrow{-C_2H_4} \diagdown Si=CH_2 \xrightarrow[CTD]{LPD} \begin{array}{l} \text{polymer} \\ \\ \text{cyclodimer as major product} \end{array}$$

LPD of 1,1-dimethyl-1-silacyclobutane (DMSCB) and 1,1-dichloro-1-silacyclobutane yields, however, mostly cyclodimer[22]. Examination of LPD of DMSCB in the presence of oxygenated olefins revealed[22] that intermediary 1,1-dimethyl-1-silaethene is efficiently trapped with MA, AC and VA. With AC and VA, only gaseous products of trapping reactions are formed, while with MA polymeric substance is a major product. We assume that similar depletion of MA and DMSCB, no significant formation of gaseous organosilicon compounds, and the i.r. (ν(C=O)) absorption of the deposit at 1760 cm^{-1} are consistent with the reaction steps assumed below:

$$Me_2Si\triangle \xrightarrow[-C_2H_4]{SF_6, MA} \left[\begin{array}{c} Me_2Si-CH_2 \\ | \quad\quad | \\ CH_2-CH \\ | \\ CO_2Me \end{array} \right] \longrightarrow (-CH_2\underset{Me}{\overset{Me}{\underset{|}{\overset{|}{Si}}}}-CH_2\underset{CO_2Me}{\overset{}{\underset{|}{CH}}}-)_n$$

Molecular weight of the polymeric deposit formed in an almost quantitative yield have rather narrow molecular weight distribution (1000-5000).

MA does not polymerize, when it is irradiated in the presence of SF$_6$, but decomposes into ethene, CO and acrolein[23]. The reaction above is, to the best of our knowledge, the first example of the co-polymerization of silaethene species. Different course of LPD and CTD of silacyclobutanes can be ascribed to high peak temperatures and the elimination of heterogeneous contributions to reaction mechanism during LPD [24,25].

4 TEA CO$_2$ LASER INDUCED REACTIONS OF SILANE WITH FLUORINATED CARBONYL COMPOUNDS

The irradiation of silane in the presence of methyl trifluoroacetate, hexafluoroacetone or trifluoroacetic acid was studied with the aim to test these reactions as a technique for production of silicon containing coatings. $CF_3CO_2CH_3$, CF_3CO_2H do not but $CF_3C(O)CF_3$ does possess an i.r. absorption band in the region of CO$_2$ laser emission. It is therefore plausible to assume that 944.2 cm^{-1} radiation induces IRMPD of silane and $CF_3C(O)CF_3$. Gaseous products formed are SiHF$_3$, SiF$_4$ and some compounds that are produced by the laser induced decomposition of the fluorinated reactants[25]. Solid products deposited on reactor walls show i.r. absorption in the region of ν(C-F), ν(Si-O), ν(C=O) and ν(Si-F) vibrations and their composition, analysed by ESCA technique, differ depending upon the relative pressure of reactants and irradiation conditions. Some representative runs are given in Table 2.

The specific feature of the reaction induced in SiH$_4$-CF$_3$C(O)CF$_3$ mixture is the formation of carbon deposit. This reaction is probably the only known gas-phase reaction of perfluorocarbons that results in the formation of carbon. Using SEM technique for analysis of the deposit, we proved the presence of fluorine in the glass substrate, which implies that the process of deposition is accompanied by the etching of the substrate by fluorine of fluorine-containing species. More work on the elucidation of the structure of the solid materials and the nature of gas-phase processes is underway.

Table 2 Laser Induced Reactions[a] of SiH_4 with CF_3CO_2H, CF_3COOCH_3 and $CF_3C(O)CF_3$

Reactants, (kPa)		Gaseous products	ESCA analysis of deposit[b]	Ref.
CF_3CO_2H (0.1-0.3)	SiH_4 (0.1-0.3)	$HSiF_3, SiF_4, COF_2,$ CH_4, C_2H_2	$Si_{0.9}CO_{1.4}F_{1.4}$	26
$CF_3CO_2CH_3$ (0.1)	SiH_4 (0.1)	$CF_2{:}CH_2, CO, COF_2,$ $CH_3F, C_2F_6, C_2F_4,$ $SiF_4, SiHF_3, CH_4$	$SiCO_{1.3}F_{0.2}$	27
$CF_3C(O)CF_3$ (0.04)	SiH_4 (0.2)	$SiHF_3, SiF_4,$ $C_2H_2, CF_2{:}CH_2,$ $C_2F_4, C_2F_6, CO,$ COF_2	$SiC_{0.6}O_{1.1}F_{0.1}$ $(Si^0$ 39 %, Si^{4+} 61 %)	27
$CF_3C(O)CF_3$ (0.2)	SiH_4 (0.1)	$SiHF_3, SiF_4,$ $C_2H_2, CF_2{:}CH_2,$ $C_2F_4, C_2F_6, CO,$ COF_2	$CO_{0.04}F_{0.2}$	27

[a] Focused radiation (energy in pulse 2 J). [b] K_2SiF_6 found on KBr reactor windows

REFERENCES

1. Y. Koga, R.M. Serino, R. Chen and P.M. Keehn, J.Phys.Chem., 1987, 91, 298.
2. J.S. Haggerty and W.R. Cannon in 'Laser Induced Chemical Processes', Steinfeld, J., Ed., Plenum, New York 1981, p. 170.
3. G. Fritz, Z.Naturforsch., 1951, 6b, 47.
4. G. Fritz, Z.Naturforsch., 1952, 7b, 207; 379.
5. D.G. White and E.G. Rochow, J.Am.Chem.Soc., 1954, 76, 3897.
6. R.N. Haszeldine, M.J. Newlands and J.B. Plumb, J.Chem.Soc., 1965, 2101.
7. S. Simeonov and J. Pola, Spectrochim. Acta, 1990, 46 A, 443.
8. Z. Papoušková, J. Pola, Z. Bastl and J. Tláskal, J.Macromol.Sci.-Chem., in the press.
9. T.F. Deutch, J.Chem.Phys., 1979, 70, 1187.
10. P.A. Longeway and F.W. Lampe, J.Am.Chem.Soc., 1981, 103, 6813.
11. D.S. Rogers, K.L. Walker, M.A. Ring and H.E. O'Neal, Organometallics, 1987, 6, 2313.
12. J. Pola in 'Carbon-functional Organosilicon Compounds', V. Chvalovský and J.M. Bellama, Eds., Plenum

Press, New York 1984.
13. J.C. Lockhart, 'Redistribution Reactions', Academic Press, New York, 1970.
14. R. Walsh, Private communication.
15. J.M. Jasinski and J.O. Chu in 'Silicon Chemistry', E.R. Corey, J.Y. Corey and P.P. Gaspar, Eds., E. Horwood, Chichester 1988, Chap. 40.
16. R. Walsh in 'Silicon Chemistry', E.R. Corey, J.Y. Corey and P.P. Gaspar, Eds., E. Horwood, Chichester 1988, Chap. 41.
17. R. Alexandrescu, J. Morjan, C. Grigoriu, Z. Bastl, J. Tláskal, R. Mayer and J. Pola, Appl.Phys., 1988, 46 A, 768.
18. L.E. Guselnikov and N.S. Nametkin, Chem.Rev., 1979, 79, 529.
19. J. Pola, V. Chvalovský, E.A. Volnina and L.E. Guselnikov, J.Organometal.Chem., 1988, 341, C 13.
20. M. Sedláčková, J. Pola, L.E. Guselnikov and E.A. Volnina, J.Anal.Appl.Pyrol., 1989, 14, 345.
21. J. Pola, E.A. Volnina and L.E. Guselnikov, J.Organometal.Chem., in the press.
22. J. Pola, D. Čukanová and M. Jakoubková, to be published.
23. J. Pola, Tetrahedron, 1989, 45, 5065.
24. J. Pola, SPIE, 1988, 1033, 482.
25. J. Pola, Spectrochim.Acta, 1990, 46 A, 607.
26. J. Pola, Z. Bastl and J. Tláskal, Infrared Phys., 1990, 30, 355.
27. Z. Papoušková, J. Tláskal, Z. Bastl and J. Pola, to be published.

Reactions of Functionalized Silenes and Cyclosilanes

Wataru Ando* and Yoshio Kabe
DEPARTMENT OF CHEMISTRY, UNIVERSITY OF TSUKUBA,
TSUKUBA, IBARAKI, 305 JAPAN

Introduction

Trimethylsilyl carbenes (**2**) generated by photolysis or thermolysis of trimethylsilyl diazomethane (**1**) form silenes (**3**) by migration of a substituent from the silicon atom to carbene center. This silyl carbene-to-silene rearrangement, reported by us 1973[1] has proved to be a convenient route to a variety of functionalized silenes. The migration of a silyl group on the silicon atom to a carbene center occurs exclusively than methyl migration (**4-5-6**). In the field of silaaromatics, silylcarbene approaches are also successful. The photolysis of cyclic silyl diazo compounds (**7**) and (**9**) resulted in silafulvene (**8**) and silabenzene (**10**) intermediates via the migration of a dienyl group to carbene center.[2]

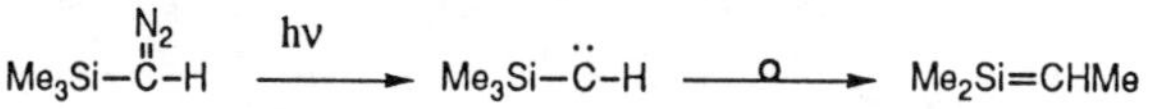

$$Me_3Si-\overset{N_2}{\underset{\|}{C}}-H \xrightarrow{h\nu} Me_3Si-\overset{..}{C}-H \longrightarrow Me_2Si=CHMe$$
$$\quad(1) \quad\quad\quad\quad (2) \quad\quad\quad\quad (3)$$

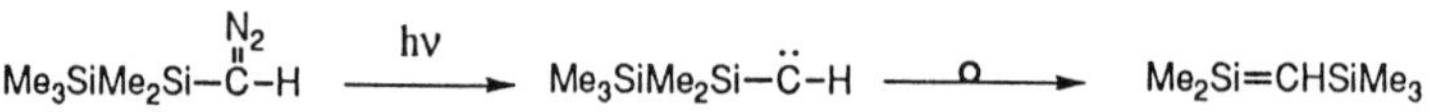

$$Me_3SiMe_2Si-\overset{N_2}{\underset{\|}{C}}-H \xrightarrow{h\nu} Me_3SiMe_2Si-\overset{..}{C}-H \longrightarrow Me_2Si=CHSiMe_3$$
$$\quad(4) \quad\quad\quad\quad (5) \quad\quad\quad\quad (6)$$

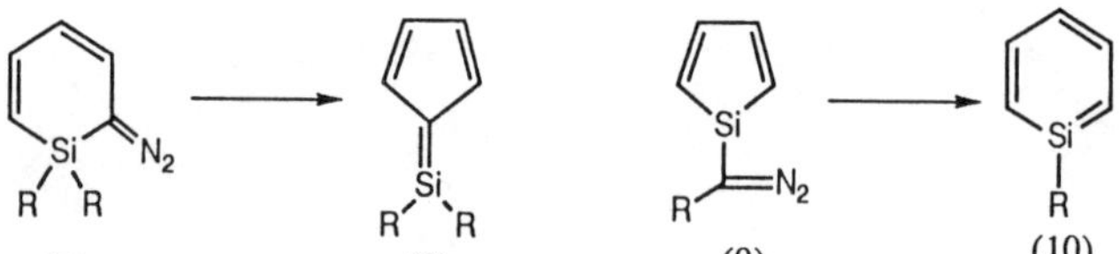

(7) → (8)   (9) → (10)

As conjugated silenes, silaacrylate (**12**) and sila-enone (**15**) intermediates were obtained by photolysis of disilanyl diazo compounds (**11**) and (**14**).[3] Interestingly, the silaacrylate (**12**) isomerizes to ketene (**13**), while sila-enone (**15**) undergoes intramolecular [2+2] cycloaddition reaction to give 1,2-silaoxetene (**16**).

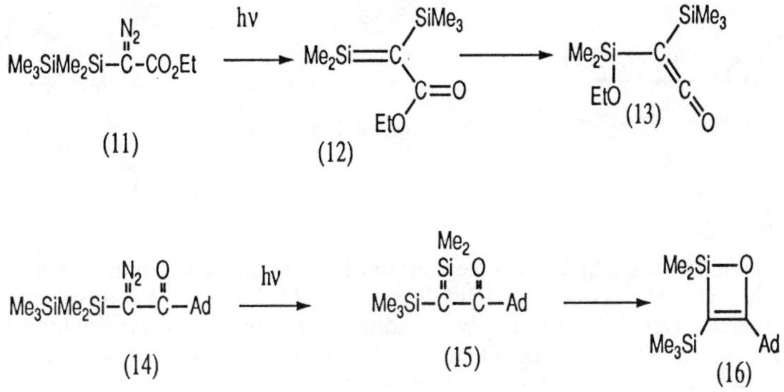

For comparative study, various silenes were isolated in 3-methyl pentane matrix at 77°K by photolysis of the corresponding silyl diazo compounds and their ultraviolet spectra measured.[4] The results are summarized in Table 1. The introduction of trimethylsilyl group on carbon results in slight red shift compared to the parent silene ($H_2Si=CH_2$, 258nm). As one might expect, considerable bathochromic shifts for conjugated silenes such as **12**, **15** and **19** have been observed.

Table 1   UV-Absorptions of Silenes

Silenes	λ_{max}/nm	Precursor
$Me_2Si=CHMe$ (**3**)	255	Me_3SiCHN_2
$Me_2Si=CHSiMe_3$ (**6**)	265	$Me_3SiMe_2SiCHN_2$
$Me_2Si=C(Me)SiMe_3$ (**17**)	274	$(Me_3Si)_2CN_2$
$Me_2Si=C(SiMe_3)_2$ (**18**)	278	$Me_3SiMe_2SiC(N_2)SiMe_3$
$Me_2Si=C(Ph)CO_2Me$ (**19**)	280	$PhMe_2SiC(N_2)CO_2Me$
$Me_2Si=C(SiMe_3)COAd$ (**15**)	284	$Me_3SiMe_2SiC(N_2)COAd$
$Me_2Si=C(SiMe_3)CO_2Et$ (**12**)	293	$Me_3SiMe_2SiC(N_2)CO_2Et$

Polysilanyl Silenes

The good migrating ability of silyl group made us pursue the project aimed at polysilanyl silenes. Firstly we have investigated the photochemical and thermal behaviour of polysilanyl diazo compounds (**20** and **21**)[5], which are accessible in 41 and 83 % yield from lithiation of trimethylsilyl diazomethane (**1**) and quenching with $Cl(SiMe_2)_nCl$ (n=2,3).

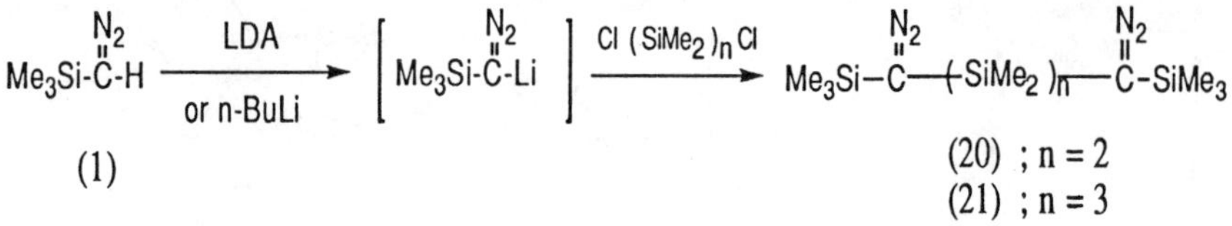

$$Me_3Si-\underset{\underset{(1)}{}}{\overset{N_2}{C}}-H \xrightarrow{\text{LDA or n-BuLi}} \left[Me_3Si-\overset{N_2}{C}-Li \right] \xrightarrow{Cl(SiMe_2)_nCl} Me_3Si-\overset{N_2}{C}-(SiMe_2)_n-\overset{N_2}{C}-SiMe_3$$

(**20**) ; n = 2
(**21**) ; n = 3

Photolysis of **20** with a high pressure mercury lamp in benzene containing excess tert-butylalcohol gave one molar ^tBuOH adducts (**22**, **23** and **24**) and two molar ^tBuOH adducts (**25**, **26**) respectively. From these products analysis it appears that dimethylsilanyl group migration and successive methyl migration to each carbene center yields the stepwise formation of unsymmetrical silenes as intermediate

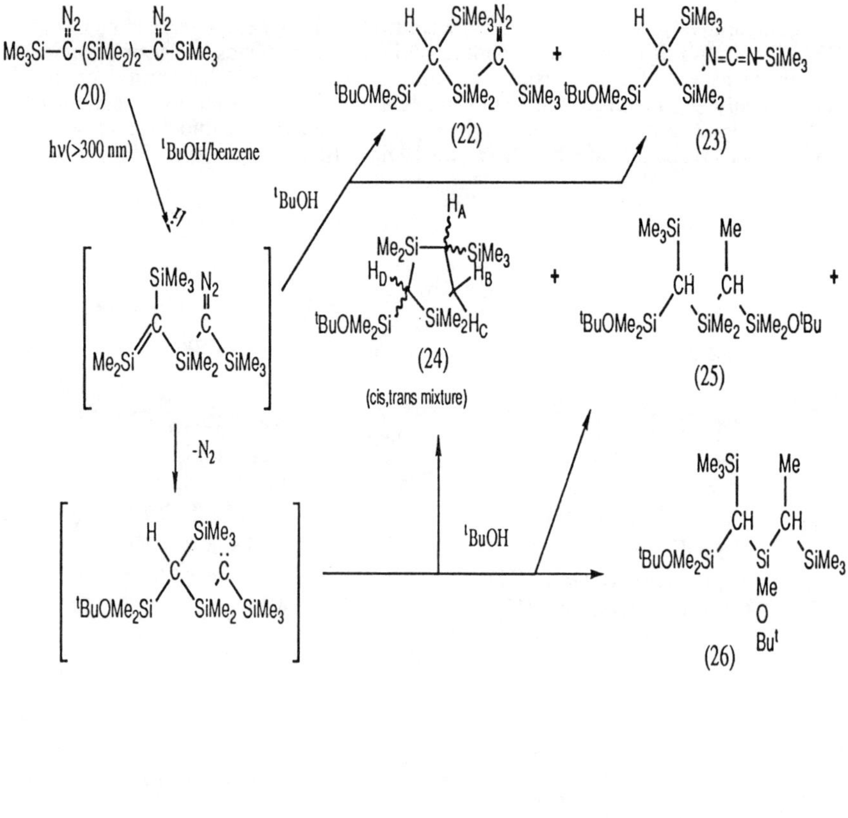

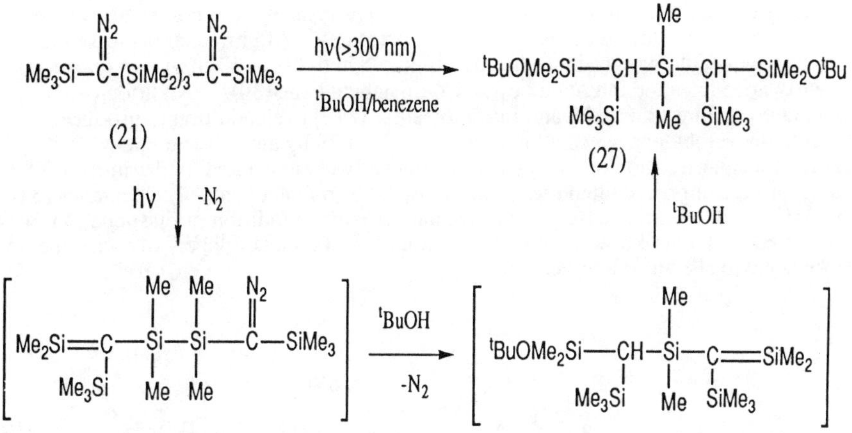

$$\underset{(21)}{Me_3Si-\overset{\overset{N_2}{\|}}{C}-(SiMe_2)_3-\overset{\overset{N_2}{\|}}{C}-SiMe_3} \xrightarrow[\text{^{t}BuOH/benezene}]{h\nu (>300\ nm)} \underset{(27)}{{}^tBuOMe_2Si-\underset{\underset{Me_3Si}{|}}{CH}-\underset{\underset{Me}{|}}{\overset{\overset{Me}{|}}{Si}}-\underset{\underset{SiMe_3}{|}}{CH}-SiMe_2O^tBu}$$

$h\nu \downarrow -N_2$ $\uparrow ^tBuOH$

$$\left[Me_2Si{=}\underset{\underset{Me_3Si}{|}}{C}-\underset{\underset{Me}{|}}{\overset{\overset{Me}{|}}{Si}}-\underset{\underset{Me}{|}}{\overset{\overset{Me}{|}}{Si}}-\overset{\overset{N_2}{\|}}{C}-SiMe_3 \right] \xrightarrow[-N_2]{^tBuOH} \left[{}^tBuOMe_2Si-\underset{\underset{Me_3Si}{|}}{CH}-\underset{\underset{Me}{|}}{\overset{\overset{Me}{|}}{Si}}-\underset{\underset{SiMe_3}{|}}{C}{=}SiMe_2 \right]$$

In contrast to the polysilanyl diazo compound (**20**), the photolysis of **21** gave rise to double dimethylsilanyl group migration to each carbene center yielding the stepwise formation of symmetrical silene as intermediate. In order to maximize

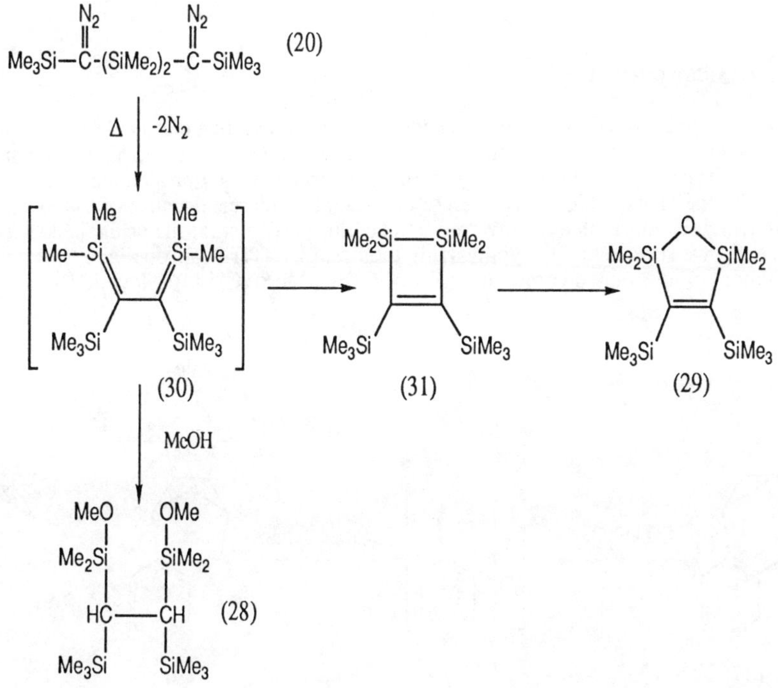

$$\underset{(20)}{Me_3Si-\overset{\overset{N_2}{\|}}{C}-(SiMe_2)_2-\overset{\overset{N_2}{\|}}{C}-SiMe_3}$$

$\Delta \downarrow -2N_2$

(30) → (31) → (29)

↓ McOH

$$\underset{(28)}{\underset{\underset{Me_3Si}{|}}{\overset{\overset{MeO}{|}}{Me_2Si}}-\underset{\underset{SiMe_3}{|}}{\overset{\overset{OMe}{|}}{SiMe_2}}}$$
HC——CH

chances of intramolecular reactions, gas phase pyrolysis of these polysilanyl diazo compounds were carried out under reduced pressure. After methanolysis of the pyrolysate of **20**, two products are identified (**28** and **29**). Each of two products clearly arise from the intermediacy of 1,4-disilabutadiene (**30**). Addition of methanol to **30** produces **28** and intramolecular [2+2] cycloaddition to produce 1,2-disilacyclobutene (**31**), which is oxidized to **29** by air. Efficient intramolecular reaction of silenes was dramatically demonstrated by thermolysis of **21** to afford the unprecedented cyclosilane; i.e. 2,4,5-trisilabicyclo[1.1.1]pentanes (**32**) in 38% yield. In this case, intramolecular criss-cross addition of bissilene (**33**) was observed. To our knowledge this represents the first example of silyl carbene to produce two silenes in a molecule.

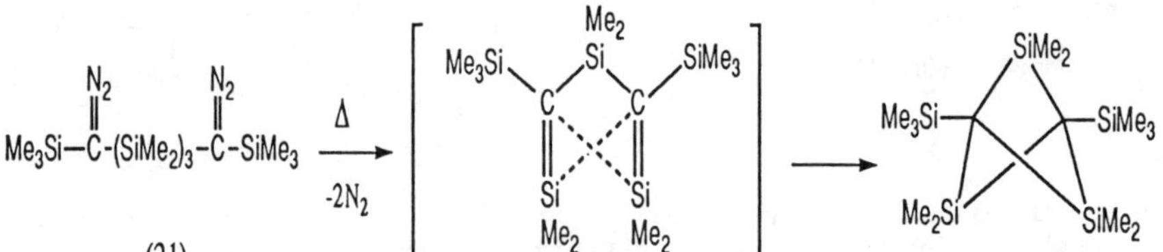

[n.1.1.]Silapropellane

After polysilanyl diazo derivatives yielded to synthesis in a classical reaction, our interest expanded towards construction of strained polysilanyl cyclosilanes in similar synthetic pathway. Our first target molecule was small ring propellane such as [n.1.1]propellanes, which have been of considerable interest because of the extremely high reactivity attributed to "inverted tetrahedral" geometry at the bridgehead carbons. Recently we succeeeded in synthesizing the first [3.1.1]Trisilapropellane (**36**) by treatment of dilithiated compound (**35**) with Cl(SiMe$_2$)$_3$Cl in 23 % yield.[6)]

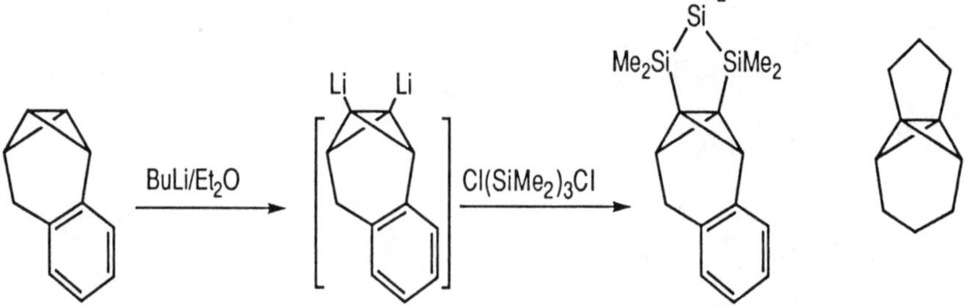

The thermal stability of **36** was found to significantly increase compared to the corresponding carbon analogue (**37**), which easily isomerized at 0°C. When a solution of **36** in benzene, sealed in a tube under a vacuum was kept at 190°C, no decomposition was observed. However under the flow pyrolysis at 450°C, **36** decomposed and finally converted to **38** in 30 % yield. On the other hand, **36** was labile to acidic contamination such as silica gel chromatography or HCl present in CHCl$_3$ and quantitatively rearranged to **39**. We rationalized the intermediate leading to **39** as the cation **41**, which is doubly stabilized by cyclopropyl and silyl substitution. The cation **41** is probably attacked by chloride ion with concomitant rearrangement to homoallyl skeleton of **39**. In order to get [2.1.1]disilapropellane,

treatment of dilithiated compound **35** with Cl(SiMe$_2$)$_2$Cl provided [5.1.1]propellane (**42**) and the ether incorporated product (**43**), which is probably derived from the reactive [2.1.1]disilapropellane (**44**) similar to carbon system (**45-46**). Although there are several mechanistic possibilities, the mechanism which we currently favour for this unique rearrangement is that 1,2-silyl shift proceeds through the intermediacy of bissilene (**47**) or biradical (**48**) to produce silyl radical (**49**). Meanwhile, the radical disproportionation of ether affording enol ether is initiated by biradicals (**48** and/or **49**), finally silyl radical (**49**) added to enol ether.

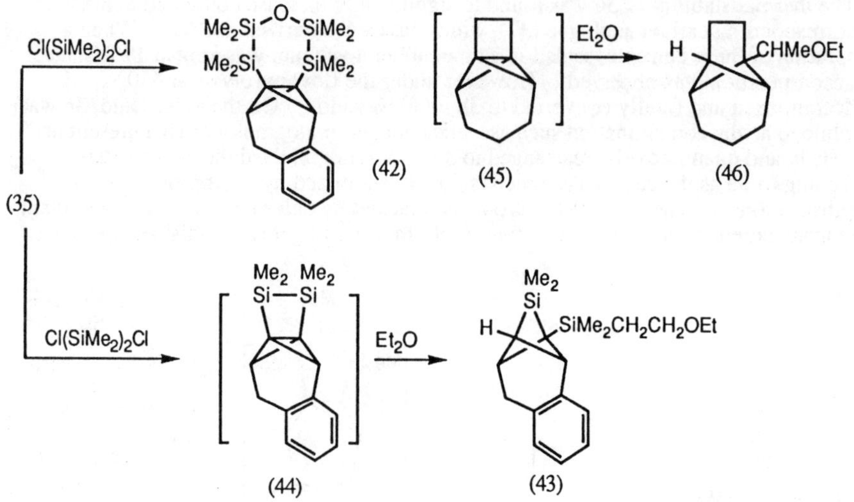

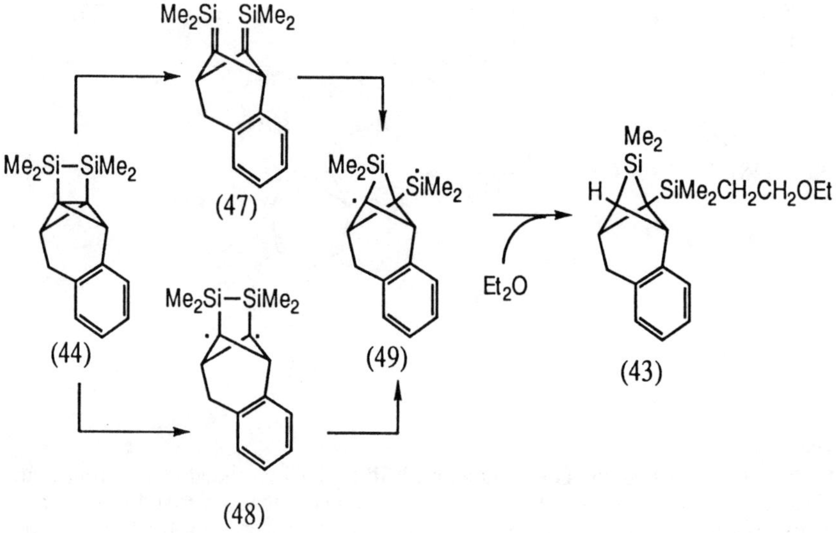

$$Et_2O \xrightarrow{(48)\text{or}(49)} Me\dot{C}HOEt \xrightarrow{\text{disproportion}} CH_2{=}CHOEt + Et_2O$$

Polysilacycloalkynes

Another target molecule was the small ring cycloalkyne which is a subject of considerable interest because of bond-angle strain at the sp carbons.
Our efforts towards construction of strained polysilanylcycloalkynes culminated in the synthesis of hexasilacyclooctyne (**50**) and pentasilacycloheptyne (**51**) respectively.[7]
Treatment of Cl(SiMe$_2$)$_6$Cl with acetylene di-Grignard reagent under dilute conditions gave **50** in 46% yield, which was stable colorless liquid.

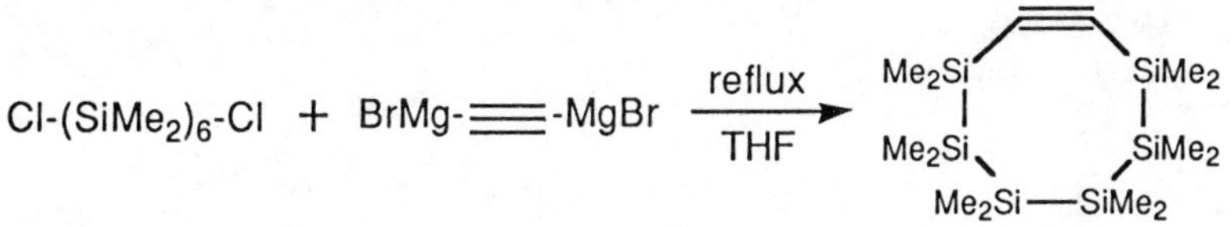

The photolysis of hexane solution of **50** with a low pressure mercury lamp in the presence of triethylsilane gave lower homologous cycloheptyne (**51**) in 22% yield as colorless crystals, which is also stable under atmospheric oxygen and moisture.   The cycloheptyne (**51**) could be also obtained by thermolysis in 18% yield.   Interestingly irradiation of **50** produced minor products (**52** and **53**).   Although several mechanistic possibilities were not excluded, the products suggest a more exotic reaction pathway in the photolysis of cyclooctyne (**50**).

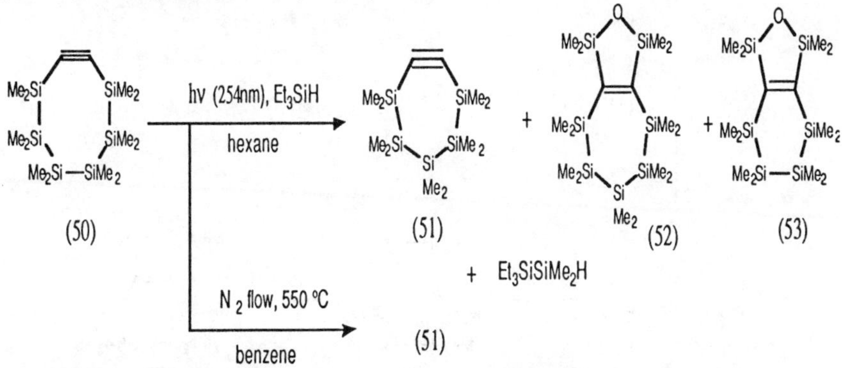

The formation of three products from **50** can be mechanistically rationalized by intermediacy of bicycloolefins (**54** and **55**).   An unprecedented rearrangement for the routes from **50** and **51** to **54** and **55** are shown in the schemes. The bicycloolefins (**54** and **55**) could undergo dimethylsilylene addition to provide **56** and **57**, which are easily oxidized to siloxanes (**52** and **53**) by air.
To put this mechanistic reasoning to a test, bicycloolefins (**58** and **59**) have been independently prepared through the reaction of polysilacycloalkynes with silylenes.   Further photolysis of **58** produced not only cyclooctyne (**50**) through silylene extrusion, but also unexpectedly ring expansion products (**60** and **61**).

A rationalization for this latter unique behaviour was that we were observing the reverse rearrangement corresponding to the conversion of **50** and **51** to **54** and **55**. Finally we have had considerable success in producing silacyclooctyne (**50**) by irradiation of **60**, where the same rearrangement does occur.

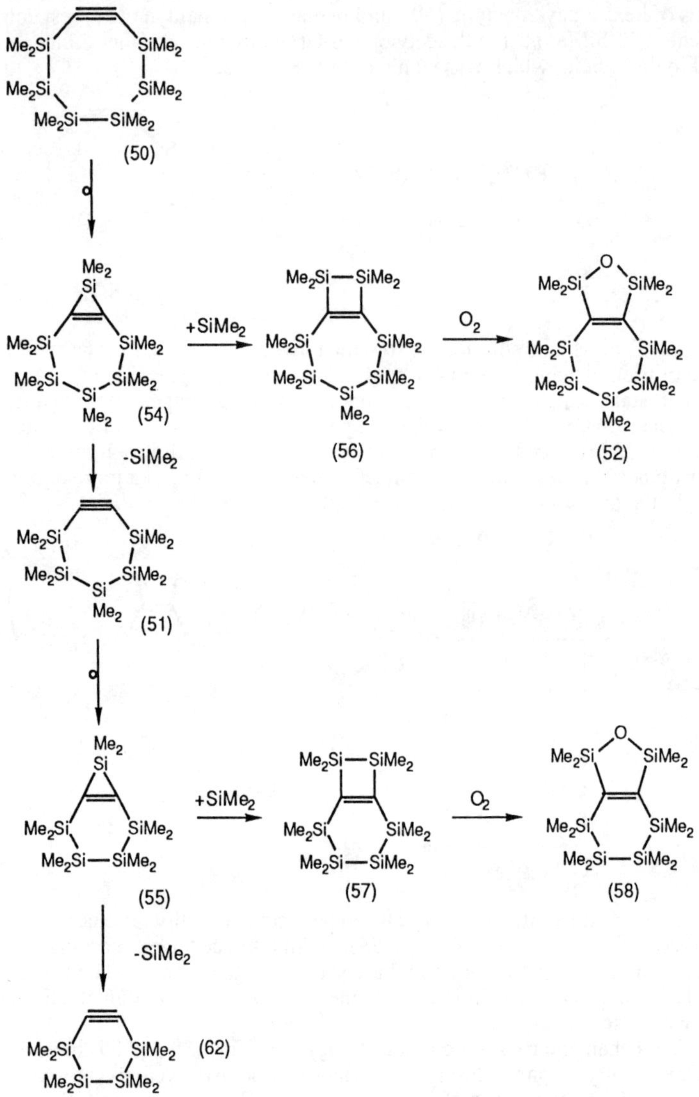

The molecular structure of **51** could be established by X-ray analysis shown in Figure 1. The Si-Si,Si-C single and C-C triple bond lengths are normal and ranged between 2.340-2.350, 1.80-1.84 and 1.22Å, respectively. However, there is an interesting alternation in the bond angle around the polysilane chain; the bond angles of C(2)-Si(1)-Si(2) (100.4°) and C(1)-Si(5)-Si(4) (101.7°) are contracted and those of Si(1)-Si(2)-Si(3) (108.7°) and Si(3)-Si(4)-Si(5) (109.9°) are normal while those of Si(2)-Si(3)-Si(4) (117.4°) are considerably expanded from the normal bond

Figure 1. ORTEP Plot of 51

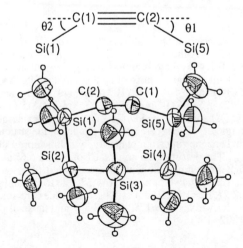

angles. The most interesting point in the structure of **51** is of course the bending of the acetylene bond from linear geometry, which is represented by the deviation angles θ1 and θ2. The bending angles of θ1 and θ2 of **51** are 20.4 and 17.8° respectively, which are comparable with those observed in the non-substituted cyclooctyne (θ=21.5°) and cyclononyne (19.8°). Apparently, the structure of **51** reveals that the magnitude of the bending in acetylene decreases, but an increase in the deformation of the polysilane chain is definitely noticeable. The observations are explained by high flexibility of the conformation in the polysilane chain. Photochemical silylene extrusion of cycloheptyne (**51**) would be expected to produce lower homologous cycloalkyne; i.e. cyclohexyne (**62**). From preliminary result, we have found the photolysis of **51** under similar conditions to those applied to **50** led to cyclohexyne which is too labile to isolate and only identified by GC-Mass spectrometry.

Polysila[n]paracyclophane

The last target molecule was [n]paracyclophane which has unique properties resulting from bending of the benzene ring. After many attempts, the reductive coupling of 1-chloro-6-parachlorodimethylsilyl- 1,1,2,2,3,3,4,4,5,5,6,6 -dodecamethylhexasilane (**63**) with sodium in refluxing toluene in the presence of [18]crown-6 produced heptasila[7]paracyclophane (**64**) in 1.6% yield[8], as colorless crystal which is stable under atmospheric oxygen and moisture even when heated to its melting point. The presence of [18]crown-6 is essential for the formation of **64** and only polymeric substances were obtained in the absence of [18]crown-6.

Cl—SiMe$_2$—⟨⟩—(SiMe$_2$)$_6$—Cl $\xrightarrow{\text{Na/18-Crown-6}}_{\text{Toluene reflux}}$

(63)   (64)

X-ray structural determination established unequivocally the molecular structure of **64** shown in Figure 2. Although the Si-Si bond lengths are normal and ranged between 2.338(5)-2.371(4)Å, there is an interesting alternation in the bond angles around polysilane chain similar to those of cycloheptyne (**51**); the bond angles of C1-Si1-Si2 (105.5(3)°) and C4-Si7-Si6 (107.4(3)°) are contracted while those of Si1-Si2-Si3 (117.7(2)°), Si2-Si3-Si4 (126.5(2)°), Si3-Si4-Si5 (116.2(2)°), Si4-Si5-Si6 (115.6(2)°) and Si5-Si6-Si7 (113.7(2)°) are considerably expanded from the normal bond angles. The most interesting point in the structure of **64** is the deformation of the benzene ring from planarity, which is represented by the deviation angles θ1 and θ2. The deformation angles θ1 of **64** are 6.6° and 4.5° and θ2 are 6.5° and 9.0°, which are comparable with those observed for octamethyltetrasila-[2,2]paracyclophane (θ1=4.3°, θ2=15.0°) and its chromium complex (θ1=2.5,2.2°, θ2=10.0,10.9°). These θ values are smaller than those of [7]paracyclophane derivatives (θ1=17°, θ2=23.5°) and those of [8]paracyclophane derivatives (θ1=9°,θ2=15°).

Figure 2. ORTEP Plot of 64

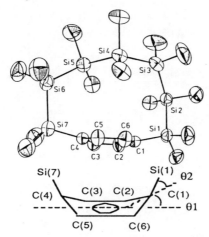

Apparently the structure of **64** also reveals that the magnitude of deformation in the benzene ring decreases, but an increase of deformation in polysilane chain is definitely noticeable. These observations are best explained by high flexibility of the conformation in polysilane chain compared to methylene chain, involving expansion of Si-Si-Si angles. In the hope of synthesizing lower homologous [n]paracyclophanes and finding out unique rearrangement, we are investigating the thermal and photochemical behaviour of **64**.

Acknowledement

The author is grateful to his collaborators who actually did the work discussed in this lecture, Dr. H. Yoshida, Dr. T. Shimizu, Dr. T. Tsumuraya, Mr. Y. Igarashi, Miss N. Nakayama, Mr. T. Suzuki and Mr. K. Kurishima.

References

1)a) W. Ando, T. Hagiwara, T. Migita, J. Am. Chem. Soc.,1973, 95, 7518.
  b) W. Ando, A. Sekiguchi, T. Migita, S. Kammula, M. Green, M. Jones, Jr., J. Am. Chem. Soc., 1975, 97, 3818.
2)a) A. Sekiguchi, W. Ando, J. Am. Chem. Soc., 1981, 103, 3579.
  b) W. Ando, H. Tanikawa, A. Sekiguchi, Tetrahedron Lett., 1983, 4245.
  c) A. Sekiguchi, H. Tanikawa, W. Ando, Organometallics, 1985, 5, 584.
3)a) W. Ando, A. Sekiguchi, T. Sato, J. Am. Chem. Soc., 1981, 103, 5573.
  b) A. Sekiguchi, W. Ando, J. Am. Chem. Soc., 1984, 106, 1486.
4) A. Sekiguchi, W. Ando, Chem. Lett., 1986, 2025.
5) W. Ando, T. Suzuki, K. Kurishima, H. Yoshida, Y. Kabe, to be published.
6) W. Ando, Y. Igarashi, Y. Kabe, N. Tokitoh, Tetrahedron Lett., 1990, 31, in press.
7) W. Ando, N. Nakayama, Y. Kabe, T. Shimizu, Tetrahedron Lett., 1990, 31, in press.
8) W. Ando, T. Tsumuraya, Y. Kabe, Angew. Chem. Int. Ed. Engl., 1990, 29, in press.

Section III: Mechanistic Organosilicon Chemistry
b) Hypervalent Silicon, Nucleophilic Substitution, and Biotransformations

Hypervalent Species of Silicon: Structure and Reactivity

R.J.P. Corriu

UNITÉ MIXTE CNRS, RHÔNE POULENC, USTL N° 44, UNIVERSITÉ DE MONTPELLIER II, PLACE E. BATAILLON, 34095 MONTPELLIER CEDEX O5, FRANCE

1 INTRODUCTION AND PRELIMINARY RESULTS *(Studies carried out by Dr. HENNER-LEARD, Dr.DABOSI, Dr. MARTINEAU)*

Our interest concerning the mechanism of nucleophilic substitution at silicon [1] led us to afford some experimental evidence for the formation of pentacoordinated silicon species as intermediates in the course of nucleophilic substitution taking place either with retention or with inversion at silicon [2,3]. The possibility for a silicon atom to extend its coordination number and to become penta or hexacoordinated is one of the most important factors in the mechanisms of reactions at silicon.

The first case in which we were faced directly with these species involved the racemisation of chlorosilanes induced by solvents. It was Sommer [4] who observed that optically active chlorosilanes were not optically stable in some solvents (acetonitrile, nitromethane, nitrobenzene). In studying this phenomenon, it appeared to us that the driving force for the racemisation was not the polarity but the nucleophilicity of the solvent [5-10]. Furthermore using very efficient agents like HMPA, DMF or DMSO, it was possible to induce racemisation using a small amount of these reagents and to perform kinetic studies. The kinetic law was second order in the racemising agent (Cat), and first order in chlorosilane. The reaction also involved a very high negative entropy of activation. The activating nucleophile (Cat) acts as a catalyst. It does not appear in the final composition of the products.

$$R_2{\cdots}\overset{R_1}{\underset{R_3}{\diagdown}}Si-Cl + Cat \longrightarrow R_2{\cdots}\overset{R_1}{\underset{R_3}{\diagdown}}Si-Cl \quad Cl-Si{\cdots}\overset{R_1}{\underset{R_3}{\diagup}}R_2$$

$$V = k\,[R_3\,Si\,Cl]\,[Cat]_0^2$$
$$-70 < \Delta S^{\ddagger} < -50 \text{ e.u.}$$
Process controlled by $\Delta S^{\ddagger}$

(d)   (l)

RACEMISATION WITHOUT SUBSTITUTION

[Cat] = HMPA, DMSO, DMF, CH_3CN, RNO_2 ...

The large negative values for the activation entropy are consistent with a highly organized transition state. The mechanism we proposed [7] involves the formation of a hexacoordinated intermediate (or transition state) during the rate determining step. This step involves attack by a second molecule of the catalyst on the pentacoordinate silicon formed in the preequilibrium.

The same mechanistic features were observed in the case of nucleophilic substitutions (hydrolysis and alcoholysis) occurring at silicon and assisted by nucleophilic catalysts. In this case also the kinetics are controlled by the activation entropy (sometimes, the reaction proceeds faster when the temperature is lowered) and the kinetic law is first order in chlorosilane, first order in alcohol and first order in activating nucleophile (Cat). A mechanism similar to the mechanism proposed in the case of the racemisation of chlorosilanes has been suggested [11].

$$R_3SiX + Nu \longrightarrow R_3Si—Nu + X^-$$

(A)

$V = k\ [\ R_3\ Si\ Cl\ ]\ [\ Nu\ ]\ [\ Cat\ ]_o$

$-40 < \Delta S^{\ddagger} < -60$ e.u.

HEXACOORDINATE INTERMEDIATE

Cat = HMPA, F^-, RCO_2^-, Cl^- etc...

This mechanism involves the formation of a pentacoordinated intermediate in a preequilibrium, followed, during the rate determining step, by nucleophilic attack of the nucleophilic reagent (Nu) on this pentacoordinated silicon. The proposed mechanistic pathway agrees with the kinetic law and the high negative entropy of activation. The mechanism has been established in the case of chlorosilanes. However, the number of cases which could be concerned by this mechanism is very large since many functional groups at silicon (Si-H, Si-OR, Si-NR$_2$, Si-Cl) can be activated either by neutral nucleophiles (HMPA, NMI, DMF) or by anionic ones (F^-, RCO_2^-, etc).

The stereochemistry of nucleophilic attack on pentacoordinate silicon has been studied only in the case of the Si-Cl bond [12] and found to be retention instead of the inversion of configuration usually observed for nucleophilic displacements on tetracoordinate chlorosilanes. In fact the precise geometry of nucleophilic attack is not known and may depend on many factors.

The mechanistic implications of these processes are the following:
- The pentacoordinated species (A) formed between the nucleophilic catalyst and the substrate must be stable and easily formed.
- (A) must react faster with nucleophiles than the starting tetracoordinate silane since the catalytic procedure results in an acceleration of the rates.

This point is very important since the silicon atom in (A) can be negatively charged in the case of coordination with anions like F^- or RCO_2^- for instance. It might thus be expected to be more crowded and less electrophilic than in the starting tetracoordinate silane.
- The rate determining step involves nucleophilic attack on a pentacoordinated silicon atom. This step implies the formation of a hexacoordinate intermediate (or transition state) in which the bond between the nucleophile and silicon is formed during the breaking of the bond between the silicon and the leaving group.

These considerations incited us to study in more detail the stability and properties of pentacoordinated silicon species, their reactivity towards nucleophiles and also the possible transformation from penta to hexacoordination at silicon. The results obtained during this research will be discussed in this paper.

2 REACTIVITY OF PENTACOORDINATE SILICON TOWARDS NUCLEOPHILES

Reactivity of Anionic Hydridosilicates.(*Studies carried out by Dr. GUERIN, Dr. HENNER*)

Anionic hydridosilicates have been prepared by coordination of KOR to $HSi(OR)_3$ without addition of 18-crown-6 [13,14]. Interestingly LiOR leads only to the substitution of the SiH bond certainly because of the stability of LiH [15].

$$HSi(OR)_3 \ + \ RO^-K^+ \ \xrightarrow{THF} \ HSi(OR)_4^- \quad (R = Me, Et, iPr, nBu, Ph)$$

Table 1 : ^{29}Si NMR DATA (a) for $HSi(OR)_3$ and $HSi(OR)_4^-K^+$

Silane	δ ppm	$^1J_{Si-H}$, Hz	Silane	δ ppm	$^1J_{Si-H}$, Hz
$HSi(OMe)_3$	-62.6	290	$HSi(OnBu)_3$	-59.2	286
$HSi(OMe)_4^- K^+$	-82.5	223	$HSi(OnBu)_4^- K^+$	-86.1	219
$HSi(OEt)_3$	-59.6	285	$HSi(OPh)_3$	-71.3	320
$HSi(OEt)_4^- K^+$	-86.2	218	$HSi(OPh)_4^- K^+$	-112.6	296
$HSi(OiPr)_3$	-63,4	285			
$HSi(OiPr)_4^- K^+$	-90.5	215			

(a) THF as solvent with a capillary containing C_6D_6 to lock the instrument

These reagents are able to react with carbonyl groups without activation. The reaction occurs very quickly at room temperature [16]. $HSi(OR)_3$ does not react at all under the same conditions. These observations provide good support for the mechanism proposed for the reduction of carbonyl groups by hydrosilanes activated by F^- and RO^- [17,18].

$$HSi(OR)_4^- K^+ \ + \ O=C{<}^{R_1}_{R_2} \ \longrightarrow \ \xrightarrow{OH_2} \ HO-CH{<}^{R_1}_{R_2}$$

R = Et $R_1=H, R_2=Ph$ 90 % $R_1=R_2=Ph$ 73 %

$$HSi(OR)_3 \ + \ O=C{<}^{R_1}_{R_2} \ \xrightarrow{\ \times\ } \ \text{No reaction}$$

These hydridosilicates appear to be very versatile reagents:
   ---They react as reducing agents by single electron transfer:

$$HSi(OR)_4^- K^+ \ + \ Ag^+ A^- \ \longrightarrow \ Ag\downarrow \ + \ Si(OR)_4 \ + \ AK \ + \ 1/2\ H_2$$

$$HSi(OR)_4^- K^+ \ + \ \underset{(CO)_2}{\underset{Fe-I}{Cp}} \ \longrightarrow \ Si(OR)_4 \ + \ \left[\underset{(CO)_2}{\underset{Fe}{Cp}}\right]_2 \ + \ KI \ + \ 1/2\ H_2$$

with intermediate $[HSi(OR)_4]^{\cdot} K^+$ and $[Cp Fe-I (CO)_2]^{\cdot-}$, $Cp Fe^{\cdot} (CO)_2$, and $H^{\cdot} \rightarrow 1/2\ H_2$.

ESR signals corresponding to the triphenyl methyl radical [19] and to the 2,6-di-t-butyl benzoquinoyl radical anion [20] have been observed when $HSi(OEt)_4^- K^+$ is treated by Ph_3CBr and 2,6-di-t-butylbenzoquinone respectively.

---They can be very quickly hydrolysed with loss of H_2 leading to silica powders or gels. These results strongly support the formation of pentacoordinate intermediates in the course of the formation of silica by sol-gel process catalysed by bases [21].

---They also react as electrophilic centers with Grignard and organolithium reagents with formation of hydrosilanes as the major products.

$$3 RMgX + HSi(OR)_4^- K^+ \longrightarrow HSiR_3 \quad (R=Ph, PhCH_2)$$

Furthermore dihydridosilicates may be prepared by the reaction of KH/18-crown-6 with $HSi(OR)_3$ [16]:

$$HSi(OR)_3 + KH/18\text{-crown-6} \longrightarrow H_2Si(OR)_3^- K^+ (18\text{-crown-6}) \quad (R=Et, iPr, nBu)$$

Table 3- ^{29}Si NMR DATA (a) for $H_2Si(OR)_3^-K^+$

Silane	δ ppm (multiplicity)	^1JSi-H, Hz	Silane	δ ppm (multiplicity)	^1JSi-H, Hz
$H_2Si(OEt)_3^- K^+$ (b)	- 81.85 (t)	217	$H_2Si(OiPr)_3^- K^+$	- 87.1 (t)	213
$H_2Si(OnBu)_3^- K^+$ (b)	- 80.8(t)	215	(c)	- 87.3(dd)	224 193

(a) THF as solvent with a capillary containing C_6D_6 to lock the instrument
(b) not isolated (c) C_6D_6 as solvent

<u>Reactivity of Pentacoordinate Organofluorosilicates and Alkoxysilicates.</u>(*Studies carried out by Dr.GUERIN, Dr. HENNER, Dr. WONG CHI MAN*)

The pentacoordinate organofluorosilicates $Ph_3SiF_2^-$ and $MePhSiF_3^-$, as their ether complexed 18-crown-6 potassium salts, have proved to be very reactive towards strong nucleophiles (RLi, RMgX, RO⁻, metal hydrides)[22]. In fact the pentacoordinate ions react more rapidly than the corresponding neutral tetravalent compounds (lacking an F⁻ ion). Semi-quantitative comparisons of the relative reactivity are shown in the following scheme :

$$\left.\begin{array}{l}Ph_3SiF_2^{-*}\\Ph_3SiF\end{array}\right\} \xrightarrow[\text{THF, RT}]{iPrMgBr} Ph_3Si\text{-}iPr$$

Relative reactivity
(penta : tetra)
10 : 1

$$\left.\begin{array}{l}MePhSiF_3^{-*}\\MePhSiF_2\end{array}\right\} \xrightarrow[\text{Et}_2\text{O, -10°C}]{iPrMgBr} MePh(F)Si\text{-}iPr$$

150 : 1

* as K^+/18-crown-6 salts

Similar results were obtained with pentacoordinated alkoxysilicates which also react faster than the tetracoordinated parent [22].

Reactivity of Organobis(benzene-1,2-diolato)silicates *(Studies carried out by Dr CERVEAU, Dr CHUIT, Dr REYE)*

The complexes $Na^+[RSi(o-OC_6H_4O)_2]^-$ (R = Me, Ph, 1-Naphthyl) were also found [23a] to be very reactive towards nucleophilic reagents such as organometallic reagents and hydrides. An excess of hydride leads to trihydrogenosilanes.
Reactions with three moles of organolithium reagent or allyl and alkynyl magnesium bromide leads to the tetrasubstituted product. However hydrosilanes can be obtained in the reactions of alkyl Grignard reagents (2 equivalents) followed by in situ reduction. A "one pot" procedure has also been reported using alkylmagnesium bromide activated by Cp_2TiCl_2 [23b]. Primary Grignard reagents lead to monohydrosilanes while secondary and tertiary Grignard reagents lead to dihydrosilanes:

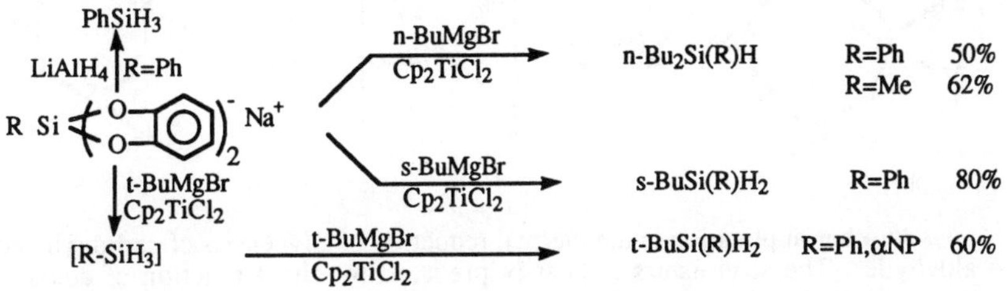

	reagent	product	R	yield
	n-BuMgBr / Cp_2TiCl_2	$n\text{-}Bu_2Si(R)H$	R=Ph R=Me	50% 62%
	s-BuMgBr / Cp_2TiCl_2	$s\text{-}BuSi(R)H_2$	R=Ph	80%
	t-BuMgBr / Cp_2TiCl_2	$t\text{-}BuSi(R)H_2$	R=Ph,αNP	60%

(with $PhSiH_3$ from $LiAlH_4$ when R=Ph; and [R-SiH_3] from t-BuMgBr / Cp_2TiCl_2)

3 UNEXPECTED REACTIVITY OF PENTACOORDINATED SILICON COMPOUNDS

Neutral Pentacoordinated Silicon Hydrides as Reducing Agents. *(Studies carried out by Dr. LANNEAU)*

An early report by Eaborn [24] suggested hydrosilatranes as possible reducing agents for benzaldehydes, in contrast to tetracoordinated hydrosilanes which do not exhibit any reactivity towards alcohols, acids or carbonyl groups.

The intramolecularly coordinated dihydrosilanes previously prepared [25] exhibit high reactivity towards alcohols, acids and carbonyl groups without any activation. Under the same conditions tetracoordinate dihydrosilanes do not react at all.

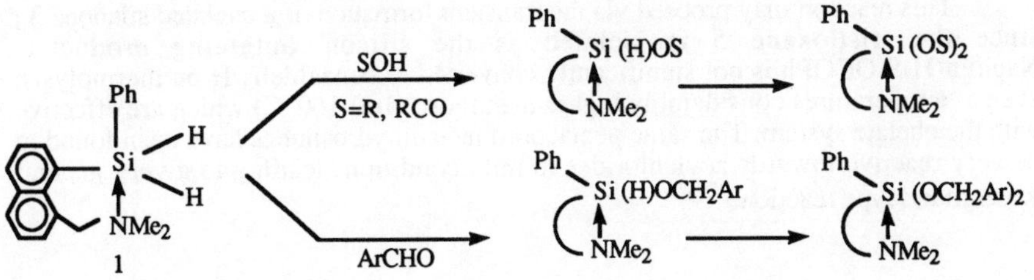

Confirmation of the high reactivity of pentacoordinate silicon compounds towards carbonyl groups was obtained in the reaction of 1 with carbon dioxide, from which the silyl ester of formic acid 2 can be isolated and characterized. The same compound was obtained from the reaction of 1 with formic acid. The thermal decomposition of this ester gave $H_2C{=}O$ and trisiloxane. When the reaction was performed in the presence of hexamethyltrisiloxane, the adduct 4 corresponding to insertion of the silanone into the Si-O bond was observed. The mechanism proposed is described in following scheme. The transient formation of silanone 3 from hypervalent silicon compound 2 is proven by formation of 4 and 5 [26].

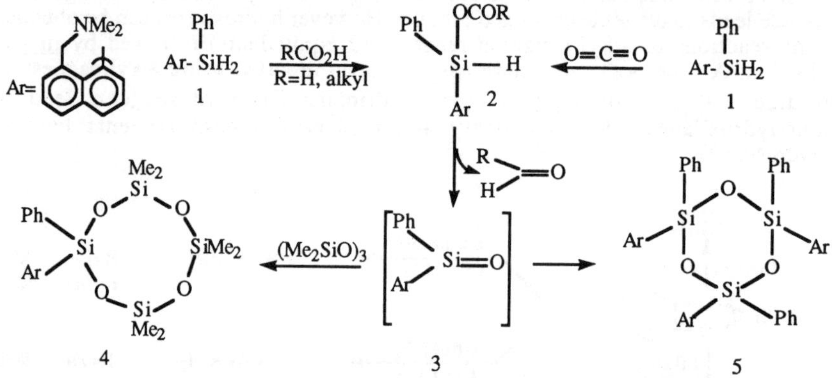
NMe2
Ar=
Ph
Ar- SiH2
1
RCO2H
R=H, alkyl
OCOR
Ph
Si— H
Ar
2
O=C=O
Ph
Ar- SiH2
1
R
H
=O
Me2
Si
O
O
Ph
Si
SiMe2
Ar
O
O
Si
Me2
4
(Me2SiO)3
Ph
Si=O
Ar
3
Ph
Ph
O
Si
Si
Ar
Ar
O
O
Si
Ar
Ph
5

Another application is the thermal reduction of silyl esters of carboxylic acids to aldehydes. The silyl esters are easily prepared by direct reaction of acids with pentacoordinated hydrosilanes. Thermal decomposition of the monocarboxylate esters provides a new route [27a] for the reduction of carboxylic acids to aldehydes:

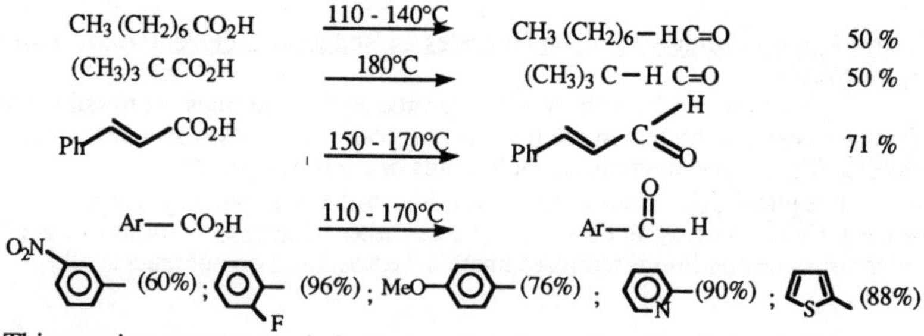
CH3 (CH2)6 CO2H
110 - 140°C
CH3 (CH2)6—HC=O
50 %
(CH3)3 C CO2H
180°C
(CH3)3 C—H C=O
50 %
Ph
CO2H
150 - 170°C
Ph
C
H
O
71 %
Ar— CO2H
110 - 170°C
Ar— C —H
O
O2N
(60%)
F
(96%)
MeO
(76%)
N
(90%)
S
(88%)

This reaction may proceed via the transient formation of a chelated silanone 3, since the trisiloxane 5 is isolated as the silicon-containing product. 1-NaphPh(H)SiOCOPh is not significantly converted to benzaldehyde on thermolysis, even at temperatures considerably higher than those (100-160°C) which are effective with the chelate system. The same pentacoordinate dihydrosilanes have been found to be very reactive towards acylchlorides in mild conditions leading to a very general Rosenmund type reaction [27b]:

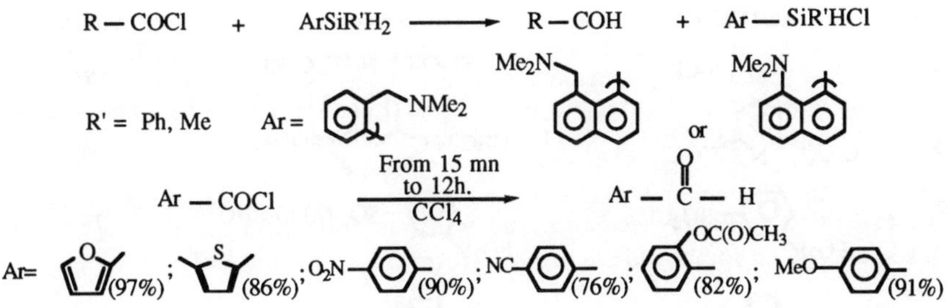

R—COCl + ArSiR'H$_2$ ⟶ R—COH + Ar—SiR'HCl

R' = Ph, Me   Ar = [2-(Me$_2$N-CH$_2$)-C$_6$H$_4$-] ; [8-(Me$_2$N)-naphthyl-]

$$\text{Ar—COCl} \xrightarrow[\text{CCl}_4]{\text{From 15 mn to 12h.}} \text{Ar—C(O)—H} \text{ or } [8\text{-Me}_2\text{N-naphthyl-C(O)H}]$$

Ar= furyl (97%); thienyl (86%); O$_2$N-C$_6$H$_4$- (90%); NC-C$_6$H$_4$- (76%); CH$_3$C(O)O-C$_6$H$_4$- (82%); MeO-C$_6$H$_4$- (91%)

$$\text{ClOC—Y—COCl} \xrightarrow[\text{CCl}_4]{\text{From 30 mn to 3h.}} \text{H—C(O)—Y—C(O)—H}$$

Y= C$_6$H$_4$: Ortho (89%), Meta (89%), Para (85%) - Y= (CH$_2$)$_8$ (85%)

Ph-CH=CHCOCl $\xrightarrow{12\text{h.}}$ Ph-CH=CHCHO (86%)

Br(CH$_2$)$_3$COCl $\xrightarrow{2\text{h.}}$ Br(CH$_2$)$_3$CHO (90%)

This method appears to be a good one for the preparation of deuterated aldehydes:

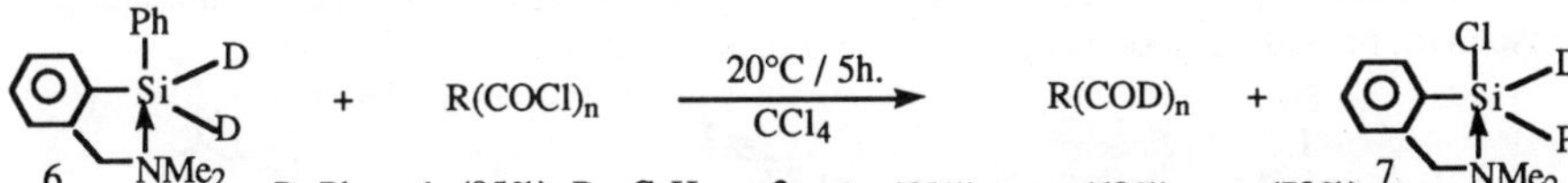

$$\underset{6}{[\text{Ph, D, D, NMe}_2\text{-aryl-Si}]} + \text{R(COCl)}_n \xrightarrow[\text{CCl}_4]{20°\text{C}/5\text{h.}} \text{R(COD)}_n + \underset{7}{[\text{Cl, D, Ph, NMe}_2\text{-aryl-Si}]}$$

R=Ph, n=1 (85%)   R= C$_6$H$_4$, n=2 ortho (65%); meta (68%); para (73%)

Reactivity towards Isocyanates and Isothiocyanates (Allenoïdes) *(Studies carried out by Dr. LANNEAU)*

The same compounds react also very efficiently with isocyanates and isothiocyanates (allenoïdes) at room temperature with reaction times from 5 minutes to one hour according to the substrate. A similar reaction was reported [28] to occur at 100°C with Pt catalyst in the case of tetracoordinate R$_3$SiH. The one pot synthesis of complex organic molecules can be performed by this way.

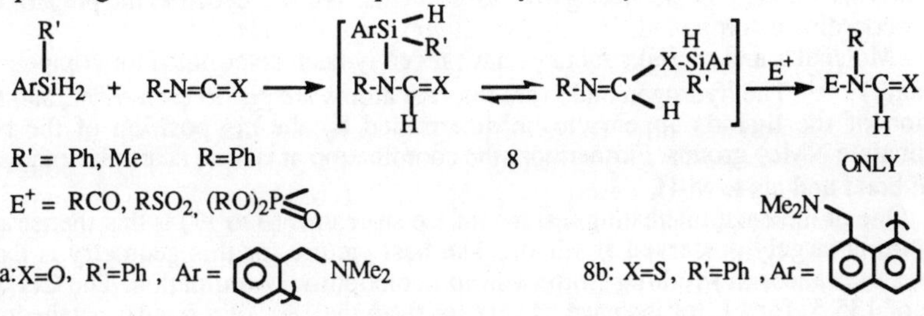

$$\text{ArSiH}_2\text{R'} + \text{R-N=C=X} \longrightarrow \left[\underset{\text{H}}{\overset{\text{ArSi-H, R'}}{\text{R-N-C=X}}} \rightleftharpoons \text{R-N=C} \overset{\text{X-SiAr, R'}}{\underset{\text{H}}{}} \right] \xrightarrow{\text{E}^+} \underset{\text{H}}{\overset{\text{R}}{\text{E-N-C=X}}}$$

R' = Ph, Me   R=Ph   8   ONLY

E$^+$ = RCO, RSO$_2$, (RO)$_2$P=O

8a: X=O, R'=Ph, Ar = [2-(Me$_2$N-CH$_2$)-C$_6$H$_4$-]   8b: X=S, R'=Ph, Ar = [8-(Me$_2$N)-naphthyl-]

Some examples of compounds prepared [29] are listed:

Silane	Acid Chloride	Product	Yield %
8 a	PhO-CS-Cl	Ph-O-CS-N(Ph)-CHO	69
8 a	ClCO-N(Ph)-COCl	OHC-N(Ph)-CO-N(Ph)-CO-N(Ph)-CHO	80
8 a	FO$_2$S-C$_6$H$_4$-SO$_2$Cl	FO$_2$S-C$_6$H$_4$-SO$_2$-N(Ph)-CHO	79
8 a	(2-thienyl)-SO$_2$Cl	(2-thienyl)-SO$_2$-N(Ph)-CHO	74
8 b	(2-furyl)-COCl	(2-furyl)-CO-N(Ph)-CHS	70
8 b	ClCO-(CH$_2$)$_8$-COCl	HCS-N(Ph)-CO-(CH$_2$)$_8$-CO(Ph)-N-CHS	72

Synthesis of stabilized silathiones. *(Studies carried out by Dr. LANNEAU and Dr. CARRE)*

The high reactivity of pentacoordinate hydrosilanes has been used for the preparation of low valent silicon species stabilized by intramolecular coordination. By reaction with molecular sulfur, CS_2 or H_2S, a silathione has been obtained and isolated, providing good support for the transient formation of silanone 3.

$$Ar-SiH_2 + S \longrightarrow Ar\underset{Ph}{\overset{S}{Si}} \qquad Ar = \text{8-(NMe}_2\text{)naphthyl}$$

The structure of the silathione determined by X-ray diffraction has been reported [30], showing a zwitterionic type structure with a very short Si-N bond(1,96Å), Si-S (2,01Å).

4 HEXACOORDINATION AT SILICON *(Studies carried out by Dr.BRELIERE, Dr. CARRE, Dr. ROYO)*

In all the hexacoordinated silicon complexes described previously, the silicon atom was surrounded by electronegative substituents. We succeeded in the preparation of hexacoordinate compounds having two silicon carbon bonds.

Molecular and crystal structures have recently been determined for compounds 9,10 and 11 [31]. The hydrogen and / (or) fluorine atoms are cis to each other, and the position of the ligands appears to be determined by the cis position of the two coordinative NMe$_2$ groups. Furthermore the coordination at silicon takes place trans to a Si-F bond and cis to Si-H.

One of the most interesting features of the structures (9 to 11) is that the tetrahedral structure is largely preserved at silicon. The best picture for this geometry is that of a bicapped tetrahedron resulting from twofold nucleophilic coordination. The C(1)SiC(2) angle of 135.5° for 11, for instance, is very far from the 180° of a regular octahedron and larger than tetrahedral angle (109°). Following the ideas of Dunitz [32], these compounds corresponding to a tetrahedral silicon atom undergoing two nucleophilic attacks, can be considered as good models for hexacoordinate intermediates (or transition states). They thus lend credence to the possible formation of these species in the course of reactions such as nucleophilic displacement at silicon with nucleophilic activation.

Dissymmetric compounds of this general structure 10 have been utilised in studies of their fluxional behaviour.

(Idealized drawings)

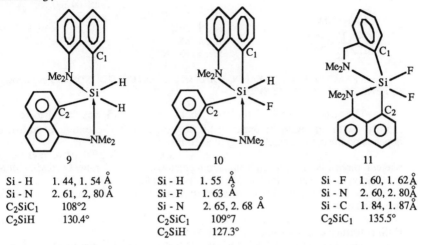

9		10		11	
Si - H	1.44, 1.54 Å	Si - H	1.55 Å	Si - F	1.60, 1.62 Å
Si - N	2.61, 2.80 Å	Si - F	1.63 Å	Si - N	2.60, 2.80 Å
C$_2$SiC$_1$	108°2	Si - N	2.65, 2.68 Å	Si - C	1.84, 1.87 Å
C$_2$SiH	130.4°	C$_2$SiC$_1$	109°7	C$_2$SiC$_1$	135.5°
		C$_2$SiH	127.3°		

Figure 1: Variable temperature ^{1}H NMR spectra for 12

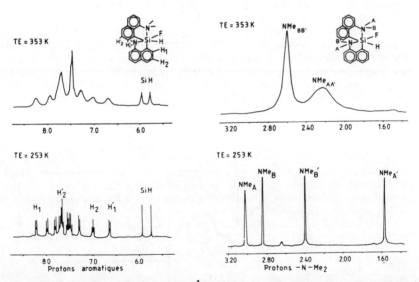

At the low temperature limit the ^{1}H NMR spectra show two diastereotopic naphthyl groups and two distinct diastereotopic pairs of methyl groups bound to nitrogen. As the temperature is raised, the two sets of naphthyl signals coalesce, and the four methyl signals are reduced to two. Each of the latter signals corresponds to two methyl groups: one on each nitrogen atom as proven by experiments double irradiation (cf. Figure 1)

Table 5- Free Energy of Activation for Isomerization of Hexacoordinate Compounds from Line-Shape Data [33]

	$\Delta G^{\ddagger}$, kcal (at 300°K)
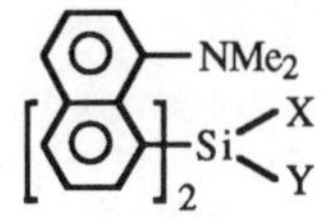 X =H , Y=F	14,7
X =H , Y=OMe	15,2
X =H , Y=Me	9,3
X =Me, Y=Ph	12,7

12 $[(\text{O}_2\text{C}_6\text{H}_4)\text{Si}(X)(Y)]_2$

5 REACTIVITY OF HEXACOORDINATE SILICON COMPOUNDS *(Studies carried out by Dr.REYE, Dr. CHUIT, Dr. CERVEAU)*

The tris(benzene-1,2-diolato) silicon complex 13 was found to react very rapidly with Grignard or organolithium reagents[34]. The extent of substitution depends on the organometallic reagent as follows:

1) When RM is an alkyl (except MeMgBr) or benzyl Grignard reagent, three silicon - carbon bonds are formed whatever the ratio 13 /RM . (MeMgBr leads only to the formation of Me4Si in good yield.)

2) When RM is an alkyllithium reagent, a mixture of tri - and tetra - organosilanes is obtained.

3) When RM is an allyl, vinyl, phenyl or alkynyl Grignard reagent, R4Si is formed directly whatever the ratio 13 /RM.

The intermediates R3Si-OC6H4OMgX obtained in the reaction of an alkyl Grignard reagent can be isolated by hydrolysis, leading to catechoxysilanes. However they can be treated directly with different reagents leading to various organosilicon compounds. Complex 13 can also be treated with a reducing agent such as LiAlH4 to give SiH4 in quantitative yield.

These reactions correspond to an alternative route for the preparation of organosilanes. The normal route implies the reduction of SiO_2 to elemental silicon which is transformed into SiX_4 or $HSiX_3$. These two compounds are the general precursors, leading to organosilanes either by hydrosilylation or by organometallic coupling reactions. In our case, the trisbenzenediolato complexes 13 can be directly prepared from silica and also from Na2SiF6 which is a very common by-product from the fertilizer industry.

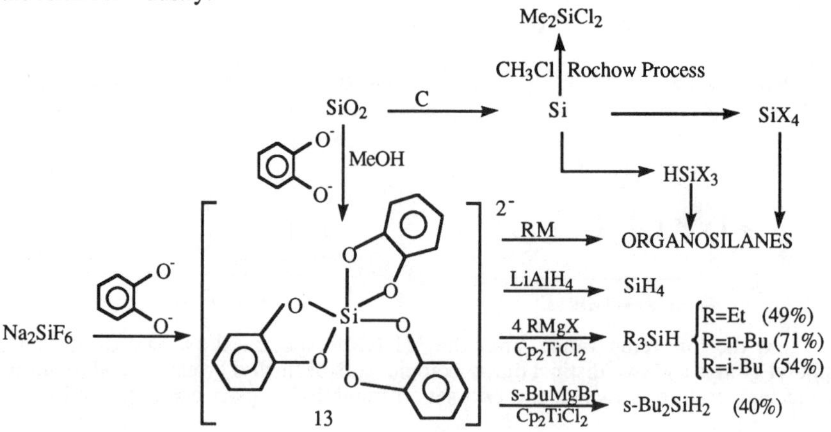

13

R₃SiOMe R=Bu (85%) — MeOLi / CoCl₂

R₃SiX R=nBu(84%) — HX, X=Br,Cl

R₃SiMPh₃ R=Et, M=Si(68%), Ge(58%) — Ph₃MLi

R₃Si–O–C₆H₄–O–MgX (13) ← RMgX

R₃SiR' R=Et, R'= Ph (60%), PhC≡C (67%) ← R'MgBr

R₃SiH R=Et (53%), Bu (68%) ← LiAlH₄

R₃Si–O–C₆H₄–OH →(MeOLi/MeOH)→ R₃SiOMe R=Et(72%), Bu(82%) [via H₃O⁺]

R₃Si–CH₂CH=CH₂ R=Et (64%) ← allylMgBr

These reactions are useful for the preparation of functional compounds [34]. The reaction of β-hydrogenated Grignard reagents activated by Cp_2TiCl_2 (Cp = cyclopentadienyl) [35] on 13 produces the trisubstituted hydrosilanes. This is an excellent way to obtain hydrosilanes from silica in two steps [23b].

CONCLUSION

In conclusion, hypervalent silicon species both penta and hexacoordinate offer great interest from both the structural and reactivity point of view. The isolation of these compounds and their very particular chemical behaviour open new possibilities in silicon chemistry and confirm the validity of the mechanistic propositions based on extension of coordination at silicon.

Acknowledgements- R. Corriu thanks very much all those colleagues who have been involved in this chemistry and whose names have been cited in the text and also all the students and Post Doctoral Fellows who have strongly contributed (Dr. BECKER, Dr. BOUDIN, J.L.BREFORT, Dr. MAZHAR, Dr.MEHTA, Dr.NAYYAR, Dr. ARYA, Dr. PERROT, Dr. PRIOU, Dr. GERBIER, Dr. de SAXCE, Dr. WONG CHI MAN, Q. J. WANG, Dr. ZWECKER). He acknowledges gratefully Dr. C. YOUNG'S assistance with the English version and thanks CNRS and Rhône Poulenc for supporting this research.

REFERENCES

1. L.H.SOMMER "Stereochemistry, Mechanism and Silicon" Mc Graw-Hill, New-York 1965
2. a) R.J.P.CORRIU,C.GUERIN, J. Organomet. Chem. 1980, 198 231 b) R.J.P.CORRIU, C. GUERIN Adv. Organomet. Chem. 1982, 20 265
3. a) R. J. P. CORRIU, C. GUERIN, J. J. E. MOREAU, Topics in stereochemistry 1984, 15, 43; b) R. J. P. CORRIU, C. GUERIN, J. J. E. MOREAU "The Chemistry of Organic Silicon Compounds" S. Pataï and Z. Rappoport Eds. John Wiley & Sons Pub.1989, p:305-370
4. a) ref 1 page 84-7 b) F. O. STARK Ph. D. Pensylvania State University 1962
5. R. CORRIU, M. LEARD, J. MASSE Bull. Soc. Chim. Fr. 1968, 2555
6. R. CORRIU, M. LEARD, J. Chem. Soc. Chem. Commun. 1971, 1087
7. R. CORRIU, M. HENNER-LEARD, J. Organomet. Chem. 1974, 64 351
8. R. CORRIU, M. HENNER-LEARD, J. Organomet. Chem. 1974, 65 C39-C41
9. F. CARRE, R. CORRIU, M. HENNER-LEARD, J. Organomet. Chem. 1970 24 101
10. R. CORRIU, M.LEARD, J. MASSE Bull. Soc. Chim. Fr. 1974, 1447

11. R. J. P. CORRIU, G. DABOSI, M. MARTINEAU, J. Organomet. Chem. 1978, 150 27; 1980, 186 25
12. R. J. P. CORRIU, G. DABOSI, M. MARTINEAU J. Chem. Soc. Chem. Commun. 1977, 649; J. Organomet. Chem. 1978, 154 33
13. R. DAMRAUER, S.E. DANAHEY Organometallics 1986, 5, 1490
14. R. CORRIU, C. GUERIN, B. HENNER, Q. WANG J. Organomet. Chem.1989, 365 C7
    B. BECKER, R. CORRIU, C. GUERIN, B. HENNER, Q. WANG J. Organomet. Chem. 1989, 359 C33
15. R. CORRIU, C. GUERIN, B. HENNER, Q. WANG unpublished results
16. B. BECKER, R. CORRIU, C. GUERIN, B. HENNER, Q. WANG J. Organomet. Chem. 1989, 368 C 25
17. R.J.P. CORRIU, R. PERZ, C. REYE Tetrahedron 1983, 39, 999 ; J. BOYER, R.CORRIU, R. PERZ, C. REYE Tetrahedron 1981, 37, 2165 ; J.Organomet. Chem.1978 148 C1 1979, 172 143; J. BOYER, R.J.P. CORRIU, R. PERZ, M. POIRIER, C. REYE Synthesis 1981, 558; R.J.P. CORRIU, R. PERZ, C. REYE, ibid. 1982, 981
18. A. HOSOMI, H. HARYASHIDA,S. KOHRA, Y. TOMINAGA,J.Chem. Soc. Chem. Commun. 1986, 1411 ,Tetrahedron Lett. 1988, 29 89
19. E. C.ASHBY, A. B. GOEL, R.N. DE PRIEST Tetrahedron Lett. 1981, 22 (38), 3729
20. D.YANG, D.D. TANNER, J. Org. Chem. 1986, 51 2267
21. R. CORRIU, D. LECLERCQ, A. VIOUX, M. PAUTHE, J. PHALIPPOU, in Ultrastructure Processing of Advanced Ceramics ; J.D. Mackenzie, D.R. Ulrich Eds; J. Wiley and Sons, New-York 1988, p. 113; V. BELOT, R. CORRIU, C. GUERIN, B. HENNER, D. LECLERCQ, H. MUTIN, A. VIOUX, Q. WANG, in Better Ceramics Through Chemistry IV, C. J. Brinker, D. E. Clark, D.R. Ulrich, Eds; Materials Research Society Pittsburgh 1990
22. R. CORRIU, C. GUERIN, B. HENNER, W.W.C. WONG CHI MAN Organometallics 1988, 7, 237 ; J.L. BREFORT, R. CORRIU, C. GUERIN, B. HENNER,W.W.C. WONG CHI MAN Organometallics, accepted for publication
23. A. BOUDIN, G. CERVEAU, C. CHUIT, R. CORRIU, C.REYE, a) Angew. Chem. Int. Ed. Engl.1986, 25, 473; b) J. Organomet. Chem. 1989, 362, 265
24. M. T. ATTAR-BASHI, C. EABORN, J. VENCL, D.R.M. WALTON J.Organomet. Chem. 1976, 117 C87
25. J.BOYER, C. BRELIERE, R. J. P. CORRIU, A.KPOTON, M. POIRIER, G.ROYO, J.Organomet. Chem. 1986, 311 C39
26. P. ARYA, J. BOYER, R. J. P. CORRIU, G. F. LANNEAU, M. PERROT J.Organomet. Chem. 1988, 346 C11
27. R.J.P. CORRIU, G. F. LANNEAU, M. PERROT Tetrahedron Lett. a) 1987, 28 ,3941 b) 1988, 29 , 1271
28. I.OJIMA, S. INABA J.Organomet. Chem. 1977, 140, 97
29. R. CORRIU, G. LANNEAU, M. PERROT,V.D. MEHTA, Tetrahedron Lett. (1990) in press
30. P. ARYA, J. BOYER, F. CARRE, R. CORRIU, G. LANNEAU, J. LAPASSET, M. PERROT, C. PRIOU Angew. Chem. Int. Ed. Engl. 1989, 28 1016
31. C.BRELIERE,F.CARRE,R.J.P. CORRIU, M.POIRIER, G. ROYO, J.ZWECKER, Organometallics 1989, 8, 1831
32. D. BRITTON, J. D. DUNITZ J. Amer. Chem. Soc. 1981,103 , 2971
33. C.BRELIERE, R.J.P. CORRIU, G. ROYO, J.ZWECKER, Organometallics 1989, 8, 1834
34. A.BOUDIN, G. CERVEAU, C. CHUIT, R. CORRIU, C.REYE, Organometallics, 1988, 7,1165
35. E.COLOMER, R.J.P.CORRIU, J. Organomet. Chem. 1974, 82 362
    R.J.P.CORRIU, B. MEUNIER J. Organomet. Chem. 1974, 65 187

Hydrogen Peroxide Oxidation of the Silicon-Carbon Bond: Mechanistic Studies

Kohei Tamao, Takashi Hayashi, and Yoshihiko Ito

DEPARTMENT OF SYNTHETIC CHEMISTRY, FACULTY OF ENGINEERING, KYOTO UNIVERSITY, KYOTO 606, JAPAN

1 INTRODUCTION

In 1983, we found that silicon-carbon bonds in functional silicon compounds such as fluoro-, chloro-, alkoxy-, and amino-silanes are readily cleaved by 30% hydrogen peroxide to form alcohols (Scheme 1).[1,2] The most characteristic feature is that the presence of at least one heteroatom (or functional group) on silicon is essential for the oxidation. Similar oxidative cleavages by other oxidants such as peracids[3] and an amine oxide[4] have also been developed.

$$R\text{-}SiX_3 \xrightarrow{H_2O_2} R\text{-}OH$$

SiX_3 = $SiMe_2H$, $SiMe_2F$, $SiMeF_2$, SiF_3, $SiMe_2Cl$, $SiMeCl_2$, $SiCl_3$, $SiMe_2(OR')$, $SiMe(OR')_2$, $Si(OR')_3$, $SiMe_2(NR'_2)$,

30% H_2O_2 / KF /($KHCO_3$) / MeOH / THF / room temp.

Scheme 1

Apart from the mechanistic details, the synthetic utilities of the hydrogen peroxide oxidation have been demonstrated by us[2,5,6] and others[3,7] in regio and stereoselective transformations of certain organosilicon compounds into oxygen-functional molecules.

We have been engaged in the mechanistic studies on the hydrogen peroxide oxidation and have examined several important factors, such as the role of fluoride ions, electronic effects, steric effects <u>etc</u>. Based on the studies, we now propose a plausible mechanism which involves pentacoordinate silicon species as the initial key intermediates and hexacoordinate silicon species in the transition state, as shown in Scheme 2.[8] The mechanism will be discussed herein.

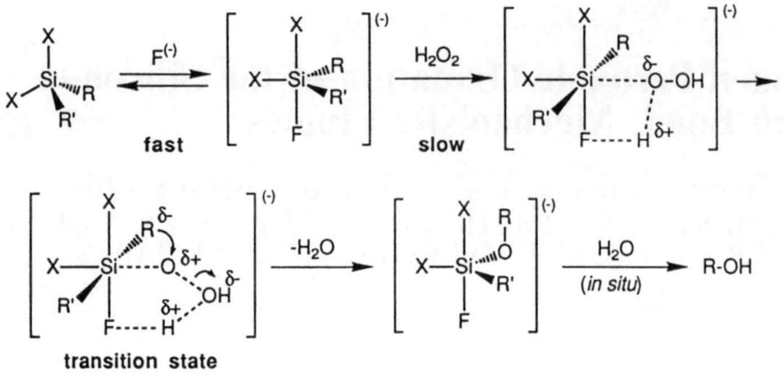

Scheme 2

2 HYDROGEN PEROXIDE OXIDATION OF TETRACOORDINATE SILANES

2.1 Reactivity Order of Organic Groups on Silicon

Relative reactivities of organic groups on silicon in diorganodialkoxysilanes have been estimated by intramolecular competition methods to show that aryl, alkenyl and long-chain alkyl groups are more readily cleaved by hydrogen peroxide than the methyl group. In particular, aryl groups are more than 10 times as reactive as the least reactive methyl group, as shown in Scheme 3. The results imply that the oxidative cleavage of the methyl group may be neglected in the initial stages of the oxidation of arylmethylsilanes, ArMeSiX$_2$, being pertinent to the competitive reactions mentioned later.

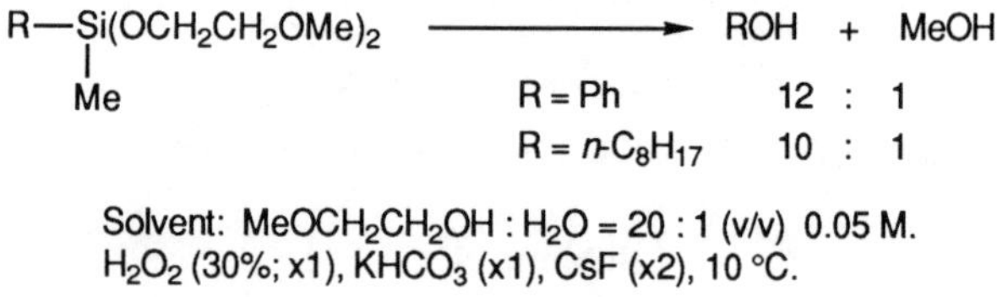

$$R-Si(OCH_2CH_2OMe)_2 \longrightarrow ROH + MeOH$$
$$\underset{Me}{|}$$

R = Ph   12 : 1
R = n-C$_8$H$_{17}$ 10 : 1

Solvent: MeOCH$_2$CH$_2$OH : H$_2$O = 20 : 1 (v/v) 0.05 M.
H$_2$O$_2$ (30%; x1), KHCO$_3$ (x1), CsF (x2), 10 °C.

Scheme 3

2.2 Number of Functional Groups on Silicon

The reactivity of organosilanes are highly dependent upon the number of functional groups on silicon. The reactivity of a series of phenylfluorosilanes, PhSiMe$_{3-n}$F$_n$, have been found in the order $n = 2 \gg n = 1 > n = 3$, as shown in Figure 1. The high reactivity of the difunctional compound may be ascribed in part to the high ability for the formation of pentacoordinate silicate species with a fluoride ion (vide infra).

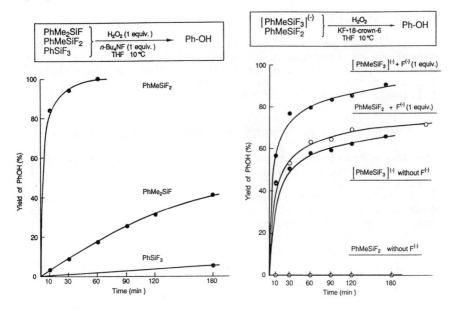

Figure 1. Reactivity of phenylfluorosilanes toward the H_2O_2 oxidation

Figure 2. Reactivity of Ph_2SiF_2 and $Ph_2SiF_3^-$ K^+·18-crown-6 in the presence or absence of F^- toward the H_2O_2 oxidation

2.3 Activation by a Fluoride Ion and Intermediary of Pentacoordinate Fluorosilicates

The presence of an extra fluoride ion is essential for the oxidation, as shown in Figure 2. Thus, while in the absence of a fluoride ion PhMeSiF$_2$ is not oxidized at all, in the presence of one equivalent of fluoride ion the oxidation proceeds smoothly to form phenol. The latter reaction curve is essentially the same as that of the pentacoordinate silicate [PhMeSiF$_3$]$^-$ in the absence of extra fluoride ion.

The quantitative formation of the pentacoordinate species [PhMeSiF$_3$]$^-$ from PhMeSiF$_2$ and n-Bu$_4$NF *in solution* has been confirmed by following the change of the methyl proton chemical shifts by ^{1}H NMR. From the titration curve the formation constant is estimated to be $K = 1.8 \times 10^3$ M^{-1}. No formation of hexacoordinate silicon species is observed in the presence of excess amounts of fluoride source.

The results suggest that the oxidation proceeds through pentacoordinate fluorosilicates as key intermediates.

2.4 Stereochemistry of Oxidation

Hydrogen peroxide oxidation proceeds with complete retention of configuration at an sp^3 carbon center.[1] The result implies that the oxidation involves a front-side attack of the oxidant, being consistent with the proposed intramolecular migration mechanism.

3 OXIDATION OF PENTACOORDINATE DIORGANOSILICATES

The above mentioned results show the important role of pentacoordinate silicates as intermediates in the hydrogen peroxide oxidation. We have then examined the reactivity of isolable pentacoordinate organosilicates. A series of diaryltrifluorosilicates have been prepared according to Damrauer's elegant method[9] (Scheme 4).

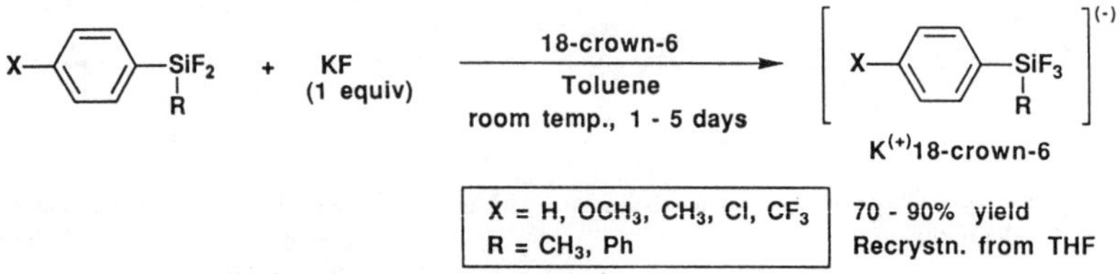

$$X\text{-}C_6H_4\text{-}SiF_2R + KF \text{ (1 equiv)} \xrightarrow[\text{Toluene, room temp., 1 - 5 days}]{\text{18-crown-6}} [X\text{-}C_6H_4\text{-}SiF_3R]^{(-)} K^{(+)}\cdot\text{18-crown-6}$$

X = H, OCH$_3$, CH$_3$, Cl, CF$_3$
R = CH$_3$, Ph

70 - 90% yield
Recrystn. from THF

Scheme 4

3.1 Kinetics

The hydrogen peroxide oxidation of $[(p\text{-}CF_3C_6H_4)PhSiF_3]^- K^+\cdot\text{18-crown-6}$ has been followed by ^{19}F NMR. Determination of the initial rates of disappearance of the silicate indicates that the oxidation rate is roughly first order with respect to the concentration of silicate and hydrogen peroxide, $-d[R_2SiF_3^-]/dt = k_2 [R_2SiF_3^-][H_2O_2]$, and $k_2 = 0.15$ M^{-1} sec^{-1} at -26 °C. The results indicate that the rate-determining step involves the interaction of hydrogen peroxide with pentacoordinate silicate species.

3.2 Electronic Effects

Electronic effects of the hydrogen peroxide oxidation have been determined by intermolecular competition reactions of arylmethyl-silanes and -silicates and by intramolecular competitions of arylphenyl-silanes and -silicates. Hammett plots between the product ratios, $XC_6H_4OH / PhOH$, and σ_p show linear relationships; a representative result is shown in Figure 3 and the whole results are summarized in Scheme 5. While in intermolecular competition reactions $\rho = +1$, intramolecular competitions afford $\rho = +0.5$ for both tetracoordinate silanes and pentacoordinate silicates.

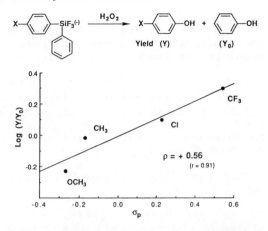

Figure 3. Hammett plots of relative reactivities between the phenyl and substituted phenyl groups in pentacoordinate silicates by intramolecular competitions. Conditions: H_2O_2 (30%, x0.5), KF (x1), 18-crown-6 (x1), $MeOCH_2CH_2OH$, 10 °C.

(4-coordinate: KF (2 equiv.), 18-crown-6 (2 equiv.), H_2O_2 (0.5 equiv.), $MeOCH_2CH_2OH$, 10 °C.
5-coordinate: KF (1 equiv.), 18-crown-6 (1 equiv.), H_2O_2 (0.5 equiv.), $MeOCH_2CH_2OH$, 10 °C.)

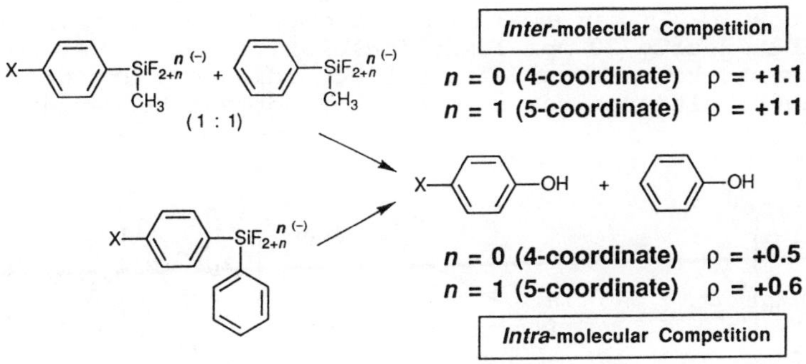

Scheme 5

Thus, electron-withdrawing groups accelerate the oxidative cleavage reaction. Two points should further be noted. First, the fact that essentially the same ρ values are obtained with tetracoordinate silanes in the presence of a fluoride ion and with pentacoordinate silicates supports the intermediary of pentacoordinate silicates. Second, the ρ value in the intermolecular competition is nearly twice as large as that in the intramolecular competition. The electronic effect in the intramolecular competition reflects that of the migration step only. The intermolecular competition, however, should involve additionally the electronic effect in the coordination step of hydrogen peroxide to the pentacoordinate silicate. The results imply that the latter is nearly equal to the former.

In the transition state, the organic group should bear a negative charge, but the rather small ρ values imply the concerted nature of the reaction. Coordination of hydrogen peroxide to pentacoordinate silicates may possibly be promoted by the hydrogen bond with an apical fluorine ligand, as shown in Scheme 2.

3.3 Electronic Effects and a Linear Relationship with ^{13}C NMR Chemical Shifts

^{13}C NMR chemical shifts of Si-*ipso* carbons, C-1 of the para-substituted phenyl ring and C-1' of the parent phenyl ring, in pentacoordinate diarylsilicates, $(XC_6H_4)PhSiF_3^-$, have been observed for the first time.[10] There are linear relationships between the Si-*ipso* ^{13}C chemical shifts and σ$^+$ values of the para substituents on the benzene ring (Figure 4), revealing that the ^{13}C chemical shifts are in turn linearly correlated with the charge densities of the Si-*ipso* carbons.[11] It may also be noted that these Si-*ipso* carbons exhibit nearly 20 ppm lower field shifts in comparison with those of the corresponding tetracoordinate diaryldifluorosilanes, although the origins have not yet been clarified.

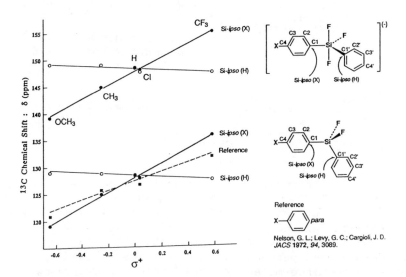

Figure 4. Hammett plots of ^{13}C NMR chemical shifts of Si-ipso carbons in diaryldifluorosilanes and diaryltrifluorosilicates

The relative reactivities of para-substituted phenyl groups in the hydrogen peroxide oxidation, mentioned above, are now found to be linearly correlated with the Si-*ipso* carbon chemical shift differences, as shown in Figure 5. Thus, the reactivity of the organic groups on silicon toward hydrogen peroxide is governed by the electron density of the reaction center, suggesting the ground-state-like transition state.

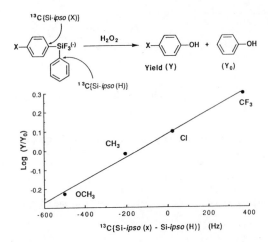

Figure 5. A linear relationship between the relative reactivity of aryl groups toward the H_2O_2 oxidation and ^{13}C NMR chemical shifts of Si-ipso carbons in diaryltrifluorosilicates

3.4 Steric Effects and Trajectory Analysis

Studies on the steric effect have provided a useful clue for the trajectory analysis in approach of hydrogen peroxide to pentacoordinate silicates.

A priori, there are three possibilities for attack by hydrogen peroxide on trigonal bipyramidal silicate RR'SiF$_3^-$ in the equatorial plane, regions A, B, and C, as shown in Scheme 6, in which the charge is omitted for clarity. Three different geometrical isomers of hexacoordinate silicon species may be formed thereby. In the hexacoordinate species formed in route A, two organic groups R and R' are <u>cis</u> to the hydroperoxy ligand and hence may migrate with equal chance to the adjacent oxygen atom to form two kinds of alcohols, ROH and R'OH, eventually. In contrast, in route B only the R' group and in route C only the R group is <u>cis</u> to the hydroperoxy ligand to form, respectively, R'OH and ROH only. We assume here that nucleophilic attack from <u>trans</u> to electronegative group (fluorine), region A in these cases, would be energetically most favorable.

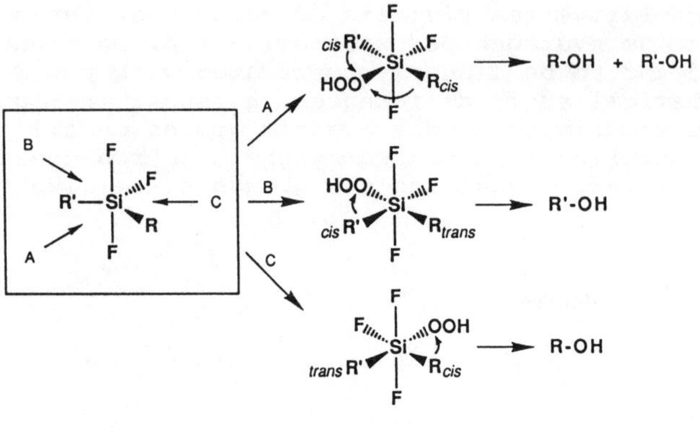

Scheme 6

Several types of diaryltrifluorosilicates containing methyl-substituted phenyl groups on silicon have been prepared in a similar manner to that described above and subjected to the hydrogen peroxide oxidation. Some representative results of the relative reactivities in the intramolecular competition reactions are summarized in Table 1.

Table 1. Relative reactivity of two aryl groups in ArAr'SiF$_3^-$K$^+$·18-crown-6 toward the H$_2$O$_2$ oxidation

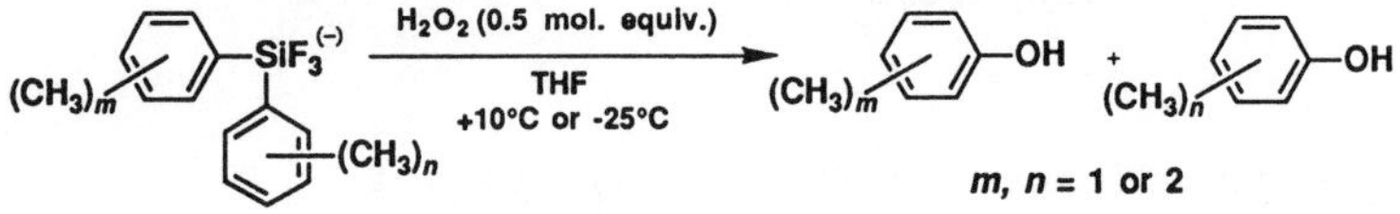

Substrate	Products (ratio)	Temp
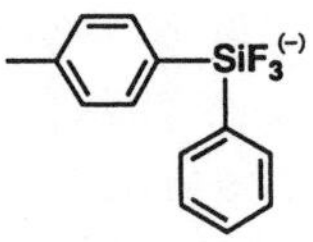 **1**	1 : 1.0	(+10 °C)
	1 : 1.1	(−25 °C)
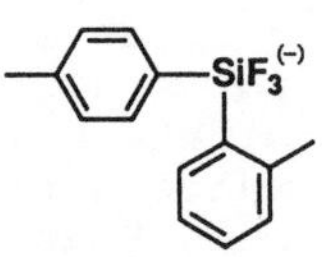 **2**	 1 : 1.5	(+10 °C)
	1 : 1.5	(−25 °C)
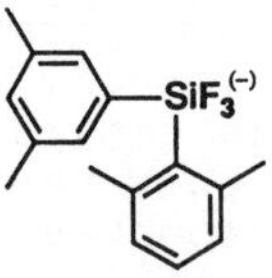 **3**	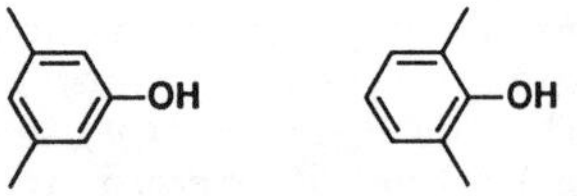 1 : 0.9	(+10 °C)
	1 : 0.4	(−25 °C)

There seem to be three tendencies. (1) Para- and meta-methyl groups exhibit little effect on the reactivity, as exemplified by compound **1**. The reactivity difference may be ascribed to the electronic effect mentioned above. (2) The o-tolyl group is more reactive than the parent phenyl or p-tolyl group, as shown by compound **2**, suggesting a steric acceleration effect. (3) The 2,6-xylyl group that contains methyl groups at both ortho positions is less reactive than the 3,5-xylyl group in compound **3**, indicating a steric hindrance effect. The reactivity difference becomes larger at lower temperatures in the last case.

The results may be explained by the following trajectory analysis based on the structural features of compounds **1 - 3**, as determined by $\underline{X}$-ray structural analyses. Thus, while in **1** the two aromatic rings are nearly perpendicular to the equatorial plane, in **2** and **3** the o-tolyl group and 2,6-xylyl group, respectively, are nearly in the equatorial plane, as shown in Figure 6.[12] In the latter cases, the ortho methyl group(s) may block the equatorial region(s). The trajectory analyses are visualized in Scheme 7. Thus, in **1** all the three regions are open for approach of hydrogen peroxide, the reactivity should be determined by the electronic effect only. In contrast, in **3** the region C may be attacked preferentially, since the other two regions A and B are blocked by the methyl groups of 2,6-xylyl group, causing steric hindrance effect. The higher selectivity observed at lower temperatures may result from the slowing down of rotation of the aryl group about the carbon-silicon bond and of pseudorotation.

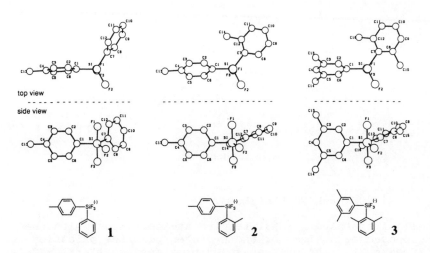

Figure 6. Top views (upper) and side views (lower) of $\underline{X}$-ray structures of representative ArAr'SiF$_3^-$K$^+$·18-crown-6. The 18-crown-6 parts are omitted for clarity.

In **2**, rotation about the o-tolyl-silicon bond would make the regions A and B half-open. In conjunction with this, the assumption (<u>vide supra</u>) of preferred attack on region A, <u>trans</u> to the equatorial fluorine ligand, may account for the observed steric acceleration.

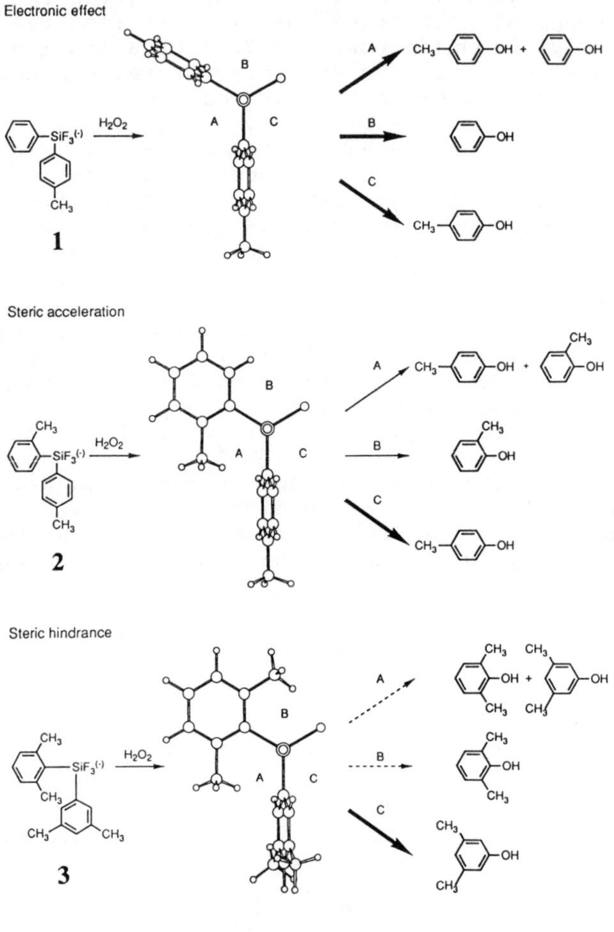

Scheme 7

4 CONCLUDING REMARKS

All of the experimental results are consistent with the proposed mechanism for the hydrogen peroxide oxidation (Scheme 2). The present study provides the first example for discussion of a structure-reactivity correlation of pentacoordinate silicates. We are grateful to Dr. Motoo Shiro of Shionogi Co., Ltd. for X-ray structure determinations of pentacoordinate silicates.

REFERENCES

1. K. Tamao, N. Ishida, T. Tanaka, and M. Kumada, *Organometallics*, 1983, *2*, 1694.
2. K. Tamao, 'Organosilicon and Bioorganosilicon Chemistry', H. Sakurai, Ed., Ellis Horwood, Chichester, 1985, p. 231.
3. I. Fleming, N. J. Lawrence, *Tetrahedron Lett.*, 1990, *31*, 3645, and references cited therein.
4. K. Sato, M. Kira, H. Sakurai, *Tetrahedron Lett.*, 1989, *30*, 4375.
5. K. Tamao, *J. Synth. Org. Chem, Jpn.*, 1988, *46*, 861.
6. K. Tamao, Y. Nakagawa, Y. Ito, *J. Org. Chem.*, 1990, *55*, 3438, and references cited therein.
7. Recent representative papers: (a) G. Stork, M. J. Sofia, *J. Am. Chem. Soc.*, 1986, *108*, 6826. (b) T. Hayashi, Y. Matsumoto, Y. Ito, *J. Am. Chem. Soc.*, 1988, *110*, 5579. (c) M. K. O'Brien, A. J. Pearson, A. A. Pinkerton, W. Schmidt, K. Willman, *J. Am. Chem. Soc.*, 1989, *111*, 1499. (d) T. H. Chan, P. Pellon, *J. Am. Chem. Soc.*, 1989, *111*, 8737. (e) M. Koreeda, I. A. George, *Chem. Lett.*, 1990, 83. (f) G. Agnel, M. Malacria, *Tetrahedron Lett.*, 1990, 25, 3555.
8. K. Tamao, T. Hayashi, Y. Ito, manuscript under preparation.
9. R. Damrauer, S. E. Danahey, *Organometallics*, 1986, *5*, 1490.
10. K. Tamao, T. Hayashi, Y. Ito, M. Shiro, manuscript under preparation.
11. G. L. Nelson, G. C. Levy, J. D. Cargioli, *J. Am. Chem. Soc.*, 1972, *94*, 3089.
12. Similar conformational dependence owing to steric hindrance has already been pointed out by Holmes and his coworkers. S. E. Johnson, J. S. Payne, R. O. Day, J. M. Holmes, R. R. Holmes, *Inorg. Chem.*, 1989, *28*, 3190 and references cited therein.

Nucleophilic Substitution Reactions of Functional Siloxanes

K. Rühlmann, U. Scheim, K. Käppler, and R. Gewald

TECHNISCHE UNIVERSITÄT DRESDEN, SEKTION CHEMIE, 8027 DRESDEN, MOMMSENSTR. 13, GERMANY

1 INTRODUCTION

Silicon compounds and silicones in particular are used in nearly all branches of industry and economy, in therapy and in the household. In many fields of application silicones are of unique importance. The production of silicones is increasing progressively from year to year.

But in spite of the growing importance of the silicones, details of the mechanisms of the reactions, which are involved in their technical production by hydrolysis of chlorosilanes and condensation of the silanols formed, are widely unknown. In addition, at the beginning of our studies there was a complete lack of inductive and steric substituent constants for siloxy groups in the literature.

Thus the first topic of our work in this field was the evaluation of these constants, which were to give us a basic tool for further kinetic and mechanistic investigations.

2 INDUCTIVE SUBSTITUENT CONSTANTS FOR SILOXY GROUPS

In 1977 Engelhardt stated a linear correlation between the ^{29}Si-NMR shifts of compounds of the type $(Me_3SiO)_3Si-X$ and the σ^*-values of X[1]. Especially the correlation with the shift values of the monofunctional Si-atoms was really good. From the data found in this publication the following equation could be derived: $\sigma^* = 0.571 \cdot \delta - 4.28$ and from the known value for the monofunctional silicon atoms of tetrakis(trimethylsiloxy)silane, the σ^*-value of the trimethylsiloxy group could be calculated to 0.35.

Of interest is the σ^*-value of the phenyl group, which was found by Engelhardt to be 0.5. That is significantly smaller than the value of 0.6 known in carbon chemistry. In this connection we recalculated the correlation between silicon-hydrogen-coupling constants and σ^*-values found by Nagai for a large number of silanes with alkyl-, aryl-, and halogenated alkyl-groups at silicon.[2] In this investigation Nagai used the value of 0.6 for a phenyl group at carbon. But if correlation is taken without phenyl groups, it becomes much better. And now from the coupling constant found by Nagai for a phenylsilane a value of 0.48 for a phenyl group at silicon was calculated. This lower σ^*-value is well intelligible, because there is a significant back donation of charge from the phenyl group to silicon.

Using the equation mentioned σ^*-values for siloxy groups could be determined from the δ-values of a series of siloxanes, prepared by us (Table 1).[3]

But at present the σ^*-values of Taft were considerably criticized, especially by Charton.[4] Thus the shift values found for alkyl and halogenated alkyl groups were correlated also with the σ_I-values of Charton. By the derived equation $\sigma_I = 0.0625 \cdot \delta - 0.44$ σ_I-values for siloxy groups could be determined starting from the δ-values (Table 1).

Table 1 ^{29}Si-NMR Shift Values δ for the monofunctional Si-atoms in $(Me_3SiO)_3Si-X$ and σ^*- and σ_I-Values for Siloxy Groups X

X	δ(ppm)	σ^*	σ_I
Me_3SiO	8.11	0.35	0.07
Cl	12.28	2.73	0.33
$ClMe_2SiO$	9.63	1.22	0.16
Cl_2MeSiO	10.50	1.71	0.22
Cl_3SiO	11.40	2.23	0.27
$ClMe_2SiOSiMe_2O$	9.08	0.90	0.13
$ClMe_2SiO(SiMe_2O)_2$	8.90	0.80	0.12
$ClMe_2SiO(SiMe_2O)_3$	8.86	0.78	0.11
HO	8.91	0.81	0.12
$HOMe_2SiO$	9.11	0.92	0.13
$(HO)_2MeSiO$	9.70	1.26	0.17
$Me_3SiOSiMe_2O$	8.51	0.57	0.09
$(Me_3SiO)_2SiMeO$	8.80	0.74	0.11
$(Me_3SiO)_3SiO$	9.08	0.90	0.13
$(Me_3SiO)_2SiMeOSiMe_2O$	8.75	0.71	0.11
$(Me_3SiO)_3SiOSiMe_2O$	8.80	0.74	0.11
$Cl_2MeSiOSiClMeO$	10.34	1.62	0.21
$Cl_2MeSiO(SiClMeO)_2$	10.27	1.58	0.20
$(Cl_2MeSiO)_2SiMeO$	10.04	1.45	0.19
Br	12.47	2.84	0.34

3 STERIC SUBSTITUENT CONSTANTS FOR SILOXY GROUPS

For the determination of steric substituent constants a suitable model reaction had to be found. As investigations on the hydrolysis reactions of chlorosiloxanes were intended, this model reaction was to be mechanistically similar to the hydrolysis reaction and easy to evaluate. We chose the acetolysis of chlorosilanes with acetic acid in acetic anhydride. First experiments showed that this reaction can be described by the equations:

$$\equiv\!Si-Cl + AcOH \longrightarrow \equiv\!Si-OAc + HCl \qquad (1)$$

$$Ac_2O + HCl \rightleftharpoons AcCl + AcOH \qquad (2)$$

By reacting chlorosilanes with different concentrations of acetic acid it was found that there is no reaction with acetic anhydride. And reactions of acetic anhydride with hydrogen chloride in acetic acid showed that the equilibrium of the second reaction is established very fast and shifted far to the right side. As an excessive amount of acetic anhydride is present and the concentration of acetic acid remains constant because of the fast establishment of the equilibrium, the acetolysis reactions should obey a pseudo first order rate law. Kinetic experiments confirmed this assumption.

The next step had to be the evaluation of ϱ, the inductive reaction constant of the acetolysis reaction. To exclude steric effects, the rate constants for a series of arylchlorodimethylsilanes were determined. Within the limit of error all the rate constants measured were equal. That means $\varrho = 0$. The constant of steric susceptibility, δ, could be obtained by plotting the lg k-values of some alkylchlorodimethylsilanes against the steric substituent

Table 2 E_s-Values for Siloxy Groups R derived from k-Values of the reaction
$RSiMe_2Cl + AcOH \longrightarrow RSiMe_2OAc + HCl$

R	$k \cdot 10^4 (l \cdot mol^{-1} \cdot s^{-1})$	E_s
Me	40	0.00
Cl	11	-0.43
Me_3SiO	11	-0.43
$Me_3SiOSiMe_2O$	10	-0.46
$Me_3SiO(SiMe_2O)_2$	7.7	-0.53
$(Me_3SiO)_2SiMeO$	4.0	-0.77
$(Me_3SiO)_2SiMeOSiMe_2O$	4.5	-0.73
$(Me_3SiO)_3SiO$	1.8	-1.04
$(Me_3SiO)_3SiOSiMe_2O$	2.6	-0.91
$ClMe_2SiO$	26	-0.14
$Cl(Me_2SiO)_2$	22	-0.18
$Cl(Me_2SiO)_3$	14	-0.33
$Cl(Me_2SiO)_4$	10	-0.46
$ClMe_2SiO\!\!\diagdown\!\!SiMeO$ $Me_3SiO\!\!\diagup$	6.4	-0.59
AcO	22	-0.18
$AcOSiMe_2O$	19	-0.25

constants of Taft.[5] These values led – in this case – to a better correlation than the $E_s(Si)$-values of Cartledge, which were published in 1983.[6] The constant of steric susceptibility δ was found to be 1.3.

Now kinetic data enabled us to calculate new steric substituent constants for some alkyl groups and furthermore for a series of siloxy groups (Table 2).[7]

4 CLEAVAGE OF SILOXANES BY HCL

The substituent constants for siloxy groups should be proved with reactions which are important for the production of silicones, e.g. the cleavage of siloxanes by HCl in dioxane. Kinetic data obtained could not be evaluated with a simple rate law. A significant induction period suggested autocatalytic effects caused by the water formed during the reaction. Further the establishment of equilibria could be observed. Thus the following mechanism was assumed:

$$\geqslant Si-O-Si \leqslant\ +\ dioxane \cdot H^+ \rightleftharpoons\ \geqslant Si-\overset{H}{\underset{+}{O}}-Si \leqslant\ +\ dioxane \quad |K(1) \quad (1)$$

$$\geqslant Si-O-Si \leqslant\ +\ H_3O^+ \rightleftharpoons\ \geqslant Si-\overset{H}{\underset{+}{O}}-Si \leqslant\ +\ H_2O \quad |K(2) \quad (2)$$

$$\geqslant Si-\overset{H}{\underset{+}{O}}-Si \leqslant\ +\ Cl^- \longrightarrow\ \geqslant Si-Cl\ +\ \geqslant Si-OH \quad |k(3) \quad (3)$$

The idea of this mechanism is that there are two species reacting with the siloxanes, adducts of hydrochloric acid with dioxane and adducts of hydrochloric acid with H_2O. Surprisingly no silanol signals could be found in the proton resonance spectra. As silanols have to be formed, their concentration must be too low to appear in the spectra under these conditions of high concentration of hydrochloric acid. From mechanism and stoichiometry the following rate law could be derived:

$$-\frac{d[\geqslant Si-O-Si \leqslant]}{dt} = (k_1 + k_2 \cdot [H_2O]^a) \cdot [\geqslant Si-O-Si \leqslant] \quad (4)$$

with $k_1 = K(1) \cdot k(3) \cdot [HCl]^b = k_1' \cdot [HCl]^b$ \quad (5)

and $k_2 = K(2) \cdot k(3) \cdot [HCl]^c = k_2' \cdot [HCl]^c$ \quad (6)

$b = 4\ ;\ c = 3$

As these reactions were only performed up to a 70% conversion of the reactants, the back reactions could be neglected in the reaction scheme as well as in the rate law. In this way the rate law could be used without being disturbed by the establishment of equilibria and it allowed determination of the rate constants for the two reactions by nonlinear regression. Table 3 shows these rate constants for substituted aryldisiloxanes and methylsiloxanes. We found a good correlation when using the value of 0.48 for the phenyl and $\sigma^*(X) + 0.48$ for the substituted phenyl groups X-Ph and further the value of 0.35 for one and 0.70 for two trimethylsiloxy groups at silicon.[8]

Table 3 Cleavage of Siloxanes by HCl in dioxane

Siloxane	$k_1' \cdot 10^6$ ($1^4 \cdot mol^{-4} \cdot s^{-1}$)	$k_2' \cdot 10^4$ ($1^4 \cdot mol^{-4} \cdot s^{-1}$)	$\Sigma\sigma^*$
$(Me_3Si)_2O$	32	8.9	0
$(pMe-PhMe_2Si)_2O$	4.7	3.7	0.62
$(PhMe_2Si)_2O$	1.9	2.5	0.96
$(pF-PhMe_2Si)_2O$	1.1	1.8	1.08
$(pCl-PhMe_2Si)_2O$	0.62	1.4	1.42
$(ClCH_2Me_2Si)_2O$	0.039	0.23	2.10
$(Me_3SiO)_2SiMe_2$	14.0	4.4	0.35
$(Me_3SiO)_3SiMe$	8.3	3.6	0.70
	$\varrho_1 = -1.4$	$\varrho_2 = -0.6$	

5 REACTIONS OF CHLOROSILANES AND CHLOROSILOXANES WITH LITHIUM SILANOLATES AND ALCOHOLATES

In order to attain a larger amount of experimental data with compounds containing siloxy groups from a reaction that can be easier performed and evaluated, chlorosilanes and chlorosiloxanes were reacted with Me_3SiOLi, $Me_2PhSiOLi$, Me_2CHOLi, and Me_3COLi. Absolute and relative rate constants could be obtained by ^1H-NMR spectroscopy, gas chromatography or by a turbidimetric method. The ϱ-values of these reactions were determined by a series of substituted aryldimethylchlorosilanes and the δ-values by different alkyldimethylchlorosilanes. In the following rate constants were determined for chlorosiloxanes and $-(\lg k_{rel} - \Sigma\sigma^* \cdot \varrho^*)$ plotted against $-\Sigma E_s$. With only one exception the δ-values obtained from chlorosiloxanes in this way are the same as those found with alkyldimethylchlorosilanes. But in all cases chlorosilanes and chlorosiloxanes form two different correlation lines as given in Figure 1 and 2.

In the case of monochloro compounds (Figure 1a, 1b and 2a) the siloxanes show an approximately tenfold reactivity in relation to that of the alkylsilanes. In contrast with dichloro and trichloro compounds the siloxanes possess only one tenth of the reactivity of the alkyl compounds (Figure 2b and 2c). Therefore the Taft equation must be written in the following form:

$$\lg \frac{k_i}{k_o} = \sigma^* \cdot \varrho^* + \delta \cdot E_s + C \qquad (7)$$

C refers in an up to now not completely clear way to the electron density at the chlorine bearing silicon atom.

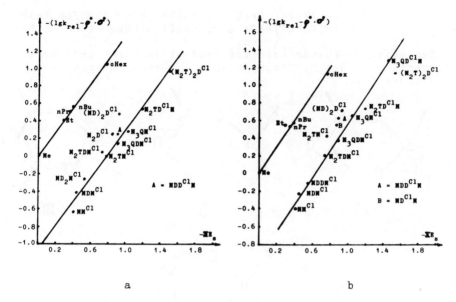

Figure 1 Reactions of alkyldimethylchlorosilanes and chlorosiloxanes with Me$_3$SiOLi (a) or Me$_2$PhSiOLi (b)

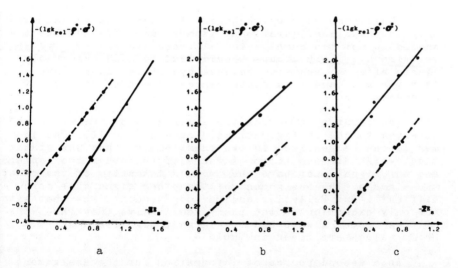

Figure 2 Reactions of alkylchlorosilanes (---) and chlorosiloxanes (——) with Me$_2$CHOLi (a) and of alkylmethyldichlorosilanes (---) and dichlorosiloxanes (——)(b) or alkyltrichlorosilanes (---) and trichlorosiloxanes (——)(c) with Me$_3$COLi

6 STEREOCHEMISTRY OF NUCLEOPHILIC SUBSTITUTION REACTIONS AT FUNCTIONAL CYCLOTRISILOXANES

The model compounds used for studies in this field were 1-substituted-3,5-bis(trimethylsiloxy)-cyclotrisiloxanes (Figure 3).

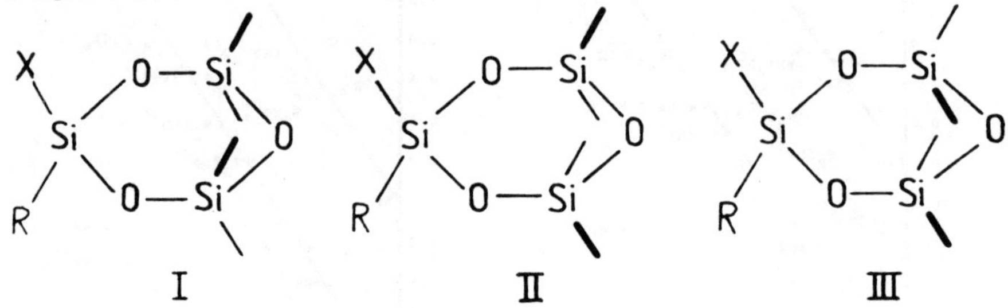

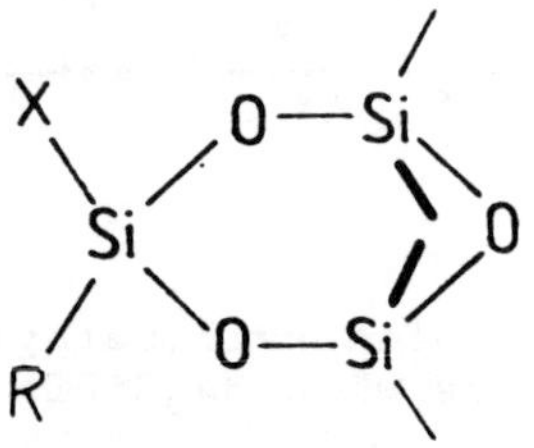

Figure 3 Stereoisomers of 1-substituted-3,5-bis(trimethylsiloxy)-cyclotrisiloxanes

These compounds were chosen: i. because they can exist only in three configurational isomers and thus NMR-spectra should be not too complicated, ii. because the fairly large trimethylsiloxy groups should lead to good observable steric effects, and iii. because cyclotrisiloxanes are fairly rigid, and thus disturbances by conformational movements should be small.

The first problem was the assignment of the signals observed to the configurational isomers I to III. The isomer II can be realized in two enantiomers, thus the statistical weight is two times of those of the other two isomers and the signal must have the twofold intensity of the other two signals. The assignment of the other signals is more difficult. But Pelletier and Harrod found in 1984 that with the only exception of the phenyl derivative $(SiMePhO)_3$ the all-cis isomer always has the largest high field shift.[9] Shift values are given in Table 4.

As a second source of information for the assignment the relative intensities of the signals should be useful. The all-cis isomers I should be present in a lower amount than could be expected from statistical reasons, because the formation of the all-cis isomers is sterically more hindered than the formation of other conformational isomers. Indeed, in all cases that isomer with the largest high field shift is present in the lowest amount (Table 5).

Thus the assignment should be fairly sure.

Table 4 ^{1}H- and ^{29}Si-NMR Shift Values of 1-substituted-3.5-bis(trimethylsiloxy)-cyclotrisiloxanes (D^X-group)

X	^{1}H-NMR shift δ(ppm)			^{29}Si-NMR shift δ(ppm)		
	I	II	III	I	II	III
Cl	0.48	0.50	0.51	-31.68	-31.76	-31.18
Br	0.585	0.600	0.615	-35.41	-35.41	-34.79
OMe	3.37*	3.39*	3.41*	-48.03	-48.21	-47.80
OiPr	1.21*	1.225*	1.24*			
OtBu	1.26*	1.28*	1.295*			
OAc	0.30	0.35	0.38	-49.86	-49.86	-48.76
OH				-48.89	-48.89	-48.29
H	0.22	0.245	0.26	-23.6	-23.74	-23.19

*methyl protons of the alkoxy group

Table 5   Relative Intensities of the Signals (Concentrations) of the Stereoisomers of 1-substituted-3,5-bis(trimethylsiloxy)-cyclotrisiloxanes

		Relative Intensities of the NMR-Signals		
R	X	I(%)	II(%)	III(%)
H	Cl	22.5	49	28.5
Me	Cl	22	51	27
Me	Br	20.5	50	29.5
Me	OH	17	53.5	29.5
Me	OMe	24.5	49.5	26
Me	OtBu	22.5	50.5	27
Me	OAc	22.5	52	25.5
Me	Bzl	24.5	50	25.5
Me	H	30	49.5	20.5
cHex	H	30	51	19
OSiMe$_3$	Cl	28	50	22
OSiMe$_3$	H	29.5	49	21.5
OSi(OSiMe)$_3$	Cl	29	50	21
OSi(OSiMe)$_3$	Me	27	53	20

Now a series of nucleophilic substitution reactions was performed. In all reactions the signals of the reaction products appear in the same order as the signals of the reactants disappear. And in all reactions the all-cis derivatives have the lowest reactivity (Table 6).[10]

So it was shown that all reactions have a retention stereochemistry. This is also true for reactions which are well known to have an inversion stereochemistry in all cases studied up to now.

Table 6 Relative Reactivities of the Stereoisomers I, II
and III in Nucleophilic Substitution Reactions

X	Nucleophile	Relative Reactivities		
		I	II	III
Cl	AcOH(Ac$_2$O)	1	3.2	10.0
Cl	AcOH(Pyr)	1	3.3	7.5
Cl	MeOH(Pyr)	1	3.0	8.6
Cl	iPrOH(Pyr)	1	2.8	7.5
Cl	tBuOH(Pyr)	1	2.3	5.2
OAc	H$_2$O	1	4.1	9.5
OAc	MeOH(Pyr)	1	2.7	6.3
OAc	iPrOH(Pyr)	1	3.1	7.5
OMe	H$_2$O	1	2.1	4.4
H	Br$_2$	1	1.8	2.2

Corriu presented two possibilities for the explanation of the steric course of nucleophilic substitution reactions.[11] First in cyclic systems with small angles at silicon the orbitals of the bonds from silicon to elements being part of the ring have a higher p-character and the bonds from silicon to the functional groups consequently have a higher s-character which should favour a retention mechanism by front side attack of the nucleophile. But the angle at Si in cyclotrisiloxanes was determined by Oberhammer to be 107.8°,[12] which nearly equals the tetrahedral angle of 109°. Second in analogy to P-compounds Si-compounds, too, should prefer transition states with the substituents of silicon that are members of the ring, in e,a-position.

We think that the sterical course is mainly determined by stereoelectronic effects. In principle there are four possibilities for the attack of a nucleophile at the functional Si-atom of the cyclotrisiloxane. Three of them should be slightly favoured by interactions between the HOMO of the nucleophile and the LUMO of the bond between Si and the functional group. In Figure 4 these versions are represented by the letters a to c. Two of these ways lead to transition states with the ring O-atoms in e,a-positions. And both of these transition states result in retention of the configuration. The last step of this reaction pathway is a Berry-pseudo-rotation, which brings the leaving group into an apical position without larger changes in the O-Si-O-angle.

As cyclic siloxanes are important intermediates in the synthesis of silicones, these stereochemical results are of interest because the stereochemistry of the substitution or condensation steps following the formation of cyclosiloxanes determines the structure of the oligo- and polysiloxanes formed.

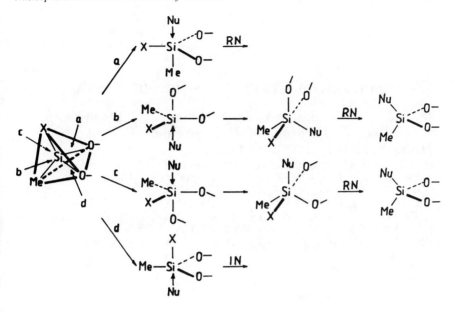

Figure 4 Steric Course of Nucleophilic Substitution Reactions with 1-substituted-3.5-bis(trimethylsiloxy)-cyclotrisiloxanes

REFERENCES

1. G. Engelhardt, H. Jancke, R. Radeglia, H. Kriegsmann, M.F. Larin, V.A. Pestunovich, E.I. Dubinskaya and M.G. Voronkov, Z.Chem., 1977, 17, 376.
2. Y. Nagai, H. Matsumoto, T. Nakano and H. Watanabe, Bull.Chem.Soc.Japan, 1972, 45, 2560.
3. U. Scheim, H. Grosse-Ruyken, K. Rühlmann and A. Porzel, J.Organomet.Chem., 1986, 312, 27.
4. M. Charton, Progr.Phys.Org.Chem., 1981, 13, 119
5. R.W. Taft, Jr., in "Steric Effects in Organic Chemistry", M.S. Newman, Ed., J. Wiley and Sons, New York, 1956, Chap.13.
6. F.K. Cartledge, Organometallics, 1983, 2, 425.
7. U. Scheim, H. Grosse-Ruyken, K. Rühlmann and A. Porzel, J.Organomet.Chem., 1985, 293, 29.
8. U. Scheim, R. Lehnert, A. Porzel and K. Rühlmann, J. Organomet.Chem., 1988, 356, 141.
9. E. Pelletier and J.F. Harrod, Organometallics, 1984, 3, 1070.
10. R. Gewald, K. Rühlmann, U. Scheim and A. Porzel, J.Organomet.Chem., 1989, 377, 9.
11. R.J.P. Corriu, C. Guerin and J.J.E. Moreau, Top.Stereochem., 1984, 15, 43.
12. H. Oberhammer, W. Zeil and G. Fogarasi, J.Mol.Struct., 1973, 18, 309.

Bioorganosilicon Chemistry – Recent Results

R. Tacke,[a] S. Brakmann,[a] M. Kropfgans,[a] C. Strohmann,[a]
F. Wuttke,[a] G. Lambrecht,[b] E. Mutschler,[b] P. Proksch,[c]
H.-M. Schiebel,[d] and L. Witte[c]

[a]INSTITUTE OF INORGANIC CHEMISTRY, UNIVERSITY OF KARLSRUHE, ENGESSERSTRAßE, D-7500, GERMANY
[b]DEPARTMENT OF PHARMACOLOGY, UNIVERSITY OF FRANKFURT, D-6000 FRANKFURT/M, GERMANY
[c]INSTITUTE OF PHARMACEUTICAL BIOLOGY, [d]INSTITUTE OF ORGANIC CHEMISTRY, TECHNICAL UNIVERSITY OF BRAUNSCHWEIG, D-3300 BRAUNSCHWEIG, GERMANY

1 INTRODUCTION

During the last thirty years, extensive research has been carried out in the field of bioorganosilicon chemistry. These studies (for a recent review, see ref. [1]) led to the development of a variety of biologically active silicon compounds, some of which have already found practical application. In contrast to the numerous publications dealing with pharmacological and toxicological properties of organosilicon drugs, reports on biotransformations of such compounds are sparse. Here, results of recent studies on the metabolism of organosilicon drugs in insects and higher organisms are presented.

As the biological properties of chiral organosilicon compounds depend on their absolute configuration, there is a need for efficient methods of preparing optically active silanes. We report here the synthesis and pharmacological properties of the enantiomers of an organosilicon muscarinic antagonist (preparation of the antipodes by a classical resolution method). In addition, biotransformations of organosilicon compounds using microbial cells and isolated enzymes are described. The aim of these studies was to investigate the biocatalyzed synthesis of optically active silanes.

2 BIOTRANSFORMATION OF ORGANOSILICON DRUGS IN HIGHER ORGANISMS AND INSECTS

Biotransformation in the rat

The silanols hexahydro-sila-difenidol (HHSiD, **1a**) and p-fluoro-hexahydro-sila-difenidol (p-F-HHSiD, **2a**)

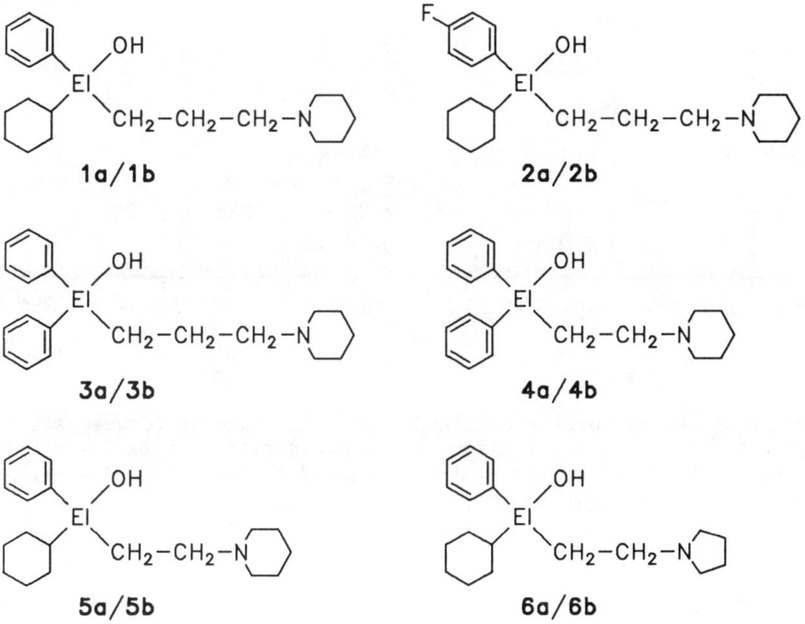

Figure 1 Chemical structure of the silicon/carbon pairs 1a/1b-6a/6b (a: El = Si; b: El = C).

are commercially available antimuscarinic agents (for synthetic aspects, see ref. [2]). Due to their interesting selectivity profiles [3], these drugs are used as tools in experimental pharmacology and physiology for the subclassification of muscarinic receptors. Here we report on the metabolic fate of these compounds and of some structurally related muscarinic antagonists.

We have studied the metabolism of 1a, 2a, sila-difenidol (3a), sila-pridinol (4a), sila-trihexyphenidyl (5a) and sila-procyclidine (6a) in the rat [4] (Figure 1). For reasons of comparison, the corresponding carbon analogues, the carbinols 1b-6b (Figure 1), were included in these studies. In addition, pharmacokinetic studies with tritiated 1a (([^{3}H]-HHSiD, tritium label in the piperidino group) were carried out [5]. All compounds were investigated as racemates.

Following oral administration of the drugs to rats, urine samples were collected, subsequently freeze-dried and extracted with organic solvents (methanol, ethyl acetate). In order to cleave phase-II metabolites, methanol extracts were incubated with glucuronidase/sulfatase and then extracted with ethyl acetate. The organic extracts obtained by these working-up procedures were then analyzed using mass spectrometric techniques (FAB-MS, GC-MS). In the FAB-MS studies, the crude extracts could be investigated directly (matrix glycerol; for typical FAB-MS spectra, see Figure 2). In

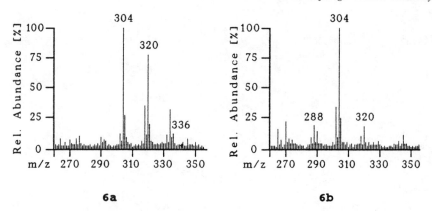

6a   **6b**

Figure 2 FAB-MS spectra obtained from urine samples (crude methanol extracts) of the rat after oral administration of **6a** (left) and **6b** (right). The marked peaks correspond to the ions $[M + H]^+$, $[M + O + H]^+$ and $[M + 2O + H]^+$.

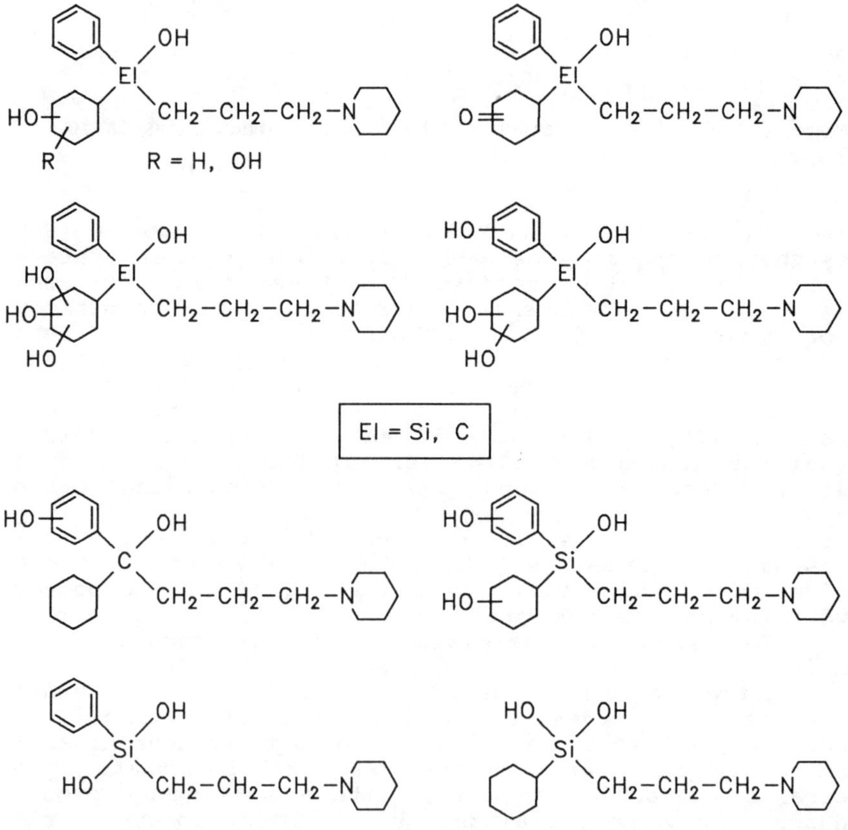

El = Si, C

Figure 3 Phase-I metabolites detected in urine samples of the rat after oral administration of **1a** and **1b**.

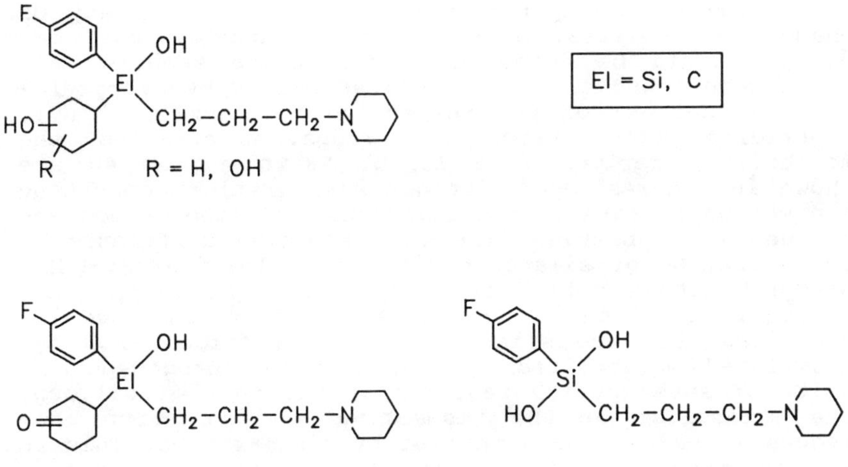

Figure 4 Phase-I metabolites detected in urine samples of the rat after oral administration of **2a** and **2b**.

El = Si, C

the case of the GC-MS experiments (EI, 40 eV), the extracts were analyzed after silylation with $Me_3SiN(Me)-C(O)CF_3$. The structures of the metabolites (Figures 3-5) were determined on the basis of characteristic fragmentation patterns in the mass spectra.

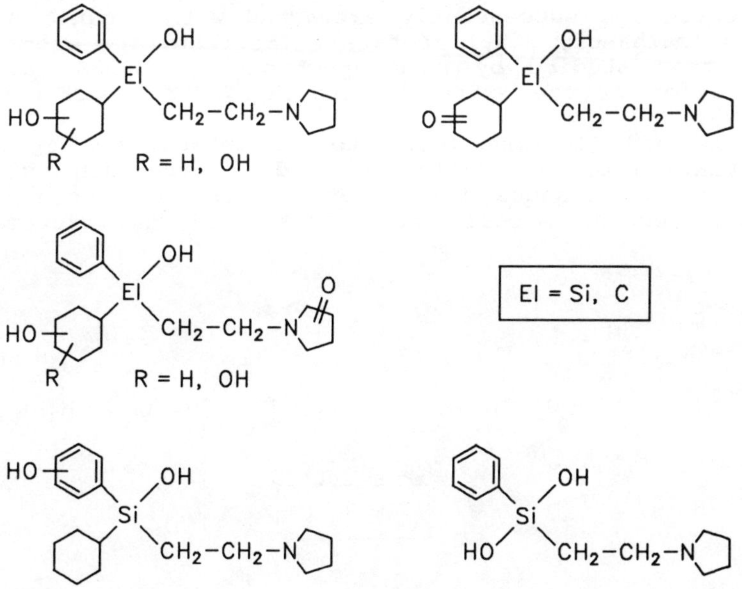

El = Si, C

Figure 5 Phase-I metabolites detected in urine samples of the rat after oral administration of **6a** and **6b**.

Using these experimental techniques, a variety of phase-I metabolites, besides the unchanged parent compounds, could be detected in the urine samples. All drugs were found to be metabolized mainly by hydroxylation (oxidation) of the respective cyclohexyl, phenyl, piperidino and/or pyrrolidino groups. As examples, the metabolites derived from **1a/1b**, **2a/2b** and **6a/6b** are shown in Figures 3-5. Distinct similarities concerning the metabolic fate of the respective silicon/carbon analogues were observed. The most striking difference is the formation of silanediols by Si-C cleavage (Si-C_6H_5, Si-C_6H_{11}). It is not clear whether this cleavage is an enzymatic reaction or not. However, there are some indications that the silanediols may be formed from hydroxylated metabolites by non-enzymatic processes. In spite of some special features in the case of silicon, the metabolism of the respective silicon/carbon analogues is mainly characterized by similarities. This is in accordance with the manifold bioisosteric relationships generally observed for structurally analogous silicon and carbon compounds [1].

Biotransformation in larvae of Spodoptera littoralis

Hexahydro-sila-difenidol (**1a**) and its carbon analogue hexahydro-difenidol (**1b**) were also studied for their metabolic fate in larvae of *Spodoptera littoralis* (Lepidoptera: Noctuidae) [6].

After administration of the drugs, as components of an artificial diet, the excrements of the insects were collected and subsequently extracted with organic solvents (methanol, ethyl acetate). The resulting extracts were then studied by mass spectrometry (GC-MS, after silylation of the extracts with $Me_3SiN(Me)C(O)CF_3$; EI, 40 eV).

1a and **1b** were found to be metabolized by hydroxylation of the cyclohexyl and/or piperidino group (Figure 6). Besides these metabolites, the unchanged parent compounds could be detected in the excrement

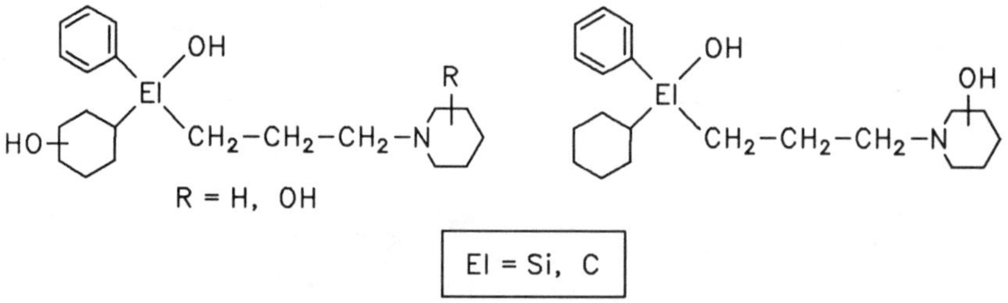

R = H, OH

El = Si, C

Figure 6 Metabolites detected in excrement samples of *Spodoptera littoralis* after administration of **1a** and **1b** as components of an artificial diet.

samples. Similarly to the results found with rats, the metabolism of **1a** and **1b** in *Spodoptera littoralis* is also characterized by distinct analogies.

3 OPTICALLY ACTIVE SILANES: PREPARATION BY CLASSICAL RESOLUTION METHODS AND BIOCATALYSIS

Optically active organosilicon drugs

As demonstrated some years ago, the enantiomers of the muscarinic antagonist sila-procyclidine (**6a**) (Fig-

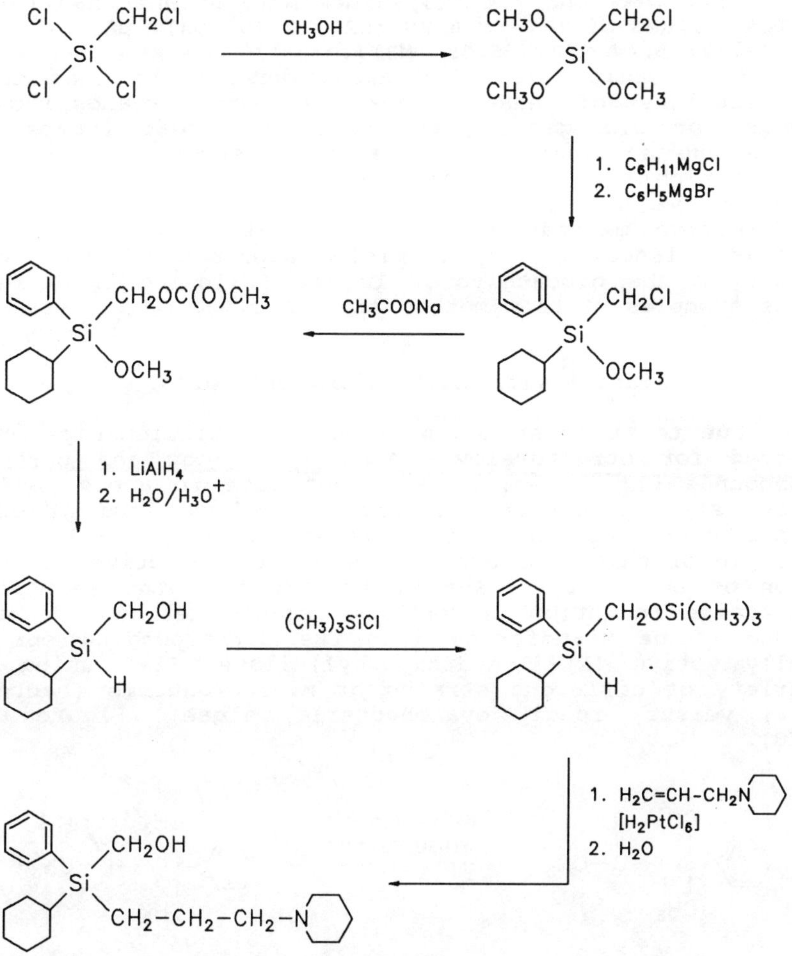

Figure 7 Synthesis of the racemic antimuscarinic agent **7** starting from Cl_3SiCH_2Cl.

ure 1) differ in their pharmacological properties [7]. The same holds true for the enantiomers of the antimuscarinic agent **7** (for the synthesis of rac-**7**, see Figure 7), recently prepared by a classical resolution method via diastereomeric salts using (R)- and (S)-1,1'-binaphthyl-2,2'-diyl hydrogen phosphate as resolving agents [8]. Functional pharmacological studies (guinea pig atrial M2 and ileal M3 receptors; agonist arecaidine propargyl ester) have shown that (-)-**7** exhibits a greater antimuscarinic potency than (+)-**7** (the sign of the optical rotations refers to solutions of the corresponding hydrochlorides in $CHCl_3$; enantiomeric purity >97% ee as determined by ^{13}C NMR experiments using $Eu(hfc)_3$ as shift reagent). (-)-**7** was found to be 5.8 (M2) and 7.4 (M3) times more potent than (+)-**7** [pA_2-values of (-)-**7**: 6.20 (M2), 7.26 (M3); pA_2-values of (+)-**7**: 5.44 (M2), 6.39 (M3)].

The results described above demonstrate that the availability of enantiomerically pure organosilicon drugs for biological purposes is of great interest. Since the application of classical resolution methods (separation via diastereomeric derivatives) is rather limited, we have been interested in the development of alternative methods for the preparation of optically active silanes. A very promising approach in this respect is the biocatalysis. In the following chapters, some examples of this method are described.

Biotransformation with whole microbial cells

Due to the distinct bioisosteric relationships observed for structurally analogous silicon and carbon compounds [1], attempts have been made to apply well-known stereoselective microbial transformations of organic compounds to their sila-derivatives [9-13]. An example of this approach is the enantioselective reduction of the acetylsilane **8**. Analogous to the enantioselective reduction of various organic ketones, **8** was found to be transformed into the corresponding optically active (R)-(1-hydroxyethyl)silane (R)-**9** using a variety of different strains of microorganisms (bacteria, yeasts, fungi, cyanobacteria, algae) (Figure 8) [10].

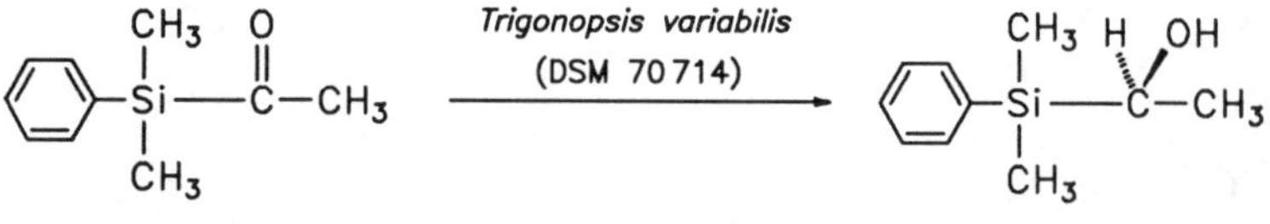

Figure 8 Enantioselective microbial reduction of **8** using various strains of microorganisms.

Biocatalyst	Enantiomeric purity of (R)-**9**
Acinetobacter calcoaceticus (ATCC 31012)	> 95 % ee
Brevibacterium species (ATCC 21860)	90 % ee
Corynebacterium dioxydans (ATCC 21766)	> 95 % ee
Candida albicans (ATCC 10231)	86 % ee
Candida humicola (DSM 70067)	90 % ee
Trigonopsis variabilis (DSM 70714)	86 % ee
Cunninghamella elegans (ATCC 26269)	94 % ee
Synechococcus leopoliensis (SAG 1402-1)	94 % ee
Chlamydomonas reinhardii (Y-1)	85 % ee

Figure 8 (continued)

Using racemic acetylsilanes as substrates, this type of enantioselective microbial transformation yields diastereomeric (1-hydroxyethyl)silanes. This is shown in Figure 9 for the microbial transformation of rac-**10** using cells of the yeast *Trigonopsis variabilis* (DSM 70714) [11,13] and the bacterium *Corynebacterium dioxydans* (ATCC 21766) [13]. The (R)-selective reduction of rac-**10** (for the synthesis of rac-**10**, see ref. [14]) leads to the diastereomeric products (R,R)-**11** and (S,R)-**11** which can be separated by column chromatography on silica gel. Conversions on a 10 g scale in a 30 l bioreactor led to the biotransformation products with yields of about 75% {*Trigonopsis variabilis* (DSM 70714): 97% ee [(R,R)-**11**], 96% ee [(S,R)-**11**]; *Corynebacterium dioxydans* (ATCC 21766): $\geq$99% ee [(R,R)-**11** and (S,R)-**11**]}.

Using stereoselective chemical reactions, these bioconversion products can be transformed into optically active synthons. For example, the (1-hydroxyethyl)silane (R,R)-**11** could be converted into the corresponding hydridosilane (S)-**12** by a three-step synthesis as outlined in Figure 10 [15]. The reactions were found to proceed with high stereoselectivity (according to preliminary studies $\geq$85%) and with retention of configuration at the silicon atom. As optically active silanes $R^1R^2R^3Si^*H$ can be transformed stereoselectively into a variety of derivatives $R^1R^2R^3Si^*R$ [16], the microbial reduction of rac-**10** can be regarded as a key step in the synthesis of optically active silanes of the type t-BuPhMeSi*R.

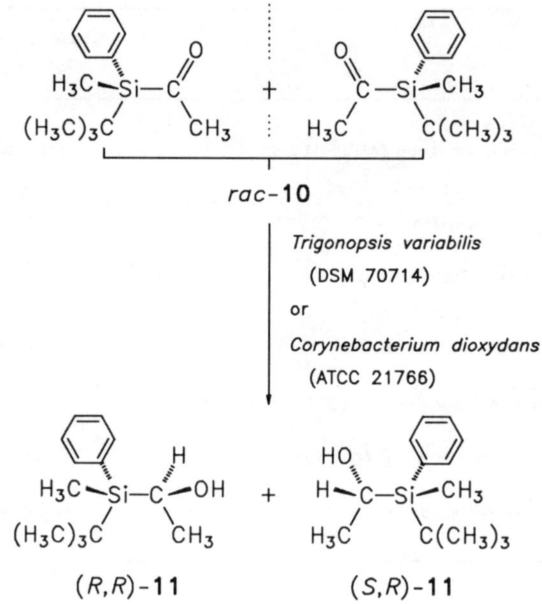

Figure 9 Enantioselective microbial reduction of *rac*-10 using cells of *Trigonopsis variabilis* (DSM 70714) or *Corynebacterium dioxydans* (ATCC 21766).

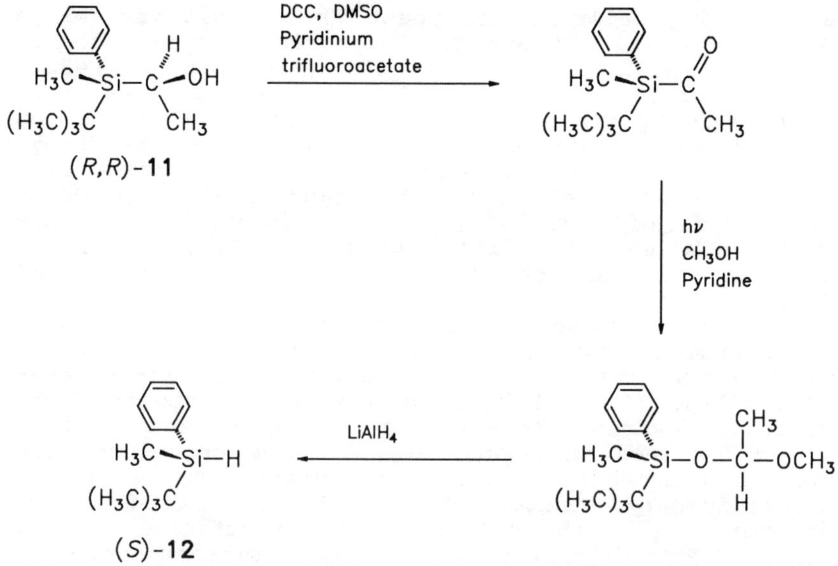

Figure 10 Stereoselective chemical transformation of (*R,R*)-11 into (*S*)-12.

Biotransformation with isolated enzymes

As may be expected from the preceding chapter, isolated enzymes can also be used as biocatalysts for the preparation of optically active silanes [17,18]. An example is the enantioselective enzymatic hydrolysis (kinetic racemate resolution) of rac-13 using immobilized penicillin-G-acylase from *Eschericia coli* 5K (pHM 12) as shown in Figure 11 [18]. Incubation of this enzyme (immobilized on Eupergit C) with rac-13 in potassium phosphate buffer/DMSO leads to the formation of the (1-aminoethyl)silane (R)-14 (96% ee, analytical scale). As racemic 14 is a sympathomimetic agent [19], pharmacological studies with the enantiomers (R)-14 and (S)-14 might be of interest.

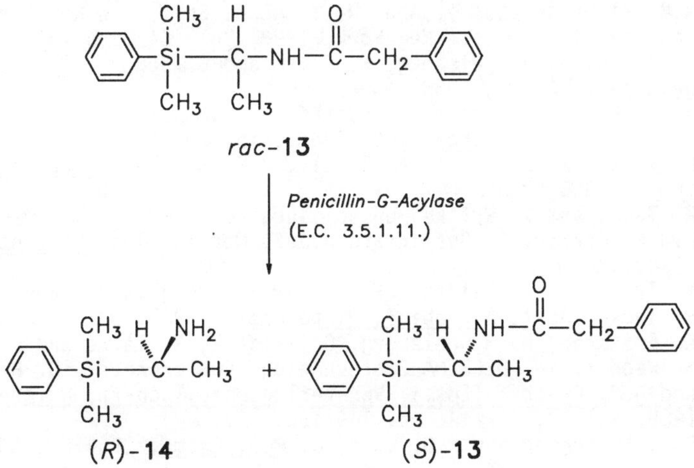

Figure 11 Enantioselective hydrolysis of rac-13 using immobilized penicillin-G-acylase (E.C. 3.5.1.11.) from *Eschericia coli* 5K (pHM).

REFERENCES

[1] R. Tacke and H. Linoh, in 'The Chemistry of Organic Silicon Compounds', S. Patai and Z. Rappoport (Eds.), John Wiley and Sons Ltd., Chichester, 1989, Part 2, pp. 1143-1206.
[2] R. Tacke, H. Linoh, H. Zilch, J. Wess, U. Moser, E. Mutschler and G. Lambrecht, <u>Liebigs Ann. Chem.</u>, 1985, 2223.
[3] G. Lambrecht, R. Feifel, M. Wagner-Röder, C. Strohmann, H. Zilch, R. Tacke, M. Waelbroeck, J. Christophe, H. Boddeke and E. Mutschler, <u>Eur. J. Pharmacol.</u>, 1989, *168*, 71.
[4] C. Strohmann, H.-M. Schiebel, L. Witte, G. Lambrecht, E. Mutschler and R. Tacke, unpublished results.
[5] N.M. Rettenmayr, J.F. Rodrigues de Miranda, N.V.M. Rijntjes, F.G.M. Russel, C.A.M. van Ginneken, C. Strohmann, R. Tacke,

G. Lambrecht and E. Mutschler, <u>Naunyn-Schmiedeberg's Arch. Pharmacol.</u>, in press.
[6] C. Strohmann, H.-M. Schiebel, L. Witte, P. Proksch and R. Tacke, unpublished results.
[7] R. Tacke, H. Linoh, L. Ernst, U. Moser, E. Mutschler, S. Sarge, H.K. Cammenga and G. Lambrecht, <u>Chem. Ber.</u>, 1987, *120*, 1229.
[8] R. Tacke, M. Kropfgans, H.-J. Egerer, G. Lambrecht and E. Mutschler, unpublished results.
[9] C. Syldatk, H. Andree, A. Stoffregen, F. Wagner, B. Stumpf, L. Ernst, H. Zilch and R. Tacke, <u>Appl. Microbiol. Biotechnol.</u>, 1987, *27*, 152.
[10] C. Syldatk, A. Stoffregen, F. Wuttke and R. Tacke, <u>Biotechnol. Lett.</u>, 1988, *10*, 731.
[11] C. Syldatk, A. Stoffregen, A. Brans, K. Fritsche, H. Andree, F. Wagner, H. Hengelsberg, A. Tafel, F. Wuttke, H. Zilch and R. Tacke, in 'Enzyme Engineering 9', H.W. Blanch and A.M. Klibanov (Eds.), Ann. N. Y. Acad. Sci., The New York Academy of Sciences, New York, 1988, Vol. 542, pp. 330-338.
[12] R. Tacke, H. Hengelsberg, H. Zilch and B. Stumpf, <u>J. Organomet. Chem.</u>, 1989, *379*, 211.
[13] R. Tacke, S. Brakmann, F. Wuttke, J. Fooladi, C. Syldatk and D. Schomburg, <u>J. Organomet. Chem.</u>, in press.
[14] R. Tacke, K. Fritsche, A. Tafel and F. Wuttke, <u>J. Organomet. Chem.</u>, 1990, *388*, 47.
[15] R. Tacke and F. Wuttke, unpublished results.
[16] R.J.P. Corriu, C. Guérin and J.J.E. Moreau, <u>Top. Stereochem.</u>, 1984, *15*, 43.
[17] K. Fritsche, C. Syldatk, F. Wagner, H. Hengelsberg and R. Tacke, <u>Appl. Microbiol. Biotechnol.</u>, 1989, *31*, 107.
[18] K. Fritsche, H. Hengelsberg, C. Syldatk, R. Tacke and F. Wagner, in 'DECHEMA Biotechnology Conferences', D. Behrens and A.J. Driesel (Eds.), VCH Verlagsgesellschaft, Weinheim, 1989, Vol. 3, Part A, pp. 149-152.
[19] R.J. Fessenden and M.D. Coon, <u>J. Med. Chem.</u>, 1964, *7*, 561.

ACKNOWLEDGEMENTS

These studies were supported by the Deutsche Forschungsgemeinschaft, the Fonds der Chemischen Industrie and the Bayer AG.

Section IV: Structural Organosilicon Chemistry and New Organosilicon Compounds

The Role of Trimethylsilyl-substituted Ligands in Co-ordination Chemistry; Bis(Trimethylsilyl)-Methylmetal Complexes

Michael F. Lappert
SCHOOL OF CHEMISTRY AND MOLECULAR SCIENCES,
UNIVERSITY OF SUSSEX, BRIGHTON BN1 9QJ, UK

1 INTRODUCTION

Trimethylsilyl-substituted ligands (Scheme 1) have featured prominently in our researches. The use of the trimethylsilyl-substituted methyls was first recognised in a 1969 patent;[1] a communication followed a year later.[2] Features of the series of ligands $CH_{3-n}(SiMe_3)_n$ are (i) the absence of β-hydrogen, and indeed of β-carbon, and (ii) the possibility of increasing the bulk of the ligand with increasing values of n; reviews were published in 1974[3a] and 1976.[3b]

The significance of such ligands was originally conceived to be that they might provide access to hitherto unknown stable transition metal alkyls or open shell main group element analogues. Up to about 1970 transition metal alkyls were considered to be inherently unstable due to the weakness of the M-C bond, but our earliest publications stressed the importance of kinetic effects (and calorimetric data showed that the bond strength proposition was invalid[4]). Thus, a common intramolecular pathway of decomposition of a metal alkyl is a β-H elimination leading to the extrusion of an alkene; the majority of the alternative decomposition routes are associative in character. Steric protection of the metal centre by a bulky alkyl group offers a means of kinetic stabilisation with regard to both intramolecular and associative processes.

This paper (and the lecture on which it is based) will focus exclusively on one of these ligands, namely $CH(SiMe_3)_2$, which henceforth will be abbreviated as R. However, the scope of this paper will largely be restricted to homoleptic complexes (i.e., compounds of type MR_n), although occasionally heteroleptic analogues, (particularly those having a neutral coligand L, i.e., of formula MR_nL_m) will be mentioned.

Further general features of the ligands indicated

$\bar{C}H_2SiMe_3$

$\bar{C}H(SiMe_3)_2$

$\bar{C}(SiMe_3)_3$ [a]

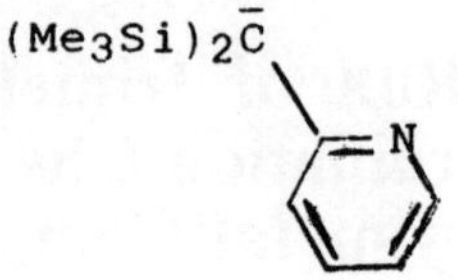

$(Me_3Si)_2\bar{C}$—pyridyl [d]

$\bar{N}(SiMe_3)_2$ [b]

$\bar{P}(SiMe_3)_2$ [c]

$\bar{S}C(SiMe_3)_3$

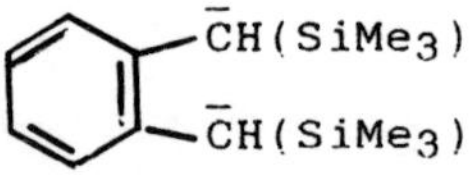

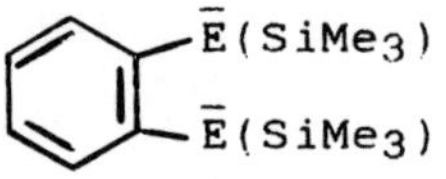

o-C$_6$H$_4$[$\bar{C}H(SiMe_3)$]$_2$

o-C$_6$H$_4$[$\bar{E}(SiMe_3)$]$_2$ [e]

n-$\bar{C}_5H_4$(SiMe$_3$) ($\bar{C}p'$) n-$\bar{C}_5H_3$(SiMe$_3$)$_2$-1,3 ($\bar{C}p''$)
n-$\bar{C}_5H_2$(SiMe$_3$)$_3$-1,2,4 ($\bar{C}p'''$) [f]

Scheme 1. Some trimethyl-substituted ligands [First used by: [a] C. Eaborn and J.D. Smith, [b] U. Wannagat, [c] G. Fritz and G. Becker, [d] C.L. Raston, [e] (for E=P) K. Issleib; [f] also P. Jutzi, for main group metal compounds]

Table 1. Some X-ray structural data on crystalline sodium alkyls

Empirical Formula	n in (NaR)$_n$	C.N.[a] of Na	<Na-C>/(Å)	Ref.
NaMe	(4), ∞	4 (6)	2.58 – 2.64 (intermol 2.76)	41
NaEt	∞	4	2.63 – 2.68	42
NaCH$_2$Ph(tmeda)	4	2 + 2N	2.65	43
NaCPh$_3$(tmeda)	∞	6 + 2N	2.64 – 3.13	44
NaR	∞	2	2.557(13)	6

[a] C.N. = Co-ordination number

Role of Trimethylsilyl-substituted Ligands in Co-ordination Chemistry   233

in Scheme 1 are that their multiplicity of methyl groups ensures that intermolecular contacts between monomolecular units are weak. Hence, the metal complexes are often hydrocarbon-soluble and volatile.

Our general objective has been to prepare unusual compounds and to study their structures and chemistry. Because of restrictions of space, there will generally only be opportunity to discuss the syntheses and structures of the complexes MR_n or their simple donor adducts. The ligand R is rather flexible and the metal alkyl MR_n, although not able to accept further R ligands for steric reasons, is often able to attract smaller ligands; in that sense, these alkyls may be regarded as models for metalloenzymes with M as the active site.

Another characteristic of R is the presence of the methine proton, which often allows the determination of $^2J(^xM-^1H)$ or $a(^1H)$, depending on whether the metal alkyl is dia- or paramagnetic. The presence of the six methyl groups in R is also useful, because for diamagnetic compounds they appear either as a singlet in the ^1H NMR spectrum or as a 1:1 doublet if the two trimethylsilyl groups are inequivalent.

The synthesis of MR_n is in principle straightforward and uses the corresponding metal chloride, t-butoxide, or a bulky phenoxide [*e.g.*, $OC_6H_2Bu^t_2$-2,6-Me-4 ($\equiv$ OAr)], together with the Grignard reagent, $MgR_2(OEt_2)$, or LiR. In general the coproduct (LiCl, LiOBu^t, or LiOAr, or a Mg analogue) is less soluble in the hydrocarbon solvent than is MR_n.

2   HOMOLEPTIC MAIN GROUP METAL ALKYLS MR_n [R = CH(SiMe$_3$)$_2$]

The homoleptic alkyls which have been structurally authenticated are summarised in Scheme 2, which also shows the co-ordination number of the metal in these compounds.

Some features of general interest are the following. (1) The alkyls of Li[5] and Na[6] are infinite polymers in the solid state, having a skeletal chain structure in which the alkali metal alternates with the α-C. Thus, there is a single alkyl bridge between successive metal atoms and the singly bridging carbon atom has the co-ordination number of 5, the alkali metal being two-co-ordinate. In the gas phase, the lithium compound is a monomer.[5]

The sodium compound,[6] Figure 1, is not isostructural with its lithium analogue. Selected X-ray data on some sodium alkyls, including (NaR)$_\infty$, are in Table 1. It will be noted that the low co-ordination number of 2 for Na in (NaR)$_\infty$ is unprecedented; furthermore, the compound is both

$(LiR)_\infty{}^a$						
$(NaR)_\infty{}^a$	$(MgR_2)_\infty{}^a$	$(AlR_2)_2$	AlR_3			
		$(GaR_2)_2$	GaR_3	$(GeR_2)_2{}^a$	$GeR_3{}^a$	$(PR_2)_2{}^a$	$(PR)_3$ or 4
		$(InR_2)_2$	InR_3	$(SnR_2)_2{}^a$	$SnR_3{}^a$	AsR_2 (gas)	$(SbR)_4$
C.N.^{b} 2	3	3	3	3	3	3 (2 for gas)	3

Scheme 2. Homoleptic main group metal alkyls MR_n [R = $CH(SiMe_3)_2$] [^{a}Monomer in C_6H_{12} and in gas; ^{b}C.N. = co-ordination number]

Role of Trimethylsilyl-substituted Ligands in Co-ordination Chemistry

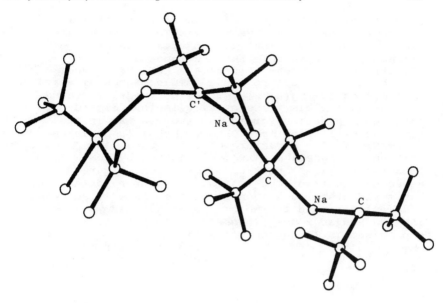

Figure 1 X-Ray structure of $[Na(\mu-R)]_\infty$ [<Na-C> 2.557(13) Å; C-Na-C' 123.1(4)°, 147.1(4)°; Na-C-Na 154.6(5), 155.1(5)°].[6]

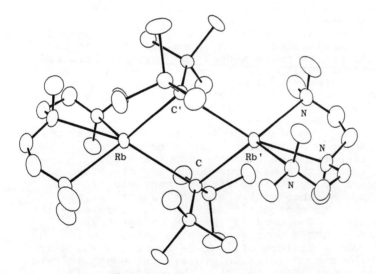

Figure 2 X-Ray structure of $[Rb(\mu-R)\{pmdeta\}]_2$ [Rb-C 3.485(8) Å, 3.361(9) Å; <Rb-N> 3.112 Å; Rb-C-Rb' 104.7(2)°; C-Rb-C' 75.3(2)°].[6]

hydrocarbon-soluble and volatile *in vacuo* at < 100 °C.

X-Ray data are not yet available on the heavier alkali metal alkyls, but the molecular geometry of the hydrocarbon-soluble [Rb(μ-R)(pmdeta)]$_2$ is illustrated in Figure 2 [pmdeta = MeN(CH$_2$CH$_2$NMe$_2$)$_2$].[6]

The usual structure for a simple magnesium alkyl is typified by [Mg(μ-Me)$_2$]$_\infty$. By contrast, we found that using the bis(trimethylsilyl)methyl ligand three-co-ordination for Mg is favoured, as in the X-ray authenticated MgR$_2$(OEt$_2$), (MgR$_2$)$_2$(μ-p-dioxane), or (MgR$_2$)$_\infty$. The structure of the crystalline homoleptic magnesium alkyl, elucidated by low temperature (15 K) neutron diffraction, Figure 3, is especially noteworthy.[7] It will be observed that in the solid state each magnesium atom, as well as being covalently bonded to two groups R, also shows a weak *intermolecular γ-methyl agostic* interaction, as emphasised in (1) using the atom numbering of

```
C(13)    H    H
    \     \  /
     Si-----C(12)····Mg'
    /      \
C(11)       H      (1)
```

Figure 3. Three-co-ordination for magnesium is unusual and was previously established in its organometallic chemistry only for the anion [Mg(CH$_2$Bu^t)$_3$]$^-$.[8] There is a single example of a two-co-ordinate crystalline alkyl, Mg[C(SiMe$_3$)$_3$]$_2$.[9]

Calcium and barium bis(trimethylsilyl)methyls have recently been made; X-ray data are to hand on the monomeric CaR$_2$(THF)$_2$.[10]

Treatment of BCl$_3$ and an excess of LiR yielded BR$_2$Cl.[11] Replacement of the final Cl$^-$ by R is not possible, nor is there a reaction between BR$_2$Cl and LiMe unless (Me$_2$NCH$_2$)$_2$ is added.[12] The mixed alkyl BR$_2$Me is noteworthy not only for being stable to disproportionation to symmetrical products, but also for its remarkable stability, as evident by its lack of reaction with methanol.[12]

Efforts have been made to establish routes to the boron analogue of a carbene, by a reductive elimination reaction using as substrate RBCl$_2$.[13] Such a fragment RB was trapped by use of either an anthracene or cyclo-octatetraene to yield (1)-(3), Scheme 3. Neither of the adducts (2) or (3) was susceptible to photolytic degradation to release the RB fragment. Compound (3) proved by variable temperature NMR spectroscopy to be not a simple adduct, but a mixture of isomers I and II, which are related by a bond rotation process for which $\Delta G^\ddagger$ was found to be 42 ± 1 kJ mol^{-1}. The difference in energy between the two structures was calculated as *ca.* 1.5 kJ mol^{-1}. As well as a bond rotation process there is also a bond shift pathway resulting in a degenerate Cope rearrangement, equation (1) (I = III).

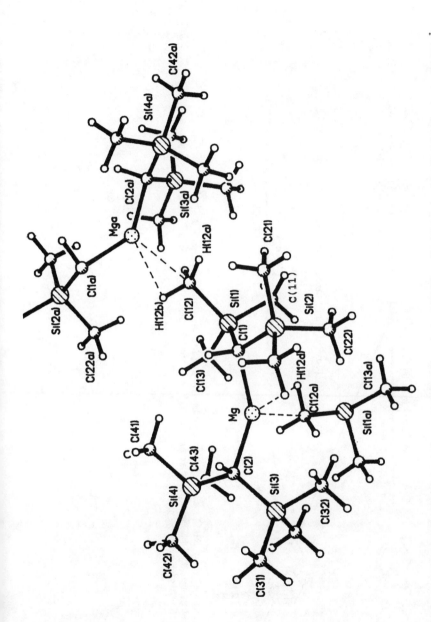

Figure 3 Neutron diffraction structure (at 15 K) of (MgR₂)∞ [Si-C(12) 1.915(8) Å, Si-C(11) 1.876(6) Å, Si-C(13) 1.879(5) Å; Mg-H(12b) 2.333(4) Å, Mg-H(12b) 2.414(5), Mg-H (12c) 2.516(4); H-C(12)-H 110.4°; Mg'···C(12)-Si 172.3(2)°.⁷

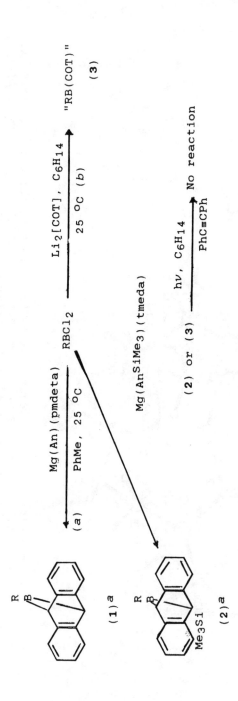

Scheme 3. RB Trapping experiments [An = anthracene, AnSiMe3 = 9-trimethylsilylanthracene, COT = cyclo-octatetraene, tmeda = $(Me_2NCH_2)_2$, pmdeta = $MeN(CH_2CH_2NMe_2)_2$]

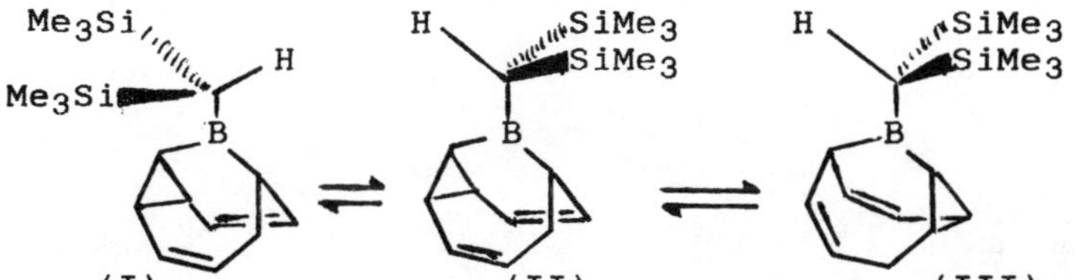

$$\text{(I)} \rightleftharpoons \text{(II)} \rightleftharpoons \text{(III)} \quad (1)$$

An interesting series of binuclear alkyls of some of the heavier group 13 elements has recently been described, equations (2),[14] (3),[15] and (4).[16] The X-ray structure of each of the compounds R_2M-MR_2 (M = Al,[14] Ga,[15] or In[16]) has been determined and shows that the central C_2MMC_2 skeleton is coplanar. Earlier it proved possible to obtain the mononuclear compounds $M'R_3$ (M' = Ga or In) which, unlike their methyl analogues, are mononuclear; the X-ray structure of the In compound shows that the InC_3 skeleton is coplanar.[17]

$$2R_2AlCl \xrightarrow{4K} R_2Al-AlR_2 \quad (2)$$

$$Ga_2Br_4 \cdot 2\text{dioxane} \xrightarrow{4LiR} R_2Ga-GaR_2 \quad (3)$$

$$In_2Br_4 \cdot 2\text{tmeda} \xrightarrow{4LiR} R_2In-InR_2 \quad (4)$$

The nature of the product of treating the divalent chloride $(MCl_2)_\infty$ (M = Ge, Sn, or Pb) with a hydrocarbyl-lithium reagent LiR' depends on the nature of the group R'. It is likely that in each case the divalent mononuclear species SnR'_2 is formed which, however, is kinetically unstable and may either associate to provide an oligomer or alternatively disproportionate, Scheme 4.

The germanium and tin compounds M_2R_4 (M = Ge or Sn) were first described in 1976;[18] the full paper on the crystalline materials appeared a decade later.[19] It was shown that the compounds are the heavy group 14 element analogues of alkenes which differ from the latter in that the central element M is in a pyramidal rather than in a trigonal planar environment. This is illustrated in Figure 4, which also shows a representation of the bonding arrangement between the two metal atoms accounting for the weakness of the MM bond. That aspect is well illustrated by the fact that in dilute cyclohexane solution the compounds are monomeric, although in a more concentrated solution of Sn_2R_4 some dimer is still present.[20] In the vapour phase the compounds exist solely as the V-shaped diamagnetic monomers.[21] In crystalline M_2R_4 the MM distance is shorter than

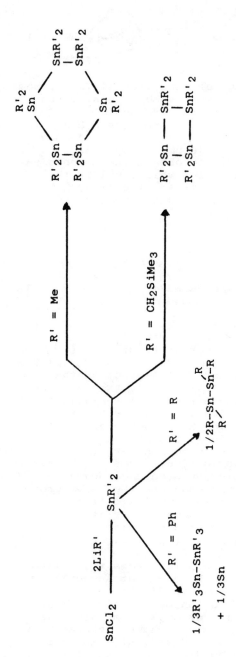

Scheme 4. Alkylation of $SnCl_2$ as a function of R'

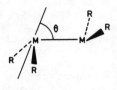

$R_2\underline{M}-\ddot{M}R_2 \longleftrightarrow R_2M-\underline{\ddot{M}}R_2$

Figure 4 Structure and bonding in M_2R_4 (M = Ge or Sn).[19]

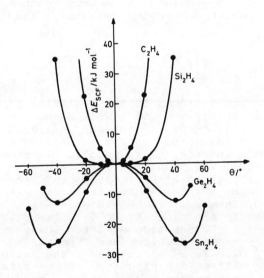

Figure 5 Variation of SCF energies for M_2H_4 with the trans-fold angle θ (M = C, Si, Ge, or Sn).[19]

in the tetrahedral form of the element M_∞. The Ge and Sn compounds are clearly related to silicon analogues, of which $Si_2(C_6H_2Me_3-2,4,6)_4$ was the first homologue.[22] The fold angle θ in these dinuclear molecules is not exclusively steric in origin, as may be judged from a plot of θ versus the SCF energy in model molecules M_2H_4, Figure 5.[19] Thus C_2H_4 has a sharp energy minimum at θ = 0, whereas for the silicon analogue the potential is much softer, and for the heavier group 14 congeners sharp minima at θ = 40° for M = Ge and θ = 46° for M = Sn are observed. The compounds M_2R_4, as well as the lead alkyl the structure of which has not yet been determined,[18] are highly coloured and thermochromic, becoming redder on heating and colourless at liquid nitrogen temperature. They have an extensive chemistry, including a versatile role in transition metal chemistry.[23] As an example of unusual reactivity, attention is drawn to $[SnR_2(\mu-O)]_2$,[24] whereas cyclostannoxanes are generally trimers. This four-membered ring compound was obtained by oxidation of Sn_2R_4 using Me_3NO.[24] It is related to $[Si(C_6H_2Me_3-2,4,6)_2(\mu-O)]_2$;[25] both compounds have short transannular M···M and O···O interactions. The cyclodistannoxane is also of interest in that its controlled hydrolysis provides the unusual compound $[(SnR_2OH)_2(\mu-O)]$ which on heating reverts to its factors.[24]

$$2MR_nCl \xrightarrow[PhMe, h\nu]{L^{Me}{}_2} 2\dot{M}R_n + [L^{Me}{}_2]Cl_2 \qquad (5)$$

The ligand $\bar{R}$ has featured in the isolation (in hydrocarbon solution) of unusual stable metal-centred radicals $\dot{G}eR_3$,[26] $\dot{S}nR_3$,[26] $\dot{P}R_2$,[27] and $\dot{A}sR_2$.[27] Each of these compounds is coloured, and was identified by characteristic ESR spectra. For example, $\dot{P}R_2$ shows a doublet of triplets, the doublet structure due to $a(^{31}P)$ and the triplet to $a(^1H)$, whereas $\dot{S}nR_3$ shows a doublet of quartets, the former arising from $a(^{119}Sn)$. The synthesis of these radicals was achieved as shown in equation (5), $L^{Me}{}_2 = [\dot{C}N(Me)CH_2CH_2\dot{N}Me]_2$. Evaporation of a red toluene solution of $\dot{P}R_2$ yields the colourless crystalline dimer P_2R_4.[27,28] The X-ray structure of the latter and the electron diffraction structure of the gaseous $\dot{P}R_2$ have been carried out by Rankin et al., and are described elsewhere in this volume.[29] From the ESR spectra it was deduced that $\dot{P}R_2$ [27] and $\dot{S}nR_3$ [26] are π- and σ- radicals, respectively; the former assignment was confirmed by the GED data.[29] The stability of the radicals is attributable to the usual kinetic effects. In the particular case of $\dot{P}R_2$, for dimerisation to occur it is necessary for an electronic and conformational re-organisation to take place prior to dimerisation.

The compounds $(PR)_n$ are believed to be cyclotrimers or -tetramers,[31] the dimers being less stable but related molecules, such as the crystalline $[As\{C(SiMe_3)_3\}]_2$ [31] and $(PAr)_{21}$ [32] are well established (Ar = $C_6H_2Bu^t_3$-2,4,6). The crystalline antimony compound $(SbR)_4$ is a cyclotetramer.[33]

3 HOMOLEPTIC d- AND f- BLOCK METAL ALKYLS MR_n [R = CH(SiMe$_3$)$_2$] AND RELATED ADDUCTS

The homoleptic alkyls which have been structurally authenticated are indicated in square brackets in Scheme 5; those shown in round brackets almost certainly have the structures indicated, as evident from spectroscopic and molecular weight data. Some syntheses involving the use of the metal chloride and an excess of LiR are shown in Scheme 6. With the lanthanide chlorides $LnCl_3$, the reaction product (containing Cl) varies in a way that may reflect the size of Ln^{3+}. Accordingly, for the synthesis of homoleptic compounds LnR_3 an alternative precursor had to be found (vide infra).

Synthetic data relating to manganese(II) compounds are shown in Scheme 7.[34] The homoleptic alkyl is a monomer in benzene and therefore quite possibly also in the solid state.[35] It is a typical high spin d^5 complex, as evident by magnetic measurements over a wide temperature range. In the gas phase, [MnR$_2$] is linear. The crystalline adducts [MnR$_2$(thf)] and [MnR$_2$(dmpe)] have been X-ray authenticated [thf = OC_4H_8, dmpe = $(Me_2PCH_2)_2$].[34]

The alkyls of d- and f- block elements MR_n are of interest as useful synthetic precursors. For example, they generally readily react with x moles of a protic compound HA to provide $MR_{n-x}A_x$ with elimination of the volatile RH, and undergo insertion reactions with many unsaturated compounds. They may also be useful R transfer reagents, as illustrated in Scheme 7 in the context of a synthesis of PbR_2.[34]

Selected structural data for some manganese(II) alkyls are collected in Table 2.

$$[Ln(OAr)_3] + 3LiR \xrightarrow{\text{pentane}} [LnR_3] + 3LiOAr \qquad (6)$$

Treatment of [Ln(OAr)$_3$] (Ar = $C_6H_3Bu^t_2$-2,6), or a simple analogue having a Me or Bu^t substituent in the 4-position, yields the homoleptic alkyl, equation 6.[36] The X-ray structures of the La and Sm alkyls have been determined and show that the molecules are pyramidal at Ln with the CLnC' angle 109.9(2)° for La or 110(1)° for Sm.[36] The mean Ln-C bond length

(ScR$_3$)	(TiR$_3$)	(VR$_3$)	[CrR$_3$]	[MnR$_2$] gas
$3d^0$	$3d^1$	$3d^2$	$3d^3$	$3d^5$
(YR$_3$)				[MnR$_2$(thf)]
$4d^0$				[MnR$_2$(dmpe)]
[LaR$_3$]		[SmR$_3$]	[UR$_3$]	
$4f^0$		$4f^5$	$5f^3$	

Scheme 5. Homoleptic d and f-block metal alkyls MR$_n$ and related compounds [R = CH(SiMe$_3$)$_2$]

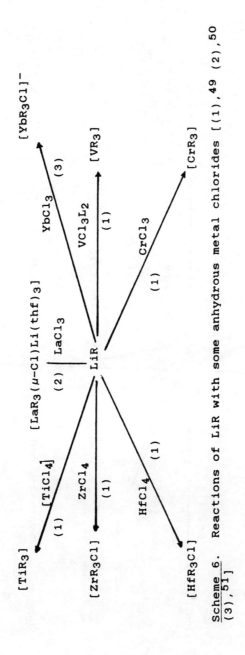

Scheme 6. Reactions of LiR with some anhydrous metal chlorides [(1),⁴⁹ (2),⁵⁰ (3),⁵¹]

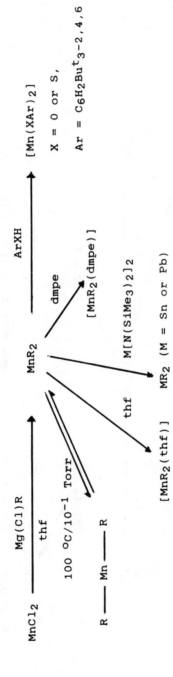

Scheme 7. Synthesis and some reactions of MnR_2 and its thf and dmpe adducts

Table 2. Selected structural data for some MnR'$_2$ complexes

R' in [(MnR'$_2$)$_n$]	n	C.N. of Mn	<Mn–C>/Å	C–Mn–C/(°)	Ref.
Np[a]	1 (G.E.D.)[a]	2	2.104(6)	180 (ass.)[a]	45
R	1 (G.E.D.)[a]		2.01(3)	180 (ass.)[a]	35
Tsi[a]	1	2	2.102(4)	180	46
Mes[a]	3	3,4	2.11(1)t[a]	122.1(4)	47
			2.20(2)b[a]		
			2.34(1)b		
[MnR$_2$(thf)]	–	3	2.06(2)	160.1(9)	34
[MnR$_2$(dmpe)]	–	4	2.17(1)	141.4(7)	34
[{MnCy(μ–Cy)}$_2$(μ–dmpe)][a]	–	4	2.12(1)t		48
			2.25(9)b		
			2.327(9)b		

[a]Np = CH$_2$Bu$^t_2$, Tsi = C(SiMe$_3$), Cy = c–C$_6$H$_{11}$, dmpe = (Me$_2$PCH$_2$)$_2$, G.E.D. = gas electron diffraction, ass. = assumed, t = terminal, b = bridging

reflects the lanthanide contraction, being 2.515(9) Å for La and 2.33(2) Å for Sm. A further feature of the crystal structures is that there are close α- and γ- agostic interactions. For example the La$\cdots$C(γ) contact is 3.121(9) Å and the corresponding Sm distance is 2.85(3) Å.

The co-ordination number of 3 for a lanthanide ion is unusually small; hence, not surprisingly, a compound [LnR$_3$] is able not only to form an adduct with a neutral donor, but also with an anionic ligand, as exemplified by equation 7, [pmdeta = MeN(CH$_2$CH$_2$NMe$_2$)$_2$].[37] The X-ray structure of this unusual mixed metal alkyl is shown in Figure 6, from which it will be noted that the bond lengths and angles from Sm to the ligand R are essentially the same as those in [SmR'$_3$].[36] Consistent with that is the placement of the Me ligand so that there is an almost linear arrangement of the SmCLi skeleton; the Sm-C(H$_3$) distance is exceptionally short, while the (H$_3$)C-Li bond length is particularly long.[37]

Alkyls of ytterbium(II) have recently been made from a corresponding phenoxide, as shown in equation (8) (Ar = C$_6$H$_2$Bu$^t_2$-2,6-Me-4).[38] Although X-ray data are not yet available, the identity of two alkyls is most clearly demonstrated by their ^{171}Yb NMR spectra in toluene-d$_8$ which shows a triplet pattern for YbR$_2$ due to $^2J(^{171}$Yb^{1}H) at 30 Hz, while for an intermediate compound Yb(OAr)R there is a doublet, $^2J(^{171}$Yb^{1}H) 30 Hz. The technique of ^{171}Yb NMR spectroscopy has recently been introduced.[39]

The homoleptic uranium(III) alkyl was made, equation (9),[40] using a technique similar to that which had been used for [LnR$_3$].[36] The X-ray structure of [UR$_3$] shows that the uranium atom is in a pyramidal environment with a relatively close U$\cdots$C(γ) contact of 3.09(2) Å compared with a mean U-C distance of 2.48(2) Å.

[LnR$_3$] + LiMe(pmdeta) $\longrightarrow$ [LnR$_3$(μ-Me)Li(pmdeta)] (7)

$$[Yb(OAr)_2(OEt_2)_2] \text{ (ref. 52)} \xrightarrow[(-LiOAr)]{LiR} YbR_2 + Yb(OAr)R \quad (8)$$

[U(OAr)$_3$] (ref. 53) + 3LiR $\longrightarrow$ [UR$_3$] + 3LiOAr   (9)

4 CONCLUSIONS

It will be noted that a wide range of lipophilic and often volatile homoleptic metal alkyls [{MR$_n$}$_m$] has now been prepared; the value of n is usually 3 for m = 1, but sometimes n is 2 or even 1. The alkyls are generally mononuclear in dilute hydrocarbon solution or in the gas phase, but in the crystal may be bi- (Ge or Sn), tri- (P), tetra - (P or Sb), or even (Li,

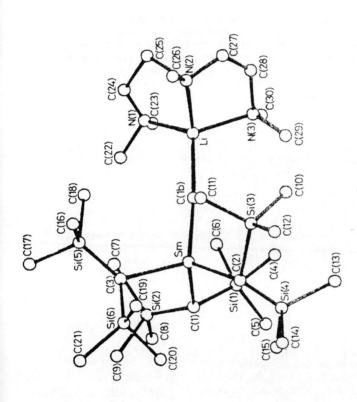

Figure 6. X-Ray structure of [SmR$_3$(μ-Me)Li(pmdeta)] [Sm-C(1b) 2.33(3) Å, C(1b)-Li 2.42(6) Å, <Sm-C(1,2, or 3) > [2.52(3) Å; C(1)SmC(2) 111(1)°, C(1)SmC(3) 104(1)°, C(2)SmC(3) 106(1)°; C(1)SmC(1b) 103(1)°, C(2)SmC(1b) 116(1)°, C(3)SmC(1b) 115(1)°].[34]

Na, or Mg) polynuclear. The co-ordination number around the metal in the crystalline state is very low, usually 3, but sometimes (Li or Na) 2. For the X-ray characterised complexes of formula [MR$_3$] the geometry at M appears to be trigonal planar for main group and d-block elements M, but tends towards tetrahedral angles for the pyramidal f-block complexes (M = La 329.7°, Sm 330°, or U).

These metal alkyls have an extensive chemistry which, by and large, it has only been possible here to touch upon.

5 ACKNOWLEDGEMENTS

I am grateful to my co-workers, whose names are listed in the bibliography, for their valuable contributions, and SERC, the Leverhulme Trust, and the EC (Stimulation Contract: ST2000335) for their support. Regarding the unpublished work, I particularly wish to acknowledge my considerable debt to Dr. W.-P. Leung.

6 REFERENCES

1. M.R. Collier, B.M. Kingston, M.F. Lappert, and M.M. Truelock, Brit. Patent 36021 (1969).
2. M.R. Collier, M.F. Lappert, and M.M. Truelock, J. Organomet. Chem., 1970, 25, C36; (CH$_2$SiMe$_3$): G. Yagupsky, W. Mowat, A. Shortland, and G. Wilkinson, J. Chem. Soc., Chem. Commun., 1970, 1369.
3. (a) P.J. Davidson, M.F. Lappert, and R. Pearce, Acc.Chem. Res., 1974, 7, 209; (b) P.J. Davidson, M.F. Lappert, and R. Pearce, Chem. Rev., 1976, 76, 219.
4. M.F. Lappert, D.S. Patil, and J.B. Pedley, J. Chem.Soc., Chem. Commun., 1975, 830.
5. J.L. Atwood, T. Fjeldberg, M.F. Lappert, N.T. Luong-Thi, R. Shakir, and A.J. Thorne, J. Chem. Soc., Chem. Commun., 1984, 1163.
6. R. Bohra, P.B. Hitchcock, M.F. Lappert, and W.-P. Leung, unpublished work.
7. P.B. Hitchcock, J.A.K. Howard, M.F. Lappert, W.-P. Leung, and S.A. Mason, J. Chem. Soc., Chem. Commun., 1990, 847.
8. E.P. Squiller, R.R. Whittle, and H.G. Richey, J. Am. Chem. Soc., 1985, 107, 432.
9. S.S. Al-Juaid, C. Eaborn, P.B. Hitchcock, C.A. McGeary, and J.D. Smith, J. Chem. Soc., Chem. Commun., 1989, 273.
10. P.B. Hitchcock, M.F. Lappert, G.A. Lawless, and B. Royo, J. Chem. Soc., Chem. Commun., 1990, 1141; Idem, unpublished work.
11. S. Al-Hashimi and J.D. Smith, J. Organomet. Chem., 1978, 153, 253.

12. G. Beck and M.F. Lappert, unpublished work.
13. A.G. Avent, M.F. Lappert, B. Skelton, C.L. Raston, L.M. Engelhardt, S. Harvey, and A.H. White, Chapter 15 in "Heteroatom Chemistry", ed. E. Block, VCH Publications, New York, 1990.
14. W. Uhl, Z. Naturforsch. Teil B, 1988, 43B, 1113.
15. W. Uhl, M. Layh, and T. Hildenbrand, J. Organomet. Chem., 1989, 364, 289.
16. W. Uhl, M. Layh, and W. Hiller, J. Organomet. Chem., 1989, 364, 139.
17. A.J. Carty, M.J.S. Gynane, M.F. Lappert, S.J. Miles, A. Singh, and N.J. Taylor, Inorg. Chem., 1980, 19, 3637.
18. D.E. Goldberg, D.H. Harris, M.F. Lappert, and K.M. Thomas, J. Chem. Soc., Chem. Commun., 1976, 261.
19. D.E. Goldberg, P.B. Hitchcock, M.F. Lappert, K.M. Thomas, A.J. Thorne, T. Fjeldberg, A. Haaland, and B.E.R. Schilling, J. Chem. Soc., Dalton Trans., 1986, 2387.
20. M.F. Lappert, Silicon, Germanium, Tin, and Lead Compounds, 1986, 9, 129; K.W. Zilm, G.A. Lawless, R.M. Merrill, J.M. Millar, and G.G. Webb, J. Am. Chem. Soc., 1987, 109, 7236.
21. T. Fjeldberg, A. Haaland, M.F. Lappert, B.E.R. Schilling, and A.J. Thorne, J. Chem. Soc., Dalton Trans., 1986, 1551.
22. M.J. Fink, M.J. Michalczyk, K.J. Haller, R. West, and J. Michl, J. Chem. Soc., Chem. Commun., 1983, 1010.
23. M.F. Lappert and R.S. Rowe, Coord. Chem. Revs., 1990, 100, 267.
24. M.A. Edelman, P.B. Hitchcock, and M.F. Lappert, J. Chem. Soc., Chem. Commun., 1990, 1116.
25. M.J. Fink, K.J. Haller, R. West, and J. Michl, J. Am. Chem. Soc., 1984, 106, 822.
26. A. Hudson, M.F. Lappert, and P.W. Lednor, J. Chem. Soc., Dalton Trans., 1976, 2369.
27. M.J.S. Gynane, A. Hudson, M.F. Lappert, P.P. Power, and H. Goldwhite, J. Chem. Soc., Dalton Trans., 1980, 2428.
28. P.P. Power, D.Phil. Thesis, University of Sussex, 1977.
29. D.W.H. Rankin, IXth International Symposium on Organosilicon Chemistry, Edinburgh, July 1990, Paper No. I.13 (this volume).
30. A.H. Cowley, J.E. Kilduff, S.K. Mehrotra, N.C. Norman, and M. Pakulski, J. Chem. Soc., Chem. Commun., 1983, 528.
31. C. Couret, J. Escudie, Y. Madaule, H. Ranivonjatovo, and J.-G. Wolf, Tetrahedron Lett., 1983, 24, 2769.
32. M. Yoshifuji, I. Shima, N. Inamoto, K. Hirotsu, and T. Higuchi, J. Am. Chem. Soc., 1981, 103, 4587.

33. H.J. Breunig, M. Ates, and M. Dräger, IXth International Symposium on Organosilicon Chemistry, Edinburgh, July 1990, Paper No. 3.22.
34. P.B. Hitchcock, M.F. Lappert, W.-P. Leung, and N.H. Buttrus, J. Organomet. Chem., 1990, 394, 57.
35. R.A. Andersen, D.J. Berg, L. Fernholt, K. Faegri, J.C. Green, A. Haaland, M.F. Lappert, W.-P. Leung, and K. Rypdal, Acta Chem. Scand., 1988, A42, 554.
36. P.B. Hitchcock, M.F. Lappert, R.G. Smith, R.A. Bartlett, and P.P. Power, J. Chem. Soc., Chem. Commun., 1988, 1007.
37. P.B. Hitchcock, M.F. Lappert, and R.G. Smith, J. Chem. Soc., Chem. Commun., 1989, 369.
38. S.A. Holmes and M.F. Lappert, unpublished work.
39. A.G. Avent, M.A. Edelman, M.F. Lappert, and G.A. Lawless, J. Am. Chem. Soc., 1989, 111, 3423.
40. W.G. van der Sluys, C.J. Burns, and A.P. Sattelberger, Organometallics, 1989, 8, 855.
41. E. Weiss, G. Sauermann, and G. Thirase, Chem. Ber., 1983, 116, 74.
42. E. Weiss and G. Sauermann, J. Organomet. Chem., 1970, 21, 1.
43. C. Schade, P. v. R. Schleyer, H. Dietrich, and W. Mahdi, J. Am. Chem. Soc., 1986, 108, 2484.
44. H. Köster and E. Weiss, J. Organomet. Chem., 1979, 168, 273.
45. R.A. Andersen, A. Haaland, K. Rypdal, and H.V. Volden, J. Chem. Soc., Chem. Commun., 1985, 1807.
46. N.H. Buttrus, C. Eaborn, P.B. Hitchcock, J.D. Smith, and A.C. Sullivan, J. Chem. Soc., Chem. Commun., 1985, 1380.
47. S. Gambarotta, C. Floriani, A. Chiesi-Villa, and C. Guastini, J. Chem. Soc., Chem. Commun., 1983, 1128.
48. G.S. Girolami, C.G. Howard, G. Wilkinson, H.M. Dawes, M. Thornton-Pett, M. Motevalli, and M.B. Hursthouse, J. Chem. Soc., Dalton Trans., 1985, 921.
49. G.K. Barker, M.F. Lappert, and J.A.K. Howard, J. Chem. Soc., Dalton Trans., 1978, 734.
50. J.L. Atwood, M.F. Lappert, R.G. Smith, and H. Zhang, J. Chem. Soc. Chem. Commun., 1988, 1308.
51. J.L. Atwood, W.E. Hunter, R.D. Rogers, J. Holton, J. McMeeking, R. Pearce, and M.F. Lappert, J. Chem. Soc., Chem. Commun., 1978, 140.
52. G.B. Deacon, P.B. Hitchcock, S.A. Holmes, M.F. Lappert, P. MacKinnon, and R.H. Newnham, J. Chem. Soc., Chem. Commun., 1989, 935.
53. W.G. van der Sluys, C.J. Burns, J.C. Huffman, and A.P. Sattelberger, J. Am. Chem. Soc., 1988, 110, 5924.

Structure and Reactivity – The Role of Bulky Silyl Groups

David W.H. Rankin
DEPARTMENT OF CHEMISTRY, UNIVERSITY OF EDINBURGH, WEST MAINS ROAD, EDINBURGH EH9 3JJ, UK

The fact that bulky groups can stabilise otherwise reactive groupings of atoms is long established, as is the collorary that reactions of compounds may be profoundly modified by inclusion of one or more large substituents. For example, the first preparation of a disilene[1] used four mesityl groups to protect the Si=Si bond.

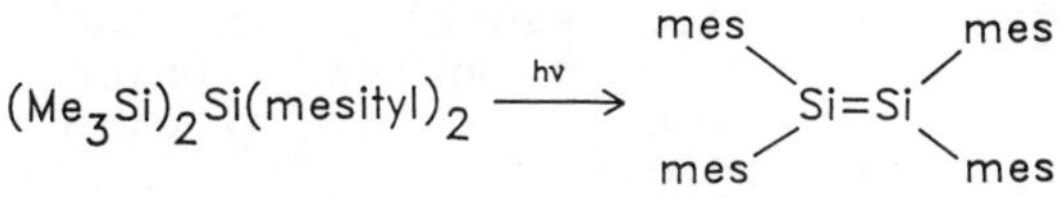

$$(Me_3Si)_2Si(mesityl)_2 \xrightarrow{h\nu} \underset{mes}{\overset{mes}{\diagdown}}Si=Si\underset{mes}{\overset{mes}{\diagup}} \quad (1)$$

An alternative route to a disilene also involved preparation of a cyclotrisilane, this time using 2,6-dimethylphenyl as the bulky ligand.[2]

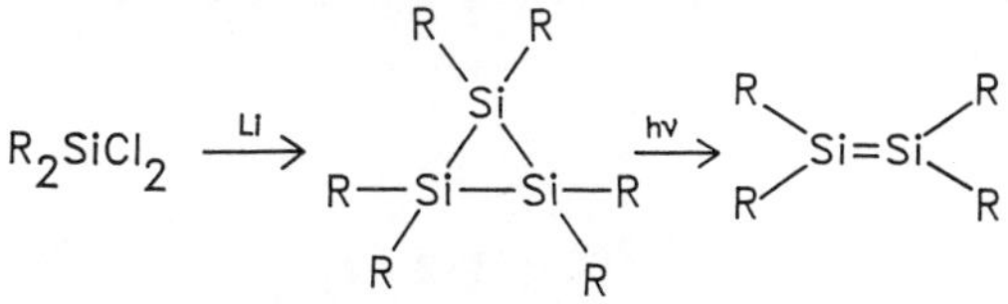

$$R_2SiCl_2 \xrightarrow{Li} \begin{array}{c} R\,R \\ Si \\ R-Si-Si-R \\ R\;\;\;R \end{array} \xrightarrow{h\nu} \underset{R}{\overset{R}{\diagdown}}Si=Si\underset{R}{\overset{R}{\diagup}} \quad (2)$$

R = 2,6-dimethylphenyl

Similarly, the first compound containing an Si=C double bond to be isolated contained both the very bulky adamantyl groups and trimethylsilyl groups[3] [Figure 1(a)].

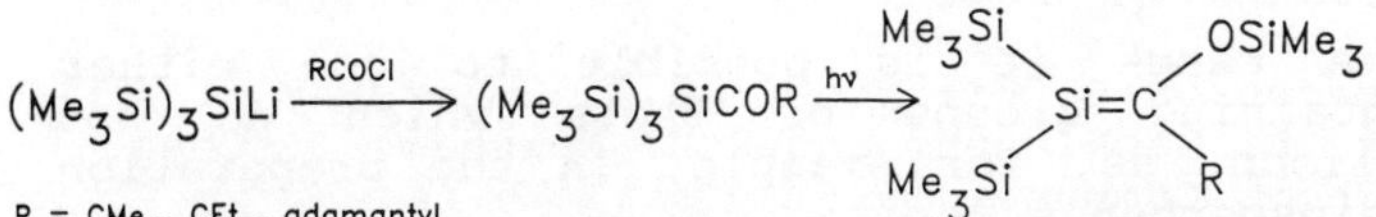

$$(Me_3Si)_3SiLi \xrightarrow{RCOCl} (Me_3Si)_3SiCOR \xrightarrow{h\nu} \underset{Me_3Si}{\overset{Me_3Si}{\diagdown}}Si=C\underset{R}{\overset{OSiMe_3}{\diagup}} \quad (3)$$

R = CMe$_3$, CEt$_3$, adamantyl

In this work, t-butyl (-CMe$_3$) and 3-ethyl-pentyl (-CEt$_3$) groups were also used, and the importance of the size of the group in stabilising the product was thus shown. It is natural to think of groups in terms of size, but shape is also important. Tetramesityl-disilene has flat groups, which must be twisted so

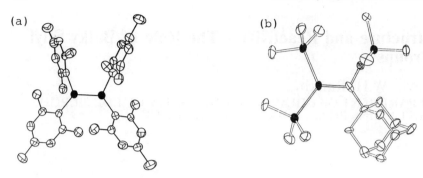

Figure 1 Structures of (a) tetramesityldisilene and (b) $(Me_3Si)_2Si = C(OSiMe_3)(adamantyl)$

that the four of them can be accommodated [Figure 1(b)].

Silicon-containing groups are often used in this way. One reason is that the size of the group can be selected, either by using different numbers of silicon-containing fragments or by choosing different alkyl substituents. For example, $-C(SiMe_3)_3$ and $-CH(SiMe_3)_2$ are often used. The shape of the latter group is particularly important, as they can pack together. They were used, for example, in the preparation of a tetralkylbiphosphine which dissociates in solution or in the gas phase to give stable dialkylphosphinyl free radicals.[4]

$$((Me_3Si)_2CH)_2PCl \xrightarrow{Na} ((Me_3Si)_2CH)_2P-P(CH(SiMe_3)_2)_2 \quad \text{solid}$$

$$\Updownarrow \quad \text{liquid}$$

$$\cdot P(CH(SiMe_3)_2)_2 \quad \text{gas} \qquad (4)$$

$$RPCl_2 \xrightarrow{Na^+ naphthalenide^-} \underset{R}{\overset{R}{\diagdown}} P=P \underset{R}{\diagdown} \qquad (5)$$

$R = C(SiMe_3)_3 ; 2,4,6-C_6H_2Bu^t_3$

In some cases it is possible to use either silicon-containing groups or those which do not contain silicon, as, for example, in the preparation of dialkyldiphosphenes.[5]

So are the silyl groups just large substituents, with fixed size and shape? No. They are more than that. They are flexible, and that is an important factor in their usefulness. One consequence of their flexibility is that they can pack together easily, just as soft objects can be packed more easily into a box than can rigid ones. The groups within a sterically crowded compound may be packed together in this way, but it is also important to remember that easily deformed molecules may well have different conformations in different phases. The second consequence is a dynamic one: the groups may change their internal structure during the course of a reaction. It is therefore possible to prepare compounds which are stabilised by protecting groups, but not so much that they cannot undergo further reactions. The protecting groups resemble a coat of mail, rather than a more restricting suit of rigid armour. And so we can prepare useful new compounds, and not just chemical curiosities which may contain some novel bonding arrangement, but are not the basis for further developments in chemistry.

The flexibility of some of these ligands is revealed by comparing their structures in ranges of compounds. We have studied several compounds containing trisyl [$-C(SiMe_3)_3$] and disyl [$-CH(SiMe_3)_2$] groups in the gas phase by electron diffraction. Examples include $(Me_3Si)_3CPH_2$ [Figure 2(a)][6] and $(Me_3Si)_3CSiCl_3$ [Figure 2(b)].[7]

Both show one feature common to almost all compounds containing an $-A(BC_3)_3$ fragment or having the $A(BC_3)_4$ arrangement: the BC_3 (here $SiMe_3$) groups are twisted by between 15 and 20° from perfectly staggered conformations. This is necessary to relieve the Me···Me interactions between neighbouring $SiMe_3$ groups. Any further twist causes other interactions to become important, and the observed angles are those minimising the total interaction energy. The angle can easily be predicted using a simple model of such a molecule. Other important structural features are revealed by comparing geometrical parameters in a range of compounds (Table 1).

In all these compounds the three methyl groups in each $SiMe_3$ group are compressed, as revealed by the $C_{inner}SiC_{Me}$ angles. In contrast, the SiCSi angles are

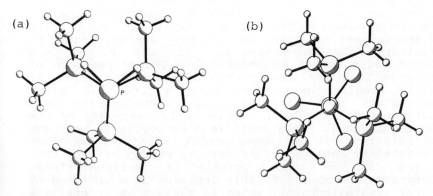

<u>Figure 2</u> Gas-phase molecular structures of (a) $(Me_3Si)_3CPH_2$, viewed along the P-C bond, and (b) $(Me_3Si)_3CSiCl_3$, viewed along the three-fold axis.

Table 1 Average bond lengths and angles in some trisyl compounds

Compound	inner Si-C pm	outer Si-C pm	SiCSi deg	CSiC(Me) deg	Ref.
RPH$_2$	194.1	188.3	106.9	114.3	6
RSiMe$_3$	193.1	189.6	109.5	113.1	8
RSiMe$_2$Ph	191.6	187.7	109.5	113.5	9
RP=PR	192	187	111.1	112.5	5
RSiCl$_3$	191.4	187.8	108.1	112.0	7
RB$_2$H$_4$R	190.4	187.9	111.2	112.7	10
<u>R</u>B(Ph)O(CH$_2$)$_4$R	190.0	186.7	110.4	113.0	11
RB(Ph)O(CH$_2$)$_4$<u>R</u>	190.0	186.7	110.8	113.0	11
RHgR	188.7	187.1	112.6	113.0	12
RH	188.8	187.3	117.5	112.8	13

variable, and corrolate strongly with the Si-C$_{inner}$ distances. It is clear that with a sterically undemanding fourth ligand, such as hydrogen, on the central carbon atom, the SiMe$_3$ groups separate widely, and the inner Si-C bonds are not much longer than in unstrained compounds. When the fourth ligand is larger, the SiMe$_3$ groups are forced closer together. They cannot compress much further internally, and so the inner Si-C bonds must become longer. However, the size of the fourth carbon substituent is not the only factor. Electronic properties must also be of importance, and when the central carbon is bound to phosphorus, the SiCSi angle may even be less than the tetrahedral angle. In this context, it would be interesting to know the structures of trisyl compounds with an electronegative substituent on the central carbon atom.

One problem in studying these compounds by electron diffraction arises from the limited information obtainable by this method. There are many atom pairs contributing to the radial distribution curve for (Me$_3$Si)$_3$CPH$_2$ (Figure 3)[6] but there are few resolved peaks. The P-C and inner and outer Si-C distances are all close to 190 pm, so they overlap. Similarly the non-bonded Si$\cdots$Si, Si$\cdots$P and some C$\cdots$C peaks cannot be separated. The consequence of this is that any experimental error in one parameter will result in compensating errors in other parameters. If the Si-C$_{inner}$ distance is overestimated, the SiCSi angle will necessarily be underestimated. This could lead to precisely the same correlation, as an experimental artefact, as that revealed by Table 1. Fortunately, we <u>can</u> be sure that there is a real correlation between these parameters, because the table includes data obtained by <u>X</u>-ray crystallography, which is subject to an entirely different set of experimental errors.

There is an even greater variation in the internal structures of disyl [CH(SiMe$_3$)$_2$] groups, because they

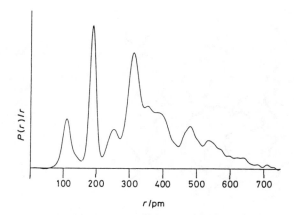

Figure 3 Radial distribution curve $\underline{P}(\underline{r})/\underline{r}$, for $(Me_3Si)_3CPH_2$.

are less internally crowded than trisyl $[C(SiMe_3)_3]$ groups. They are also more difficult to study by electron diffraction, because they have lower symmetry. The ranges of some parameters, taken from published data for gaseous GeR_2 and SnR_2,[14] crystalline Ge_2R_4 and Sn_2R_4,[15] gaseous LiR,[16] solid (polymeric) LiR,[17] gaseous PR_2 and crystalline P_2R_4,[14] are shown in Figure 4.

We will return to the last two compounds later, but first we should consider what we mean by crowded and strained molecules, for they are not the same thing. The molecule $N_2(SiMe_3)_4$ (Figure 5)[18] is crowded, in that there is very little vacant space within the structure, for which all the torsion angles can be predicted, to within about 1°, simply by making a ball and stick model. But it is not strained. The atoms can all be accommodated without forcing the molecule to have any unusual bond lengths or angles.

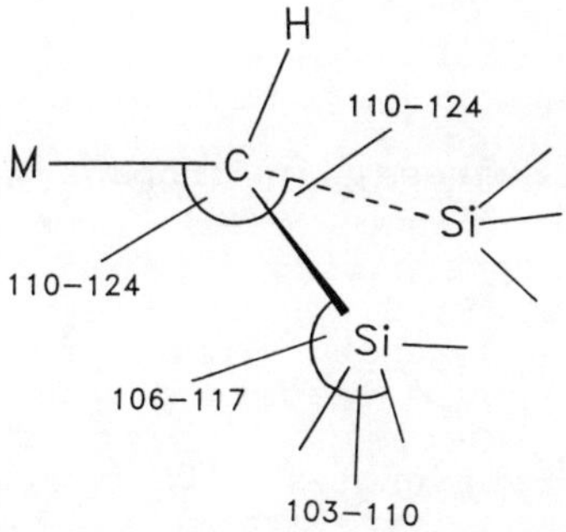

Figure 4 Ranges of angles (in degrees) observed in disyl groups

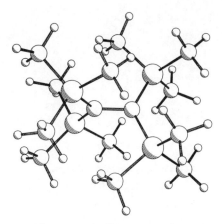

Figure 5   The gas-phase molecular structure of $N_2(SiMe_3)_4$

By considering some simple model compounds, we can see the origins of steric strain. The bond angles at the central atoms of propane, dimethylsilane and disilylmethane are shown in Figure 6. Only in the last one is the angle widened much above the tetrahedral angle, 109.5°, so we can see that two silicon atoms, on the ends of two Si-C bonds separated by 110°, would be slightly strained. But not much. We can put three or four silyl groups onto one carbon atom, without stretching the Si-C bonds significantly at all (Figure 7).

With t-butyl or trimethylsilyl groups the situation is very different. Both $SiH_2Bu^t_2$ and

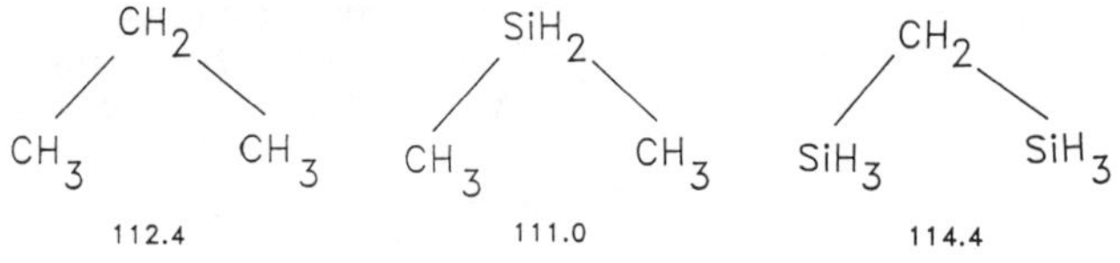

Figure 6   Angles (in degrees) in propane and related compounds.

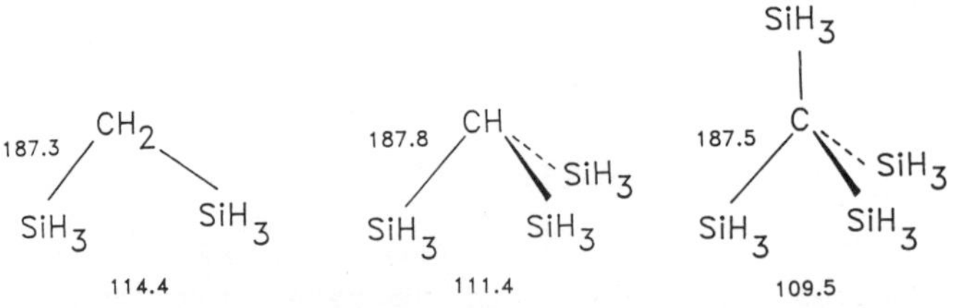

Figure 7   Bond lengths (in pm) and angles (in degrees) in silyl-substituted methane derivatives.

```
      SiH₂                    CH₂
     /    \                  /    \
   CMe₃   CMe₃            SiMe₃   SiMe₃
        125.1                  123.2
```

Figure 8 Angles (in degrees) in SiH₂But₂ and CH₂(SiMe₃)₂

CH₂(SiMe₃)₂ are substantially distorted (Figure 8), although with only two large groups, such distortions can easily take place. So if three or four trimethylsilyl groups are bound to one carbon atom, and the central angles must be reduced towards 109.5°, the molecules must be very strained (Figure 9), and this is shown by substantial lengthening of the inner Si-C bonds. Similarly with t-butyl groups on silicon there is crowding, but here the problem is even more severe (Figure 9). Even with three groups, the Si-C bonds are very much lengthened, and SiBu$^t_4$ is unknown. This is not surprising, as even in SiPr$^i_4$, in which the four substituents can pack together quite well, the Si-C bond is lengthened to 192 pm. However, PBu$^t_4{}^+$, which is isoelectronic with SiBu$^t_4$, is known, so it may yet be possible to make the silicon compound. We should also note here that a silicon with one t-butyl and two methyl groups is not strained at all (Figure 10) and so the SiBu^{t}Me₂ group should be a useful ligand, which can easily be compressed if necessary.

The structure of CH₂(SiMe₃)₂, with an angle of 123° between the Si-C bonds, shows that the two SiMe₃ groups ideally will move away from one another. What happens if this is not possible, because there are several -CH(SiMe₃)₂ groups compressing one another? The angle must be reduced - down to 112° in some

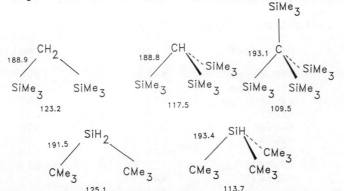

Figure 9 Bond lengths (in pm) and angles (in degrees) in trimethylsilyl-substituted methanes and t-butyl silanes.

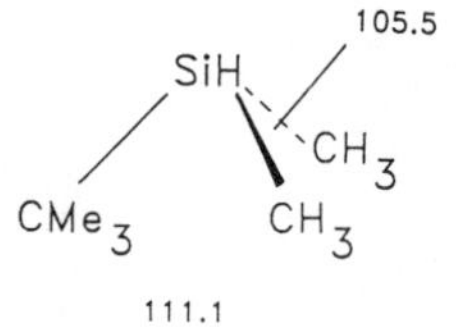

```
              105.5
       SiH   /
         \  /
      /   \/ `CH₃
     /    /\
  CMe₃  CH₃
        111.1
```

Figure 10 Angles (in degrees) in SiBu^{t}HMe$_2$

compounds. In the case of P$_2$[CH(SiMe$_3$)$_2$]$_4$ we can find out more about the effects of compression, because this molecule dissociates into stable phosphinyl radicals. In solution, we get both forms, with slow interconversion - but we cannot determine the structures. X-ray crystallography gives us the solid structure (P$_2$R$_4$), and electron diffraction gives the gas-phase structure (·PR$_2$), but not very well, because of the low symmetry, and parameters are strongly correlated. What we want to know is where the energy to break the P-P bond comes from. Is this bond very weak, so that the energy needed to break it is very small? No. The P-P bond length is 230 pm, compared with about 220 pm for a normal, unstrained P-P bond. It is by considering the structure of a solid phase (Figure 11) that we see the answer to the problem. The whole structure is distorted, with the two PCSi angles in each -CH(SiMe$_3$)$_2$ group very different from one another, with one about 112° and the other close to 124°. Similarly, the C$_{inner}$SiC$_{Me}$ angles are on average some 6° more than the C$_{Me}$SiC$_{Me}$ angles, but again there is very great variation, and the CSiC angles range from 103 to 117°. In addition, all the P-P, P-C and Si-C$_{inner}$ bonds are stretched. Table 2 summarises all the distortions. It is clear that there is potential energy stored in 13 bonds, and in 14 atoms which have deformed configurations, so that the energy is distributed over most of the molecule.

Table 2 Distortion of P$_2$[CH(SiMe$_3$)$_2$]$_4$

No.*	Parameter	Average	Normally
1	r(P-P)/pm	231	220
4	r(P-C)/pm	189	183
8	r(Si-C)/pm	191	187
4	<PPC/°	106	100
2	<CPC/°	103	100
4	<PCSi/°	112	110
4	<PCSi/°	124	110
4	<SiCSi/°	112	110
48	<CSiC/°	103-117	110
		(std. dev. 3.3)	

*   Number of bonds or angles of each type in the molecule

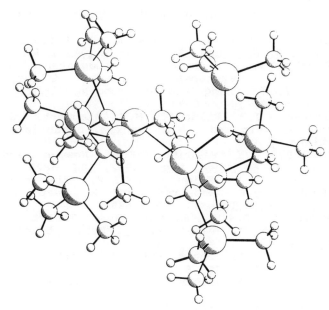

Figure 11 The structure of $P_2[CH(SiMe_3)_2]_4$ in the crystalline phase

When the biphosphine dissociates into two radicals, the overall conformation changes (Figure 12). That change of conformation is shown by models to be impossible by simple rotation about P-C bonds. It must involve a complex, concerted motion of many atoms, a process which is inevitably slow. But it is only possible at all if the groups are very flexible. Once the conformation of the radical is reached, most of the strain due to overcrowding is released, and so we can see how the P-P bond energy is provided.

Thus the chemistry of this particular system is critically dependent on the flexible nature of the silyl group. In designing compounds we need to consider the dynamic aspects of geometry, and not think only in terms of static, rigid structures.

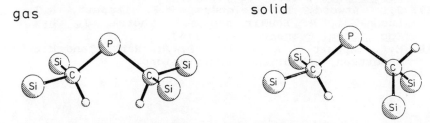

Figure 12 Conformations of the disyl groups in solid $P_2[CH(SiMe_3)_2]_4$ and gaseous $P[CH(SiMe_3)_2]_2$

1. R. West, M.J. Fink and J. Michl, Science, 1981, 214, 1343.
2. S. Masamune, Y. Hanzawa, S. Murakami, T. Bally and J.F. Blount, J. Am. Chem. Soc., 1982, 104, 1150.
3. A.G. Brook, S.C. Nyburg, F. Abdesaken, B. Gutekunst, G. Gutekunst, R.K.M.R. Kallury, Y.C. Poon, Y.-M. Chang and W. Wong-Ng, J. Am. Chem. Soc., 1982, 104, 5667.
4. A.H. Cowley, J.E Kilduff, M.F. Lappert, G. Gundersen, R. Seip and D.W.H. Rankin, unpublished observations.
5. A.H. Cowley, J.E Kilduff, T.H. Newman and M. Pakulski, J. Am. Chem. Soc., 1982, 104, 5820.
6. A.H. Cowley, J.E Kilduff, E.A.V. Ebsworth, D.W.H. Rankin, H.E. Robertson and R. Seip, J. Chem. Soc., Dalton Trans., 1984, 689.
7. D.G. Anderson, D.W.H. Rankin, H.E. Robertson, A.H. Cowley and M. Pakulski, J. Mol. Struct., 1989, 196, 21.
8. B. Beagley, R.G. Pritchard and J.O. Titiloye, J. Mol. Struct., 1988, 176, 81.
9. C. Eaborn, P.B. Hitchcock and P.D. Lickiss, J. Organomet. Chem., 1981, 221, 13; P.B. Hitchcock, J. Organomet. Chem., 1982, 228, C83.
10. P. Paetzold, L. Geret and R. Boese, J. Organomet. Chem., 1990, 385, 1.
11. C. Eaborn, N. Retta, J.D. Smith and P.B. Hitchcock, J. Organomet. Chem., 1982, 235, 265.
12. F. Glockling, N.S. Hosmane, V.B. Mahale, J.J. Swindall, L. Magos and T.J. King, J. Chem. Res., 1977, (S) 116, (M) 1201.
13. B. Beagley and R.G. Pritchard, J. Mol. Struct., 1982, 84, 129.
14. T. Fjeldberg, A. Haaland, B.E.R. Schilling, M.F. Lappert and A.J. Thorne, J. Chem. Soc., Dalton Trans., 1986, 551.
15. D.E. Goldberg, P.B. Hitchcock, M.F. Lappert, K.M. Thomas, T. Fjeldberg, A. Haaland and B.E.R. Schilling, J. Chem. Soc., Dalton Trans., 1986, 2387.
16. T. Fjeldberg, M.F. Lappert and A.J. Thorne, J. Mol. Struct., 1985, 127, 95.
17. J.L. Atwood, T. Fjeldberg, M.F. Lappert, N.T. Luong-Thi, R. Shakir and A.J. Thorne, J. Chem. Soc., Chem. Commun., 1984, 1163.
18. M.F. Lappert, I.A. Irving, D.W.H. Rankin and H.E. Robertson, unpublished observations.

Sterically Overloaded Organosilicon Compounds

N. Wiberg
INSTITUT FÜR ANORGANISCHE CHEMIE, UNIVERSITÄT
MÜNCHEN, GERMANY

1 INTRODUCTION

Tri-*tert*-butylsilyl groups $^tBu_3Si$ (**"supersilyl"**) are especially suitable components of *sterically overloaded silicon compounds* due to their extreme space requirement, their remarkable inertness, and their good accessibility *). The study of supersilyl compounds is of interest because substitution of "normal loaded" silyl groups (e.g. Me_3Si groups) by supersilyl groups leads to silicon compounds with unusual *bond lengths* and *angles*. Besides that, the group displacement is associated with *unexpected stability* and *activity*.

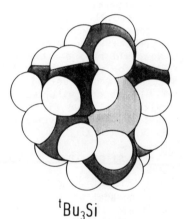

$^tBu_3Si$

("supersilyl")

Pioneers in the field of supersilyl compounds are M. Weidenbruch et al. (ref.1) who synthesized supersilyl derivatives of hydrogen, hydrogen halides, hydrogen pseudohalides, water, ammonia, hydrazoic acid. During the last few years we have successfully isolated new supersilyl compounds of main group elements. The compounds are mostly somewhat difficult to synthesize. As will be shown, they can be prepared by three alternatives which involve transfer of positively and negatively polarized supersilyl groups or supersilyl radicals.

---

* Concerning the $(Me_3Si)_3C$ ("trisil") group, which is as bulky as the $(Me_3C)_3Si$ ("supersilyl") group, and the $(Me_3Si)_2CH$ group, refer to the contributions of D. W. H. Rankin, and of M. F. Lappert in this volume.

2 PREPARATION AND PROPERTIES OF SOME SUPERSILYL COMPOUNDS

2.1 SYNTHESES WITH POSITIVELY POLARIZED SUPERSILYL GROUPS.

Some of the silylating agents used by us are compounds 1-3:

$^t Bu_3 Si-Hal$ $^t Bu_3 Si-OSO_2 CF_3$ $^t Bu_3 Si-N=N-NH-Si^t Bu_3$
   1   **2**   **3**

Their syntheses involve the silane $^t Bu_3 SiH$, formed by the reaction of SiF_4 with $Li^t Bu$ (ref.1,2). It undergoes halogenation to form the halides 1 and reacts with $CF_3 SO_3 H$ to form the triflate 2. On the other hand, $^t Bu_3 SiBr$ is transformed to triazene derivative 3 by multiple step reactions (see below).

The nucleophilic substitution at silicon in supersilyl compounds is somewhat slow due to steric factors. The *halides* 1 therefore, are very poor silylating agents (ref.1). It is mostly the hard and small nucleophiles like *azide* N_3^- or *hydroxide* OH^- which can lead to the desired products in reasonable time. On the other hand, *amide* NH_2^- do not undergo any reaction even in boiling toluene. In fact, the supersilyl amide was first prepared by the reduction of the azide (ref.3).

A stronger silylating agent, the *triflate* 2, obtained recently by us, reacts with *amide* NH_2^- to form $^t Bu_3 SiNH_2$. With *phosphide* PH_2^-, we could also synthesize $^t Bu_3 SiPH_2$ for the first time. The same holds for other *element hydride anions*.

3 acts as a stronger silylating agent than even the ester (see below). It reacts with *water* H_2O or *sulphane* H_2S to form the sterically overloaded disiloxane 4 or disilsulphane 5. On *heating*, it transforms into a disilazane 6:

                                                                     H
$^t Bu_3 Si-O-Si^t Bu_3$ $^t Bu_3 Si-S-Si^t Bu_3$ $^t Bu_3 Si-N-Si^t Bu_3$
         4   **5**   **6**

X-ray structural analysis (ref.4) shows that the first two compounds are linear and the latter has almost a linear central framework. Thus, the two supersilyl groups not only lead to a linear structure on oxygen but also to a linear stucture on sulphur, although it is well-known that the angle opening in sulphane requires much more energy than in water. Even when the angle opening in disilazane 6 is strongly hindered by the hydrogen, bound to the nitrogen atom, the supersilyl groups bring about surprisingly large bond angles and the compound has almost a T-type structure on nitrogen.

2.2 SYNTHESES WITH SUPERSILYL RADICALS

The radical $^t Bu_3 Si\cdot$ can be generated either by the *reduction* of supersilyl halides or by the *oxidation* of metal supersilyls (scheme 1):

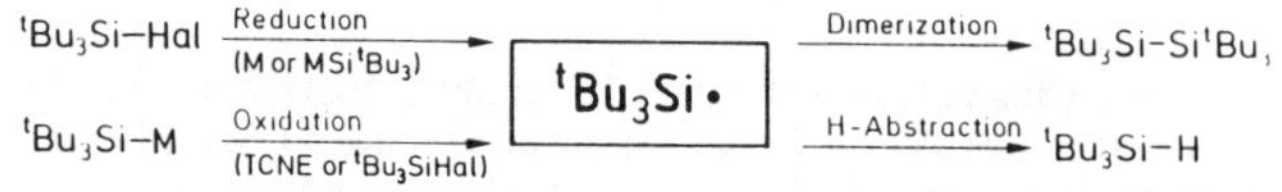

$^t Bu_3Si-Hal \xrightarrow[\text{(M or MSi}^t Bu_3)]{\text{Reduction}}$ $^t Bu_3Si\cdot$ $\xrightarrow{\text{Dimerization}} {}^t Bu_3Si-Si^t Bu_3$

$^t Bu_3Si-M \xrightarrow[\text{(TCNE or }^t Bu_3SiHal)]{\text{Oxidation}}$ $\xrightarrow{\text{H-Abstraction}} {}^t Bu_3Si-H$ (1)

It has not so far been characterized by esr spectroscopy and its intermediate existence follows from chemical investigations only. It appears that the life-time of the radical is short. Consequently its concentration remains very small. Our results show that the follow-up reactions of the radical in the absence of radical traps are primarily the dimerization at lower temperatures and hydrogen-abstraction at higher temperatures.

Supersilyl halides can be reduced with alkali metals in organic solvents. The rate of reaction increases in the order: supersilyl fluoride, chloride, bromide, and iodide. The same is true for the series lithium, sodium, potassium, rubidium, and cesium as reducing agents, or heptane, dibutyl ether, and tetrahydrofuran as solvents. The reduction temperature decreases in the same direction. This way, the tendency to form $^tBu_3SiH$ decreases and the tendency to form $^tBu_3Si-Si^tBu_3$ increases.

However, the reaction is more complex than shown in scheme (1), because the radicals, thus formed, can react with excess alkali metals to form alkali metal supersilyls. Partly these remain as such and partly they react with the supersilyl halides to re-build supersilyl radicals or react with the solvent to form $^tBu_3SiH$.

The formation of the disilane $^tBu_3Si-Si^tBu_3$ ("*disupersilyl*") follows from the low temperature oxidation of $NaSi^tBu_3$ with tetracyanoethylene. After many unsuccessful attempts by others, we could isolate the desired disilane by this method (ref.5). The compound forms colourless crystals, melting at 162°C, and is soluble in organic media. It shows a Si-Si bond distance of 270 pm which is significantly greater than the Si-Si bond distance in all the disilanes known so far (ref.5). The bond distance is almost 16% longer than that in disilane or hexamethyldisilane.

The disilane is quite an inactive substance at room temperature. Its reactivity starts above 60°C with the formation of supersilyl radicals in small equilibrium concentration which can combine with the given reactants. Some of the reactions, studied by us (ref. 6), are summarized in scheme (2) and shall be discussed in a clockwise direction.

$$\begin{array}{c} ^tBu_3Si-H \quad\quad\quad ^tBu_3Si-S_n-Si^tBu_3 \\ (n=2-6) \end{array}$$

scheme (2)

The disilane thermolyses in *mesitylene* at 100°C in a first order reaction to form quantitative amounts of $^tBu_3SiH$. The activation energy of this reaction, which can more or less be equated with the Si-Si dissociation energy, is 160 kJ/mol. This is significantly less than the normal Si-Si bond energy of about 285 kJ/mol. *Halogens* react with the disilane around 60°C to form halosilanes but *sulphur* reacts only at 130°C to form colourless bis(silyl)polysulphanes. The same is true for *selenium* and *tellurium* which give red coloured bis(silyl)polyselanes and blue coloured bis(silyl)polytellanes. Although no bis(silyl)monosulphane is formed as yet, bis(silyl)monoselane and bis(silyl)monotellane are observed. White *phosphorus* reacts with the disilane at 110°C to form a mixture of silylated polyphosphanes from which one compound has the composition shown in scheme (2). In view of the expected steric hindrance, it forms only one geometric isomer. *Nitrogen monoxide* reacts with the disilane to form bis(silyl)hyponitrite, and *butadiene* reacts at 100°C to form oligomeric to polymeric products. The value of n is small in excess of $^tBu_3Si-Si^tBu_3$, but large in shortage. In the last case the disilane acts as a polymerisation starter. It leads to a colourless glass-like product, sparingly soluble in organic media.

2.3 SYNTHESES WITH NEGATIVELY POLARIZED SUPERSILYL GROUPS

In general, the syntheses of metal silyls from halosilanes and alkali metals causes difficulties. In fact, the expected metal silyls are formed as intermediates which immediately react further with halosilanes to form disilanes (scheme 3):

$$
\begin{array}{rcl}
R_3SiHal + 2M & \longrightarrow & R_3SiM + MHal \\
R_3SiHal + R_3SiM & \longrightarrow & R_3Si-SiR_3 + MHal \\
\hline
2R_3SiHal + 2M & \longrightarrow & R_3Si-SiR_3 + 2MHal
\end{array} \quad (3)
$$

As an exception, the reaction of $^tBu_3SiBr$ or $^tBu_3SiI$ with *sodium* in ether at 80°C stops at formation of $NaSi^tBu_3$ (ref.3,7,8). In this way, the latter is obtainable in almost quantitative amounts. The use of potassium, rubidium, or cesium leads to increasing amounts of $^tBu_3Si-Si^tBu_3$ and $^tBu_3SiH$, so that the yield of metal supersilyls decreases significantly in the same order. *Lithium* and *magnesium* prove to be almost inactive towards even $^tBu_3SiI$ under all conditions. The preparation of a lithium supersilyl, or bromomagnesium supersilyl, or magnesium bis(supersilyl) follows from the reaction of sodium supersilyl with lithium chloride or magnesium dibromide in tetrahydrofuran (ref.6).

"*Sodium supersilyl*", obtained by the reaction of $^tBu_3SiBr$ with sodium in tetrahydrofuran (THF), crystallizes out in the form of yellow needles from a solution of pentane at -78°C. Its ^1H-, ^{13}C-, ^{29}Si- and ^{23}Na-NMR studies indicate the constitution $^tBu_3Si-Na(THF)_2$. In spite of many attempts, it has not so far been possible to obtain crystals suitable for X-ray analysis (ref.6). The same is true for magnesium silyls (ref.6).

Some of the reactions of $NaSi^tBu_3$, studied by us (ref.6), are summarized in scheme 4 and shall be taken-up in a clockwise direction:

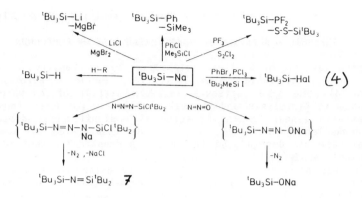

The parent supersilyl anion $^tBu_3Si^-$ acts as a strong base and reacts with *proton active reagents* HR to form $^tBu_3SiH$. At higher temperatures it reacts with tetrahydrofuran itself. Still more active are the rubidium and especially cesium supersilyls which deprotonate even hydrocarbons. Much less reactive are lithium and especially magnesium supersilyls. Besides that, $^tBu_3Si^-$ acts as a nucleophile and displaces halide in element halides such as *lithium chloride, magnesium bromide, chlorobenzene, trimethylchlorosilane, phosphorus trifluoride* and *sulphur monochloride*. These form supersilyl derivatives shown on the top row of scheme 4. In cases where the outgoing halogen is less negatively polarized and/or is sterically protected, then the nucleophilic attack of silyl anion occurs frequently on halogen. For example, the reaction with *bromobenzene, phosphorus trichloride,* or *di-tert-butylmethylsilyl iodide* leads to supersilyl halides.

Since the anion $^tBu_3Si^-$ is isoelectronic with $^tBu_3P$, therefore, it shows the reactivity of this phosphane and likewise adds on to the double bonded system of *dinitrogen oxide* or *di-tert-butylsilylene chloride azide*. The adducts are, of course, labile and decompose with the elimination of dinitrogen. In the latter case, a silicon derivative with SiN double bond is formed (ref.8). It was with this substance that we could first isolate a silaneimine and determine its structure (see below).

With that I wish to move on from the preparation of the supersilyl compounds to the *unusual stability* and *reactivity* of these substances. Under this category of silicon compounds *silenes* are especially important.

3 SUPERSILYL COMPOUNDS WITH UNUSUAL STABILITY AND REACTIVITY

3.1 SUPERSILYL SUBSTITUTED UNSATURATED SILICON COMPOUNDS

A typical example of a silene, stabilized by a supersilyl group is the light yellow silaneimine **7** (ref.8), mentioned before (cf. scheme 4). The X-ray strucure analysis (ref.9) of **7**, which is unstable to hydrolysis and thermally stable up to 80°C, shows an essentially linear Si=N-Si skeleton with significantly different SiN bond lengths. The geometry of the unsaturated Si atom is planar. The SiN double bond (1.568 A) is drastically shorter than SiN single bonds (1.64 - 1.78 A).

It has also been possible for us to establish a supersilyl substituted silanephosphimine **8** with SiP double bond, corresponding to silaneimine **7** (ref.10). In addition, a supersilyl substituted silaethene **9** with SiC double bond could be proved by us (ref.11). Interestingly the supersilyl substituted silaneimine **10** is metastable in diethyl ether solution which, once again, is attributable to the exceptional bulkiness and stabilizing action of supersilyl groups (ref.3).

$^tBu_2Si=P(Si^tBu_3)$ $Me_2Si=C(SiMe_3)(Si^tBu_3)$ $Me_2Si=N(Si^tBu_3)$
8   **9**   **10**

As another category of supersilyl stabilized compounds, I shall now discuss supersilyl triazenes.

3.2 BIS(SUPERSILYL)TRIAZENE

Scheme 5 presents the reaction of $NaSi^tBu_3$ with $^tBu_3SiN_3$. The reagents combine to form a yellow crystalline adduct, sodium bis(supersilyl)triazenide, which undergoes *protonation* with MeOH, *methylation* with Me_2SO_4, and *stannylation* with Me_3SnCl (ref.7,12). The protolysis product, *"bis(supersilyl)triazene"*, is a light yellow crystalline substance, melting at 140°C. Its X-ray determination (ref.12) shows a planar zigzag conformated framework of the heavier atoms SiNNNSi with a short and a long NN bond.

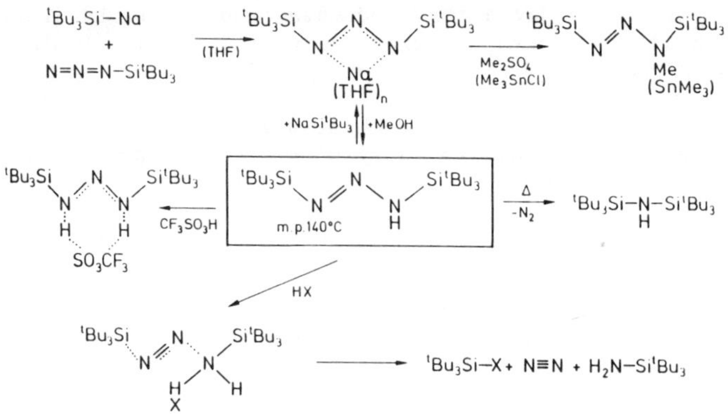

(5)

With $^tBu_3Si-N=N-NH(Si^tBu_3)$, it was possible for us to isolate for the first time a triazene, substituted by silyl groups only. Such triazenes are normally unstable to nitrogen elimination and hence not isolable. Because of excessive bulkiness, the supersilyl derivative loses dinitrogen very slowly only at 150°C in a first order reaction. The activation energy of *thermolysis* is 140 kJ/mol. The compound can be *deprotonated* with $NaSi^tBu_3$ and *protonated* with CF_3SO_3H (ref.13). The amino nitrogen can, of course, be protonated with acids such as hydrogen chloride, or water, or hydrogen sulphide. But this leads to the subsequent decomposition of the compound as shown in scheme 5 (ref.13). This way, as already mentioned, bis(supersilyl)triazene acts as a strong silylating reagent.

4 CONCLUSION

Now, as discussed in connection with the formation of unsaturated silicon compounds and silyltriazenes, supersilyl groups can lead to compounds which are normally extremely labile. This is also likely for many other exciting silicon compounds:

(i) If, for example, $^tBu_3Si-SiCl_3$, obtainable easily from $SiCl_4$ and $NaSi^tBu_3$, is reacted with sodium in dibutylether, then a mixture of products is formed along with precipitation of NaCl (ref.14). After methanolic work up, the mixture of products can be separated by preparative HPLC. The different fractions are under investigation and show simple NMR spectra. The mass spectral studies indicate the build-up of compounds like $(^tBu_3SiSiH_2)_2$ or $(^tBu_3SiSiH)_3$. The X-ray structure analysis of $(^tBu_3SiSiH_2)_2$ (ref.15) shows a zigzag conformated SiSiSiSi framework with SiSi bond lengths of ca. 235 pm and SiSiSi bond angles of ca. 120°.

(ii) It is known that aluminum can be transported in contact with aluminum trichloride at higher temperatures. The gaseous phase contains *aluminum monochloride* AlCl. It has been found (ref.16) that quenching can lead to a metastable solution of this chloride in a mixture of toluene and ether. Addition of sodium supersilyl gives a violet-red solution, and sodium chloride precipitates out. Our investigations show that a mixture of "*supersilyl aluminum(I)* compounds" $(^tBu_3SiAl)_n$ are formed. It was possible to isolate a dark red fraction by sublimation. The mass and NMR spectral studies of this fraction indicate tetrameric supersilyl aluminum $(^tBu_3SiAl)_4$ with a Al_4 frame work (ref.17).

REFERENCES

1. M. Weidenbruch, W. Peter, Angew. Chem., Int. Ed. Engl. **14** (1975) 642; M. Weidenbruch, H. Pesel, W. Peter, R. Streichen, J. Organometal. Chem. **141** (1977) 9; M. Weidenbruch. H. Pesel, Z. Naturforsch. **33b** (1978) 1465; M. Weidenbruch, H. Flott, B. Ralle, Z. Naturforsch. **38b** (1983) 1062.
2. N. Wiberg, K. Schurz, J. Organometal. Chem. **341** (1988) 145.
3. L. H. Sommer, P. M. Nowakowski, J. Organometal. Chem. **178** (1979) 95.
4. N. Wiberg, E. Kühnel, K. Schurz, H. Borrmann, A. Simon, Z. Naturforsch. **43b** (1988) 1075.
5. N. Wiberg, H. Schuster, A. Simon, K. Peters, Angew. Chem., Int. Ed. Engl. **25** (1986) 79.
6. N. Wiberg, H. Schuster, preliminary communication.
7. N. Wiberg, G. Fischer, P. Karampatses, Angew. Chem., Int. Ed. Engl. **23** (1984) 59.
8. N. Wiberg, K. Schurz, Chem. Ber. **121** (1988) 581.
9. G. Reber, J. Riede, N. Wiberg, K. Schurz, G. Müller, Z. Naturforsch. **44b** (1989) 786; N. Wiberg, K. Schurz, G. Reber, G. Müller, J. Chem. Soc., Chem. Commun., **1986**, 591.
10. N. Wiberg, H. Schuster, Chem. Ber., in print.
11. N. Wiberg, Th. Passler, preliminary communication; cf. N.Wiberg, G. Wagner, J. Riede, G. Müller, Organometallics **6** (1987) 32.
12. N. Wiberg, P. Karampatses, E. Kühnel, M. Veith, V. Huch, Z. Anorg. Allg. Chem. **562** (1988) 91.
13. N. Wiberg, E. Kühnel, preliminary communication.
14. N. Wiberg, Th. Passler, preliminary communication.
15. N. Wiberg, Th. Passler, K. Polborn, preliminary communication.
16. M. Tacke, H. Schnöckel, Inorg. Chem. **28** (1989) 2896.
17. C. Dohmeier, Th. Passler, H. Schnöckel, N. Wiberg, preliminary communication.

Structure Systematics of Di- and Oligosilanes

László Párkányi
CENTRAL RESEARCH INSTITUTE FOR CHEMISTRY,
HUNGARIAN ACADEMY OF SCIENCES, PO BOX 17, H-1525-
BUDAPEST, HUNGARY

Abbreviations:

Ad = adamantyl
Cp = cyclopentadienyl
ED = gas electron diffraction
Ind = indenyl (C_9H_7)
Me = methyl
Mes = mesityl
MW = microwave spectroscopy
Ph = phenyl
t–Bu = tertiary–butyl

1 INTRODUCTION

Three-dimensional structural parameters of molecules containing Si–Si linkages have long been in the focus of interest. A new impetus of structural research was given by the potential use of the new class of polysilane polymers in modern industrial applications. The electron delocalization along the Si–Si σ–framework and the special photochemistry render polysilanes unique polymers.

The present number of X–ray structures may be estimated about 130. 116 structures were registered with the Cambridge Crystallographic Database [1] in 1989; a few gas–phase structures determined by ED and two MW structures are also available. Si–Si bond distances observed with the highest frequency are clustered at 2.358 Å. The Si–Si distance in elemental silicon is 2.3515(1) Å [2] compares with this value.

2 CHAIN MOLECULES

2.1 Compounds containing Si_2 units

2.1.1 *Disilenes*

The silicon atoms and their three substituent atoms in disilenes (Table 1) lie in an approximate plane: a slight pyramidalization is observable at the sp^2–hybridized silicon atoms. According to *ab initio* calculations [6] the fold-angle (defined as the angle formed by the X=X (X=C,Si,Ge) vector and the H-X-H plane) for ethylene has a well defined potential energy minimum at $\Theta=0°$, disilene has a broad energy

Table 1. Structural parameters of disilenes

			$R^1R^2Si=SiR^1R^2$		
Cmpnd.	R^1	R^2	d(Si-Si), Å	Θ, deg.	Ref.
1.1	X	X	2.140(3)	1	3
1.2	t–Bu	Mes	2.143(1)	4	4
1.3	Mes	Mes	2.160(1)	18	4

X = 2,6–diethylbenzene

Table 2. Si–Si distances (Å) in disilanes

$$R^1R^2R^3Si-SiR^4R^5R^6$$

Cmpnd.	R^1	R^2	R^3	R^4	R^5	R^6	d(Si–Si)		Ref.
2.1	H	H	H	F	F	F	2.319(5)	[MW]	7
2.2	F	F	F	F	F	F	2.324(6)	[ED]	8
2.3	Cl	Cl	Cl	Cl	Cl	Cl	2.32(3)	[ED]	9
2.4	H	H	H	H	H	H	2.331(3)	[ED]	10
2.5	H	H	H	H	H	F	2.332(5)	[MW]	11
2.6	A	Me	Me	Me	Me	A	2.335(2)		12
2.7	Me	Me	Me	Me	Me	Me	2.340(9)	[ED]	13
2.8	Me	Me	B	Me	Me	B	2.341(3)		14
2.9	H	Ph	Ph	H	Ph	Ph	2.343(1)		15
2.10	H	Br	Br	H	Br	Br	2.348(19)	[ED]	16
2.11	D	Q		D	Q		2.349(1)		17
2.12	Me	Me	Me	Ph	Ph	Ph	2.354(1)		18
2.13	H	Mes	Mes	H	Mes	Mes	2.356(6)		15
2.14	H	Ph	Ph	Ph	Ph	Ph	2.357(1)		19
2.15	H	X	X	H	X	X	2.365(1)		20
2.16	Me	Y		Y		Me	2.378(3)		21
2.17	H	t-Bu	t-Bu	H	Mes	Mes	2.397(1)		22
2.18	Ph	Ph	Ph	Ph	Ph	Ph	2.519(4)		23
2.19	t-Bu	t-Bu	t-Bu	t-Bu	t-Bu	t-Bu	2.697(3)		24

A = S–Si(O-t-Bu)$_3$
D = α-naphtyl
B = [C$_5$H$_4$]–Fe(CO)$_2$Me
Q = [cyclobutadienyl]
X = cyclohexyl

$$Y = \begin{array}{c} Me_2Si-SiMe_2 \\ \diagdown SiMe_2 \\ Me_2Si-SiMe_2 \end{array}$$

minimum and only small energy is required to a slight pyramidalization of 5–20°.

Comparing **1.3** with the **2.13** disilane (Table 2) a decrease of the Si–Si bond length is observed ($\Delta = 0.196$ Å). Though this difference is greater than that for the C–C distances in the analogous aryl-substituted pair of Ph$_2$HC–CHPh$_2$ (1.540) and Ph$_2$C=CPh$_2$ (1.355, $\Delta = 0.185$ Å) [4], on a percentage basis the bond contraction for the silicon compounds is smaller by 8–9% than for carbon ones (ca. 12%). The π-component seems to be smaller for silanes than for the carbon compounds, which is possibly due to the the less efficient overlap of the $3p$ orbitals.

2.1.2 Disilanes

The Si–Si distances (Table 2) seem to be affected by electronic and steric (repulsive) interactions. The Si–Si distance in elemental silicon (2.352 Å) is equivalent to the values observed for **2.11–2.12**. The shortest Si–Si distances were reported for compounds with electronegative and/or small substituents. He[1] photoelectron spectroscopy revealed [8] that bonding electrons in the Si–Si bond are more tightly bound in **2.2** than in **2.4**. This is shown by the slightly elongated Si–Si bond in **2.4**.

Bulky substituents in **2.13–2.19** are responsible for the rather long Si–Si bonds. The longest bond reported is 2.697 Å (**2.19**). Steric repulsion also affects the length of the Si–C bonds: strain may be relieved by the deformation of these bonds (d(Si–C) = 2.00 Å, **2.19**). Rather short Si–C(sp^3) bonds are also frequently reported. Such short bonds are found, for example, for **2.7** in the range of 1.857–1.873 Å (the expected value is 1.94 Å). One explanation for short methyl–silicon bond distances might be based on the concept of electron transfer from methyl groups to silicon d-orbitals [13]. High thermal motion also causes shrinkage of observed bond

Table 3. Mean Si–Si bond distances (Å) in trisilanes

$$R^1R^2R^2Si-SiR^3R^4-SiR^1R^2R^2$$

Cmpnd.	R^1	R^2	R^3	R^4	$\bar{d}$(Si-Si)		Ref.
3.1	Me	Me	Me	Me	2.325(12)	[ED]	29
3.2	Cl	Cl	Cl	Cl	2.329(7)	[ED]	30
3.3	A	Me	Me	Me	2.340(1)		12
3.4	Me	Me	B		2.344(1)		32
3.5	H	1/2Q	Q		2.348(7)		33
3.6	Me	1/2Y	Me	Me	2.356(1)		21
3.7	Me	Me	Br	X	2.369(1)		34
3.8	Me	Me	Ph	Z	2.396(1)		31
3.9	Me	t-Bu	t-Bu	t-Bu	2.593(1)		45
3.10	I	t-Bu	t-Bu	t-Bu	2.613(32)		35

A= S–Si(O–t–Bu)₃

B= ≡C⟨Ad / OSiMe₃⟩

Q= (anthracenyl)

X= (fluorenyl with Me₃Si)

Y= Me₂Si–SiMe₂ / Me₂Si–SiMe₂ / SiMe₂

Z= ⟨cyclopropyl with t-Bu, t-Bu, t-Bu⟩

lengths obtained from X-ray structure analysis. X-ray Si–C(sp^3) bonds tend to be somewhat longer, however, than those determined by ED or MW.

The crystal structure determinations of an isomeric pair of germyl-silanes related to **2.12** were also performed. These are the first X-ray structure determinations of a compound with Ge–Si bond (the Si—Ge distance is 2.357(4) Å by an MW study [5] of H₃Ge–SiH₃, **2.20**). While the Ge–Si bond distance in Ph₃Ge–SiMe₃ (**2.21**) is 2.384(1) Å [26], this bond is 2.394(1) Å long in Me₃Ge–SiPh₃ (**2.22**) [27]. The conformation of **2.12**, **2.21** and **2.22**, owing to isomorphism and the three-fold crystallographic symmetry axis, is completely identical, the lengthening of the Ge–Si bond therefore is not attributable to steric repulsion. We suggest that the elongation of the Ge–Si bond is due to the expansion of the bonding orbitals of germanium caused by the relatively electron donating methyl groups in conjunction with the contraction of the corresponding orbitals of silicon due to the presence of relatively electron withdrawing phenyl groups. The overall result is a reduced orbital overlap of the bonding orbitals between the Si and Ge atoms for **2.22**.

The Ge–Ge bond in digermane (**2.23** [44]) is 2.403 Å. The Ge–Ge bond is shorter in hexaphenyldigermane (**2.24** [25], 2.437(2) Å) than the Si–Si bond in hexaphenyl-disilane (**2.18**). The Ge–Ge bond length in hexa–t–butyl digermane (**2.25** [28]) is 2.710(4) Å.

Usually a mixture of various conformations of disilanes exists in gas phase. The ED analysis of **2.4** showed that no single conformation state can be assigned to the observations and even free rotation about the Si–Si bond cannot be ruled out.

The solid state conformations of disilanes were studied for **2.9**, **2.13** and **2.15** by X-ray structure analysis and molecular mechanical calculations. While aryl substitution shows preference for *anti–*, alkyl substitution results in *gauche* conformation at room temperature. **2.17** with mixed-type substituents also adopts *anti* conformation. **2.12** is in the intermediate conformation between fully staggered and eclipsed form. Because of the low rotation barrier about the Si–Si bond, the extent of rotation is easily affected by packing forces in the crystal.

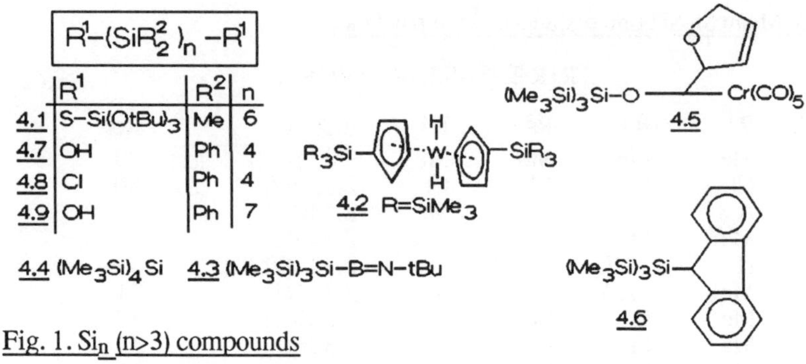

Fig. 1. Si$_n$ (n>3) compounds

2.2 Trisilanes

The steric demand of the substituents plays a decisive role in affecting the length of the Si–Si bond in trisilanes (Table 3).

Compounds **3.1**, **3.2**, **3.3** and **3.6** have the same substituents as **2.7**, **2.3**, **2.6** and **2.16**, thus they are directly comparable. The Si–Si bond is generally longer in disilanes than in the corresponding trisilanes. The central silicon atom in trisilanes has silyl substituents which form longer bonds than the organic groups they have replaced. The steric strain in trisilanes is relieved with respect to disilanes.

2.3 Si$_n$ (n > 3) chain molecules

The backbones of these compounds (Fig. 1, Table 4) may constitute linear (**4.1**, **4.7**–**4.9**) or branched (**4.2**–**4.6**) Si–Si chains.

The (mean) Si–Si bond length is practically independent of the degree of oligomerization (n) in the **4.1** (n=6), **3.4** (n=3) and **2.6** (n=2) homologous series. The length of the middle bond in compound **4.1** is the longest (2.349 Å) while the lengths of the end-bonds are identical with that observed in the disilane (2.335(2) Å). The average Si–Si bond distances are slightly longer in the tris-trimethylsilyl moieties. The longest mean Si–Si distances are found in the **4.7**–**4.9** perphenylated derivatives which may be attributable to the steric demand of the bulky phenyl groups though the influence of the interaction of the aryl π–electron system and the d-orbitals of the silicon atoms cannot be excluded entirely. **4.7** is quasi-cyclic: an intramolecular hydrogen bond is formed by the OH groups. All OH groups of both α,ω–diols participate in hydrogen bonding, yet the **4.9** heptasilane-diol has longer Si–Si bonds. Mean central and chain-end bonds are nearly the same in the tetrasilane-diol but they are considerably different in the heptasilane-diol. While the Si–Si bonds adjacent to the hydroxyl groups are of identical length with those in the tetrasilane, the averages of the middle bonds are significantly longer. The opposite

Table 4. Si–Si bond lengths (Å) for Si$_n$ (n>3) chain molecules

Cmpnd.	$\bar{d}$(Si–Si)	Central bonds	Terminal bonds	Si–X	X	Ref.
4.1	2.338(6)	2.349(7)	2.335(6)	2.138(5)	S	12
4.2	2.342(3)					36
4.3	2.350(4)					37
4.4	2.361(3)				[ED]	38
4.5	2.363(1)					39
4.6	2.366(6)					40
4.7	2.370(3)	2.373(2)	2.369(8)	1.648(5)	O	41,42
4.8	2.381(8)	2.374(2)	2.389(2)	2.077(2)	Cl	43
4.9	2.389(1)	2.400(4)	2.368(1)	1.643(8)	O	42

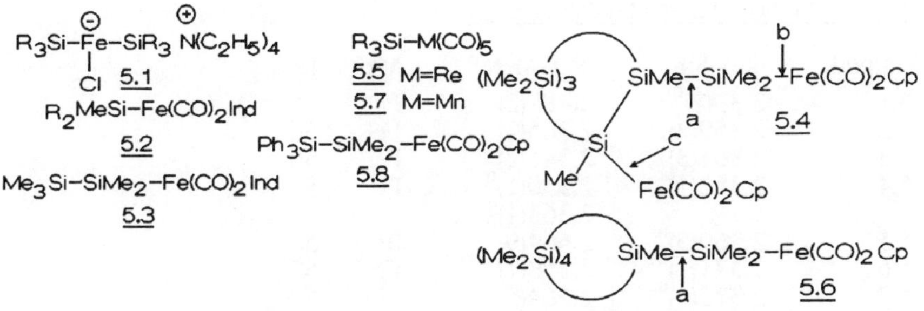

$R_3Si-\overset{\ominus}{Fe}-SiR_3$ $\overset{\oplus}{N(C_2H_5)_4}$ $R_3Si-M(CO)_5$
| **5.1**   **5.5** M=Re
Cl   **5.7** M=Mn
$R_2MeSi-Fe(CO)_2Ind$
5.2 $Ph_3Si-SiMe_2-Fe(CO)_2Cp$
                                        5.8
$Me_3Si-SiMe_2-Fe(CO)_2Ind$
5.3

$(Me_2Si)_3$ $SiMe\underset{a}{-}SiMe_2\overset{b}{\underset{\downarrow}{-}}Fe(CO)_2Cp$
                                                        5.4
             $\overset{Si}{\underset{Me}{\diagdown}}\overset{c}{\diagup}Fe(CO)_2Cp$

$(Me_2Si)_4$ $SiMe\underset{a}{\underset{\downarrow}{-}}SiMe_2-Fe(CO)_2Cp$
                                                    5.6

Fig. 2. Compounds containing Si-transition metal bonds $(R=SiMe_3)$

is observed in the **4.8** α,ω-dichloro compound, where the terminal bonds are longer than the middle ones. A rationale for this phenomenon might be an oxygen-silicon interaction. The central Si-Si bond in **4.7** is wedged between two Si-Si-O moieties, but longer chains (e.g. in **4.9**) isolate such interacting moieties.

Torsion angles indicate that the Si-Si framework of linear oligosilanes is rather flexible.

2.4 Transition metal compounds

The interest in silyl-transition metal complexes (Fig. 2, Table 5) stems in part from their catalytic properties in the reactions of organosilicon compounds, such as hydrosilylation and formation of silicon-silicon bonds. The oligo and polysilyl derivatives of the $(\eta^5-C_5H_5)Fe(CO)_2$-system have been shown to exhibit unique range of deoligomerizations, molecular rearrangements and migrations. Such chemistry is not always duplicated with other metal systems.

The particular importance of the $(\eta^5-C_5H_5)Fe(CO)_2$-system has initiated spectroscopic investigations (Mössbauer and IR [53] and ^{29}Si NMR [54]). It has been shown that there is little retrodative π-bonding between the metal atom and the silicon atom in complexes of the type $(\eta^5-C_5H_5) Fe(CO)_2 -SiR_3$.

A ^{29}Si NMR spectroscopic study of the **2.12** disilane and the **5.8** iron-complex revealed that the silyl group acts as a σ-donor towards Fe. This effect is demonstrated by the changes in the molecular geometry. The Si-Si bond is longer by 0.018 Å than that in the 'free ligand' **2.12** [Si-Si 2.354(1) Å]. NMR and X-ray investigations were also performed for the the analogous germylsilyl pair of compounds **2.25** and $Ph_3Ge-SiMe_2-Fe(CO)_2Cp$ (**5.13**) [26]. A comparison of the interatomic distances involving Si, Ge and Fe indicates that the σ-donor property of the silicon atom in **5.13** is more pronounced than in **5.8**. The lengthening of the Ge-Si bond (+0.021 Å) is accompanied by a significant shortening of the Si-Fe bond (-0.021 Å) with respect to **5.8**. The replacement of a triphenylsilyl group by a triphenylgermyl group in the iron complexes permits a greater removal of electron density by the iron atom from the bonded silicon atom (i.e. the germyl-silyl complex is a better σ-donor than the disilyl complex).

The Si-Si and Fe-Si bond distances in **5.6** compare to those in **5.8**. Two different Si-Fe bonds are present in the related derivative **5.4**. The Si-Fe and the Fe substituents are in 1,2-positions to each other. The iron atom directly bound to the five-membered ring maintains a longer Si-Fe bond than the one terminating the Si-Si chain. The longest bond is between the silyl and iron substituted ring silicon atoms. This is in agreement with the concurrent transfer of electron density from Si-Si bonds to the metal centers.

Further evidence for the σ-donor property of the silyl group is provided by the

Table 5. Si–Si and Si–metal distances (Å)

Cmpnd.	d(Si–Si)	d(Si–Metal)	Metal	Ref.
5.1	$2.33(2)^m$	2.489(2)	Fe	55
5.2	2.359(2)	2.365(2)	Fe	56
5.3	2.361(3)	2.341(2)	Fe	56
5.4	$2.362(1)^a$	$2.350(1)^b$ $2.363(1)^c$	Fe	57
5.5	$2.369(5)^m$	2.664(9)	Re	58
5.6	$2.371(1)^a$	2.350(1)	Fe	57
5.7	$2.372(6)^m$	2.564(5)	Mn	59
5.8	2.373(1)	2.346(1)	Fe	60
5.9	$2.374(5)^m$	2.669(1)	Mo	61
5.10	$2.378(1)^m$	2.378(1)	Fe	56
5.11	2.393(1)	2.365(2)	Fe	56
5.12	2.396(5)	2.378(2)	Fe	14

^{m} mean value ^{a,b,c} cf. Fig 2.

structure determination of the di–iron complex **5.12**. The steric effect of the bulky terminal groups bound to the disilane was studied for **2.8** [(Me(CO)$_2$Fe(η^5-C$_5$H$_4$)-SiMe$_2$)$_2$]. Significant lengthening of the Si–Si bond is observed in **5.12** with respect to **2.8** which does not have direct Si–Fe bond. The lengthening of the Si–Fe bonds is also noted with respect to **5.8**. This dual elongation of both Si–Si and Si–Fe seems to reflect the problem of electron donation to two iron atoms: neither Fe atom is bonded as strongly as in the monoiron complex. NMR and structural data indicate that two transition metal atoms bonded to a disilane moiety in 1,2-positions have a capacity to weaken the Si–Si bonds.

We performed X-ray analyses of a number of complexes of the general type (η^5-C$_9$H$_7$)Fe(CO)(L)–Si$_n$, Si$_n$=Si$_2$Me$_5$, 2-Si$_3$Me$_7$, L = CO, PPh$_3$ [**5.2**, **5.3** L=CO; **5.10**, **5.11** L=PPh$_3$]. The phosphine ligands alter the capacity of the silyl ligands to accept electron density from the Fe atom via π-bonding. The phosphine increases the electron density upon the iron atom, since it is a more powerful electron donor but less capable π-acceptor than the CO group it has replaced. The crystal structures of the monosilyl derivatives Me$_3$Si–Fe(CO)$_2$(L)Ind, L=CO, (**5.13**) and L=PPh$_3$ (**5.14**) have also been determined [56]. Si–Fe distances are significantly longer in the **5.10**, **5.11** and **5.14** [Fe-Si: 2.339(1) Å] phosphine-substituted complexes than those in the dicarbonyl derivatives **5.2**, **5.3** and **5.13** [Fe-Si: 2.325(3) Å]. Within the group of dicarbonyl complex molecules the Si–Fe bond increases regularly with the silyl bulk. The Si–Si bond distances, however, do not follow this pattern of incremental change. They are practically identical in **5.2** and **5.3** but the Si–Si bond

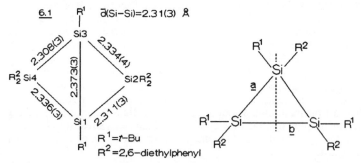

Fig. 3. Si–Si bond lengths in **6.1** (Å) and the general formula of cyclotrisilanes (Table 6.)

Table 6. Si–Si bond distances (Å) in cyclotrisilanes

Cmpnd.	R^1	R^2	d(Si–Si)	a	b	Ref.
6.2*	X	X	2.380(8)	2.386(9)	2367(8)	69
6.3 (cis,trans)	Y	Y	2.407(17)	2.424(2)	2.373(?)	70
6.4 (cis,trans)	Mes	t–Bu	2.416(2)	2.404(9)	2.441(2)	71
6.5 (cis,cis)	Z	t–Bu	2.421(1)	2.422(3)	2.418(3)	72
6.6 (cis,cis)	Mes	t–Bu	2.428(2)	2.426(2)	2.431(2)	71
6.7	t–Bu	t–Bu	2.510(1)	2.510(1)	2.510(1)	73

X = neopentyl   Y = 2,6–dimethylphenyl   Z = cyclohexyl
*Two molecules/asymmetric unit

distance difference between **5.11** and **5.10** may be due to the different steric requirements of the substituents.

The metal–Si bonds in the **5.5** Re and **5.7** Mn complexes are longer than those in the corresponding monosilane compounds Me₃Si–M [Δ(Re) = 0.065, Δ(Mn) = 0.067 Å] which possibly reflects the difference in the bulk of the silyl substituents.

The **5.1** complex has a three–coordinate iron center. The average Fe–Si bond distances are the longest yet observed for a Si–Fe complex and the average Si–Si bond distance is the shortest among the structures listed in Table 5.

3 RING SYSTEMS

3.1 Cyclotrisilanes [(Si)₃]

Cyclotrisilanes (Fig. 3, Table 6) are strained rings stabilized by bulky substituents. **6.1** (Fig. 3) is the first uncharged tetrasilabicyclo compound [64]. The facile ring inversion of this compound occurs about the rather long central bond. Other Si–Si bonds in this ring system are normal or even short, when the sizes of the substituents and the ring size are considered. Structural studies of the carbon analogs (e.g. [68]) revealed that the middle bond length widely varies with the dihedral angle formed by the two three–membered rings implying great flexibility of this system. The dihedral angle formed by the two halves of **6.1** is 121°. The non–bonded Si...Si distance across the ring is 3.475 Å.

The mean Si–Si bond in the **6.7** cyclotrisilane is shorter by 0.093 Å than the mean bond lengths in the linear cyclotrisilanes **3.9**–**3.10**.

Of the **6.4** (*cis,trans*) and **6.6** (*cis,cis*) compounds, the *cis,cis*–isomer has approximate three–fold symmetry. The individual Si–Si bonds are longer than those in the *cis,trans*–isomer. The latter possesses a non–crystallographic mirror–plane.

A characteristic geometric feature of some cyclotrisilanes is the presence of an approximate two–fold axis (**6.2**, **6.3**), or a mirror–plane (**6.4**). **6.7** has a genuine crystallographic three–fold (D₃) symmetry, and **6.5** has an approximate one. The shape of the cyclotrisilane is best described either as an isosceles triangle, or as an equilateral triangle (**6.5** and **6.7**).

The structure of the per–*t*–butylated cyclotrigermane has also been published [28] (Ge–Ge: 2.563(1) Å, Ge–Ge–Ge: 60.0° [three–fold symmetry]).

Disilaoxiranes, disilathiiranes and disilacyclopropanes. Disilaoxiranes, –thiiranes and –cyclopropanes are three–membered rings with one heteroatom. In disilaoxiranes [48–50] a significant amount of double bond character is retained in the Si–Si bond. The Si–Si bond length is intermediate between the Si=Si double and Si–Si single bonds, the configuration is planar at the silicon atoms. The molecular structure may rather be described as one having some of the character of the oxirane and the π–complex structure. The Si=Si double bond seems to act as a powerful electron donor and the oxygen atom as an electron acceptor in the π–complex.

Table 7. Si–Si distances (Å) and dihedral angles (°) in cyclotetrasilanes

Cmpnd.	R^1	R^2	R^3	R^4	R^5	R^6	R^7	R^8	$\bar{d}$(Si–Si)	A/B	Ref
7.1	t-Bu	Cl	t-Bu	Cl	t-Bu	Cl	t-Bu	Cl	2.35(3)	?	74
7.2	Me	Me	Me	Me	Me	Me	Me	Me	2.363(4)	0	77
7.2	Me	Me	Me	Me	Me	Me	Me	Me	2.364(4)	29.4 [ED]	78
7.3	Me	t-Bu	Me	t-Bu	Me	t-Bu	Me	t-Bu	2.376(1)	36.8	79
7.4	Ph	Ph	Ph	Ph	Ph	Ph	Ph	Ph	2.378(3)	12.8	80
7.5	F	A	F	A	F	A	F	A	2.391(3)	35.0	81
7.6	B	B	B	B	Q	Q	Q	Q	2.406(15)	39.4	82
7.7	H	t-Bu	X	X	OH	t-Bu	X	X	2.409(12)	36.6	83
7.8	t-Bu	Y	t-Bu	Y	t-Bu	Y	t-Bu	Y	2.444(1)	34.0	72

A = (tetramethylcyclobutenyl) B = i-propyl Q = neopentyl X = 2,6-diethylphenyl Y = cyclohexyl

(silicon atoms are more tetrahedral than in the oxirane and the Si–Si bonds are also longer) but still a significant amount of double bond character is present in the disilane bonds. The hybridization of silicon in these compounds is rather sp^2–p than sp^3.

3.2 Cyclotetrasilanes [(Si)$_4$]

Compound **7.2** (Table 7) was studied by both ED and X-ray diffraction techniques. The conformations are different in the two phases, though the Si–Si bond lengths are not. Other Si–Si bond lengths in this class of compounds, at least in part, seem to be dependent on substituent size. **7.8** is the cyclotetrasilane (cis-cis-trans) analogue of the **6.5** (cis-cis) cyclotrisilane. The mean Si–Si bond length is longer in **7.8**.

A conformational feature of the four-membered Si$_4$ ring is the degree of folding of the two half-rings about one of the diagonals. The minimal energy conformation is probably the puckered ring. The only known planar cyclotetrasilane ring is that of **7.2** (the molecule has a crystallographic inversion center). A review on the organic analogs [84] showed that the dominant conformation for the cyclobutane-rings is the puckered one. The mean dihedral angle for simple substituted cyclobutanes is 26(3)°, unless crystal packing forces/crystallographic symmetry requires the rings to be planar. No structures have been published, however, with intermediate degree of puckering (i.e. in the range of 1–18°). The steric or electronic nature of the substituents has only limited effect on ring puckering. The mean dihedral angle for cyclotetrasilanes is 35.4(6)° greater by ca. 9° than that in cyclobutanes (if data for **7.2** and **7.4** is omitted). The low degree of puckering in the perphenylated compound **7.4** seems to be somewhat special because of the intermediate range of puckering.

Substituents on puckered cyclotetrasilane rings are oriented in axial-equatorial fashion. Axial Si–C bonds are longer by 0.025 Å than equatorial ones in **7.3**, though they are practically equal in **7.4**.

A number of crystal structure determinations has been carried out recently on strained silyl polycycles [47,74].

Table 8. Mean Si–Si distances (Å) in cyclopentasilanes

Cmpnd.	d̄(Si–Si)	Ref.	Cmpnd.	d̄(Si–Si)	Ref.
8.1	2.342(3) [ED]	76	8.4	2.358(1)	75
8.2	2.346(2)	75	5.4	2.362(4)$_1$	57
8.3	2.353(13)	88	8.5	2.362(4)^1	88
5.6	2.353(4)	57		2.366(12)^2	
			8.6	2.362(9)	89

^{1} room temperature data
^{2} low temperature data

3.3 Cyclopentasilanes [(Si)$_5$]

The shortest Si–Si bonds are measured in cyclopentasilane in gas–phase by ED (Fig. 4, Table 8). Studies of cyclopentasilane **8.1** and disilane (**2.4**) showed that the Si–Si bond is distinctly longer in the ring than in the disilane (Δ = 0.011 Å). The bond length difference for cyclopentane [92] and ethane [63] is exactly the same. The longest bond is observed in the perphenylated compound **8.6**, longer by 0.017 Å than in the **7.4** perphenylated tetrasilane. Of the two cyclopolymethylene substituted rings (pentaspiro-cyclopentasilanes, 'rotanes'), the **8.4** cyclopentamethylene substituted ring has longer mean Si–Si bonds (*i.e.* the steric demand of the cyclopentamethylene rings is greater than that of the cyclotetramethylene rings).

All cyclopentasilanes have puckered rings. No conformational species could be assigned for **8.1**: the molecule, similarly to cyclopentane, undergoes dynamic pseudorotation. In other rings, ring conformations are normally intermediate between C$_s$ (envelope) and C$_2$ (half–chair) (the only exception is **5.6** which is of pure envelope conformation). **8.4** has the most puckered ring.

3.4 Cyclohexasilanes [(Si)$_6$] and (Si)$_n$ rings (n > 6)

Average Si–Si bond lengths for **9.1**–**9.4** (Fig. 5, Table 9) are practically identical. All the six–membered rings are of regular chair conformations (**9.3** has a dominant chair conformation). Owing to longer bonds between the ring–constituent atoms, the ring is less flattened than that in cyclohexane [62]. The **A** ring of **9.4** is chair while ring **B** is neither chair nor boat. The overall conformation is a compromise to avoid close Me...Me contacts. Wide Si–Si–Si bond angles (mean: 116.2°) were observed in the seven–membered permethylated ring (**9.2**). This compares to the trend noted for cycloalkanes which also exhibit expanded C–C–C angles relative to strain–free values.

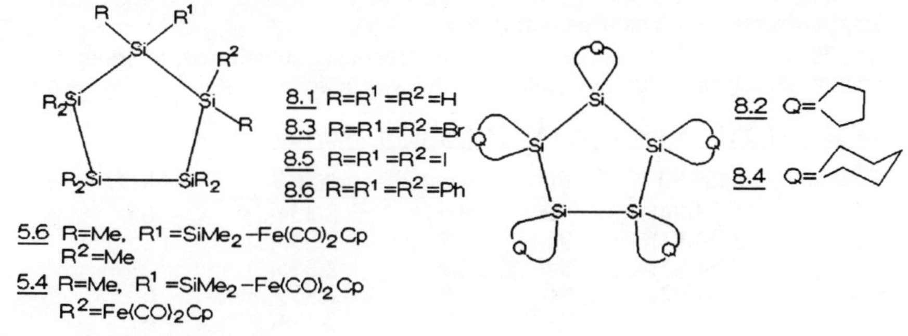

8.1 R=R^1=R^2=H
8.3 R=R^1=R^2=Br
8.5 R=R^1=R^2=I
8.6 R=R^1=R^2=Ph

5.6 R=Me, R^1=SiMe$_2$–Fe(CO)$_2$Cp
R^2=Me
5.4 R=Me, R^1=SiMe$_2$–Fe(CO)$_2$Cp
R^2=Fe(CO)$_2$Cp

8.2 Q=⬠
8.4 Q=⬡

Fig. 4. Cyclopentasilanes

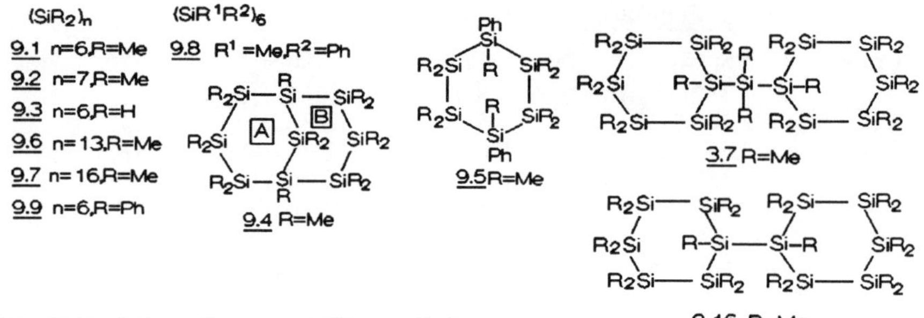

Fig. 5. Cyclohexasilanes and $(Si)_n$ $(n>6)$ rings

- **9.1** n=6,R=Me
- **9.2** n=7,R=Me
- **9.3** n=6,R=H
- **9.6** n=13,R=Me
- **9.7** n=16,R=Me
- **9.9** n=6,R=Ph
- **9.8** R^1=Me,R^2=Ph
- **9.4** R=Me
- **9.5** R=Me
- **3.7** R=Me
- **2.16** R=Me

Slightly longer bonds are found in **3.7**, **9.5**–**9.7**, **2.16** and **9.8**. These examples are characterized by higher degree of oligomerization, greater ring size or the presence of phenyl substituents. Of the isolated bicycles **3.7** and **2.16**, the latter is more crowded. The greater number of phenyl substituents in **9.8** with respect to **9.5** results in slightly elongated Si–Si bonds. The degree of oligomerization also seems to affect the length of the Si–Si bonds in permethylated macro-rings. Going from the seven-membered ring to the thirteen-membered one, an elongation of 0.014 Å is observed, well beyond the 3σ significance level. The conformations of both macro rings are governed by maximizing Me...Me interactions. Si–Si bonds are the longest in the perphenylated cyclohexasilane **9.9**, either because of the substituent bulk, or the interaction of the aryl π–electron system with the polysilane framework.

The mean Ge–Ge bond length in the related germane **9.10**, [52]) is 2.457(1) Å. The bond length differences between **2.18** and **9.9**, and **2.23** and **9.10** are Δ(Si–Si) = 0.125 Å, and Δ(Ge–Ge)=–0.020 Å, respectively. Contrary to the silicon derivatives, the mean Ge–Ge bond in the cyclohexagermane is longer than in the digermane.

The oxygen–containing heterocycle decaphenyl-1-oxa-cyclohexasilane [51] has two different types of Si–Si bond distances. The longer distances are between silicon atoms adjacent to oxygen [(Si)Si–Si(O): 2.390(1) Å], the shorter ones are wedged between Si–Si bonds [(Si)Si–Si(Si): 2.373(3) Å]. The six-membered ring is of boat conformation (^{1,4}B).

4 CONCLUDING REMARKS

Taking 2.352 Å as the length of the "normal" Si–Si bond, Pauling's bond order (PBO) can be calculated [$c \cdot \log(PBO) = (d_l - d_n)$, $c = 0.6$, and d_l is the actual bond length, d_n is the "normal" bond length [46]]. Disilane bond lengths cover the widest range of bond distances; PBO's range from 0.2–0.3 to 1.20–1.30.

The Si–Si bond distances in homocyclic rings are as expected: the mean bond lengths decrease with increasing ring size.

There is no complete series of homogenously substituted compounds with known structural data available due to synthetic and stability problems. A

Table 9. Mean Si–Si distances (Å) for $(Si)_n$ (n=6, n<6) rings

Cmpnd.	$\bar{d}$ (Si–Si)	Ref.	Cmpnd.	$\bar{d}$(Si–Si)	Ref.
9.1	2.338(4)	90	9.6	2.354(2)	65
9.2	2.340(1)	91	9.7	2.355(2)	65
9.3	2.342(5) [ED]	87	2.16	2.355(2)	21
9.4	2.343(2)	86	9.8	2.359(2)	66
3.7	2.346(4)	21	9.9	2.394(1)	67
9.5	2.350(3)	85			

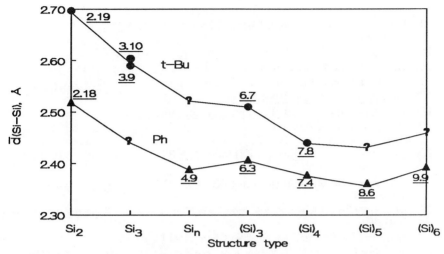

Fig. 6. Average Si-Si distances for bulky (phenyl and t-butyl) substituents

comparison of the most complete homologous series (silanes, methyl-, phenyl-, and *t*-butylsilanes) shows characteristic features as the function of substituent bulk. Si-Si distances in silanes are independent of the sizes and shapes of the molecules (only a slight increase of the bond distances is detected with increasing degree of oligomerization). Methylsilanes show a drop of the Si-Si bond distance at trisilanes. The longest Si-Si bonds are found in small ring size compounds. For bulky substituents the Si-Si bonds are longer (Fig. 6). In the perphenylated and per–*t*–butylated series disilanes seem to be the sterically most congested, and with increasing chain length the mean Si-Si bond distance decreases. A slight increase (a local maximum) of the Si-Si bonds is observed in three-membered rings. A shallow minimum is found at the five-membered rings followed by a slight increase of the average Si-Si bond length in higher–membered ring compounds.

REFERENCES

1. F.H.Allen,O.Kennard,R.Taylor,Acc.Chem.Res., 1983,16,146.
2. J.Donohue,"The Structure of the Elements",Wiley,New York,N.Y.,1974.
3. S.Masamune,S.Murakami,J.T.Snow,H.Tobita,D.J.Williams,Organometallics, 1984,3,333.
4. M.J.Fink,M.J.Michalczyk,K.J.Haller,R.West,J.Michl,Organometallics,1984, 3,793.
5. A.P.Cox,R.Varma,J.Chem.Phys., 1967,46,2007.
6. T.Fjeldberg,A.Haaland,M.F.Lappert,B.E.R.Schilling,R.Seip,A.J.Thorne, J.Chem.Soc.Chem.Commun, 1982,1408,(and references therein).
7. J.Pasinski,S.A.McMahon,R.Beaudet,J.Mol.Spectr., 1975,55,88.
8. D.W.H.Rankin,A.Robertson,J.Mol.Struct.,1975,27,438.
9. J.Haase.,Z.Naturforsch., 1973,28A,542.
10. B.Beagley,A.R.Conrad,J.M.Freeman,J.M.Monaghan,B.G.Norton,G.C. Holywell,J.Mol.Struct., 1972,11,371.
11. A.P.Cox,R.Varma,J.Chem.Phys., 1966,44,2619.
12. W.Wojnowski,B.Dreczewski,K.Peters,E.-M.Peters,H.G.von Schnering, Z.Anorg.Allg.Chem., 1986,540,271.

13. B.Beagley,J.J.Monaghan,T.G.Hewitt,J.Mol.Struct.,1971,8,401.
14. K.H.Pannell,J.Cervantes,L.Párkányi,F.Cervantes-Lee,J.Organomet.Chem.,To be published
15. S.G.Baxter,K.Mislow,J.F.Blount,Tetrahedron,1980,36,605.
16. H.Thomassen,K.Hagen,R.Stolevik,K.Hassler,J.Mol.Struct.,1986,147,331.
17. O.A.D'yachenko,Yu.A.Sokolova,L.O.Atovmyan,N.V.Ushakov,Izv.Akad. Nauk SSSR,Ser.Khim.,1984,1314.
18. L.Párkányi,E.Hengge,J.Organomet.Chem.,1982,235,273.
19. W.Wojnowski,K.Peters,E.-M.Peters,H.G.von Schnering,B.Becker, Z.Anorg.Allg.Chem.,1987,553,287.
20. S.G.Baxter,D.A.Dougherty,J.P.Hummel,J.F.Blount,K.Mislow,J.Am.Chem. Soc.,1987,100,7795.
21. K.Hassler,F.K.Mitter,E.Hengge,C.Kratky,U.G.Wagner,J.Organomet.Chem., 1987,333,291.
22. M.Weidenbruch,K.Kramer,K.Peters,H.G.von Schnering,Z.Naturforsch.,1985, B40,601.
23. N.Kleiner,M.Dräger,J.Organomet.Chem.,1984,270,151.
24. N.Wiberg,H.Schuster,A.Simon,K.Peters,Angew.Chem.,Int.Ed.Engl.,1986, 25,79.
25. M.Dräger,L.Ross,Z.Anorg.Allg.Chem.,1980,460,207.
26. L.Párkányi,K.H.Pannell,C.Hernandez,J.Organomet.Chem.,1983,301,145.
27. K.H.Pannell,R.N.Kapoor,R.Raptis,L.Párkányi,V.Fülöp,J.Organomet.Chem.,. 1990,384,41.
28. M.Weidenbruch,F.-T.Grimm,M.Herrndorf,A.Schäfer,K.Peters,H.G.von Schnering,J.Organomet.Chem.,1988,341,353.
29. A.Almenningen,T.Fjeldberg,E.Hengge,J.Mol.Struct.,1984,112,239.
30. A.Almenningen,T.Fjeldberg,J.Mol.Struct.,1981,77,315.
31. M.J.Fink,D.B.Puranik,Organometallics,1987,6,1809.
32. S.C.Nyburg,A.G.Brook,F.Abdesaken,G.Gutekunst,W.Wong-Ng,Acta Crystallogr.,1985,Sect.C41,1632.
33. J.Y.Corey,L.S.Chang,E.R.Corey,Organometallics,1987,6,1595.
34. U.Schubert,C.Steib,J.Organomet.Chem.,1982,238,C1.
35. M.Weidenbruch,B.Blintjer,K.Peters,H.G.vonSchnering, Angew.Chem.,Int.Ed.Engl.,1986,25,1129.
36. U.Schubert,A.Schenkel,J.Müller,J.Organomet.Chem.,1985,292,C11.
37. M.Haase,U.Klingebiel,R.Boese,M.Polk,Chem.Ber.,1986,119,1117.
38. L.S.Bartell,Inorg.Chem.,1970,9,2436.
39. U.Schubert,M.Wiener,F.H.Kohler,Chem.Ber.,1979,112,708.
40. A.Rengstl,U.Schubert,Chem.Ber.,1980,113,278.
41. L.Párkányi,H.Stüger,E.Hengge,J.Organomet.Chem.,1987,333,187.
42. Yu.E.Ovichinnikov,V.E.Shklover,Yu.T.Struchkov,V.V.Dement'ev,T.M. Frunze,B.A.Antipova,J.Organomet.Chem.,1987,335,157.
43. V.V.Korshat,Yu.E.Ovchinnikov,V.V.Dement'ev,V.E.Shklover, Yu.T.Struchkov,T.M.Frunze,Dokl.Akad.NaukSSSR,1987,293,140.
44. B.Beagley,J.J.Monaghan,J.Trans.Farad.Soc.,1970,66,2745.
45. M.Weidenbruch,B.Flintjer,K.Kramer,K.Peters,H.G.vonSchnering, J.Organomet.Chem.,1988,340,13.
46. L.Pauling,"The Nature of the Chemical Bond",Cornell University Press,New York,N.Y.,1984.

47. Matsumoto,H.Miyamoto,N.Kojima,Y.Nagai,M.Goto,Chem.Lett.,1988,629.
48. H.B.Yokelson,A.J.Millevolte,G.R.Gillette,R.West,J.Am.Chem.Soc., **1987,** 109,6865.
49. R.West,D.J.DeYoung,K.J.Haller,J.Am.Chem.Soc., **1985,**107,4942.
50. S.Masamune,S.Murakami,H.Tobita,D.J.Williams,J.Am.Chem.Soc.,1983,105, 7776.
51. L.Párkányi,E.Hengge,H.Stüger,J.Organomet.Chem.,1983,251,167.
52. M.Dräger,L.Ross,D.Simon,Z.Anorg.Allg.Chem.,1980,466,145.
53. K.H.Pannell,J.J.Wu,G.J.Long,J.Organomet.Chem.,1980,186,85.
54. K.H.Pannell,A.R.Bassindale,J.Organomet.Chem.,1982,229,1.
55. D.M.Roddick,T.D.Tilley,A.L.Rheingold,S.J.Geib,J.Am.Chem.Soc.,1987,109, 945.
56. Unpublished results.
57. T.J.Drahnak,R.West,J.C.Calabrese,J.Organomet.Chem.,1980,198,55.
58. M.C.Couldwell,J.Simpson,W.T.Robinson,J.Organomet.Chem, 1976,107,323 (and references therein).
59. B.K.Nicholson,J.Simpson,W.T.Robinson,J.Organomet.Chem.,1973,47,403.
60. L.Párkányi,K.H.Pannell,C.Hernandez,J.Organomet.Chem.,1983,252,127.
61. M.H.Chisholm,H-T.Chiu,K.Folting,J.C.Huffman,Inorg.Chem.,1984,23,4097.
62. M.Davis,O.Hassel,ActaChem.Scand.,1963,17,1181.
63. L.S.Bartell,H.K.Higginbotham,J.Chem.Phys.,1965,42,851.
64 R.Jones,D.J.Williams,Y.Kabe,S.Masamune,Angew.Chem.,Int.Ed.Engl.,1986, 25,173.
65. F.Shafiee,K.J.Haller,R.West,J.Am.Chem.Soc.,1986,108,5478.
66. S.-M.Chen,L.D.David,K.J.Haller,C.L.Wadsworth,R.West,Organometallics, 1983,2,409.
67. M.Dräger,K.G.Walter,Z.Anorg.Allg.Chem.,1981,479,65.
68. M.Eisenstein,F.L.Hirshfeld,ActaCrystallogr.,1983,Sect.B39,61.
69. H.Watanabe,M.Kato,T.Okawa,Y.Nagai,M.Goto,J.Organomet.Chem.,1984, 271,225.
70. S.Masamune,Y.Hanzawa,S.Murakami,T.Bally,J.F.Blount J.Am.Chem.Soc., 1982,104,1150.
71. J.C.Dewan,S.Murakami,J.T.Snow,S.Collins,S.Masamune, J.Chem.Soc.,Chem.Comm.,1985,892.
72. M.Weidenbruch,K.-L.Thom,S.Pohl,W.Saak,J.Organomet.Chem.,1987,329, 151.
73. A.Schäfer,M.Weidenbruch,K.Peters,H.G.vonSchnering,Angew.Chem., Int.Ed.Engl.,1984,23,302.
74. Y.Kabe,M.Kuroda,Y.Honda,O.Yamashita,T.Kawase,S.Masamune, Angew.Chem.,Int.Ed.Engl.,1988,27,1725.
75. C.W.Carlson,K.J.Haller,Xing-HuaZhang,R.West,J.Am.Chem.Soc.,1984,106, 5521.
76. Z.Smith,H.M.Seip,E.Hengge,G.Bauer,ActaChem.Scand.,1976,Sect.A32,697.
77. C.Kratky,H.G.Schuster,E.Hengge,J.Organomet.Chem.,1983,247,253.
78. V.S.Mastryukov,S.A.Strelkov,L.V.Vilkov,M.Kolonits,B.Rozsondai, H.G.Schuster,E.Hengge,J.Mol.Struct.(in press).
79. C.J.Hurt,J.C.Calabrese,R.West,J.Organomet.Chem.,1975,91,273.
80. L.Párkányi,K.Sasvári,I.Barta,ActaCrystallogr.,1978,Sect.B34,883.
81. P.Jutzi,U.Holtmann,H.Bogge,A.Müller,J.Chem.Soc.,Chem.Comm.,1988,305.
82. H.Matsumoto,K.Takatsuna,M.Minemura,Y.Nagai,M.Goto, J.Chem.Soc.,Chem.Comm.,1985,1366.

83. S.Masamune,Y.Kabe,S.Collins,D.J.Williams,R.Jones,J.Am.Chem.Soc.,1985, 107,5552.
84. F.A.Cotton,B.A.Frenz,Tetrahedron,1974,30,1587.
85. K.Kumar,M.H.Litt,R.K.Chadha,J.E.Drake,Can.J.Chem.,1987,65,437.
86. W.Stallings,J.Donohue,Inorg.Chem.,1976,15,524.
87. Z.Smith,A.Almenningen,E.Hengge,D.Kovar,J.Am.Chem.Soc.,1982,104, 4362.
88. C.Kratky,E.Hengge,H.Stüger,A.L.Rheingold,ActaCrystallogr.,1985, Sect.C41,824.
89. L.Párkányi,K.Sasvári,J.-P.Declercq,G.Germain,ActaCrystallogr.,1978, Sect.B34,3678.
90. H.L.Carrell,J.Donohue,ActaCrystallogr.,1972,Sect.B28,1566.
91. F.Shafiee,J.R.Damewood Jr.,K.J.Haller,R.West,J.Am.Chem.Soc.,1985,107, 6950.
92. W.J.Adams,H.J.Geise,L.S.Bartell,J.Am.Chem.Soc.,1970,92,5013.

Unusual Low Valent and Multiply Bonded Silicon Compounds from Cyclopropenyl Silanes

Mark J. Fink
DEPARTMENT OF CHEMISTRY, TULANE UNIVERSITY, NEW ORLEANS, LOUISIANA 70118, USA

Compounds possessing a low valent or multiply bonded silicon atom have attracted the attention of synthetic, mechanistic, and theoretical chemists alike.[1] These interesting molecules have a rich chemistry which is just beginning to be explored. Our research efforts have focused on a number of low valent and multiply bonded silicon compounds derived from cyclopropenyl silanes. The combination of high ring strain and unsaturation in these molecules have enabled us to generate the first examples of silacyclobutadienes, silatetrahedranes, and cyclopropenylsilylenes.

Silacyclobutadienes

Silacyclobutadiene (**I**) is a molecule of considerable interest due to its potential anti-aromatic character and possible interconversions to unusual isomers such as silatetrahedrane(**II**), cyclopropenylsilylene(**III**) and silylenecyclopropene(**IV**).

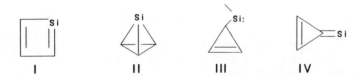

Early theoretical studies of **I** suggest that this molecule should be planar with strongly alternating C=C and Si=C bonds.[2,3] The large anti-aromatic destabilization energy of 49.1 kcal/mole predicted for **I** is somewhat less than that of cyclobutadiene itself (67.9 kcal/mol).[3] The triplet state of **I** is very low lying, only 5 kcal/mole over the ground state singlet. With respect to possible decomposition products, **I** is predicted to be 33 kcal/mole more stable than **II** and 44.8 kcal/mole more stable than the dissociation products acetylene and silyne. More recent calculations show that isomers **I**, **III**, and **IV** are essentially isoenergetic.[4]

Our approach to preparing silacyclobutadienes is based on the known rearrangements of cyclopropenyl nitrenes[5] and carbenes[6] to give azetes and cyclobutadienes, respectively. This ring expansion rearrangement may in principle occur for derivatives of cyclopropenyl silylenes to give silacyclobutadienes. Since silacyclobutadienes would be predicted to be highly reactive due to the presence of a Si=C bond, sterically protected derivatives of these species were sought in the hope that stable examples of these compounds may be obtained.

Two highly substituted photochemical precursors of cyclopropenylsilylenes, **1** and **2**, have been obtained by the reaction of tri-$\underline{t}$-butylcyclopropenium tetrafluoroborate with polysilyl lithium reagents.[7,8]

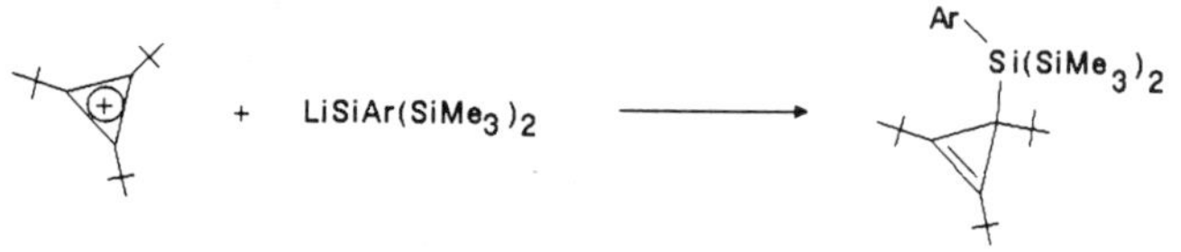

1: Ar = mesityl (Mes)
2: Ar = 2,4,6 – trisopropylphenyl (Trip)

These reactions are high yield and give stable covalent products. $\underline{X}$-ray crystal structures have been obtained for both of these compounds.[7,8] The structures reveal unusually long bonds to the central silicon atom which become more pronounced with increasing bulk of the aromatic substituent.

Photolysis of these trisilanes in a hydrocarbon solvent containing EtOH gives the expected products from the insertion of the silylene into the O-H bond of the EtOH.[9] However, this reaction is complicated by both secondary photolysis of the product and a competitive photolysis of the trisilane precursor. These photoreactions give rise to radical products derived from homolytic bond cleavage of the silicon-cyclopropene bond.

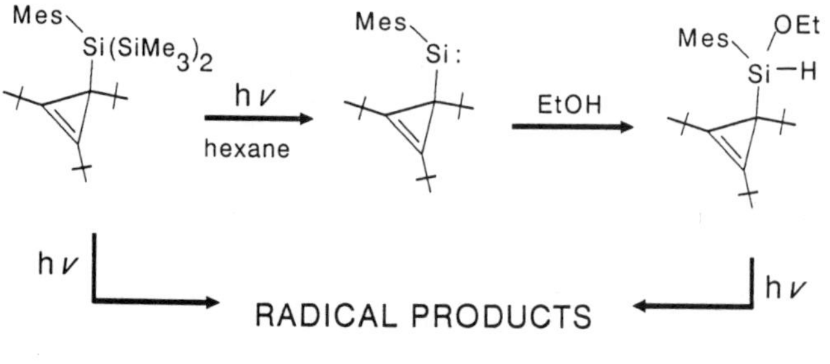

No evidence for isomers of cyclopropenylsilylenes (which should be reactive with EtOH) is seen in these experiments. The absence of thermal rearrangement is in accord with theoretical calculations which show that the two most likely rearrangement products of cyclopropenylsilylene, **I** and **IV**, are isoenergetic.

The possibility for photoisomerization of the cyclopropenylsilylenes was examined. The 254 nm photolysis of cyclopropenyltrisilanes **1** and **2** in 3-methylpentane (3-MP) matrices at 77K gives high yields of the yellow cyclopropenylsilylenes **3** and **4**. These silylenes are distinguished by prominent absorption bands with λ_{max} near 450 nm. Silylenes **3** and **4** can be efficiently converted to silacyclobutadienes **5** and **6** by photolysis into the visible absorption band of the silylene.[10] The benzosilacyclobutenes **7** and **8**, arising from the insertion of the silylene into a benzylic C-H bond of the aromatic substituent, are observed as minor products.

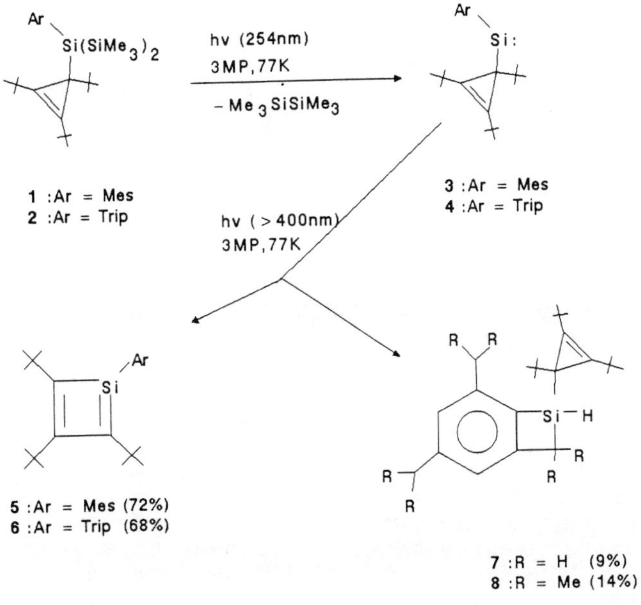

Whereas the benzosilacyclobutenes are stable isolable compounds, the silacyclobutadienes are extremely reactive. The existence of the silacyclobutadienes in matrix is confirmed by trapping reactions which give stable silacyclobutenes. These silacyclobutenes arise exclusively from addition of the trapping reagents across the Si=C bond. Shown below are some typical reactions of the silacyclobutadiene **5**.

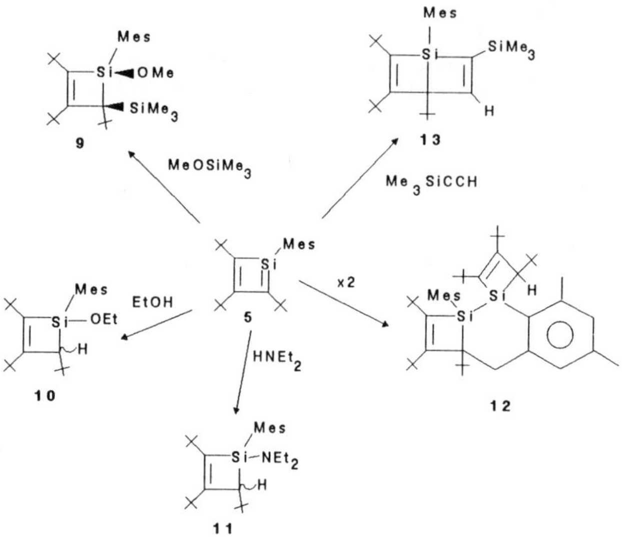

The additions of MeOSiMe$_3$, EtOH, and Et$_2$NH across the Si=C bond proceed in a regiospecific manner typical of silenes. The addition of MeOSiMe$_3$ is stereochemically syn as indicated by an X-ray crystal structure of the corresponding silacyclobutene adduct (9).[10] In contrast, trapping experiments with either EtOH or HNEt$_2$ result in the formation of two diastereomeric silacyclobutenes. This observation is readily explained by photoisomerization of an initially formed syn addition product.

Since the trapping experiments require 20-30 successive photochemical events to generate useful amounts of trapping products, the silacyclobutenes may also undergo photolysis. The two diastereomeric silacyclobutenes resulting from the addition of EtOH to **5** have been separated by chromatography. Independent photolysis of either silacyclobutene isomer under the same conditions of the trapping experiments gives a photostationary mixture which is nearly the same as found in the trapping experiments. Although this experiment shows that the two silacyclobutenes photochemically interconvert, it does not necessarily show that the addition is stereospecific. This is demonstrated by the trapping of silacyclobutadiene **5**, with EtOH after one photolysis cycle, thereby avoiding photolysis of the trapping product. In this case, only one diastereomeric silacyclobutene is produced, presumably the syn isomer. The mechanism for photoisomerization may be explained from the photochemical ring opening of the silacyclobutene to give an acyclic siladiene, followed by thermal ring closure. Woodward-Hoffmann rules predict that the newly formed silacyclobutene should have the opposite stereochemistry.

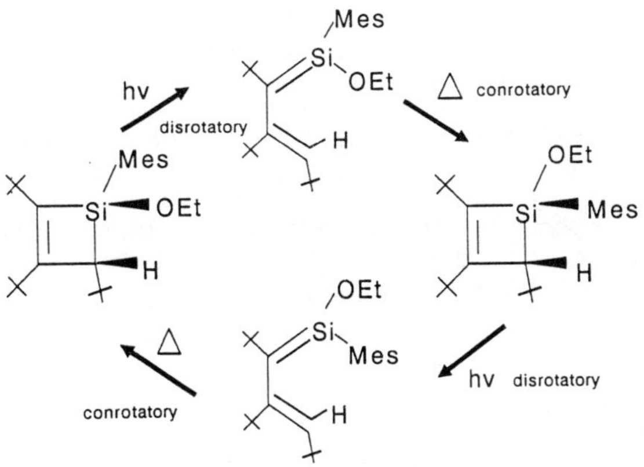

The reaction of **5** with Me₃SiC≡CH is a case of π addition to the Si=C bond. The resultant adduct (**13**) is the first stable example of the Dewar form of silabenzene. The molecule is thermally stable at room temperature although reacts slowly with oxygen.

In the absence of trapping reagents an unusual dimeric product (**12**) is obtained in nearly quantitative yield. An X-ray crystal structure of this molecule has been obtained.[11] The mechanism of its formation is discussed in a later section.

The reaction chemistry of the more hindered silacyclobutadiene (**6**) has also been examined. Although this molecule reacts analogously with EtOH and Me₃SiC≡CH, no reaction is observed with MeOSiMe₃. This is presumably due to the higher degree of protection of the Si=C bond in this analog. The Dewar silabenzene formed by the reaction with Me₃SiC≡CH is not only thermally stable but unreactive to oxygen under ambient conditions.

The Mechanism of Cyclopropenylsilylene Isomerization

Two possibilities for the photoisomerization of cyclopropenylsilylenes to silacyclobutadienes are likely. One mechanism is a simple sigmatropic shift of a cyclopropenyl C-C single bond to the divalent silicon atom. Another is an internal π addition of the silylene to give a silatetrahedrane intermediate which ultimately decomposes to the silacyclobutadiene.

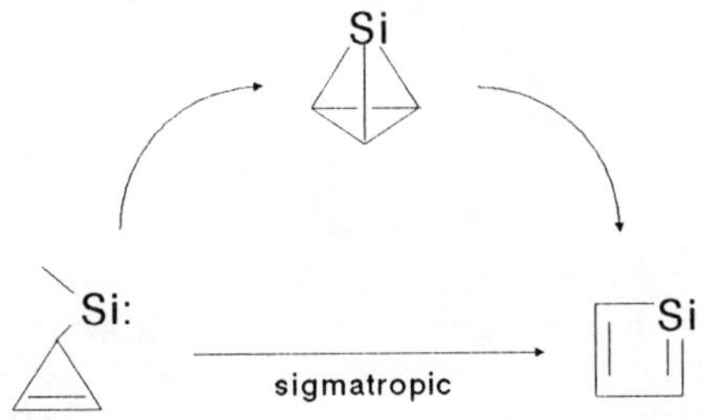

These two mechanisms may be distinguished by labelling experiments in which one of the substitituents on the cyclopropene ring of the silylene is different from the remaining two. Two isomeric cyclopropenyltrisilanes **14** and **15** have been isolated from the reaction shown below.[12] The t-pentyl groups serve as labels to distinguish the carbon atoms in the original cyclopropene ring. Each trisilane can be photolyzed to a corresponding silylene such that the label retains its position on the cyclopropene ring.

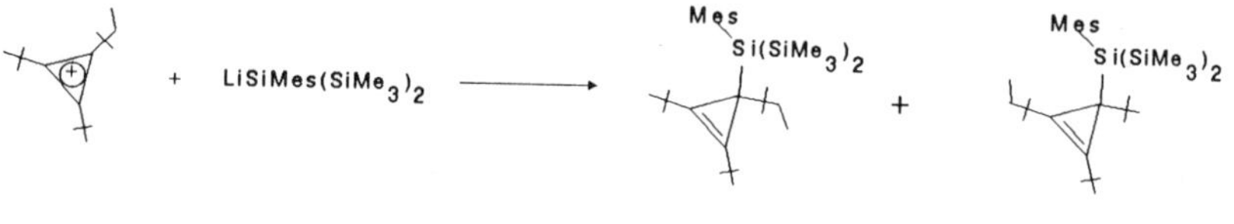

If we consider the case of the cyclopropenylsilylene in which the label is on the allylic carbon atom, simple sigmatropic shifts (path A, below) would produce two potentially tautomeric silacyclobutadienes, both with the label attached to a carbon atom vicinal to the ring silicon.

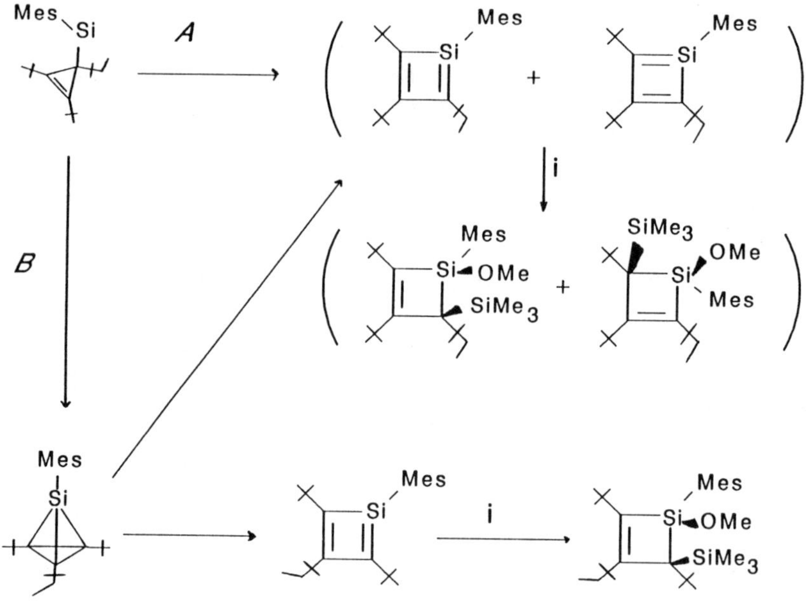

i = MeOSiMe$_3$

On the other hand, the formation and decomposition of an initial silatetrahedrane intermediate (path B) would produce three silacyclobutadienes. Two of these silacyclobutadienes have their label on a vicinal carbon atom, and are identical to the silacyclobutadienes produced by path A. A third unique silacyclobutadiene has the label on the ring carbon transannular to the silicon. If breaking of the Si-C bonds in the silatetrahedrane is equally likely then one-third of the silacyclobutadienes produced would have their label on the transannular carbon atom.

Since the silacyclobutadienes generated in the labelling experiments are unstable, trapping reactions are necessary. The stereospecific trap MeOSiMe$_3$ was used since there exists a one-to-one correspondence between the stable silacyclobutene adducts and the silacyclobutadiene which is trapped.

Independent photolysis of the two isomeric cyclopropenyltrisilanes **14** and **15** to their corresponding silacyclobutadienes, followed by trapping with MeOSiMe$_3$, results in the formation of identical silacyclobutene products. One silacyclobutene (32% yield) has a transannular t-pentyl group; the other (68% yield) has the t-pentyl group on a vicinal carbon atom. Since both isomeric cyclopropenyl silylenes give identical trapping products, a common intermediate is implied. The product distribution is wholly consistent with an intermediate silatetrahedrane. Attempts to intercept the silatetrahedrane with chemical traps have been unsuccessful to date implying that the intermediate silatetrahedrane must have a very short lifetime even at 77K in 3-MP glass.

Thermal Reactions of Silacyclobutadiene Analogs

Perhaps the most interesting chapter in the chemistry of silacyclobutadiene analogs is their thermal reactions. As mentioned earlier for the mesityl derivative **5**, a single dimeric product is isolated in greater than 95% yield. This compound is now believed to occur via a diradical pathway involving the initial formation of a Si-Si bond from the dimerization of two silacyclobutadienes (scheme below). The initially formed diradical **16** then undergoes an internal hydrogen atom abstraction to afford the new diradical species **17**. Ring closure of this ultimate diradical gives the observed dimer,**12**.

Direct evidence for the intermediate diradicals **16** and **17** is obtained by monitoring the thermal decomposition of the silacyclobutadiene by epr spectroscopy.[13] Upon annealing a 3-MP glass containing **5**, a strong signal indicative of two randomly oriented triplet species appears. A "half-field" transition confirms the triplet state. The zero field splitting parameters of the two triplets have been derived. The major triplet species has a $|D/hc|$ = .0243 cm^{-1} and $|E/hc|$ = 0; the minor triplet has $|D/hc|$ = .0138 cm^{-1} and $|E/hc|$ = .00047 cm^{-1}.

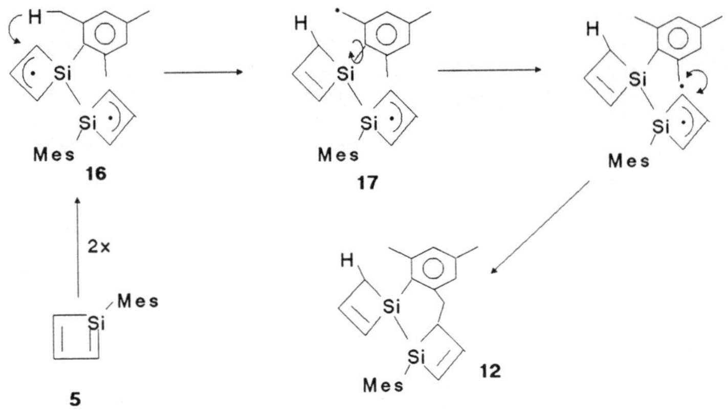

Careful thawing of the 3-MP glass containing these two diradicals results in the loss of the initially larger triplet signal with the concomitant increase of the initially smaller signal. We therefore assign the triplet with $|D/hc|$ = .0243 cm^{-1} as the initial bisallyllic diradical **16** and the triplet with $|D/hc|$ = .0138 cm^{-1} as the isomeric diradical **17**. A calculated $|D/hc|$ of .027 cm^{-1} has been obtained for the AM1 optimized structure of the parent diradical of **16** using a point dipole model. Both diradicals disappear rapidly above 250K.

The thermal decomposition of the Trip substituted silacyclobutadiene (**6**) is an interesting case of competitive intramolecular and bimolecular processes. Similar to the mesityl analog, the annealing of a glass containing **6** results in the appearance of a triplet spectrum, although in this case from only one species. The triplet has a $|D/hc|$ = .0247 cm^{-1} and $|E/hc|$ = .0009 cm^{-1}. The $|D/hc|$ value is very similar to **16** and is assigned as the analogous bisallylic diradical(**18**). This diradical is extremely stable with a half-life of at least 14 hrs at room temperature!

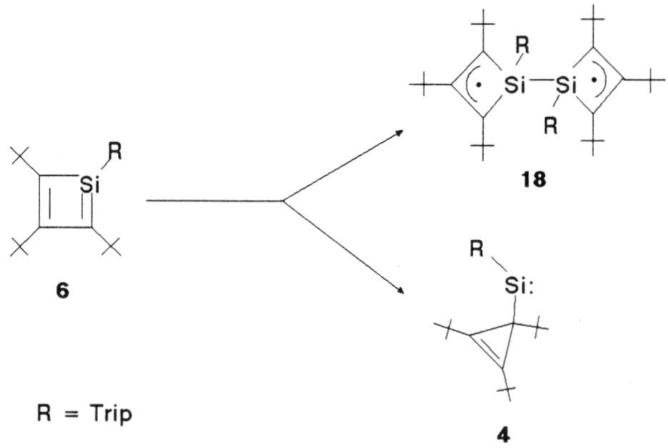

Based on spin counting and trapping experiments, dimerization accounts for only 40% of the decomposition of the silacyclobutadiene **6**. The remaining silacyclobutadiene decomposes via a novel ring contraction isomerization to afford the cyclopropenylsilylene **4** at temperatures above 150K. The transformation may be conveniently monitored by uv-vis spectroscopy.[10] This observation clearly shows that the cyclopropenylsilylene is thermodynamically more stable than the silacyclobutadiene; a result in contrast to the behavior of cyclopropenyl carbenes and nitrenes. Since theoretical calculations predict that the parent cyclopropenylsilylene and silacyclobutadiene are isoenergetic , it is likely that relief of steric strain is the driving force for the transformation of **6** to **4**.

A Stable Silylene

The cyclopropenylsilylene **4** is remarkably stable in hydrocarbon solutions at low temperature. At 200K, this silylene is persistent in hexane solutions and reacts only sluggishly with EtOH at this temperature. The half-life of **4** in hexane at 270K is approximately 15 minutes.In contrast to typical organosilylenes which dimerize in solution to give intermediate disilenes, no evidence for these intermediates is found.[10]

The extraordinary stability of **4** is attributed to a extremely high degree of steric protection provided by the bulky aromatic and cyclopropenyl substituents. An X-ray crystal structure of the water insertion product has been obtained.[14] This molecule is probably a good model for the configuration of these large substituents about the silicon center. Shown below (Figure 1) are space filling graphs which depict the steric protection of the silicon center. One side of the molecule is completely shielded and only a small cleft in the front of the molecule exposes the silicon atom. It is no wonder that the silylene is unable to dimerize in the typical fashion to give a disilene intermediate.

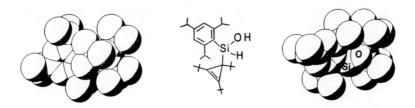

Figure 1: Space Filling Plots of Water Insertion Product of **4**

The exact pathway for the decomposition of **4** is not yet fully understood. Significantly, no products from C-H insertion reactions are observed which implies that the silylene neither reacts with the hydrocarbon solvent nor does it undergo an intramolecular C-H insertion into one of its substituents. This result is therefore encouraging to the prospects that a silylene stable to room temperature may be eventually synthesized.

Acknowledgements

I wish to thank all my collaborators and coworkers: Dhananjay Puranik, Marie Johnson, Jason Gee, Gary McPherson and William Howard. I also thank the Tulane Committee on Research, the Dow Corning Corporation , and the Louisiana Board of Regents for financial assistance.

REFERENCES

1. G. Raabe and J. Michl, Chem. Rev., 1985, 85, 419
2. M.S. Gordon, J.C.S. Chem. Commun., 1980, 1131
3. M.E. Colvin and H. Schaefer III, Faraday Sym. Chem. Soc., 1984, 19, 39
4. G.W. Schriver, M.J. Fink and M.S. Gordon, Organometallics, 1987, 6, 1977
5. U.-J. Volgelbacher, M. Regitz and R. Mynott, Angew. Chem. Int. Ed. Eng., 1986, 25, 842.
6. a.) S. Masamune, M. Nakamura, M. Suda and H. Ono, J. Amer. Chem. Soc., 1973, 95, 8481. b.) G. Maier, Angew. Chem. Int. Ed. Eng., 1988, 27, 309
7. D.B. Puranik, M.P. Johnson and M.J. Fink, Organometallics, 1989, 8, 770
8. M.P. Johnson, D.B. Puranik and M.J. Fink, Acta. Cryst. C., in press
9. M.J. Fink and D.B. Puranik, Organometallics, 1987, 6, 1809
10. D.B. Puranik and M.J. Fink, J. Amer. Chem. Soc., 1989, 111, 5951
11. D.B. Puranik, M.P. Johnson and M.J. Fink, J.C.S. Chem. Commun., 1989, 706
12. D.B. Puranik, Ph. D. Thesis, Tulane University, 1989
13. J. Gee, W.A. Howard, G.L. McPherson and M.J. Fink, manuscript in progress
14. D. B. Puranik, M.P. Johnson and M.J. Fink, unpublished work

Transition-metal Complexes of Reactive Silicon Intermediates

T.D. Tilley, B.K. Campion, S.D. Grumbine, D.A. Straus, and R.H. Heyn

CHEMISTRY DEPARTMENT, D-006, UNIVERSITY OF CALIFORNIA AT SAN DIEGO, LA JOLLA, CALIFORNIA 92093-0506, USA

1 INTRODUCTION

Transition-metal complexes are well established as useful catalysts for reactions of silicon compounds. Much of this chemistry is based on empirical observations, and surprisingly little is known about how transformations of silicon species occur at a metal center. Synthetic, reactivity, and mechanistic studies on new transition-metal silicon complexes have contributed significantly to a description of the chemical properties of silicon-based fragments within the coordination sphere of a metal.[1,2] Interest in metal-mediated transformations has naturally focused attention on complexes of "activated" silicon species, and the last few years have seen much progress in the isolation and study of transition-metal complexes of reactive silicon intermediates (e.g. silylenes, SiR_2; silenes, $R_2C=SiR'_2$; disilenes, $R_2Si=SiR'_2$; silanones, $R_2Si=O$, etc.). Such complexes have often been invoked in mechanistic proposals, but presently there is little direct evidence for their involvement in catalytic processes.[1,2]

The first reports of well-characterized silylene complexes, isolated as donor adducts, appeared in 1987. Our route is based on electron-rich transition-metal fragments for stabilization of the silylene ligand, and removal of a silicon-bound group.[3] The complex $Cp^*(PMe_3)_2RuSiPh_2OTf$ ($Cp^* = \eta^5-C_5Me_5$; $OTf = OSO_2CF_3$) contains a weakly associated triflate group, as characterized by its solution behavior, the molecular structure, and a down-field ^{29}Si NMR shift of 112.39 ppm. In acetonitrile solution, the triflate group is completely displaced to give $[Cp^*(PMe_3)_2RuSiPh_2(NCMe)]^+$. Dynamic NMR studies have shown that in dichloromethane, the latter species dissociates its acetonitrile to produce the base-free silylene complex $[Cp^*(PMe_3)_2Ru=SiPh_2]^+$.[3b] Zybill and Müller reported syntheses of $(CO)_4FeSi(O^tBu)_2(S)$ (S = HMPA, THF) via reaction of $Fe(CO)_4^{2-}$ with $(^tBuO)_2SiCl_2$, and have subsequently reported similar solvent-complexed silylene derivatives.[4] Tobita, Ogino, and coworkers have described another type of donor-stabilized silylene complex, $Cp^*(CO)FeSiMe(OMe)(\mu\text{-}OMe)SiMe_2$, obtained from a photochemically induced migration in $Cp^*(CO)_2FeSiMe_2SiMe(OMe)_2$.[5]

A transition-metal silene complex was first proposed by Pannell to explain the conversion of $Cp(CO)_2FeCH_2SiMe_2H$ and PPh_3 to $Cp(CO)(PPh_3)FeSiMe_3$.[6] More recently an analogous intermediate, $Cp^*(CO)Fe(H)(\eta^2\text{-}CH_2=SiMe_2)$, was observed spectroscopically at low temperature.[7] Berry and Procopio have presented strong evidence that the osmium silene complex $(PMe_3)_3Os(H)(D)(\eta^2\text{-}CH_2=SiMe_2)$ is an intermediate in catalytic H/D exchange between benzene-d_6 and alkylsilanes.[8] We recently communicated the first examples of isolated silene complexes, $Cp^*(PR_3)Ru(H)(\eta^2\text{-}CH_2=SiR'_2)$ (R = ^{i}Pr, Cy; R' = Me, Ph)[9] and $Cp^*(PMe_3)Ir(\eta^2\text{-}CH_2=SiPh_2)$.[10] Related η^2-disilene complexes (of platinum,[11] molybdenum,[12] and tungsten[12]) have also recently been reported by West and Berry, and interesting new structures possessing bridging disilene ligands[13] indicate that "reactive" silicon fragments may display a range of bonding patterns.

2 BASE-FREE SILYLENE COMPLEXES OF RUTHENIUM

Encouraged by the kinetic data suggesting that $[Cp^*(PMe_3)_2Ru=SiPh_2]^+$ was relatively stable in solution,[3b] we began a search for transition-metal and silicon substituents that might allow isolation of a base-free silylene complex. Work by Lambert and coworkers indicated that thiolate groups have a stabilizing influence on silylenium ions (SiR_3^+).[14] We therefore investigated the use of thiolate groups in stabilizing cationic silylene complexes, which would also contain a three-coordinate silicon center.

The tri(thiolato)silyl complexes $Cp^*(PMe_3)_2RuSi(SR)_3$ (**1**, R = p-tol; **2**, R = Et) are readily prepared by an established procedure from $Cp^*(PMe_3)_2RuCH_2SiMe_3$ and the appropriate silane $HSi(SR)_3$.[3b] From these silyl complexes, the triflate derivatives **3** and **4** are obtained via abstraction of a thiolate group by Me_3SiOTf (equation 1). X-ray crystallography established that

$$Cp^*(PMe_3)_2RuSi(SR)_3 \xrightarrow[-Me_3SiSR]{Me_3SiOTf} Cp^*(PMe_3)_2RuSi(SR)_2OTf \qquad (1)$$

3, R = p-tol
4, R = Et

3 possesses long, covalent Si-O(triflate) bonds.[15] Solution NMR and infrared spectroscopy are also consistent with covalent triflate structures for these compounds. The ^{29}Si NMR shifts for **3** and **4** (δ 77.14 and 86.05, respectively, in dichloromethane-d_2) are not unusual for transition-metal silyl complexes.[1] Infrared $\nu(SO_3)$ vibrational modes for covalently bound triflate are observed for solid samples and in dichloromethane solution (1358-1367 cm^{-1}).[16]

When **3** and **4** are dissolved in acetonitrile, triflate is displaced to produce $[Cp^*(PMe_3)_2RuSi(SR)_2(NCMe)]^+OTf^-$ silylene complexes. Infrared spectra of these solutions reveal only the presence of ionic triflate (for **3**, $\nu(SO_3)$ = 1269 cm^{-1}; for **4**, $\nu(SO_3)$ = 1268 cm^{-1}). In dichloromethane, compound **4** undergoes a dynamic process that exchanges the inequivalent methylene protons of the -SEt groups. Thus, these hydrogens appear as a single quartet (δ 2.88 at 23°C) down to -70°C. In the less polar solvent toluene-d_8, the exchange is slowed dramatically, so that a coalescence temperature of 21°C is observed ($\Delta G^\ddagger_{294K}$ = 14.9 ± 0.3 kcal mol^{-1}). This solution behavior is consistent with an exchange mechanism involving dissociation of triflate to form [Cp*(PMe$_3$)$_2$Ru=Si(SEt)$_2$]$^+$, and then return of triflate to the opposite face of the silylene ligand.[17]

Efforts to isolate the base-free silylene complexes [Cp*(PMe$_3$)$_2$Ru=Si(SR)$_2$]$^+$ were based on replacement of triflate by a less-coordinating anion. Compounds 3 and 4 react with NaBPh$_4$ in dichloromethane to precipitate NaOTf and form 5 and 6, which were crystallized from dichloromethane-diethyl ether (equation 2). Analytical and spectroscopic data show that the crystals

$$Cp^*(PMe_3)_2RuSi(SR)_2OTf \xrightarrow[-NaOTf]{+NaBPh_4} \left[Cp^*(PMe_3)_2Ru=Si\begin{matrix} SR \\ SR \end{matrix} \right]^+ BPh_4^- \quad (2)$$

5, R = p-tol
6, R = Et

contain no solvent. The ^{29}Si NMR spectrum of 5 at 23°C exhibits a remarkably low-field resonance at δ 250.6, which sharpens to a well-defined triplet at -80°C (δ 259.4, J_{SiP} = 34 Hz). For 6, a broad peak at δ 264.4 was observed at -60°C. Based on correlations between ^{13}C and ^{29}Si NMR data,[18] silylene complexes are expected to exhibit low-field ^{29}Si NMR shifts, since terminal carbene ligands generally give rise to ^{13}C NMR shifts in the range 240-370 ppm.[19] Note that the ^{13}C NMR shift for the carbene carbon in [Cp(CO)$_2$Ru=C(SMe)$_2$]PF$_6$ is δ 285.3.[20] Alternative structures for 5 and 6 possessing bridging silylene ligands have been ruled out based on molecular weight data and the fact that the silicon of 6 is coupled to only two phosphorus nuclei. A solution molecular weight determination for 6 (isopiestic method, dichloromethane) gave 990, compared to a calculated value of 982 for the ion pair. Finally, 5 reacts with acetonitrile cleanly to produce the adduct {Cp*(PMe$_3$)$_2$RuSi[S(p-tol)]$_2$(NCMe)}$^+$BPh$_4^-$, which has been crystallographically characterized.[15]

Attempts are currently underway to obtain X-ray quality crystals of 5, 6 and related silylene derivatives. Such complexes should provide a wealth of information concerning the reactivity of transition-metal silylene complexes.

3 SILENE COMPLEXES OF RUTHENIUM

Our approach to the synthesis of ruthenium silene complexes is based on β-hydrogen transfer from silicon in a Ru-CH$_2$SiR$_2$H complex, and stabilization of the silene ligand by a sterically hindered, electron-rich metal center. The key starting materials are the 16-electron complexes Cp*(PR$_3$)RuCl (R = ^{i}Pr, Cy = cyclohexyl),[21] which are alkylated according to equation 3 to

$$\text{Cp*}(PR_3)\text{RuCl} + \text{ClMgCH}_2\text{SiR'}_2\text{H} \longrightarrow \underset{\begin{array}{c}\text{7, R = }^i\text{Pr, R' = Me}\\\text{8, R = Cy, R' = Me}\\\text{9, R = }^i\text{Pr, R' = Ph}\\\text{10, R = Cy, R' = Ph}\end{array}}{\text{[Cp*Ru(PR}_3\text{)(H)(SiR'}_2\text{CH}_2\text{)]}} \quad (3)$$

give the silene complexes **7-10** directly. Apparently these reactions proceed via transient, 16-electron alkyls Cp*(PR$_3$)RuCH$_2$SiR'$_2$H which undergo β-hydrogen elimination to give the observed silene complexes. An intermediate observed by ^{31}P NMR spectroscopy (δ 43.26) during the low-temperature (-70°C) formation of **9** may be Cp*(^{i}Pr$_3$P)RuCH$_2$SiPh$_2$H. The ^{29}Si NMR spectrum of **9** consists of a multiplet centered at δ 6.14. This value is intermediate between those found in stable silenes (δ 40-50)^{22} and silacyclopropanes (~ δ -60).^{23} The largest apparent coupling constant observed was 21 Hz, which may be attributed to $^2J_{PSi}$ or $^2J_{HRuSi}$. The ^{13}C NMR shift for the silene carbon of **9** (δ -29.04) is upfield of the shift for the analogous ethylene complex Cp(PPh$_3$)RuH(η^2-CH$_2$CH$_2$) (δ 20.73). The $^1J_{CH}$ coupling constant for the coordinated =CH$_2$ group of **9** (143.3 Hz) is intermediate between values for free ethylene (156.2 Hz) and methane (125.0 Hz),^{24} suggesting significant sp^2 character for the silene carbon atom.

The crystal structure of **9** revealed Si-CH$_2$ bond distances of 1.78 (2) and 1.79 (2) Å for the two independent molecules.^{9} These distances seem to reflect partial C=Si double bond character, since Si-C single bond distances normally range from 1.87 to 1.91 Å. As expected, the Si-CH$_2$ distance in **9** is somewhat longer than Si=C double bond distances observed for free silenes (1.764 (3)^{25} and 1.702 (5)^{26} Å). The C-Si-C angles about **9** add to 344°, which falls between the expected values for sp^2 (360°) and sp^3 (329°) hybridization.

Although **7** and **8** are isolable, they are thermally unstable and decompose in solution at room temperature over ca. 1 hour. In contrast, the 1,1-diphenylsilene complexes **9** and **10** are quite stable and can be stored at room temperature in the solid state for months. In benzene solution, **9** and **10** thermally decompose (over ca. 24 h) to the products shown in equations 4

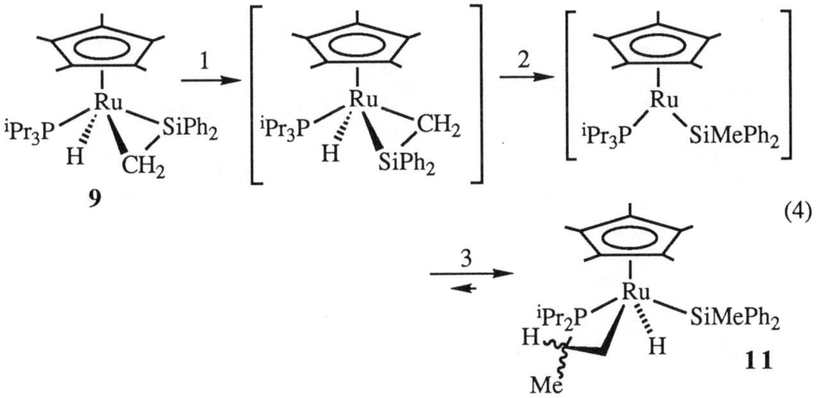

$$\text{9} \xrightarrow{1} [\text{Cp*(}^i\text{Pr}_3\text{P)Ru(H)(CH}_2\text{SiPh}_2\text{)}] \xrightarrow{2} [\text{Cp*(}^i\text{Pr}_3\text{P)Ru–SiMePh}_2] \quad (4)$$

$$\xrightarrow{3} \text{11}$$

and 5. The conversion of **9** to **11** involves exclusive migration of hydride to the carbon atom, as demonstrated by deuterium-labeling. The reaction obeys a first-order rate law, rate = k_{obs}[9], with $\Delta G^{\ddagger}$ = 23±3 kcal mol^{-1} and $\Delta S^{\ddagger}$ = -17±3 eu. The three-step process of equation 4 accounts for the observed transformation. The hydrogen migration step appears to be rapid and irreversible, since k_H/k_D = 1.00. Also for the deuteride **9-d**, deuterium does not "wash" into the methylene hydrogen positions during the course of the reaction. We believe that the third step, cyclometallation of the phosphine ligand, is rapid and characterized by an equilibrium that lies far to the right. Evidence for this comes from related studies which indicate that, in general, 16-electron Cp*(PR$_3$)Ru(silyl) species are quite unstable, and also that reactions of **11** go via the intermediate Cp*(P^{i}Pr$_3$)RuSiMePh$_2$ (<u>vide infra</u>). Another relevant observation is that no intermediates are observed to build up during the conversion of **9** to **11**. The rate-limiting step therefore appears to be rearrangement of **9** to a structure with the hydride ligand in a position to migrate to the silene carbon. This process could involve rotation about the Ru-silene bond or a pseudorotation of the complex. The rate of this rotation should then have a marked influence on the stability of silene complexes of the type Cp*(PR$_3$)Ru(H)(η^2-CH$_2$SiR'$_2$), with high barriers to rotation corresponding to more stable complexes. This is consistent with the observed ordering of stabilities for silene complexes with substituents R/R': Cy/Ph > ^{i}Pr/Ph > Cy/Me > ^{i}Pr/Me >> Me/Me.

The thermal decomposition of **10** (equation 5) involves a similar process, except that the silyl complex containing metallated phosphine (**12**) is unstable and eliminates silane to produce the β-hydrogen elimination product **13**. The lower stability of **12** compared to **11** could be due at least partially to greater steric crowding in the former species. Thermolysis of the deuteride **10-d** resulted in the clean formation of HSi(CH$_2$D)Ph$_2$. Compound **13** has been completely characterized, and its formulation as an olefin complex is based on comparisons of NMR data to similar complexes.[27]

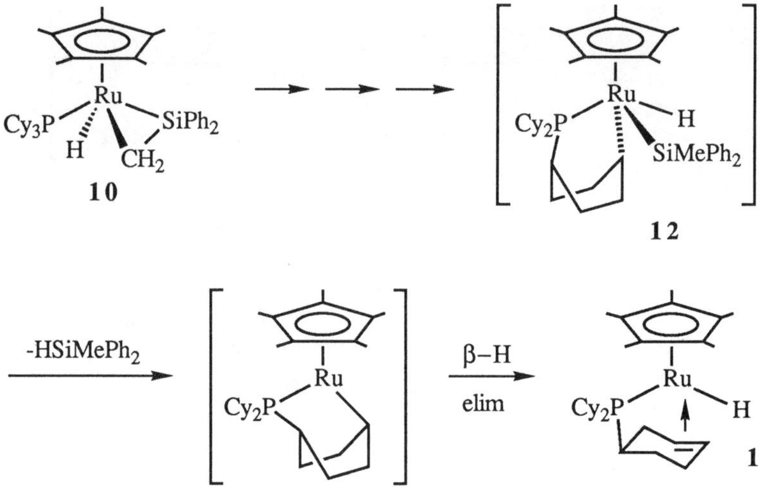

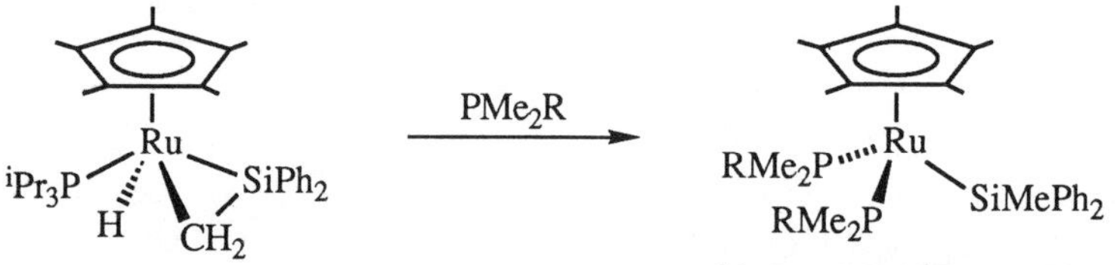

Silene complex **9** reacts cleanly with sterically unhindered phosphines to produce free P^{i}Pr$_3$ and the bis(phosphine) silyl complexes **14** or **15** (equation 6). Under a variety of reaction conditions, these were the only ruthenium-containing products observed, and no evidence for the alkyls Cp*(RMe$_2$P)$_2$RuCH$_2$SiHPh$_2$ was found. The reactions were followed by ^{31}P NMR spectroscopy under pseudo-first-order conditions using excess PMe$_2$R. The linearity of plots of ln[**9**] vs. time established first-order dependence in **9** for each case. Plots of [PMe$_2$R] vs. k_{obs} were linear but did not intercept the origin. Therefore a simple bimolecular process (rate = k_{obs}[**5**][PMe$_2$R]) is not operating, and a more accurate description of the mechanism involves competing first- and second-order pathways, rate = k [**9**] + k'[**9**][PMe$_2$R]. This behavior may be explained in terms of Scheme 1. From the plots of [PMe$_2$R] vs. k_{obs}, the first-order rate constant corresponds closely to the rate constant observed for the first-order thermal decomposition of **9** ($k_{rot} = 6 \times 10^{-5}$ s^{-1}). This component of the reaction therefore probably involves trapping of the 16-electron silyl complex which is generated by migration of hydride to carbon (cf. mechanism for thermolysis, equation 4).

The pathway giving rise to second-order behavior is explained by a rapid pre-equilibrium involving migration of hydride to the silene silicon (Scheme 1). This could lead to a

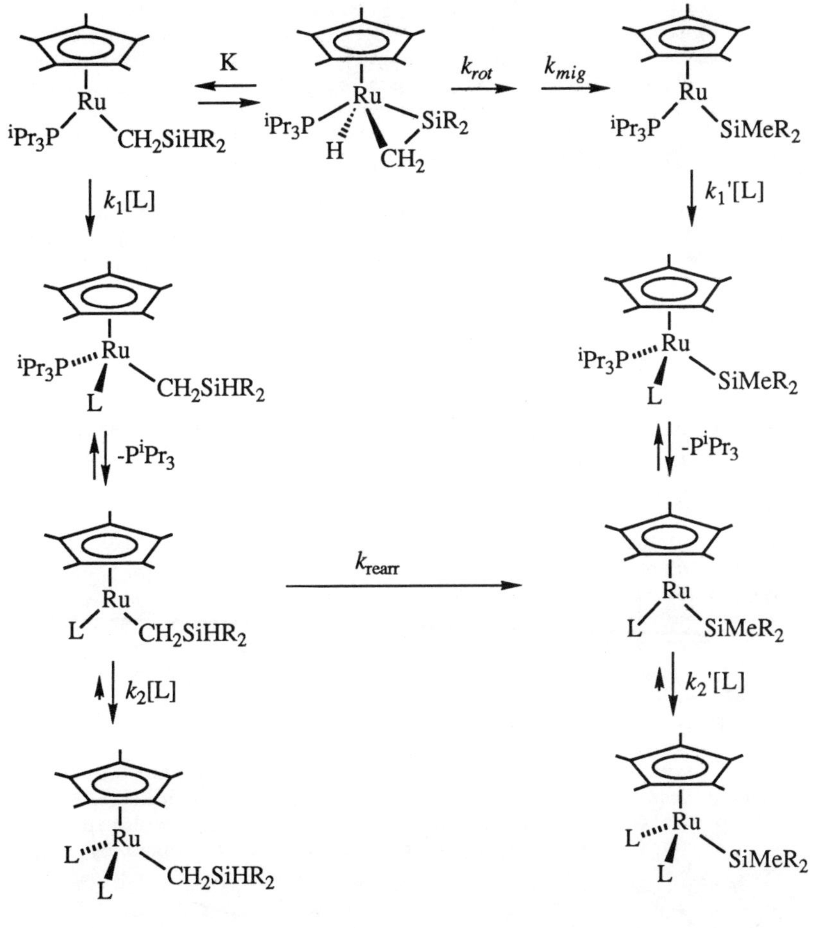

Scheme 1

second-order rate term and a rate law of the form rate = k_{rot}[9] + Kk_1[9][PMe$_2$R]. These kinetic results therefore imply that **9** can react via two hydrogen-migration manifolds involving the 16-electron intermediates Cp*(^{i}Pr$_3$P)RuCH$_2$SiHPh$_2$ and Cp*(^{i}Pr$_3$P)RuSiMePh$_2$. However, if this is true one might have expected to find some Cp*(RMe$_2$P)$_2$RuCH$_2$SiHPh$_2$ complexes among the reaction products. To learn more about such a species, Cp*(PMe$_3$)$_2$RuCH$_2$SiDPh$_2$ (**16-d**) was prepared from Cp*(PMe$_3$)$_2$RuCl and

ClMgCH$_2$SiDPh$_2$. Compound **16-d** is stable indefinitely at room temperature, but heating to 90°C for 2 hours led to quantitative conversion to the silyl Cp*(PMe$_3$)$_2$RuSi(CH$_2$D)Ph$_2$ (**14-d**). The conditions required for this reaction indicate that the rate-determining step is dissociation of PMe$_3$.[28] Therefore compound **16** is probably stable under the reaction conditions of equation 6, and if the mechanism of Scheme 1 is correct, then Cp*(PMe$_3$)RuCH$_2$SiHPh$_2$ rearranges to Cp*(PMe$_3$)RuSiMePh$_2$ much faster than it is trapped by PMe$_3$ ($k_{rearr} > k_2[L]$). Consistent with this, the reaction of **9** with neat PMe$_3$ gives both **14** and **16** in a 4:1 ratio. Thus, apparently only at extremely high PMe$_3$ concentration can the $k_2[PMe_3]$ term compete with k_{rearr}. Our inability to observe Cp*(PMe$_3$)Ru(H)(η^2-CH$_2$=SiPh$_2$) during the conversion of **16** to **14** illustrates the stability imparted to **7-10** by the bulky phosphine ligands PCy$_3$ and P^{i}Pr$_3$.

Reactions of **9** with carbon monoxide (equation 7) can also be explained via the mechanism of Scheme 1. The ratio of the two products observed (**17** and **18**) depends on the reaction conditions

$$\text{[structure 9: }^i\text{Pr}_3\text{P, Ru, H, CH}_2\text{, SiR}_2\text{]} \xrightarrow{CO} \text{[structure 17: }^i\text{Pr}_3\text{P, Ru, OC, CH}_2\text{SiHR}_2\text{]} + \text{[structure 18: }^i\text{Pr}_3\text{P, Ru, OC, SiMePh}_2\text{]} \quad (7)$$

employed. Under 15 psi of CO, a 1:2 ratio of **17** to **18** is obtained. When the pressure of CO is increased to 80 psi, a 5:1 product ratio is observed. Thus, it appears that at relatively low CO concentrations, k_{rot} and $k_1[CO]$ are comparable. At higher concentrations of CO, $k_1[CO]$ dominates and we observe predominate formation of **17**. Thermolysis of **17** (100°C, 20 min) results in quantitative conversion to **18**, presumably by way of the silene complex Cp*(CO)Ru(H)(η^2-CH$_2$=SiPh$_2$).

The reactivity of **9** toward hydrogen parallels that observed for carbon monoxide, in that the observed products reflect hydride migration to both ends of the silene ligand (equation 8). Trapping of Cp*(^{i}Pr$_3$P)RuCH$_2$SiHPh$_2$ by hydrogen is followed by

$$\text{[structure 9]} \xrightarrow[\text{-HSiMePh}_2]{H_2} \text{[structure 19: }^i\text{Pr}_3\text{P, Ru, H, H, H]} + \text{[structure 20: }^i\text{Pr}_3\text{P, Ru, H, H, SiMePh}_2\text{]} \quad (8)$$

elimination of HSiMePh$_2$ and, via oxidative addition of H$_2$, formation of the known trihydride Cp*(^{i}Pr$_3$P)RuH$_3$ (**19**).[29] Hydrogen also appears to trap Cp*(^{i}Pr$_3$P)RuSiMePh$_2$ to produce the new silyl hydride complex **20**. Therefore as expected, the ratio of **19** to **20** is dependent upon the hydrogen pressure. Under 10 psi of hydrogen, the **19:20** ratio is 3:2, whereas under 50 psi of hydrogen the ratio increases to 10:1. The product mixtures obtained can be pressurized with hydrogen without changing the ratio of **19** to **20**, indicating that **20** is not a precursor of **19**. Also, hydrogenation of Cp*(^{i}Pr$_3$P)Ru(D)(η^2-CH$_2$SiPh$_2$) (**9**-d) produced only DSiMePh$_2$ and Cp*(^{i}Pr$_3$P)RuH$_2$[Si(CH$_2$D)Ph$_2$] (**20**-d), but no HSi(CH$_2$D)Ph$_2$ (by ^{2}H NMR spectroscopy). These results show that **20** is stable to hydrogenation, unlike the intermediate Cp*(^{i}Pr$_3$P)RuH$_2$(CH$_2$SiHPh$_2$) which readily eliminates HSiMePh$_2$. An independent synthesis of **20** is based on the quantitative reaction of **11** with hydrogen. The latter reaction was suggested by the clean conversions of **11** to **14** and **18**, by reactions with PMe$_3$ and CO, respectively.

4 SILENE COMPLEXES OF IRIDIUM AND RHODIUM

The above studies with isolated ruthenium silene complexes clearly demonstrate that reactions occur via hydride migration to the silene ligand to produce reactive 16-electron intermediates. In search of other reactivity modes for silene ligands, we attempted to prepare a silene complex without a hydride ligand. The successful synthesis of such a complex is based on a system containing an alkyl ligand that can reductively eliminate with hydride after β-elimination in a M-CH$_2$SiHR$_2$ complex occurs.

The iridium silene complex **21** was synthesized from Cp*(PMe$_3$)Ir(Me)Cl[30] by the reaction of equation 9 in 56% isolated

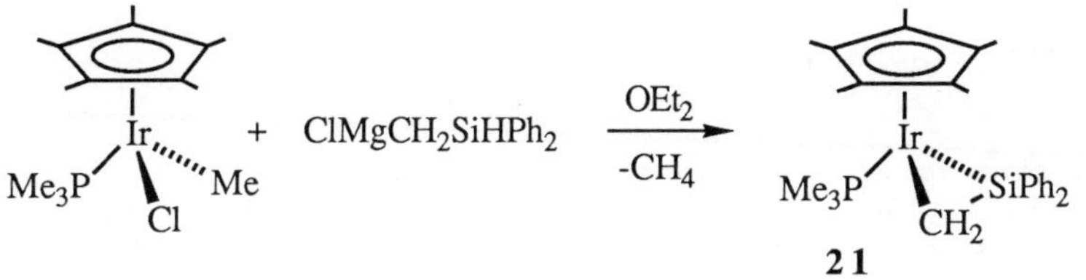

$$\text{Cp*(PMe}_3\text{)Ir(Me)Cl} + \text{ClMgCH}_2\text{SiHPh}_2 \xrightarrow[-\text{CH}_4]{\text{OEt}_2} \text{Cp*(PMe}_3\text{)Ir(CH}_2\text{SiPh}_2\text{)} \quad (9)$$

21

yield. The minor products Cp*(PMe$_3$)Ir(Me)H and Cp*(PMe$_3$)Ir(Me)SiMePh$_2$ were also identified in the reaction mixture. The analogous reaction with ClMgCH$_2$SiDPh$_2$ produced CDH$_3$ and no observable CH$_4$ (by ^{1}H NMR spectroscopy). Following the same procedure, we have also prepared Cp*(PMe$_3$)Ir(η^2-CH$_2$=SiMe$_2$) (**22**) and Cp*(PMe$_3$)Rh(η^2-CH$_2$=SiPh$_2$) (**23**).

The NMR spectra for **21** are consistent with the structure shown in equation 9. For example, inequivalent methylene protons and phenyl groups are observed by ^{1}H NMR spectroscopy. The silene carbon of **21** resonates at δ -33.37 in the ^{13}C NMR spectrum. The $^1J_{CH}$ coupling constants for the silene carbon of **21** (142.3 Hz) and the ethylene carbons of Cp*(PMe$_3$)Ir(η^2-CH$_2$=CH$_2$) (150.6 Hz)[31] are surprisingly comparable, and suggest substantial sp^2 character.[32] The $^1J_{CH}$ coupling constant for the recently reported silene complex Cp$_2$W(η^2-CH$_2$=SiMe$_2$) is 137 Hz.[33]

The molecular structure of **21** is shown in Figure 1. The Si-CH$_2$ bond length of 1.810 (6) Å is close to the analogous distances observed in **9** (1.79 (2) Å) and $Cp_2W(\eta^2\text{-}CH_2=SiMe_2)$ (1.800 (8) Å).[33] π-Bonding between iridium and the silene ligand apparently results in a relatively short Ir-Si bond[34] (2.317 (2) Å), and concomitantly a Ir-CH$_2$ bond that is somewhat elongated (2.189 (8) Å).[30] The nonplanarity of the diphenylsilene ligand can be described by angles between the Si-CH$_2$ bond and the SiC$_2$ and CH$_2$ planes. In **21** these angles are 17.3° for the CH$_2$ group and 39.2° for the SiPh$_2$ group, which can be compared to angles of 19 and 29° for the methylene groups of $Cp^*(Ph_3P)Rh(\eta^2\text{-}CH_2=CH_2)$.[35] In **9**, the SiPh$_2$ group is tilted 36.2° (average) away from the ruthenium. The sum of the C-Si-C angles in **21** (341°) suggests some sp^2 character at silicon, but slightly less than was observed in **9** (343°).

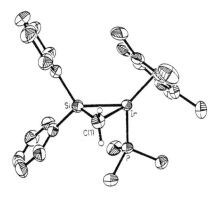

Figure 1 ORTEP view of **21**

The absence of a ligand that can readily migrate to the silene of **21** results in a dramatic difference in the chemical properties of **21** and **9**. For example **21** is quite robust, exhibiting no decomposition when heated to 140°C for several days, or when irradiated with UV irradiation for 8 hours. Note that the related ethylene complex $Cp^*(PMe_3)Ir(\eta^2\text{-}CH_2=CH_2)$ rearranges upon photolysis to $Cp^*(PMe_3)Ir(H)(CH=CH_2)$.[31] Complex **21** reacts cleanly with methanol via cleavage of the Ir-Si bond to produce $Cp^*(PMe_3)Ir(H)(CH_2SiOMePh_2)$ (**24**). Methyl iodide reacts with **21** in a similar manner to afford $Cp^*(PMe_3)Ir(I)(CH_2SiMePh_2)$ (**25**). For comparison, Werner has shown that $Cp(^iPr_3P)Ir(\eta^2\text{-}CH_2=CH_2)$ loses ethylene upon reaction with MeI to give $Cp(^iPr_3P)Ir(Me)I$.[36]

CONCLUSION

The first stable terminal silylene and η^2-silene complexes have been isolated and studied. Heteroatom-stabilization may be important in allowing isolation of the silylene complexes **5** and **6**. Reactivity studies with the ruthenium silene complexes indicate that steric factors play a large role in determining thermal stability. Reactions of $Cp^*(R_3P)Ru(H)(\eta^2\text{-}CH_2=SiR'_2)$

complexes always occur via initial migration of hydride to the carbon or silicon of the silene ligand to produce transient 16-electron intermediates. X-ray crystallography and NMR spectroscopy seem to reveal a surprising degree of double bond character for the silene ligands. Investigations of transition-metal complexes of reactive silicon intermediates is a rapidly growing area that should greatly impact the use of transition metals in silicon chemistry.

ACKNOWLEDGMENT

We would like to acknowledge support of this work by the National Science Foundation. T.D.T. thanks the Alfred P. Sloan Foundation for a research fellowship (1988-90). We thank Professor Berry for communicating results to us prior to publication.[33]

REFERENCES

1. T. D. Tilley, 'The Chemistry of Organic Silicon Compounds', S. Patai and Z. Rappoport (eds.), Wiley, New York, 1989, Chapter 24, p. 1415.
2. T. D. Tilley, *Comments Inorg. Chem.*, 1990, *10*, 49.
3. (a) D. A. Straus, T. D. Tilley, A. L. Rheingold, and S. J. Geib, *J. Am. Chem. Soc.*, 1987, *109*, 5872. (b) D. A. Straus, C. Zhang, G. E. Quimbita, S. D. Grumbine, R. H. Heyn, T. D. Tilley, A. L. Rheingold, and S. J. Geib, *J. Am. Chem. Soc.*, 1990, *112*, 2673.
4. (a) C. Zybill and G. Müller, *Angew. Chem. Int. Ed. Eng.*, 1987, *26*, 669. (b) C. Zybill, D. L. Wilkinson, and G. Müller, *Angew. Chem. Int. Ed. Eng.*, 1988, *27*, 583. (c) C. Zybill and G. Müller, *Organometallics*, 1988, 7, 1368. (d) C. Zybill, D. L. Wilkinson, C. Leis, and G. Müller, *Angew. Chem. Int. Ed. Eng.*, 1989, *28*, 203.
5. K. Ueno, H. Tobita, M. Shimoi, and H. Ogino, *J. Am. Chem. Soc.*, 1988, *110*, 4092. (b) H. Tobita, K. Ueno, M. Shimoi, and H. Ogino, *J. Am. Chem. Soc.*, 1990, *112*, 3415.
6. K. H. Pannell, *J. Organomet. Chem.*, 1970, *21*, P17.
7. C. L. Randolph and M. S. Wrighton, *Organometallics*, 1987, *6*, 365.
8. D. H. Berry and L. J. Procopio, *J. Am. Chem. Soc.*, 1989, *111*, 4099.
9. B. K. Campion, R. H. Heyn, and T. D. Tilley, *J. Am. Chem. Soc.*, 1988, *110*, 7558.
10. B. K. Campion, R. H. Heyn, and T. D. Tilley, *J. Am. Chem. Soc.*, 1990, *112*, 4079.
11. (a) E. K. Pham and R. C. West, *J. Am. Chem. Soc.*, 1989, *111*, 7667. (b) E. K. Pham and R. West, *Organometallics*, 1990, *9*, 1517.
12. D. H. Berry, J. Chey, H. S. Zipin, and P. J. Carroll, *J. Am. Chem. Soc.*, 1990, *112*, 452.
13. (a) E. A. Zarate, C. A. Tessier-Youngs, and W. J. Youngs, *J. Am. Chem. Soc.*, 1988, *110*, 4068. (b) E. A. Zarate, C. A.

Tessier-Youngs, and W. J. Youngs, *J. Chem. Soc., Chem. Commun.*, 1989, 577.
14. J. B. Lambert, W. J. Schulz, Jr., J. A. McConnell, and W. Schilf, *J. Am. Chem. Soc.*, 1988, *110*, 2201.
15. S. D. Grumbine, D. A. Straus, R. H. Heyn, T. D. Tilley, and A. L. Rheingold, manuscript in preparation.
16. G. A. Lawrance, *Chem. Rev.*, 1986, *86*, 17.
17. A related process has been identified for Cp(NO)(PPh$_3$)ReGePh$_2$OTf: K. E. Lee and J. A. Gladysz, *Polyhedron*, 1988, 7, 2209.
18. G. A. Olah and L. D. Field, *Organometallics*, 1982, *1*, 1485.
19. W. A. Herrmann, *Adv. Organomet. Chem.*, 1982, *20*, 159.
20. J. R. Matachek and R. J. Angelici, *Inorg. Chem.*, 1986, *25*, 2877.
21. B. K. Campion, R. H. Heyn, and T. D. Tilley, *J. Chem. Soc., Chem. Commun.*, 1988, 278.
22. A. G. Brook, F. Abdesaken, G. Gutekunst, and N. Plavac, *Organometallics*, 1982, *1*, 994.
23. (a) D. Seyferth, D. C. Annarelli, M. L. Shannon, J. Escucle, and D. P. Duncan, *J. Organomet. Chem.*, 1982, *225*, 177. (b) D. Seyferth, D. C. Annarelli, S. C. Vick, and D. P. Duncan, *J. Organomet. Chem.*, 1980, *201*, 179.
24. E. Breitmaier and W. Voelter, '^{13}C NMR Spectroscopy', Verlag Chemie, New York, 1978, p. 96.
25. A. G. Brook, S. C. Nyburg, F. Abdesaken, B. Gutekunst, G. Gutekunst, R. K. M. R. Kallury, Y. C. Poon, Y.-M. Chang, and W. Wong-Ng, *J. Am. Chem. Soc.*, 1982, *104*, 5667.
26. N. Wiberg, G. Wagner, and G. Müller, *Angew. Chem. Int. Ed. Eng.*, 1985, *24*, 229.
27. T. Arliguie, B. Chaudret, F. Jalon, and F. Lahoz, *J. Chem. Soc., Chem. Commun.*, 1988, 998.
28. H. E. Bryndza, P. J. Domaille, R. A. Paciello, and J. E. Bercaw, *Organometallics*, 1989, *8*, 379.
29. T. Arliguie and B. Chaudret, *J. Chem. Soc., Chem. Commun.*, 1986, 985.
30. J. M. Buchanan, J. M. Stryker, and R. G. Bergman, *J. Am. Chem. Soc.*, 1986, *108*, 1537.
31. P. O. Stoutland and R. G. Bergman, *J. Am. Chem. Soc.*, 1988, *110*, 5732.
32. C. A. Tolman, A. D. English, and L. E. Manzer, *Inorg. Chem.*, 1975, *14*, 2353.
33. T. S. Koloski, P. J. Carroll, and D. H. Berry, submitted to *J. Am. Chem. Soc.*
34. J. S. Ricci, Jr., T. F. Koetzle, M.-J. Fernandez, P. M. Maitlis, and J. C. Green, *J. Organomet. Chem.*, 1986, *299*, 383.
35. W. Porzio and M. Zocchi, *J. Am. Chem. Soc.*, 1978, *100*, 2048.
36. M. Dziallas, A. Höhn, and H. Werner, *J. Organomet. Chem.*, 1987, *330*, 207.

Decamethylsilicocene: Synthesis, Structure, and Some Chemistry

P. Jutzi
FACULTY OF CHEMISTRY, UNIVERSITY OF BIELEFELD, 4800 BIELEFELD, GERMANY

1 INTRODUCTION

In the chemistry of germanium, tin, and lead it is possible to stabilize the oxidation state +2 by π-complex formation. In this context, arene, carbollyl, and cyclopentadienyl systems have been shown to be important as π-ligands.[1] Concerning the cyclopentadienyl chemistry, the introduction of the pentamethylcyclopentadienyl ligand has allowed the synthesis of many thermally stable, but still highly reactive, complexes. In this class of compounds, the decamethylmetallocenes play an important role.

So far, no inorganic or organometallic species with divalent silicon stable under normal conditions are known. To change this situation, π-complexation seemed to us to be an interesting tool. Here we describe the synthesis, structure, and bonding of decamethylsilicocene, the first stable compound with silicon in the oxidation state +2, together with some results concerning the chemistry of this complex.

2 SYNTHESIS OF DECAMETHYLSILICOCENE (1)

Owing to the fact that no substrates with divalent silicon are available, it was necessary to start with compounds containing tetravalent silicon and to try the synthesis of **1** by a reductive elimination process. In the light of our experience in germanium and tin chemistry,[2] the bis(pentamethylcyclopentadienyl)silicon dihalides seemed to be suitable compounds for this purpose. Their synthesis and structure has been described only recently.[3]

In the reaction of dichlorobis(pentamethylcyclopentadienyl)silane with naphthalene-lithium, -sodium, or -potassium, the desired compound **1** is obtained in conjunction with some elemental silicon (equation 1).[4] The undesired formation of silicon is avoided by using the dibromobis(pentamethylcyclopentadienyl)silane as the substrate and anthracene-potassium as the reducing agent (equation 2). Recent experiments have shown that the dichlorosilane is cleanly reduced to **1** with decamethylsamarocene (equation 3).[5]

$$(Me_5C_5)_2SiCl_2 \xrightarrow{+2\ MC_{10}H_8,\ -2\ MCl,\ -2\ C_{10}H_8} (Me_5C_5)_2Si\ (\mathbf{1}) \quad (1)$$

$$(Me_5C_5)_2SiBr_2 \xrightarrow{+2\ KC_{14}H_{10},\ -2\ KBr,\ -2\ C_{14}H_{10}} \mathbf{1} \quad (2)$$

$$(Me_5C_5)_2SiCl_2 \xrightarrow{+2\ (Me_5C_5)_2Sm,\ -2\ (Me_5C_5)_2SmCl} \mathbf{1} \quad (3)$$

Crystallisation from n-pentane gives colourless crystals of **1**, which are soluble in all common aprotic organic solvents. Compound **1** is monomeric in benzene solution, stable in air for short periods of exposure, but sensitive against hydrolysis; it melts at 171°C without decomposition.

3 STRUCTURE AND BONDING OF DECAMETHYLSILICOCENE (1)

The NMR data of **1** are in accord with a more or less symmetrical π-structure. Thus, only one signal is observed for all ring carbons and all ring methyl groups in the ^{13}C and ^1H NMR spectrum. In the ^{29}Si NMR spectrum, a resonance at very high field strength ($\delta=-398$ ppm) is observed. In comparison with the high field chemical shift in the ^{119}Sn NMR spectrum of decamethylstannocene or in the ^{207}Pb NMR spectrum of decamethylplumbocene, this proves the π-structure of **1**. The ^{29}Si chemical shift **1** is of the highest value so far observed for an organosilicon compound.

In the mass spectrum of **1** (EI and CI), the molecular ion is not observed; the fragment with the highest mass (m/z=163) corresponds to the $Me_5C_5Si^+$ ion. These observations round off those of cyclovoltammetric studies and demonstrate that the $(Me_5C_5)_2Si^+$ radical cation is rather

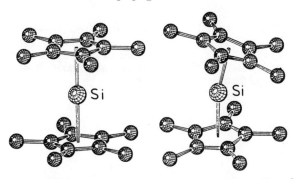

Figure 1 Crystal structure of **1** (**1a** left, **1b** right)

unstable in the gas phase and in solution and easily loses a Me_5C_5 radical.[4]

An X-ray crystal structure investigation reveals that two geometrical isomers, **1a** and **1b**, are present in the unit cell in the ratio 1:2 (see figure 1). Isomer **1a** is isostructural with decamethylferrocene, the silicon lone pair is not stereochemically active. Isomer **1b** is of a bent-metallocene type structure with an interplane angle of 25° and with pentamethylcyclopentadienyl rings asymmetrically bonded in a staggered conformation; the silicon lone pair can be stereochemically active. The Si-C separations are equidistant in **1a** [2.42(1) Å], but different in **1b** [ranging from 2.323(7) to 2.541(7) Å]. The distance between the silicon atom and the cyclopentadienyl ring centroids is 2.11 Å in **1a** and 2.12 Å in **1b**. Space-filling models clearly indicate the interplanar angle in **1b** to be of the largest possible value. According to GED studies, **1** has a bent-metallocene type structure in the gas phase with an interplanar angle of 22.3(12)°.[4]

Further informations about the bonding in silicocene stem from the He(I) photoelectron (PE) spectrum of **1** and from calculations on structural models of the parent compound $(C_5H_5)_2Si$. In the PE spectrum, the bands at 6.7 and 8.1 eV are assigned to ionisations from π-MOs and the band at 7.5 eV to the lone pair at silicon. In Figure 2, the highest occupied MOs of a silicocene molecule with D_{5d} symmetry are illustrated. The HOMOs $3e_{1g}$ are localized mainly on the ligands and are thus nonbonding, while the MO $4a_{1g}$ is an antibonding combination of the 3s atomic orbital on silicon and the $2p\pi$-MOs of the ligands.

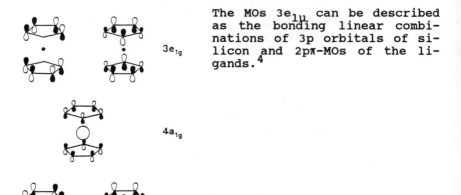

The MOs $3e_{1u}$ can be described as the bonding linear combinations of 3p orbitals of silicon and $2p\pi$-MOs of the ligands.[4]

Figure 2 Schematic drawing of the highest occupied MOs of $(C_5H_5)_2Si(D_{5d})$.

4 CHEMISTRY OF DECAMETHYLSILICOCENE (1)

In the following chapters, our preliminary results concerning the reactions of **1** are summarized.

Protonation of 1:

With protic substrates of the type HX, attack at the silicon lone pair is the preferred pathway. As a result, oxidative addition products are obtained in the reaction of **1** with hydrogen halides, carboxylic acids, trifluoromethanesulphonic acid, aliphatic amines, aromatic alcohols and thiols, as described in equation 4.[6]

$(Me_5C_5)_2Si$ (**1**) + HX --> $(\sigma\text{-}Me_5C_5)_2Si(H)X$ (**2-10**)   (4)

Table 1 ^{29}Si NMR Data of **2-10**.

Compound	2	3	4	5	6	7
X	F	Cl	Br	$C_2H_5CO_2$	CF_3CO_2	CF_3SO_3
$\delta\ ^{29}Si$ [ppm]	14.1	16.0	15.0	0.8	7.6	-3.6
$^1J_{SiH}$ [Hz]	229	224	232	224	228	249

Compound	8	9	10
X	$CH_3(H)N$	$(p\text{-}CH_3)C_6H_4O$	$(p\text{-}CH_3)C_6H_4S$
$\delta\ ^{29}Si$ [ppm]	1.6	1.0	18.8
$^1J_{SiH}$ [Hz]	201	211	215

The products **2-10** possess four functional groups, three of which are different $(Me_5C_5,H,X)_2$. In Table 1, these compounds are characterized by their ^{29}Si NMR data.

In the reaction of **1** with tetrafluoroboric acid, the elimination of pentamethylcyclopentadiene and boron trifluoride is observed; the cyclotetrasilane $(Me_5C_5SiF)_4$ (**11**) is the final product.[6] Presumably, the short-lived ionic species $Me_5C_5Si^+BF_4^-$ is formed as the first intermediate; in subsequent steps, fluoride abstraction leads to the highly reactive silylene Me_5C_5SiF, which dimerizes to the unstable disilene $Me_5C_5(F)Si=Si(F)Me_5C_5$ or, as argued by Apeloig,[7] to the fluorine-bridged silylene $Me_5C_5Si(\mu F)_2SiC_5Me_5$, which is of comparable energy. Finally, the isolable cyclotetrasilane **11** is formed (see Scheme 1). An X-ray analysis of **11** shows a folded four-membered ring of silicon atoms with all-trans orientation of pentamethylcyclopentadienyl and fluoride ligands.

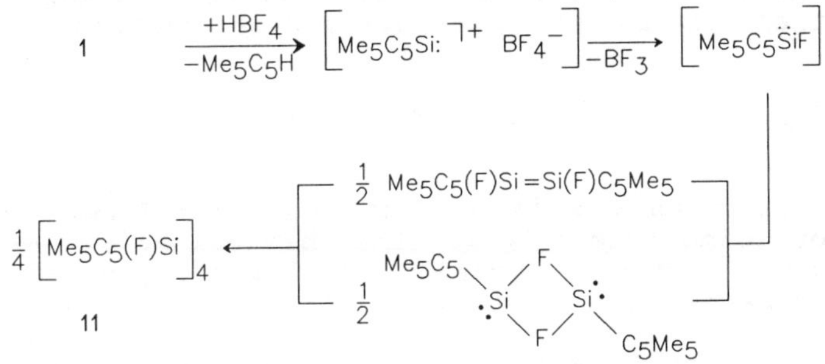

Scheme 1 Reaction of **1** with HBF_4.

Evidence for the generation of silylenium ions of the type $(Me_5C_5)_2SiH^+\ X^-$ has been obtained in the reaction of **1** with two equivalents of trifluoromethanesulphonic acid or aromatic ortho-diols.[8] The argument for the presence of silylenium ions is based on NMR data, on molecular-weight determinations, and on conductivity measurements. Some degree of π-bonding between the pentamethylcyclopentadienyl ligands and the silicon atom seems to stabilize the ionic structures; they are presented in Figure 3.

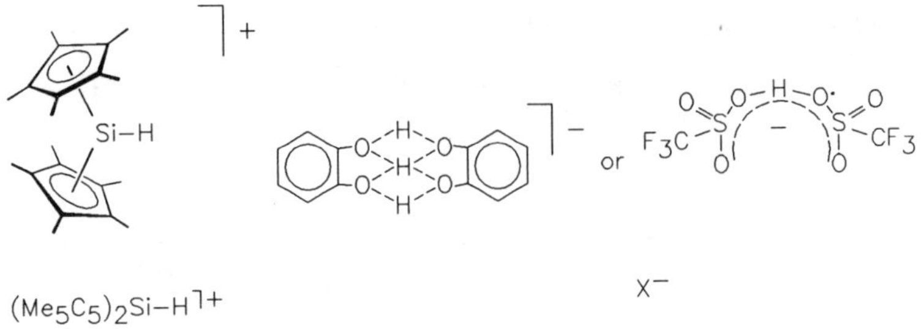

Figure 3 The $(Me_5C_5)_2SiH^+$ cation and its counteranions

Gold(I)chloride Complexes of 1.

The easy accessibility of the lone-pair in **1** prompted us to use this compound as a ligand in transition-metal chemistry. The AuCl fragment seemed to be a promising metal component due to its small steric requirements and to the isolobal relationship $H^+ \longleftrightarrow AuCl$. In fact, the complex [bis(pentamethylcyclopentadienyl)-silanediyl]gold(I) chloride **12** is formed as a deep-red, extremely air- and moisture-sensitive powder in the reaction of **1** with carbonylgold(I) chloride (equation 5).[9]

$$(Me_5C_5)_2Si + COAuCl \rightarrow (\sigma\text{-}Me_5C_5)(\pi\text{-}Me_5C_5)Si\text{-}AuCl + CO$$
$$\mathbf{1} \hspace{4cm} \mathbf{12} \hspace{2cm} (5)$$

The ^{29}Si resonance is shifted from $\delta=-398$ ppm to $\delta=+82,8$ ppm on going from **1** to **12**. The chemical shift value of **12** is in a region typical of low-coordinated silicon (silaethenes or disilenes). The ^1H and ^{13}C DNMR spectrum of **12** discloses the interesting feature that the two pentamethylcyclopentadienyl ligands are differently bonded to the silicon atom: One is σ-bonded and undergoes slow sigmatropic rearrangements, the other is π-bonded. The bonding in **12** resembles that found in Fischer-type carbene complexes: electron density is transferred to the silicon atom mainly by the π-bonded pentamethylcyclopentadienyl ligand.

By simple addition of a donor molecule, the silylene complex **12** is easily transformed to the corresponding donor-stabilized silylene complex (ylide-type complex). Thus, reaction of **12** with tert.butyl isonitrile or pyridine yields the compounds **13** or **14** in form of colourless or brown crystals (equation 6). The synthesis of **13** and

14 is also feasible starting from decamethylsilicocene (**1**) and tert.butyl isonitrile- and pyridine-gold(I) chloride, respectively (equation 6).[9]

$$\mathbf{12} \xrightarrow{+L} (\sigma\text{-}Me_5C_5)_2\overset{\uparrow}{Si}\text{-}AuCl \qquad (6)$$
$$\mathbf{1} \xrightarrow{+LAuCl} \quad \mathbf{13,14} \quad L$$

	13	14
L	tBuNC	Pyridine

Interestingly, both pentamethylcyclopentadienyl ligands in these complexes are now σ-bonded and fluxional. The ^{29}Si signals appear at +71.2 (**13**) and +54.3ppm (**14**). Thus, the electron density at silicon is not changed drastically on going from **12** to **13,14**. The electron donation from the donor molecule is nearly compensated by the electron loss from the π-σ rearrangement of a pentamethylcyclopentadienyl ligand.

Boron and Phosphorus Trihalides as Electrophiles in the Reaction with Decamethylsilicocene (1)

In the reaction of **1** with the boron trihalides BCl_3, BBr_3, and BI_3 the borylsilanes **15-17** are formed in high yields (see equation 7).[10] The reaction sequence leading to these compounds has not been elucidated so far; presumably, insertion of **1** into a boron-halogen bond is followed by dyotropic rearrangements of pentamethylcyclopentadienyl and halogen ligands. As expected, the ^{1}H and ^{13}C NMR spectra of **15-17** exhibit a different fluxionality for silicon- and boron-bonded cyclopentadiene systems. The quadrupole moment of the boron nuclei prevents the observation of ^{29}Si resonances and of $^1J_{BSi}$ coupling constants. In Table 2, the ^{11}B resonances of **15-17** are compared with those of some known borylsilanes.[11]

$$(Me_5C_5)_2Si + BHal_3 \rightarrow \sigma\text{-}Me_5C_5(Hal)_2Si\text{-}B(Hal)\sigma\text{-}C_5Me_5 \qquad (7)$$
$$\mathbf{1} \qquad\qquad\qquad\qquad \mathbf{15\text{-}17}$$

Table 2 ^{11}B NMR Data of Borylsilanes

Compound	15	16	17	$(Me_2N)_2BSiMe_3$	$(Ph_2N)_2BSiPh_3$
Hal	Cl	Br	I		
δ^{11}B[ppm]	-72.4	-73.1	-75.2	-54.4	-53.3

The borylsilanes are thermally unstable. Thus, **16** decomposes in boiling trichloromethane to pentamethyl-

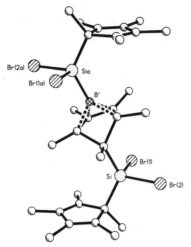

cyclopentadienyldibromoborane and a colourless compound of the composition $(Me_5C_5)_3Si_2BBr_4$ (**18**). An X-ray analysis of **18** reveals the presence of an arachno-BC_4 fragment, as portrayed in Figure 4.[10]

<u>Figure 4</u>   Structure of **18** in the crystal.

Analogous to the reaction with boron trihalides, decamethylsilicocene (**1**) reacts with the phosphorus trihalides PCl_3 and PBr_3 under final formation of the silylphosphanes **19** and **20** (see equation 8).[8]

$(Me_5C_5)_2Si + PHal_3 \rightarrow \sigma-Me_5C_5(Hal)P-Si(Hal)_2\sigma-C_5Me_5$ (8)
    1   **19,20**

In the ^{29}Si and ^{31}P NMR spectra (see Table 3), the $^1J_{SiP}$ coupling is characteristic of the presence of a silicon-phosphorus bond. Due to the chiral phosphorus atom in **19** and **20**, the number of signals for the pentamethylcyclopentadienyl ligands in the ^1H and ^{13}C NMR spectra is enhanced, so that an assignment is difficult.

<u>Table 3</u> ^{29}Si and ^{31}P NMR Data of **19** and **20**.

Compound	Hal	$\delta\ ^{29}Si(ppm)$	$\delta\ ^{31}P(ppm)$	$^1J_{SiP}$ (Hz)
19	Cl	10.3 d	96.6 *	128
20	Br	0.4 d	81.4 *	151

*=Doublets as satellites

Further Oxidative Additions at Decamethylsilicocene (1)

In the reaction of **1** with halogen molecules as well as with aryl disulfides, splitting of the relevant element-element bond is observed, and the corresponding oxidative-addition products **21 - 23** (see equations 9 and 10) are formed. Similarly, the heavier methyl halides add oxidatively to form the compounds **24** and **25** (see equation 11).[10]

$$1 + Hal-Hal \longrightarrow (\sigma-Me_5C_5)_2SiHal_2 \qquad (9)$$
$$21 (Hal=Br) \quad 22 \text{ (Hal=I)}$$

$$1 + RS-SR \longrightarrow (\sigma-Me_5C_5)_2Si(SR)_2 \qquad (10)$$
$$23 \; R=(p-Me)C_6H_4$$

$$1 + CH_3-Hal \longrightarrow (\sigma-Me_5C_5)_2Si(CH_3)Hal \qquad (11)$$
$$24 (Hal=Br) \quad 25 (Hal=I)$$

Dehalogenation with simultaneous formation of bis(pentamethylcyclopentadienyl)dihalogenosilanes takes place in the reaction of **1** with geminal or vicinal dihalogeno compounds. Some examples of redox processes of this type are given in equations 12,13,14, and 15.[8,10] Thus, bis-(pentamethylcyclopentadienyl)dichlorostannane is reduced in a clean reaction to decamethylstannocene. In the reaction with vicinal dihalogeno compounds, species with element-element double bonds arise. Thus, from the meso- or d,l-form of dibromosuccinic acid dimethyl ester, the corresponding ester of fumaric acid is formed, whereas <u>trans</u>-1,2-dibromocyclohexane is converted to cyclohexene. In phosphorus chemistry, the 1,2-diiodo-1,2-bis(pentamethylcyclopentadienyl)diphosphane is transferred quantitatively to bis(pentamethylcyclopentadienyl)diphosphene.

$$(\sigma-Me_5C_5)_2SnCl_2 \xrightarrow[-(Me_5C_5)_2SiCl_2]{1} (\pi-Me_5C_5)_2Sn \qquad (12)$$

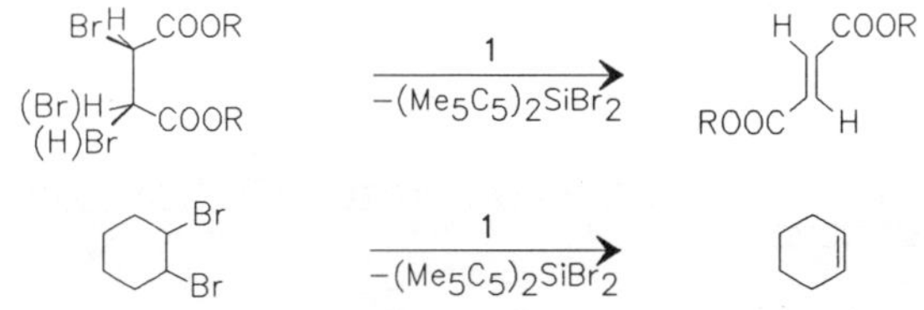

$$\begin{array}{c} Br\,H\quad COOR \\ (Br)H{\diagup}COOR \\ (H)Br \end{array} \xrightarrow[-(Me_5C_5)_2SiBr_2]{1} \begin{array}{c} H\quad COOR \\ ROOC\quad H \end{array} \qquad (13)$$

$$\underset{Br}{\overset{Br}{\bigcirc}} \xrightarrow[-(Me_5C_5)_2SiBr_2]{1} \bigcirc \qquad (14)$$

$$Me_5C_5(I)P-P(I)C_5Me_5 \xrightarrow[-(Me_5C_5)_2SiI_2]{1} Me_5C_5{\diagdown}P=P{\diagup}^{C_5Me_5} \qquad (15)$$

Reaction of 1 with sulphur, selenium and tellurium

A σ-bonded pentamethylcyclopentadienyl ligand at silicon is very bulky. For this reason, we investigated whether simple combination of 1 with electron-sextet fragments X might offer an approach to kinetically stabilized π-systems of the type $(Me_5C_5)_2Si=X$. At first, we studied the reaction of 1 with sulphur, selenium, and tellurium sources (X=S,Se,Te).[12] Compound 1 reacts with sulphur to give the dithiadisiletane **26** (equation 16); the conceivable silathione intermediate, $(Me_5C_5)_2Si=S$, has not been detected so far. In the reaction with tri-n-butylphosphane selenide, the corresponding diselenadisiletane **27** is formed. Here it is possible to trap the intermediate silaselone, $(Me_5C_5)_2Si=Se$, by [2+4] cycloaddition with 2,3-dimethylbutadiene (equation 17). The reaction with tri-n-butylphosphane telluride results in the formation of the tritelluradisilole **28** (equation 18). Intermediates have not yet been detected. The molecular structure of **28** has been determined by X-ray analysis. The functionality offered by the pentamethylcyclopentadienyl ligands makes the compounds **26** - **28** very interesting preparatively.

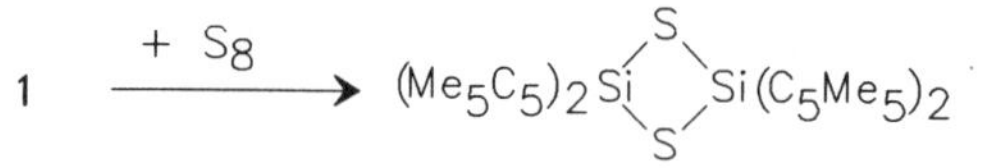

$$1 \xrightarrow{+ S_8} (Me_5C_5)_2Si\underset{S}{\overset{S}{\diamond}}Si(C_5Me_5)_2 \quad \mathbf{26} \tag{16}$$

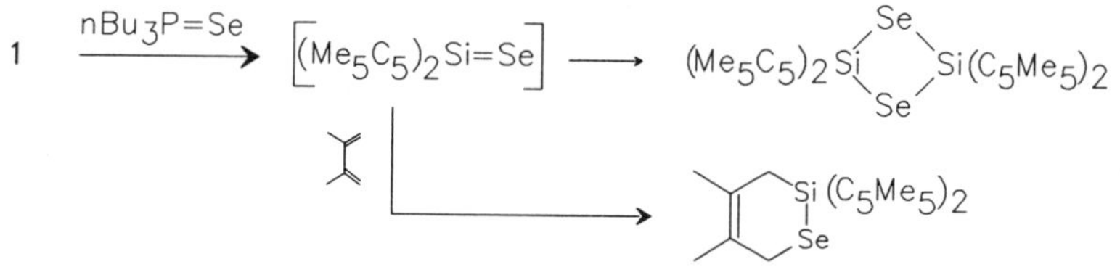

$$1 \xrightarrow{nBu_3P=Se} \left[(Me_5C_5)_2Si=Se\right] \longrightarrow (Me_5C_5)_2Si\underset{Se}{\overset{Se}{\diamond}}Si(C_5Me_5)_2 \quad \mathbf{27} \tag{17}$$

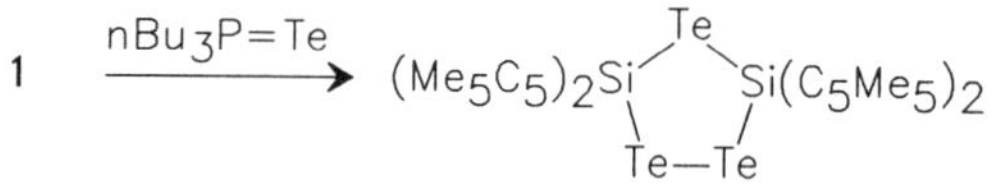

$$1 \xrightarrow{nBu_3P=Te} (Me_5C_5)_2Si\underset{Te-Te}{\overset{Te}{\diamond}}Si(C_5Me_5)_2 \quad \mathbf{28} \tag{18}$$

[1+2] Cycloadditions of 1 with compounds containing multiply bonded carbon

[1+2] Cycloaddition takes place in the reaction of 1 with carbon dioxide, carbon disulphide, and phenyl isocyanate.[10] The resulting three-membered ring systems are highly reactive and stabilise themselves either by dimerization or by CO or PhNC extrusion. In the latter cases, the resulting π-bonded species $(Me_5C_5)_2Si=X$ (X=O,S) find different routes for their further stabilisation. The observed reaction pathways, which lead to new funtionalized heterocycles, are described in scheme 2.

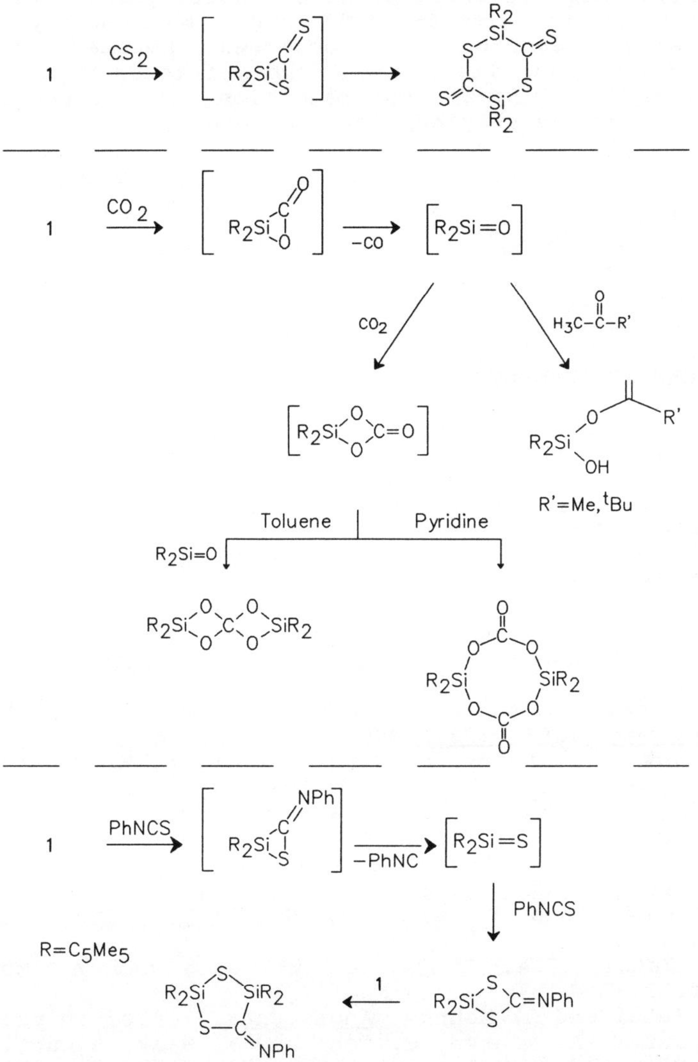

Scheme 2 Reaction of **1** with CO_2, CS_2, and PhNCS.

Alkynes with electron-withdrawing substituents under reaction with **1** form silacyclopropenes.[10] Thus, the compounds **29** and **30** could be prepared in high yields by reaction of **1** with acetylenic dicarboxylic ethylester and phenyl(sulfonyl)trimethylsilylacetylene, respectively. ^{29}Si resonances of these strained heterocycles are in the expected region (δ=-66.6(**29**) and -65.1(**30**) ppm). A surprising reaction is observed between **1** and hexafluorobutyne. Presumably, the resulting compound **31** is formed by [4+2] cycloaddition and further C-F insertion of the silylene intermediate.

	R	R'
29	COOMe	COOMe
30	SiMe$_3$	SO$_2$Ph

ACKNOWLEDGEMENTS

Sincere thanks are extended to my co-workers who have taken part in this work. Generous support by the Deutsche Forschungsgemeinschaft, the Fonds der Chemischen Industrie, and the University of Bielefeld is gratefully acknowledged as well.

REFERENCES

1. P. Jutzi, <u>Adv.Organomet.Chem.</u>, 1986, **26**, 217.
2. P. Jutzi and B. Hielscher, <u>Organometallics</u>, 1986, **5**, 1201.
3. P. Jutzi, D. Kanne, M.B. Hursthouse and A.J. Howes, <u>Chem.Ber.</u> 1988, **122**, 1299.
4. P. Jutzi, D. Kanne, C. Krüger, R. Blom, R. Gleiter and I. Hyla-Kryspin, <u>Chem.Ber.</u>, 1989, **122**, 1629.
5. W.J. Evans, T.A. Ulibarri and P. Jutzi, <u>Inorg.Chim.Acta</u>, 1990, **168**, 5.
6. P. Jutzi. U. Holtmann, A. Bögge and A. Müller, <u>J.Chem.Soc.,Chem.Commun.</u>, 1988, 305.
7. J. Maxka and Y. Apeloig, <u>J.Chem.Soc.,Chem.Commun.</u>, 1990, 737.
8. P. Jutzi, E.A. Bunte, U. Holtmann and A. Möhrke, unpublished.
9. P. Jutzi and A. Möhrke, <u>Angew.Chem.</u>, 1990, in press.
10. P. Jutzi, A. Möhrke, S. Pohl and W. Saak, unpublished.
11. H. Nöth and G. Höllerer, <u>Chem.Ber.</u>, 1966, **99**, 2197.
12. P. Jutzi, A. Möhrke, A. Müller and H. Bögge, <u>Angew.Chem.</u>, 1989, **101**, 1527; <u>Angew.Chem.Int.Ed.Engl.</u>, 1989, **28**, 1518.

Section V: Organic Synthesis Using Silicon

Stereocontrol in Organic Synthesis Using Silicon Compounds

Ian Fleming
UNIVERSITY CHEMICAL LABORATORY, LENSFIELD ROAD, CAMBRIDGE CB2 1EW, UK

1 INTRODUCTION

We have been engaged for several years in a study of stereocontrol using silicon-containing compounds. The essence of the method is to use an electrophilic reaction on a double bond adjacent to a stereogenic centre carrying a silyl group, a carbon group, and hydrogen (**1**).[1] We suggest, in outline, that this relatively good, genuinely open-chain stereocontrol stems from the sum of four mutually supporting influences. (i) The lowest energy conformation is usually well controlled to be close to that depicted in **1**, because only the hydrogen atom comfortably eclipses (or partly eclipses) the double bond. (ii) The lowest energy conformation is also likely to be close to the reactive conformation, because the Si-C bond is able to overlap with the π-bond, activating it towards electrophilic attack. (iii) The top and bottom surfaces of the π-bond are well differentiated: the silyl group is large relative to most carbon groups, and (iv) electronically the polarisation of the Si-C bond will also encourage attack by electrophiles from the lower surface of the π-bond. The only serious limitation that has emerged so far arises when the stereogenic centre has as its carbon substituent only a small group like methyl, and when this coincides with the substituent on the double bond *cis* to the stereogenic centre being no larger than a hydrogen atom. In this situation, the alternative conformation (**2**) is well populated, and attack on it will take place on the upper surface. Examples of this problem can be found in osmylation and epoxidation reactions,[2] in a nitrile oxide cycloaddition,[3] and in a Simmons-Smith reaction,[4] all of which show an approximately 2:1 preference for reaction to take place in the sense (**2**). However, although hydroboration with diborane is similarly unselective, hydroboration with 9-BBN is very selective in the sense (**1**).[5] Diels-Alder cycloadditions are very selective in the sense (**1**) with *N*-phenylmaleimide but in the sense (**2**) with dimethyl acetylenedicarboxylate.[6] When the substituent on the stereogenic centre is larger than methyl and/or when the substituent *cis* to the stereogenic centre is larger than a hydrogen atom, these reactions become relatively well behaved in the sense (**1**).[2,4,5] Of course, we cannot be certain that attack in the sense (**2**) actually takes place in this conformation, anymore than we can be certain that attack taking place in the sense (1) takes place in that conformation. We are currently investigating this point by examining in detail the stereochemistry of the electrophilic substitution of allylsilanes, but for the moment, it seems reasonable to assume that this is where the problem lies.

In spite of this limitation, stereochemical control based on this one idea can be applied in several different ways with correspondingly different outcomes. In this lecture I want to demonstrate how we have used it, in each of its manifestations, in an incomplete synthesis of ebelactone-a, which has seven stereogenic centres and a trisubstituted double bond. Our major claim with this work is that we shall be able to control all of the stereochemical problems with this single device.

The three aspects of this method of stereocontrol that we have used are: the transposition of chiral information from C-1 to C-3 in electrophilic substitution reactions of allylsilanes (3),[2,3,7,8] the setting up of stereogenic centres with a 1,3 relationship using the hydroboration of allylsilanes (4),[5] and the setting up of stereogenic centres with a 1,2 relationship by the alkylation of enolates having a β-silyl group (5).[9,10] To these three methods, we add the possibility of converting a phenyldimethylsilyl group into a hydroxyl with retention of configuration (6).[11] Perhaps most striking about these methods of stereocontrol is the sense in which the word "control" really means control: with each of them, we are able to control relative stereochemistry in whichever sense is wanted. In the electrophilic substitution of an allylsilane (3), a stereospecifically *anti* reaction allows, in principle, the new centre to be created in either sense by three methods: (i) control of the absolute configuration of the stereogenic centre, which is then relayed to the new centre, (ii) choice of which group shall be resident on C-3 of the allylsilane unit and which shall be the electrophile E+, and (iii) choice of double bond geometry. The last method also works for the hydroboration reaction (4), where we have demonstrated that the *E*-allylsilane (7) gives the 1,3-diol derivative (8) with an *anti* relationship of substitutents on the carbon chain, and the Z-allylsilane (9) gives the 1,3-diol derivative (10) with a *syn* relationship.[5] Furthermore, we have shown that the electrophilic attack on enolates (**5**) is also amenable to control in either sense, since alkylation of an enolate (**11** → **12**) gives the *anti* relationship of substituents, whereas

11 **12** 97:3 88% **13** 84%

14 **15** 91:9 **16** 63%

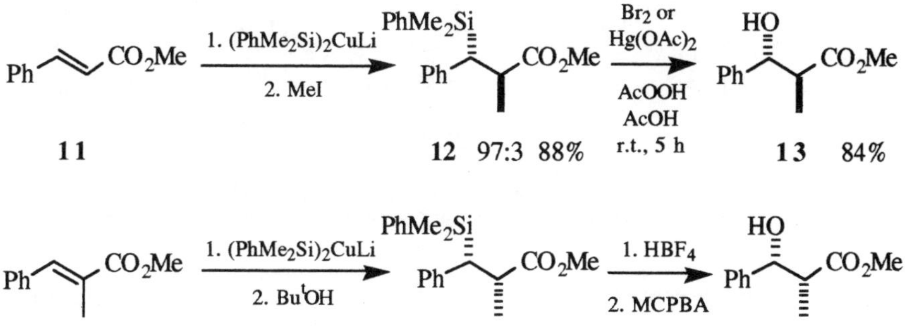

protonation of the enolate (**14** → **15**) gives the *syn* relationship.[9] In each case, the conversion of the silyl to the hydroxy group gives β-hydroxy esters (**13** and **16**, respectively).[11,12]

2 EBELACTONE-A

Ebelactone-a (**17**)[13] is an esterase inhibitor, but from the synthetic point of view it presents a much more substantial challenge than the targets that we have tackled hitherto, such as tetrahydrolipstatin.[14] Although incomplete, our work on the synthesis of ebelactone-a has led us to seek and find solutions to all the stereochemical problems. Our approach has been to divide the molecule into three parts, two electrophilic

17

18 **19** **20**

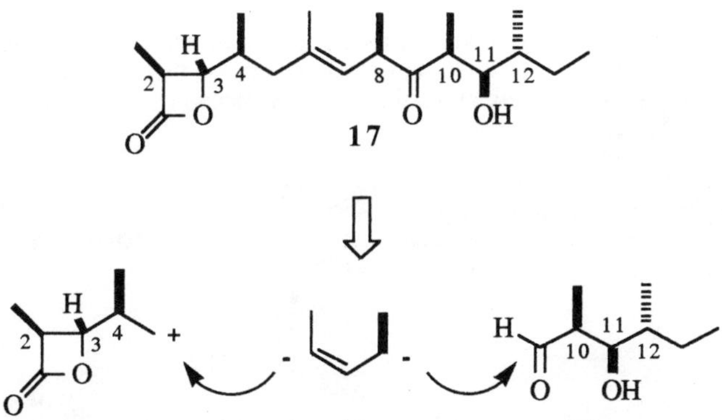

fragments (**18** and **20**) and one doubly nucleophilic fragment (**19**). The left-hand electrophile (**18**) might be derivable from a *meso* diester (**23**), if the diastereotopic ester groups can be differentiated in some way. Since there are several methods for dealing with this sort of problem, we have not looked at it yet, but have concentrated simply on setting up the three stereogenic centres with the appropriate relationship, which is that of an alkylation reaction, as in the model sequence (**11 → 12**). The first methylation step (**21 → 22**) was very highly diastereoselective, and the second (**22 → 23**) was less so,

$$MeO_2C\text{-}CH(SiMe_2Ph)\text{-}CH_2\text{-}CO_2Me \xrightarrow[\text{2. MeI}]{\text{1. LDA}} MeO_2C\text{-}CH(SiMe_2Ph)\text{-}CHMe\text{-}CO_2Me \xrightarrow[\text{2. MeI}]{\text{1. LDA}} MeO_2C\text{-}CMe(SiMe_2Ph)\text{-}CHMe\text{-}CO_2Me$$

21   **22** 97% >95:5   **23** 60% 88:12

but the diester (**23**) gave a crystalline anhydride (**24**), which was clearly a *meso* compound. There was some risk that these reactions might have induced epimerisations α to the ester groups, leading to the, presumably more stable, all-*trans meso* anhydride. However, no epimerisations had taken place in this transformation, as was readily apparent when we converted the minor component of the second alkylation step into the

23 $\xrightarrow[\text{2. Ac}_2\text{O, 100°, 1.25 h}]{\text{1. LiI, lutidine, reflux, 8 h}}$ **24** 66% $\xrightarrow[\text{quinuclidine, r.t., 1.5 h}]{\text{BnOH}}$ (±)-**25** 91%

corresponding anhydride, which was clearly different from the anhydride (**24**). For a model reaction for the differentiation of the two carbonyl groups, we opened the anhydride (**24**) with benzyl alcohol to give the racemic mono-ester (**25**), which could be selectively reduced at the carboxylic acid group using borane:THF to give the corresponding primary alcohol. It is easy to imagine how this or some related reaction can be used to enter the optically active series, and how an aldehyde or halide derived from the alcohol might function as the component (**18**).

Turning to the other electrophilic component (**20**), we can see a hydroxy group β to a carbonyl group, with the stereogenic centres C-10 and C-11 related in the sense of the protonation sequence (**14 → 15**), but the other pair, C-11 and C-12, present a new problem with many possible solutions. Our first efforts in this direction were stereochemically very successful: conjugate addition of our silyl-cuprate reagent to the unsaturated lactone (**26**), followed by alkylation, gave only, as far as we could tell, the lactone (**27**).[15] Although the C-10 centre has been put in by alkylation not protonation, it is in the correct sense, because alkylation is taking place in a cyclic structure constrained

26 $\xrightarrow[\text{2. MeI}]{\text{1. (PhMe}_2)_2\text{SiCuLi}}$ **27** 71% $\not\rightarrow$ **28**

to have the conformation (**2**). Unfortunately, we have been unable, so far, to reduce this lactone to the aldehyde (**28**). In particular, the methylene group adjacent to the oxygen atom is too hindered to be attacked by any of the powerful nucleophiles like selenide, that normally open lactones by alkyl-oxygen cleavage,[16] and approaches based on reduction of the lactone carbonyl have been similarly unfruitful.

To solve this problem, we turned to hydroboration: as well as giving 1,3 control in the sense (7 → 8) and (9 → 10), we already knew that it could be used to control 1,2-related centres. Thus the hydroboration-oxidation (29 → 30) was highly diastereoselective,[5] and we now find that the model allylsilane (31) undergoes

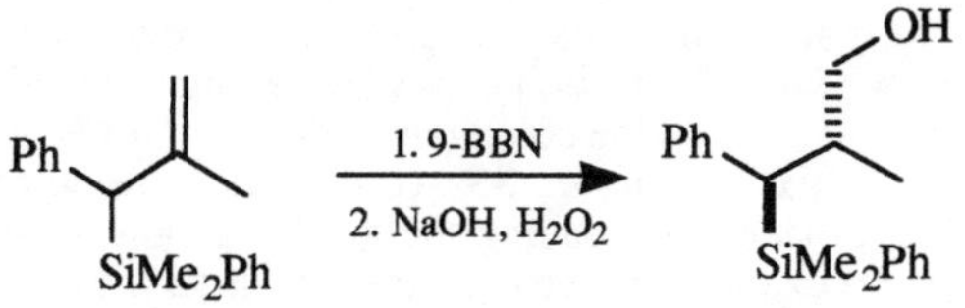

Ph—⟨SiMe$_2$Ph⟩—CH$_2$ (**29**) →[1. 9-BBN; 2. NaOH, H$_2$O$_2$]→ Ph—CH(OH)—CH(SiMe$_2$Ph)—CH$_3$ (**30**)   96%   >99:1

hydroboration-oxidation with similar selectivity giving only the lactone (**33**). We can confidently extrapolate this reaction to a substrate having an ethyl group in place of the methyl, but we still face the problem of avoiding the formation of a lactone. We really want the boron atom in the hydroboration product (**32**) to be replaced by a proton, not an

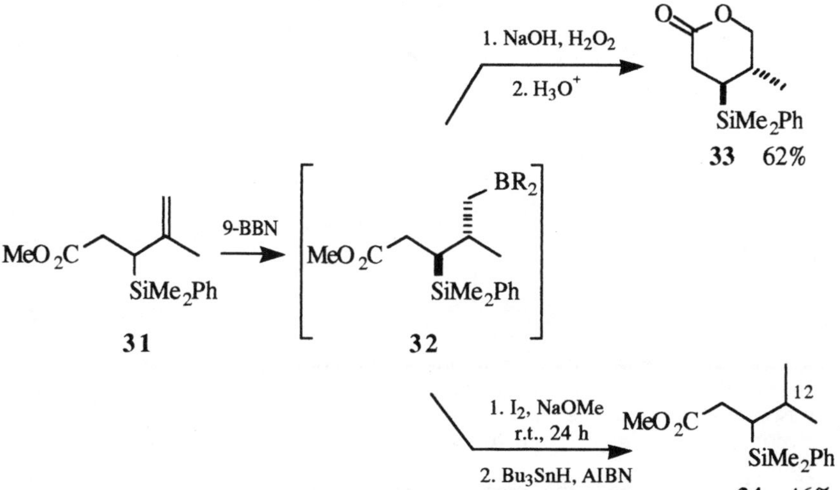

MeO$_2$C—⟨SiMe$_2$Ph⟩—CH$_2$ (**31**) →[9-BBN]→ [MeO$_2$C—CH(BR$_2$)—CH(SiMe$_2$Ph)—CH$_3$] (**32**)

→[1. NaOH, H$_2$O$_2$; 2. H$_3$O$^+$]→ lactone (**33**) 62%

→[1. I$_2$, NaOMe, r.t., 24 h; 2. Bu$_3$SnH, AIBN, THF, reflux, 16 h]→ MeO$_2$C—CH$_2$—CH(SiMe$_2$Ph)—CH$_3$ (C-12) (**34**) 46%

oxygen atom. Although this kind of process is well known using acids like propionic acid, we have had more success with a two step sequence: iododeboronation[17] followed by reduction of the iodide[18] giving the ester (**34**). This particular compound no longer has a stereogenic centre at C-12, but the sequence (31 → 34) is a good model for the problem that we want to solve. Furthermore, we already know that absolute control of

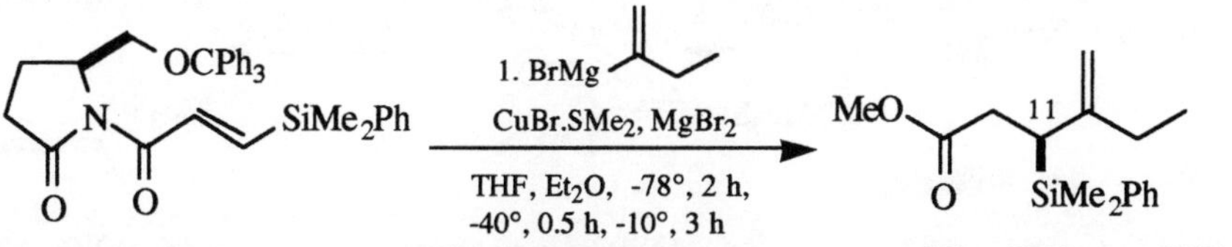

(**35**) →[1. BrMg—CH$_2$—C(=CH$_2$)—Et, CuBr.SMe$_2$, MgBr$_2$, THF, Et$_2$O, -78°, 2 h, -40°, 0.5 h, -10°, 3 h; 2. MeOLi, THF, r.t., 5 d]→ (**36**)   88% e.e. 32%

the stereo-chemistry at C-11 will be fairly easy, since the conjugate addition reaction (35 → 36) is substantially selective in the sense we want, which happens to be epimeric to the result we had in the tetrahydrolipstatin synthesis. Once again, we are able to emphasise that each of our methods is able to give either epimeric relationship; in this case control from the choice of double bond geometry in the enone precursor (*cis* in the tetrahydropipstatin synthesis and *trans* in **35**) leads to control in the absolute configuration of the silicon-substituted stereogenic centre. With the method for controlling the C-11 and C-12 relationship settled, we can now look ahead to the C-11 and C-10 relationship. Normally, since this is a "protonation relationship" in the sense of the reaction (**14** → **15**), we would control it easily by having the methyl group on C-10 already in the molecule when we do the conjugate addition. This is not possible here, because a substituent on this carbon in **35** would interfere with the absolute stereocontrol. One might consider introducing the methyl group by alkylation of the ester (**36**). This would have exactly the wrong stereochemistry, but regenerating the enolate and protonating it under kinetic control should correct the stereochemistry. Unfortunately, we have so far been unable to generate an ester enolate when there is a substituent at the α position, and we will have to use a sequence that we set up some time ago[10,19] precisely to solve this kind of problem. Methylenation of the β-silylbutyrate ester (**37**) by several methods gave the ester (**38**), and phenylthio conjugate addition-protonation gave the ester (**39**), although in this case only with low stereoselectivity in

the protonation sense. However, we know that protonation of an enolate with a methyl group on the stereogenic centre is less selective than with larger groups; protonation with an isopropyl group on the stereogenic centre (**41** → **42**),[9] for example, gives very high selectivity, which augurs well for what we will be wanting to do in the ebelactone synthesis.

We now come to the central and most intriguing problem: how to join the three pieces together, and what will be the doubly nucleophilic synthon (**19**)? Our plan here is to use an optically active allenylsilane (**45**), which will react in the general sense (**3**), *anti* to the silicon, to set up the correct absolute configuration at the isolated stereocentre C-8.

We have already established how best to prepare this type of allenylsilane,[20] and we have now used the method in the optically active series, starting from the known propargyl alcohol (**43**),[21] to prepare the allenylsilane (**45**). To improve the enantiomeric purity of the propargyl starting material, we have used the sulphonate ester (**44**), which crystallises, but we are still left with some uncertainty about the optical purity of the allenylsilane. The problem is that cuprates, although they are known[22] to react in a stereospecifically *anti* manner, are also known[23] to racemise allenes, and there is no direct and easy way by which we can tell what the optical purity of our allene is. The best we have been able to do is to treat the allenylsilane (**45**) with isobutyraldehyde and a Lewis acid, and to measure the optical purity of the product (**46**), which proved to be

45 **46** 86% 60% e.e.

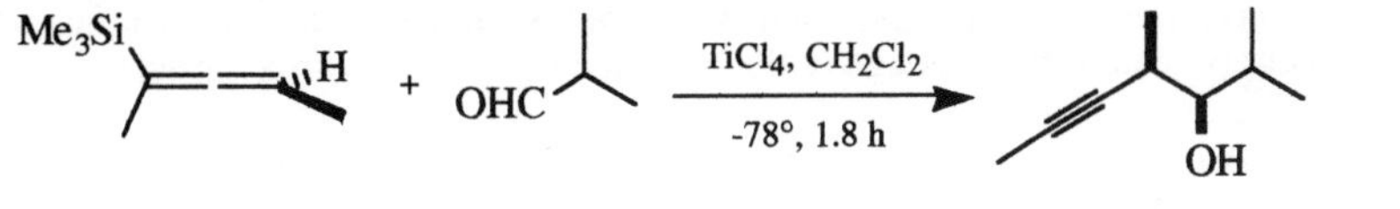

less (60% e.e.) than that of the sulphonate (**44**)(92% d.e.) from which it had been prepared. However, the diasteroselectivity for the formation of the product (**46**), assigned by analogy with Danheiser's work,[24] with the substituents *syn* on the backbone rather than *anti*, is high, and this has further consequences in our favour. When we come to combine the allenylsilane (**45**) with the aldehyde (**29**), the temporary stereogenic centre C-9 will be controlled from both components in the same sense: the allenylsilane will react in the same sense as in the model series (**45** → **46**), and the *syn* relationship between C-9 and C-10 in the product (**47**) corresponds to that expected of

45 **29** **47**

syn relationship of Me and OH groups normal

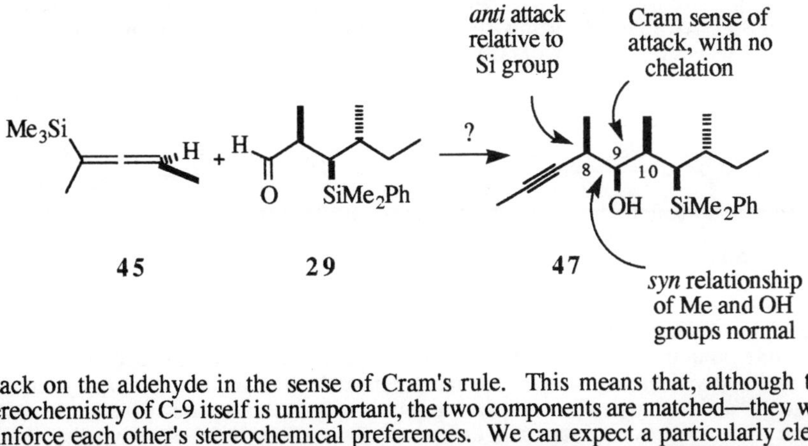

attack on the aldehyde in the sense of Cram's rule. This means that, although the stereochemistry of C-9 itself is unimportant, the two components are matched—they will reinforce each other's stereochemical preferences. We can expect a particularly clean reaction when we come to do it, and any enantiomeric impurity in our allene may be less

(±)-**45** (±)-**48** **49** 73:18:9:0 50%

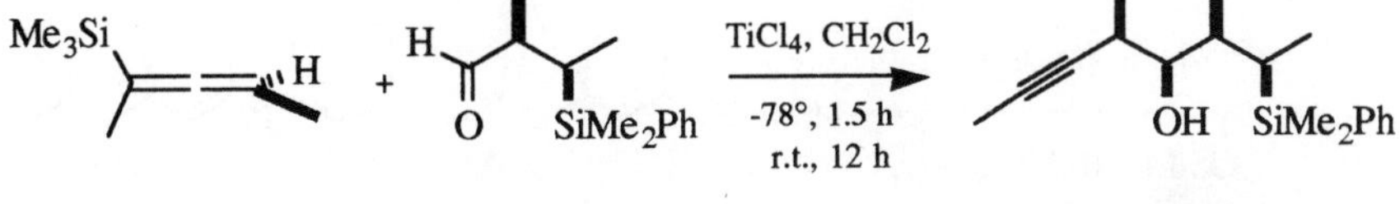

important than it would otherwise be, because the enantiomer of the allene (**45**) will not react as efficiently with the aldehyde (**29**). As a check on this point, we have carried out the model reaction (**45** + **48** → **49**), in which *both components were racemic*, and obtained, to be sure, a mixture of diastereoisomers, but with *one major component* (**49**) making up 73% of the mixture.

The success of this operation is dependent upon the phenyldimethylsilyl group masking the C-11 hydroxy group until this step is complete. If the hydroxy group is already unmasked, it will induce chelation control in the step (**45** + **29**) and the two components would then be stereochemically mismatched. However, looking ahead, there will be a problem in converting the silyl group to a hydroxy once the C-6,C-7 double bond is in place, because that reaction is dependent upon an electrophile's attacking the phenyl group on the silicon faster than it attacks anything else. A disubstituted and especially a trisubstituted double bond would be more nucleophilic than a silylated benzene ring, and will defeat our efforts. The unmasking must be the next step. We have found that a triple bond is capable of surviving the unmasking (**49 → 50**), but here we can take advantage of the presence of the C-9 hydroxy group to remove

the phenyl group under mild conditions. In the benzylation step in our tetrahydrolipstatin synthesis,[14] we had found it necessary to use acid-catalysed conditions, because of the easy loss of the phenyl group when we used sodium hydride. Now we shall be able to use sodium hydride deliberately, and the only problem then is to preserve the distinction between the two hydroxy groups before the migration of C-11 from silicon to oxygen takes place. As a model for this step, we have carried out the sequence (**51 → 52 → 53**). This leaves the problem of creating the second of the nucleophilic centres, in order

to join the two fragments (**18** and **19**). Our plan here is to carry out a silyl-cupration of the triple bond of a suitably modified and protected version of the alcohol (**47**). We know that disubstituted triple bonds react stereoselectively *syn* with our silyl-cuprate reagent,[25] and we have now tested the regioselectivity of this reaction when the two ends of the acetylene are as substantially different as they will be here—the silyl ether (**54**) of the acetylenic alcohol (**46**) reacts very selectively to place the silyl group at

the methyl-substituted end of the bond, and the product (**55**) reacts with iodine to give the iodide (**56**), in a reaction that is far from satisfactory as yet, since we get a large amount of protodesilylation, whether we protect the alcohol group or not, but is stereospecific in giving retention of double bond geometry. A vinyl iodide is obviously a suitable precursor for an organometallic nucleophile, but there are not too many ways of using a vinyl-metal to make a bond with an alkyl halide. The alternative, to use the intermediate (**18**) as the nucleophile, provides several methods, and we have carried out the conversion of the iodide (**57**) to the alkene (**58**) as a model for such a procedure, but

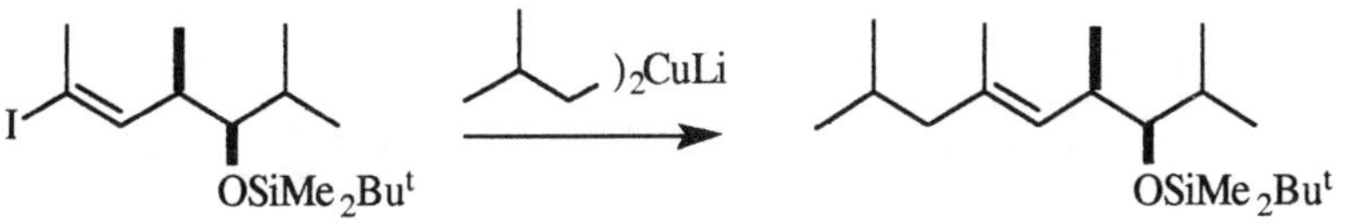

we are not sanguine about this idea—it will probably not be easy to create a nucleophilic species out of the "real" reagent corresponding to **18**; there will surely be too many problems of protection for the β-lactone group. Accordingly, we have to face the prospect that the organometallic nucleophile corresponding to the sum of the components (**19** and **20**) may have to be treated with an aldehyde in a reaction of the type (**59** + **60** → **61**), whereupon we face a new problem: how to convert the allylic alcohol (**61**) to the alkene (**62**), with retention of the position of the double bond, and the selective

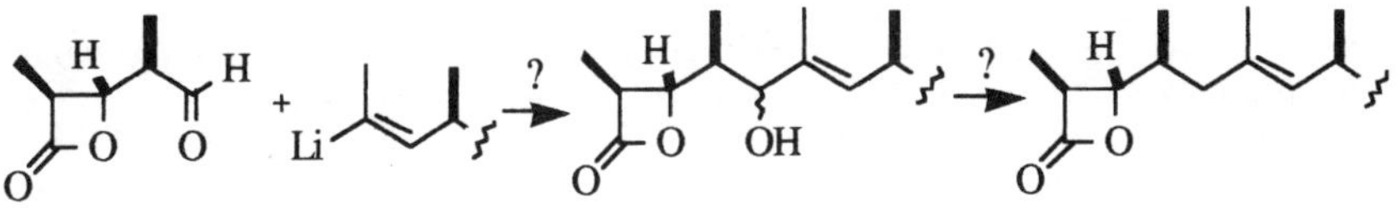

formation (or retention) of the E configuration. Once again, we have developed a model sequence that solves precisely this problem using silicon chemistry. Trisubstituted double bonds are selectively set up in the sense that we want here by the protodesilylation of allylsilanes, as in the reaction (**63** → **64**), when the substituent on the stereogenic centre carrying the silyl group is branched.[26] Furthermore, we have a

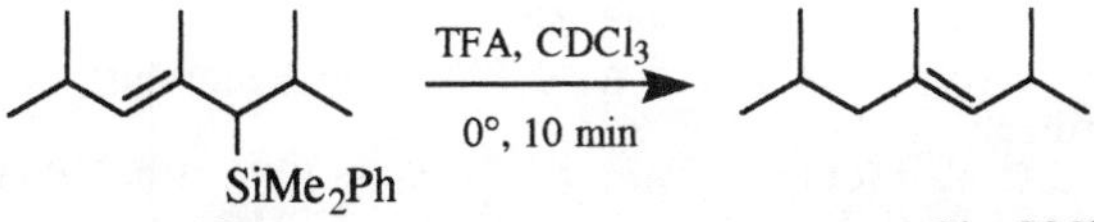

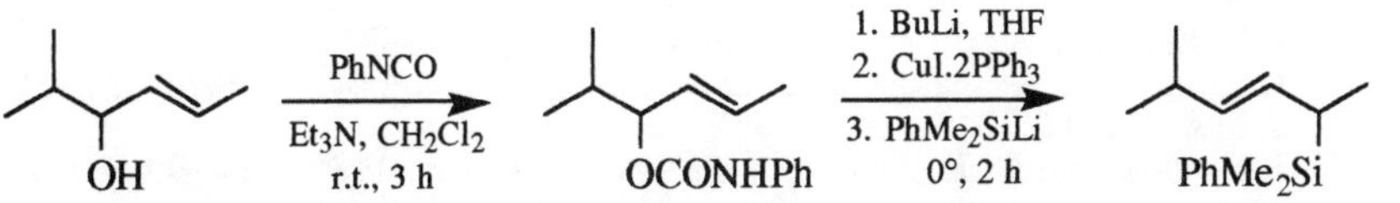

method by which an unsymmetrical allylic alcohol can reliably give the allylsilane with the necessary allylic shift in functionality, if we use the urethane-cuprate protocol,[27] as illustrated in the reaction (**65** → **66** → **67**).

With all the methods in place, we have now to put them into practice; perhaps I can finish by summarising the methods by which the stereocontrol will be achieved, if all goes according to plan! The single diagram (**68**) says it all, but I should like to emphasise that all of the methods that we plan to use could be adapted to give the

anti attack to Si on an allenylsilane

hydroboration of an allylsilane

alkylation of a β-silyl enolate

protodesilylation of an allylsilane

conjugate addition with a chiral auxiliary

protonation of a β-silyl enolate

68

opposite stereochemical outcome. Although reactions with high levels of stereoselectivity or stereospecificity are now common, stereo*control*, in the sense that I have illustrated it in this lecture, is rare.

The co-workers responsible for the new work described here are: Sarah Archibald (the reactions 31 → 33 and 34) supported by the SERC, Jérôme Bazin (the reactions 21 → 23 → 25, 43 → 45 → 46, 51 → 53, and 54 → 56), supported by Rhône Poulenc, David Parker (21 → 23), supported by the SERC, Ken Takaki (the reactions 45 → 46, 54 → 56, and 57 → 58), supported by a Ramsey Memorial Fellowship, and Anne Ware (the reactions 26 → 27, 31 → 33, 35 → 36, 45 + 48 → 49, and 49 → 50), supported by the SERC. I am grateful to these people and to the supporting bodies.

This article was originally printed in Pure Appl. Chem., 1990, **62**, 1879.

© 1990 IUPAC

REFERENCES

1. I. Fleming, *Pure Appl. Chem.*, 1988, **60**, 71; I. Fleming, D. Higgins, and A. K. Sarkar, in 'Selectivities in Lewis Acid Promoted Reactions', ed. D. Schinzer, Kluwer, Dordrecht, 1989, p. 265.
2. E. Vedejs and C. K. McLure, *J. Am. Chem. Soc.*, 1986, **108**, 1094.
3. D. P. Curran and B. H. Kim, *Synthesis*, 1986, 312-315.
4. I. Fleming, A. K. Sarkar, and A. P. Thomas, *J. Chem. Soc., Chem., Commun.*, 1987, 157.
5. I. Fleming and N. J. Lawrence, *Tetrahedron Lett.*, 1988, **29**, 2077.
6. I. Fleming, A. K. Sarkar, M. J. Doyle, and P. R. Raithby, *J. Chem. Soc., Perkin Trans. 1*, 1989, 2023.
7. I. Fleming and N. K. Terrett, *J. Organomet. Chem.*, 1984, **264**, 99; I. Fleming and N. K. Terrett, *Tetrahedron Lett.*, 1984, **25**, 5103; H.-F. Chow and I. Fleming, *ibid.*, 1988, **29**, 1825.
8. T. Hayashi, M. Konishi, H. Ito, and M. Kumada, *J. Am. Chem. Soc.*, 1982, **104**, 4962; T. Hayashi, M. Konishi, and M. Kumada, *ibid.* 4963-4965; T. Hayashi, H. Ito, and M. Kumada, *Tetrahedron Lett.*, 1982, **23**, 4605; T. Hayashi, M. Konishi, and M. Kumada, *J. Chem. Soc., Chem. Commun.*, 1983, 736; T. Hayashi, K. Kabeta, T. Yamamoto, K. Tamao, and M. Kumada, *Tetrahedron*

Lett., 1983, **24**, 5661; T. Hayashi, Y. Okamoto, K. Kabeta, T. Hagihara, and M. Kumada, *J. Org. Chem.*, 1984, **49**, 4224; T. Hayashi, M. Konishi, Y. Okamoto, K. Kabeta, and M. Kumada, *ibid.*, 1986, **51**, 3772; T. Hayashi, Y. Matsumoto, and Y. Ito, *Organometallics*, 1987, **6**, 884; H. Wetter and P. Scherer, *Helv. Chim. Acta*, 1983, **66**, 118; G. Wickham and W. Kitching, *Organometallics*, 1983, 2, 541; G. Wickham, D. Young, and W. Kitching, *Organometallics*, 1988, 7, 1187; W. Kitching, K. G. Penman, B. Laycock, and I. Maynard, *Tetrahedron*, 1988, **44**, 3819; G. Procter, A. T. Russell, P. J. Murphy, T. S. Tan, and A. N. Mather, *Tetrahedron*, 1988, **44**, 3953; S. R. Wilson and P. A. Zucker, *J. Org. Chem.*, 1988, **53**, 4682; V. G. Matassa, P. R. Jenkins, A. Kümin, L. Damm, J. Schreiber, D. Felix, E. Zass, and A. Eschenmoser, *Isr. J. Chem.*, 1989, **29**, 321.
9. I. Fleming, J. H. M. Hill, D. Parker, and D. Waterson, *J. Chem. Soc., Chem. Commun.*, 1985, 318.
10. I. Fleming and J. D. Kilburn, *J. Chem. Soc., Chem. Commun.*, 1986, 305.
11. I. Fleming and P. E. J. Sanderson, *Tetrahedron Lett.*, 1987, **28**, 4229.
12. I. Fleming, R. Henning, and H. Plaut, *J. Chem. Soc., Chem. Commun.*, 1984, 29.
13. Isolation: H. Umezawa, T. Aoyaga, K. Uotani, M. Hamada, T. Takeuchi, and S. Takahashi, *J. Antibiotics*, 1980, **33**, 1594; structure: K. Uotani, H. Nagamowa, S. Kondo, T. Aoyagi, and H. Umezawa, *ibid.*, 1982, **35**, 1495.
14. I. Fleming and N. J. Lawrence, *Tetrahedron Lett.*, 1990, **31**, 3645.
15. I. Fleming, N. L. Reddy, K. Takaki, and A. C. Ware, *J. Chem. Soc., Chem. Commun.*, 1987, 1472.
16. R. M. Scarborough, Jr. and A. B. Smith, III, *Tetrahedron Lett.*, 1977, 4361; D. Liotta and H. Santiesteban, *ibid.* 4369.
17. N. R. De Lue and H. C. Brown, *Synthesis*, 1976, 114.
18. For a review, see W. P. Neumann, *Synthesis*, 1987, 665.
19. J. D. Kilburn, *Ph.D. Thesis*, Cambridge, 1986, p. 13.
20. I. Fleming, K. Takaki, and A. P. Thomas, *J. Chem. Soc., Perkin Trans. 1*, 1987, 2269.
21. H. C. Brown and G. G. Pai, *J. Org. Chem.*, 1982, **47**, 1606, following M. M. Midland, D. C. McDowell, R. L. Hatch, and A. A. Tramontano, *J. Am. Chem. Soc.*, 1980, **102**, 867.
22. G. Tadema, R. H. Everhardus, H. Westmijze, and P. Vermeer, *Tetrahedron Lett.*, 1978, 3935.
23. A. Claesson and L.-I. Olsson, *J. Chem. Soc., Chem. Commun.*, 1979, 524.
24. R. L. Danheiser, D. J. Carini, and C. A. Kwasigroch, *J. Org. Chem.*, 1986, **51**, 3870.
25. I. Fleming, T. W. Newton, and F. Roessler, *J. Chem. Soc., Perkin Trans. 1*, 1981, 2527.
26. I. Fleming and D. Higgins, *J. Chem. Soc., Perkin Trans. 1*, 1989, 206.
27. I. Fleming and A. P. Thomas, *J. Chem. Soc., Chem. Commun.*, 1985, 411.

Stereocontrol by Means of Organosilanes: Some Observations on Chemoselectivity, Stereoselectivity, and Enantioselectivity in Allylsilane Reactions

Cristina Nativi,[b] Alfredo Ricci,[a,b] and Maurizio Taddei*[a]

[a]DIPARTIMENTO DI CHIMICA ORGANICA 'UGO SCHIFF', UNIVERSITÀ DI FIRENZE, VIA G.CAPPONI 9, I-50121 FIRENZE, ITALY
[b]CENTRO CNR COMPOSTI ETEROCICLICI, VIA G.CAPPONI 9, I-50121 FIRENZE, ITALY

The trialkylsilyl group is a powerful auxiliary for organic synthesis. It has been used in organic synthesis, for more than 25 years, as an efficient controller of the chemoselectivity and stereoselectivity of reactions.[1] A trialkylsilyl group exerts an electron donating effect on the organic skeleton to which it is bonded, it is an excellent leaving group and it is sufficiently large to exert a steric influence, so that organosilanes are generally good nucleophiles.[2]

Allylsilanes are a class of compounds in which all the above attributes are combined.[3] They react with electrophiles in the presence of activators and the stereochemistry of the reaction products is often influenced by the steric hindrance around silicon. The possibility of employing a triorganosilyl group to control stereochemistry in such reactions has prompted attempts to perform enantioselective reactions with allylsilanes (and more generally with organosilanes) using one ligand bonded to the silicon as the enantiocontroller.

This paper reports all the steps of our approach to this problem, namely the exploration of the chemoselectivity of this class of organosilylated reagents, then the examination of all the possibilities of stereocontrolled reactions and finally the design of a suitable series of ligands.

Chemoselectivity

Trialkylallylsilanes are good nucleophiles. They can be prepared from the corresponding allylmetal (Li, Na, K, Mg) and electrophilic trialkylsilyl groups (such as R_3SiCl), but also by reaction of an electron deficient system (such as an allyl acetate) with nucleophilic trialkylsilyl groups (such as silylcuprates), or by hydrosilylation and other radical-related procedures.[4] The transformation of allylmetals into allylsilanes is an example of strong nucleophiles that can be converted into derivatives which are less reactive but able to react with weaker electrophiles if the latter reactivity is enhanced by an additive.[5] In this case the reaction products may be the same as those coming from the reaction with allylmetals.

By the term chemoselectivity we mean that allylsilanes react with different selectivity than the corresponding allylanion of Li, Mg or Cu.[6] The activation of the electrophile is performed, in this case, with a Lewis acid, so that the reaction occurs in an acidic rather than in a basic medium, as it does with allylmetal derivatives. It is not the main purpose of this paper to explain the electronic factors which control these reactions, for which the reader is referred to the excellent recent report of Dunogues, Fleming and Smithers on allylsilanes and vinylsilanes.[7] Among the large number of functional groups that react with allylsilanes (the said authors report at least 30 different classes of compounds) we wish to point out some representative examples of reactions between allylsilanes and electrophiles giving different products than the reaction performed with the analogous allylmetal.

Allylsilanes react with aldehydes in the presence of a Lewis acid to give the corresponding homoallyl alcohols, the same products that can be obtained from allyllithium and aldehydes.[8] The presence of a strong Lewis acid, such as AlCl$_3$ can determine a further condensation of the homoallyl alcohol with one equivalent of the aldehyde, to give the tetrahydropyranic system (1) which can be obtained in excellent yields working with two equivalents of the aldehyde.[9]

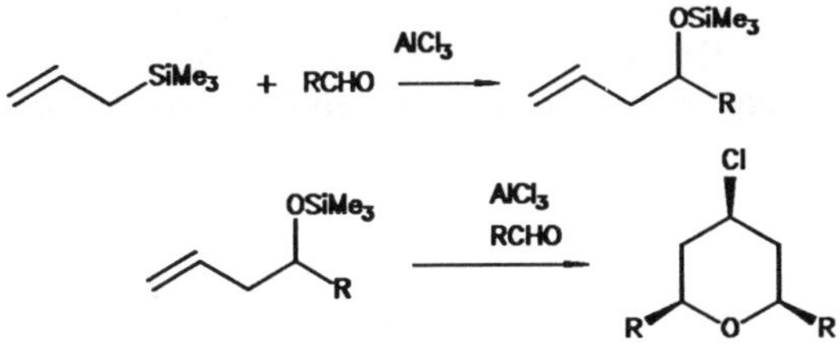

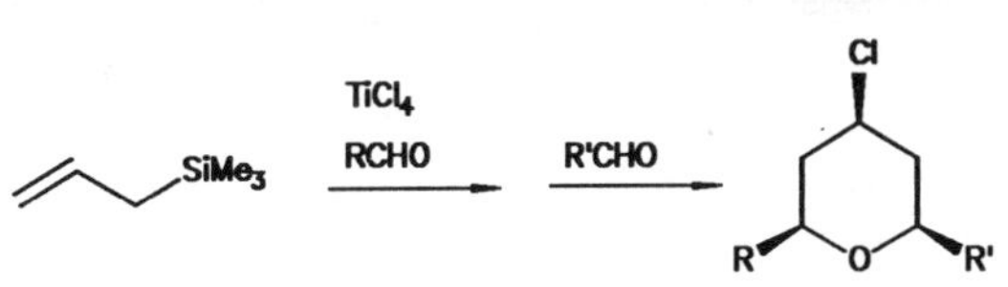

Different substituents may subsequently be introduced in positions 2 and 6 by appropriate addition of the reagents (first addition of one equivalent of aldehyde and Lewis acid followed by one equivalent of a different aldehyde). Substituted allylsilane (3) gave the pentasubstituted tetrahydropyran (4) with complete control of the stereochemistry.[10]

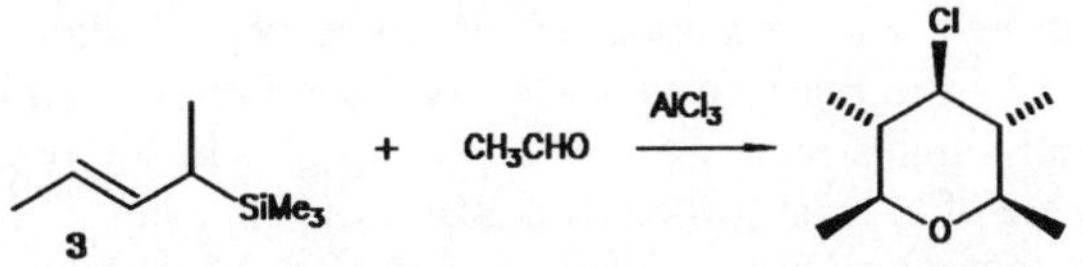

By this technique we prepared (5), a constituent of the perfume material civet, a glandular secretion of the civet cat (*Viverra civetta*).[11]

The presence of a Lewis acid allows the condensation of several allylsilanes with thioacetals, compounds that generally lose the acetalic proton in the presence of allyllithium or magnesium derivatives. Reaction of allylsilanes and phenylthioacetals in the presence of $AlCl_3$ leads to unsaturated phenylsulphides, which may be regarded as versatile intermediates in the stereocontrolled synthesis of polyenic molecules, like the diene (6).[12]

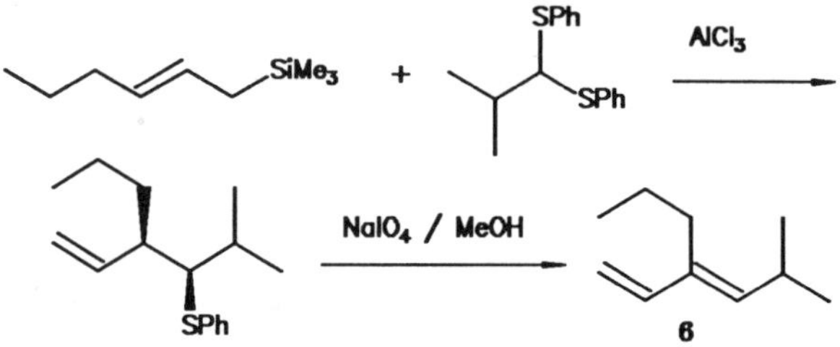

Peculiar differences in chemoselectivity are obtained in reactions of allylsilanes with lactones. Allyllithium is too basic to react with the carbonyl centre, giving mainly enolate (7); allylcuprates (or allylmagnesium bromide in the presence of Cu(I) salts) attack the carbonyl centre producing the allyl keto alcohol (8);[13] allylsilanes in the presence of $BF_3.Et_2O$ give allylation at the CH_2 α to the oxygen with formation of the unsaturated acid (9);[14] unsaturated lactones can moreover undergo conjugate addition to give the polyunsaturated acid (10).[15]

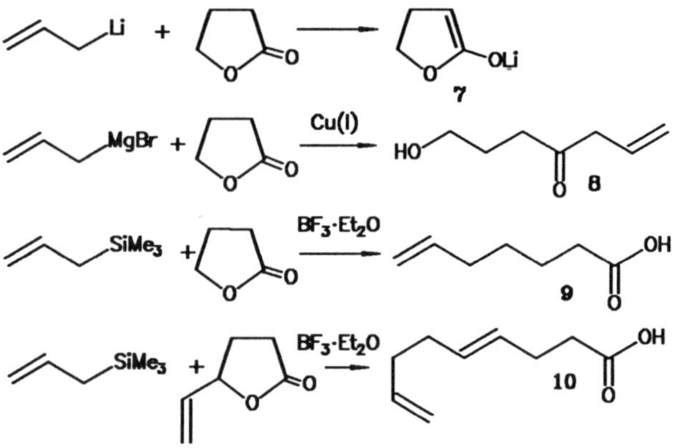

The high chemoselectivity in the reaction of allylsilanes with iminium ions allows the preparation of several kinds of unsaturated nitrogen containing compounds. Among the large number of iminium ions studied, a significant example is given by the reaction of 4-acetoxy-2-azetidinones and allylsilanes. 4-Acetoxy-2-azetidinones react well with allylcuprates in the presence of the NH at the expense of at least two equivalents of the cuprates[16]. In the case of allylsilanes, the generation of the iminium ion allowed the use of the N-substituted 4-acetoxy-2-azetidinone (11) with a group on the nitrogen which can be employed in the further synthesis of the second ring, as for example in the formal synthesis of thienamicine (12) described in the following scheme.[17]

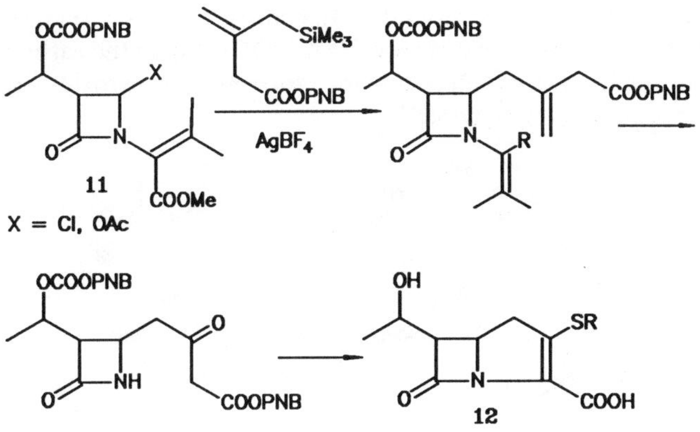

Stereoselectivity

One of the main reasons for the development of allylsilane chemistry in the last decade has been the high level of stereocontrol that can be obtained in reactions with electrophiles. The most interesting characteristics in the stereochemical outcome of the condensation of allylsilanes and aldehydes are as follows:

1) The electrophiles generally react in γ position with respect to the silicon (reaction with allylic shift).

2) The electrophiles generally attack the π system generally from the opposite side with respect to the silicon.

The correct knowledge of the factors that control these two aspects of allylsilane reactivity was extremely important in our target to use silicon as a bulky chiral hydrogen in organic synthesis. A simplified picture of the reaction of allylsilanes with electrophiles is reported in the following scheme.

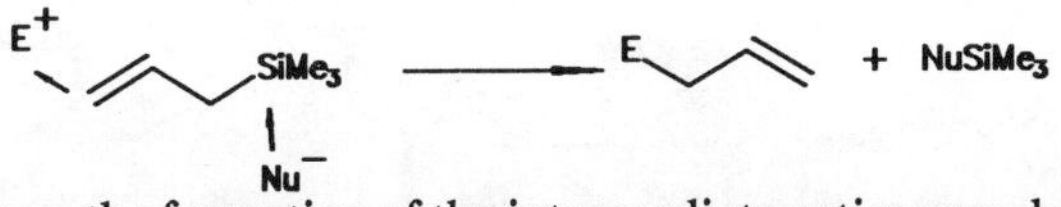

In this process the formation of the intermediate cation was demonstrated[18] and it can also be warranted by the stabilization of this carbocation by hyperconjugative effects.[19] From this picture it emerges that allylsilanes react with electrophiles with allylic shift. It is a rule but sometimes stereochemical factors allow this rule to be violated.

A case in which the silyl group does not exert good regiochemical directing effects is when, for example, the silyl group remains in the final product. Moreover, when working with a hindered allylsilane, one should consider the possibility of first having the attack of the nucleophile on silicon and then the elimination of the silyl group by electrophilic substitution on the *ipso* carbon (SE_1 or SE_2 mechanism).

We demonstrated that allylsilane (**13**), coming from β-ionone, reacts with larger electrophiles (like the complex CH3CHO/TiCl4) on the carbon carrying the silicon, while smaller electrophiles (such as the peroxidic oxygen of MCPBA) attack the double bond.[20]

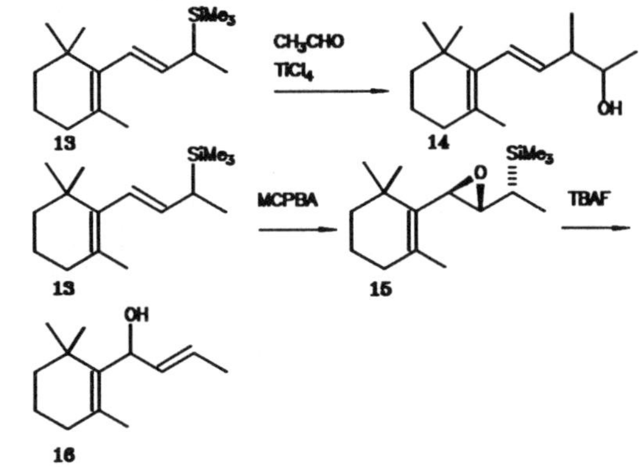

A molecular model of (**13**) showed that the double bond of the allylsilane is particularly hindered, so that probably CH3CHO- TiCl4 prefers to give, first the attack of the Cl⁻ at the silicon, then it gives the electrophilic substitution on the C sp^3 carbon. Using this reaction we prepared a component of the sandalwood perfume (**14**)[20], and starting from (**16**), β-damascone and β-damascenone were obtained in good yields.[21]

The condensation of a substituted allylsilane with a carbonyl compound is a straightforward reaction in which two new stereocentres are formed. The mode of approach of the two π systems has been extensively studied and it has been demonstrated that the stereochemical course of this reaction is extremely sensitive to the structure of the allylsilanes employed, including the stereochemistry around the double bond. *E*-Allylsilanes generally react with aldehydes[22] (or acetals)[23] in the presence of TiCl4 giving mostly the *syn* product.

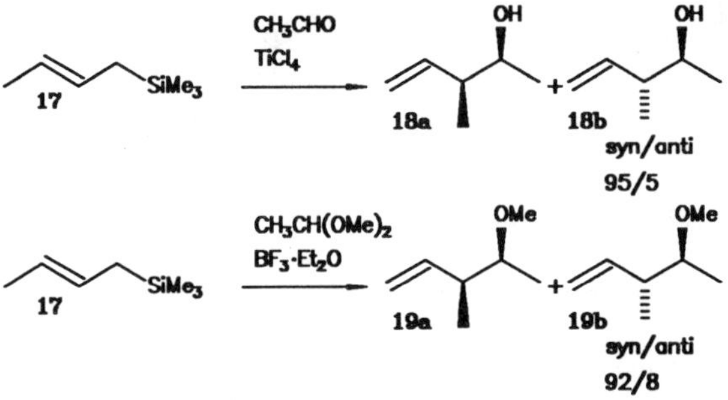

In the case of the aldehydes two possible transition states have been postulated to explain such stereochemistry[24] (in our opinion an analogous explanation can be adopted in the reaction of acetals). The structures postulated for the transition states for the *E* isomers are reported in the following scheme.

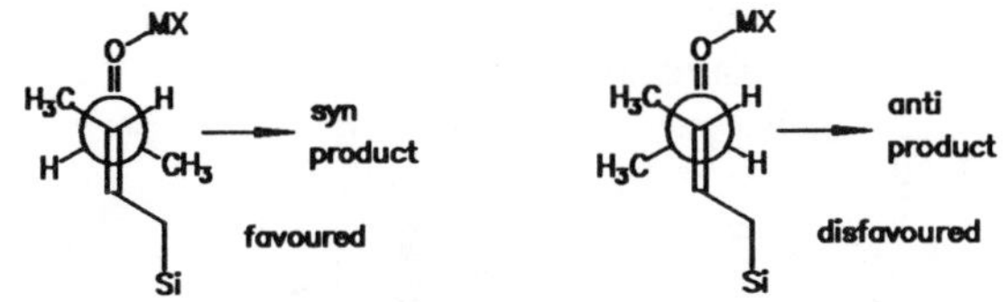

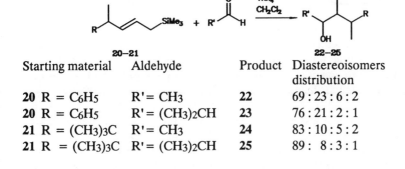

ANTIPERIPLANAR TS

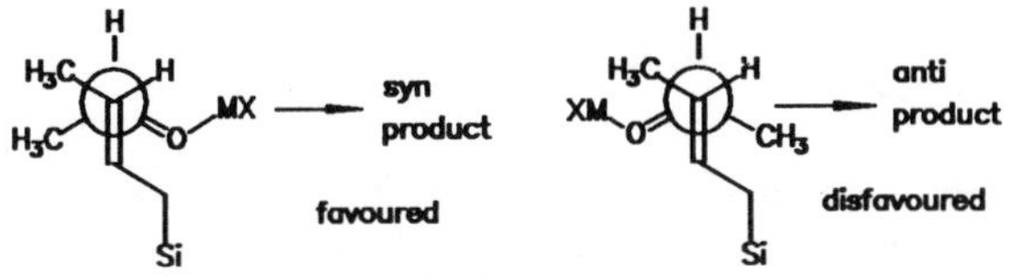

SYNCLINAL TS

We wanted to investigate the effect of a stereocentre close to the double bond in position 4 of the allylsilane chain. The π system of the allylsilane now has two diastereotopic faces and the level of stereocontrol can be related to the substituents of the stereocentre. The results are reported in Table I

Table I Condensation of allylsilanes **20-21** with aldehydes

Starting material	Aldehyde	Product	Diastereoisomers distribution
20 R = C_6H_5	R' = CH_3	22	69 : 23 : 6 : 2
20 R = C_6H_5	R' = $(CH_3)_2CH$	23	76 : 21 : 2 : 1
21 R = $(CH_3)_3C$	R' = CH_3	24	83 : 10 : 5 : 2
21 R = $(CH_3)_3C$	R' = $(CH_3)_2CH$	25	89 : 8 : 3 : 1

The reaction mostly gives one of the four possible diastereoisomers. In the case of compound (**24**), we were able to isolate the major diastereoisomer and to assign the stereochemistry as 2R(2R*,3R*,4R*) after oxidation of the double bond and ^{1}H NMR analysis of the 1,3,4-dioxanone (**24b**).

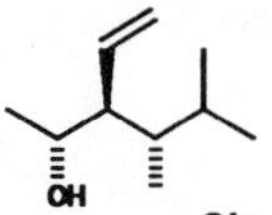

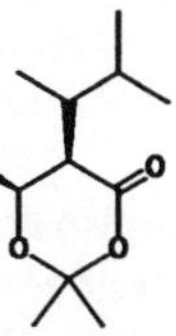

The above results show that it is possible, in principle, to control the stereochemistry of the reaction by the presence of a stereocentre in position 4. The relative stereochemistry of the approach of the C=O π system is well explained through the antiperiplanar transition states (Hayashi-Kumada transition state) and the selection between the diastereotopic faces of the C=C π

system of the allylsilane, probably controlled through a Felkin-Anh type transition state (TS of type A favored against B).

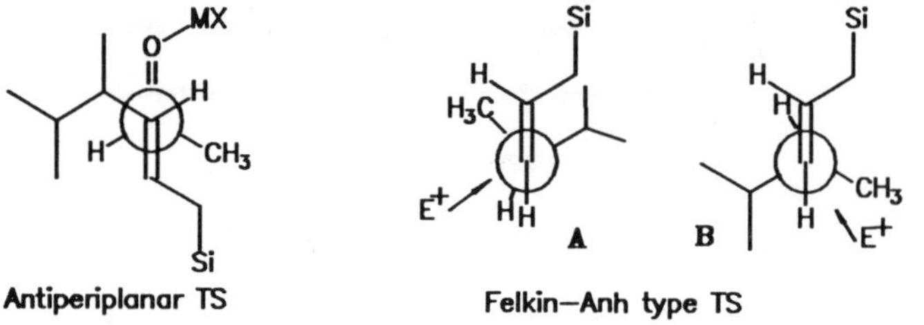

Antiperiplanar TS   Felkin–Anh type TS

We also tried to have a TS with a chelation control using the 4- hydroxy allylsilane (**26**) prepared as described in the following scheme[26], but it turned out to be unsuitable for condensation with aldehydes or acetals in the presence of a Lewis acid, giving mainly 1-4 elimination.[27]

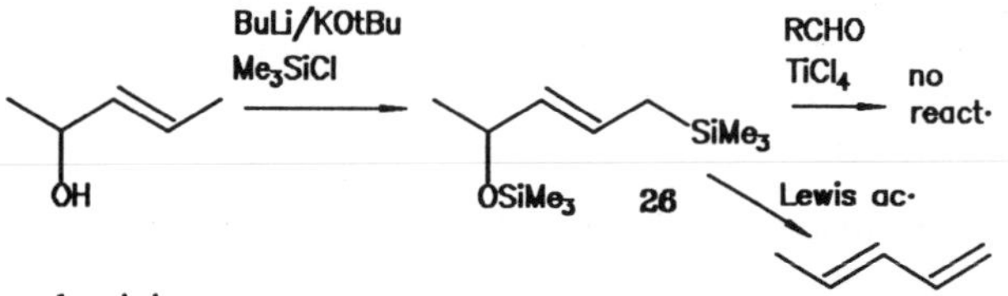

Enantioselectivity

The possibility of enantioselection in reactions that employ a "chiral silicon" has attracted the attention of many research groups. The possible approaches are to use a chiral non-racemic silicon (Si-centered optically active organosilane) or an enantiotopic ligand bonded to the silicon (C-centered optically active organosilane).

The first approach was realized by Brook in the early seventies with an enantioselective carboxylation of a lithium-benzylsilane carrying a chiral non-racemic silylate group.[28] Paquette prepared several vinylsilanes, like those reported in the picture, and tried different reactions on the double bond, concluding that chirality transfer from silicon to carbon may be generally inefficient in the absence of a template effect.[29] Later Larson and his group tried to use chiral silicon as a hindered chiral proton in reactions in which the silyl group was not directly involved.[30] Good results in enantioselection were recently achieved by cycloaddition of the thioacylsilane (**27b**).[31]

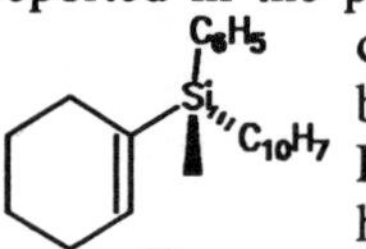

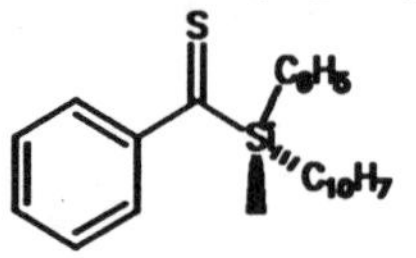

Our original approach to the problem was to synthesize a product of general formula R*Me$_2$SiCl where R* is an enantiotopic ligand. We tried to have the R* group bonded to the silicon through a carbon atom so that the auxiliary would be more stable to acidic and hydrolytic conditions. This C-atom would also be a non-racemic stereocentre. We set out to form a new C-Si bond in an enantiocontrolled way.

First attempts to perform a silylcupration on the derivatives **(28a)** and **(28b)** of myrtenol were disappointing, because the unwanted compound **(29)** was always the major product.[32]

28a,b **29** **30**

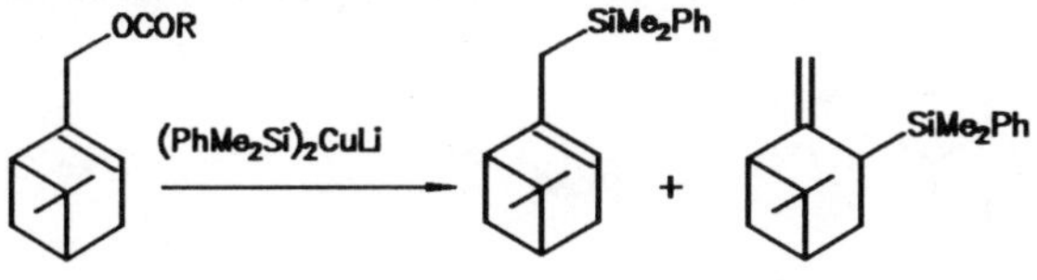

28a R = CH_3 **28b** R = NHPh

Silylcupration of myrtenal **(31)**, on the other hand, gave the β-silylaldehyde

31 **32** **33**

(32) which was further reduced to alcohol **(33)** and transformed, after two other steps, into the allylsilanes **(40- 42)**.[33]

33

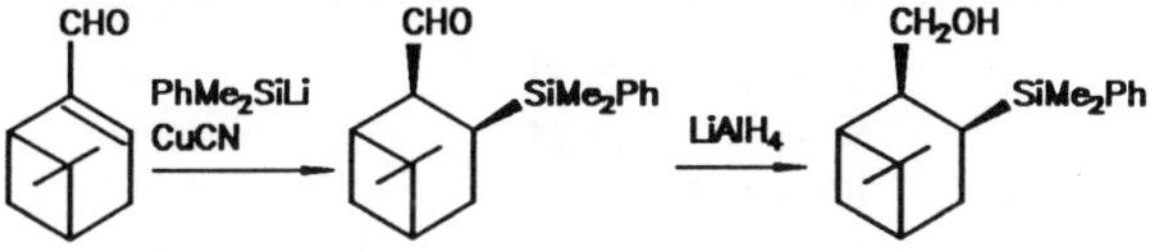

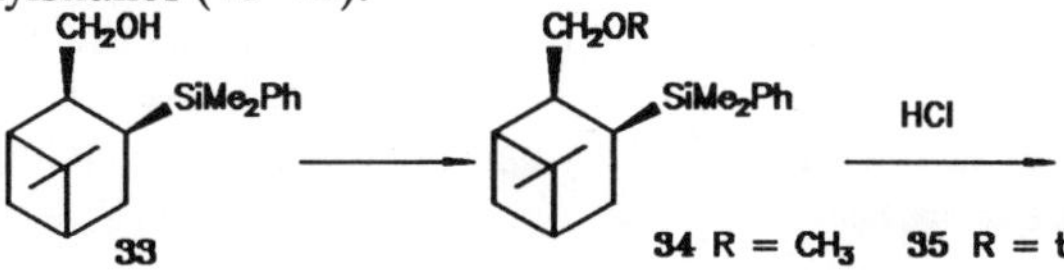

34 R = CH_3 **35** R = tBu
36 R = CONHPh

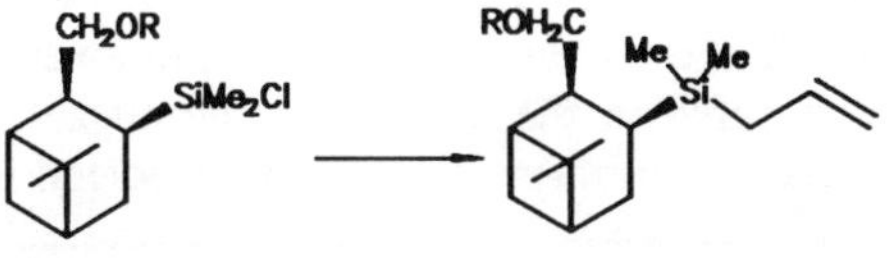

37 R = CH_3 **38** R = tBu
39 R = CONHPh

40 R = CH_3 **41** R = tBu
42 R = CONHPh

Allylsilane **(45)** was prepared in the same way, as reported in the scheme.[32]

33 **43**

44 **45**

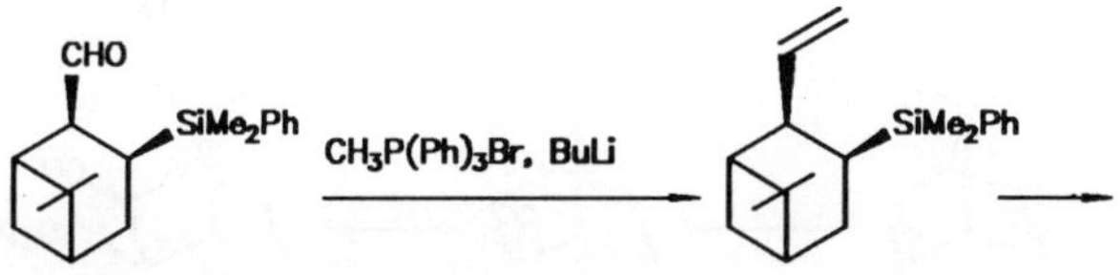

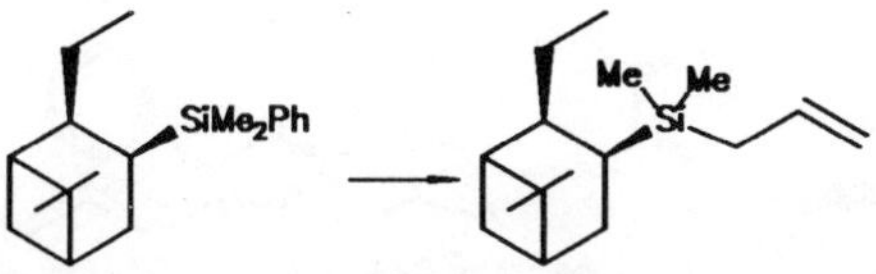

Products **(40-42)** and **(45)** reacted with aldehydes in the presence of Lewis Acids. Using the reaction between **(40)** and butyraldehyde as model, we discovered that the reaction was rather complex and that the distribution of the reaction products could change in different reaction conditions.[34] Under optimized reaction conditions, the effects on varying the different ligands on the silicon were found to be as reported in Table II.

Table II. Condensation of allylsilanes (**40-42**) and (**45**) with aldehydes

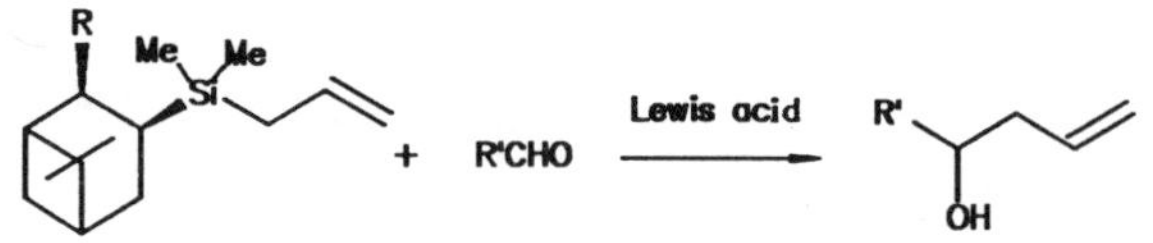

Allylsilane	Aldehyde	Yield	e.e.	Absolute Config.
R = OMe	R' = C3H7	50%	46%	R
R = OMe	R' = C6H5	45%	21%	S
R = OMe	R' = (CH3)2CH	41%	52%	R
R = OtBu	R' = (CH3)2CH	30%	50%	R
R = OCONHC6H5	R' = (CH3)2CH	—	—	—
R = C2H5	R' = (CH3)2CH	65%	12%	S

From the observation of these results we concluded that the presence of a heteroatom in the chain would have an effect in directing the enantioselectivity of the reaction, but it lowered the chemical yield. However, a fully hydrocarbonic ligand, as in the case of allylsilane (**45**), enhanced yields but decreased enantiomeric excess.

The second generation of C-centered optically active organosilanes was based on the observation that the ligands previously employed probably could not differentiate enough between the π faces of the allylsilane system. We prepared a derivative based on a bornyl skeleton as reported in the following scheme.

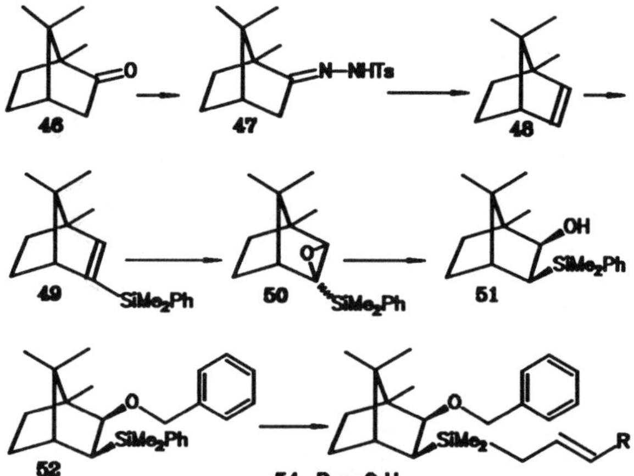

Bornene (**48**) was metallated using BuLi/KOtBu in pentane and reacted with PhMe2SiCl. Epoxidation of vinylsilane (**49**) was stereoselective giving the *exo*-epoxide (**50**) which was further reduced with LiAlH4 to alcohol (**51**) and subsequently protected as benzyl derivative (**52**). Exchange of the phenyl group with Cl gave the chlorosilane (53), which was used to prepare allylsilanes (54-56).[35] The absolute configuration of product 53 was proved by transformation of (52) into the bornyldiol derivative (57).

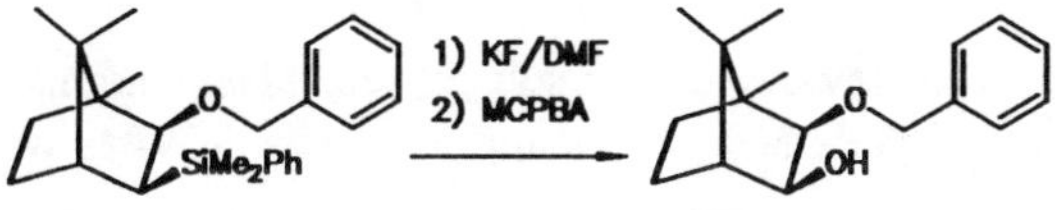

Allylsilanes (54-56) were treated with m-chloroperbenzoic acid followed by TBAF to give allylic alcohols (58-60). The results are presented in Table III.[35]

Table III. Preparation of optically active allyl alcohols (58-60)

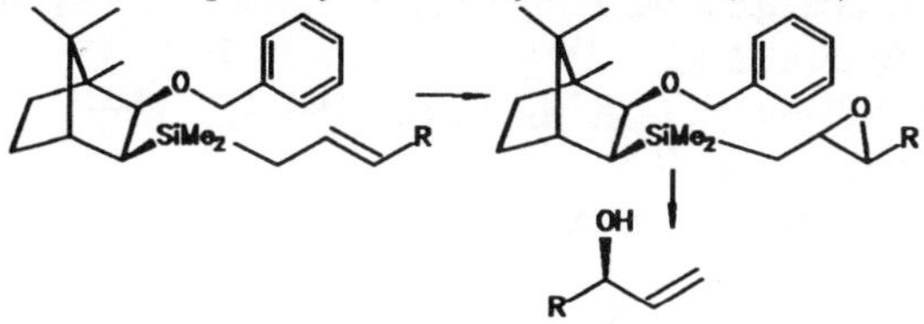

Starting material	Alcohol	Yield	e.e.	Absolute config.
54 R = C4H9	58	62%	76%	R
55 R = C6H5	59	59%	32%	S
56 R = (CH3)2CH	60	60%	87%	R

Alcohol (60) was obtained in high yields and with good enantiomeric purity. The enhancement of the e.e. value is correlated with the major hindrance of the R' group. This is consistent with a model which shows a rigid system with difficulties of rotation around the possible conformationally free bonds in the allylsilane chain. It is noteworthy to see that this is the chemical synthesis which gives alcohol (60) with the highest e.e. value, since this product does not undergo kinetic resolution using the Sharpless reagent.[36]

The above results, together with those of other authors,[37] bear the hypothesis that the ligands to the silicon accomplish the function of the enantiodiscrimination around the π allylic system especially with a substrate that has few possibilities of free rotations around σ bonds.

We think that in the future the use of organosilanes in the synthesis of enantiomerically pure compounds will become more and more popular, and probably reagents other than allylsilanes will be the subject of similar studies.[38] It will be of focal importance to find simple new reactions involving silicon, possibly with a cyclic or symmetric transition state, and to design simple auxilaries to bond to the silicon to perform enantiocontrol.

The number of papers on the use of organosilanes in organic synthesis is increasing very quickly and we hope that in the near future silylated chiral auxiliary (possibly commercially available)[40] will be widely used in enantiomerically pure compound synthesis.

References and Notes

1) For the most recent reviews on the subject see: "The Chemistry of Organic Silicon Compounds", S.Patai and Z.Rappoport eds. J.Wiley and Son, 1989, vol.1-2.
2) The trimethylsilyl group is a σ donor and a p acceptor. W.F.Reynolds, G.H.Hamer, A.R.Bassindale, J.Chem.Soc., Perkin Trans. 2, 1977, 971.
3) D.Schinzer, Synthesis, 1988, 263. G.G.Furin, O.A.Vayazankina, B.A.Gostevsky, N.S.Vyazankin, Tetrahedron, 1988, 44, 2675.
4) L.Birkofer, O.Stuhl, in ref.1, vol.1, 655.
5) Y.Yamamoto, S.Hatsuya, J.Yamada, J.Org.Chem., 1990, 55, 3118.
6) M.Schlosser, Pure App.Chem. 1988, 60, 1627. B.H.Lipshutz, E.L.Ellsworth, T.J.Siahaan, J.Am.Chem.Soc. 1989, 111, 1581. B.H. Lipshutz, R.Crow, S.H.Dirock, E.L.Ellsworth, J.Am.Chem.Soc. 1990, 112, 4063.
7) I.Fleming, J.Dunogues, R.Smithers, "The electrophilic substitution of alkylsilanes and vinylsilanes" in Organic Reactions, J.Wiley and Son, 1989, 37, 57.
8) A.Hosomi, H.Sakurai, Tetrahedron Lett. 1976, 16, 1295.
9) L.Coppi, A.Ricci, M.Taddei, Tetrahedron Lett. 1987, 28, 973.
10) A.Mordini, G.Palio, A.Ricci, M.Taddei, Tetrahedron Lett. 1988, 29, 4991.
11) L.Coppi, A.Ricci, M.Taddei, J.Org.Chem., 1988, 53, 911.
12) A.Mann, A.Ricci, M.Taddei, Tetrahedron Lett. 1988, 29, 6175.
13) T.Fujisawa, T.Sato, M.Kawashima, K.Naruse, K.Tamai, Tetrahedron Lett. 1982, 23, 3583 and references therein.
14) C.Cassinelli, *"Tesi di Laurea"* (Thesis), University of Florence, 1982.
15) T.Fujisawa, M.Kawashima, S.Ando, Tetrahedron Lett. 1984, 25, 3213.
16) S.Hanessian, D.Desilets, J.L.Bennani, J.Org.Chem. 1990, 55, 3098.
17) M.Aratani, K.Sawada, M.Hashimoto, Tetrahedron Lett. 1982, 23, 3921.
18) I.Fleming, J.A.Langley, J.Chem.Soc., Perkin Trans.1, 1981, 1421. G.D.Hartman, T.G.Traylor, Tetrahedron Lett. 1975, 939.
19) J.B.Lambert, Tetrahedron, 1990, 46, 2677.
20) E.Azzari, C.Faggi, N.Gelsomini, M.Taddei, Tetrahedron Lett. 1989, 30, 1106.
21) E.Azzari, C.Faggi, N.Gelsomini, M.Taddei, J.Org.Chem. 1990, 55, 1106.
22) T.Hayashi, K.Kabeta, I.Hamachi, M.Kumada, Tetrahedron Lett. 1983, 28, 2865.
23) H.Hosomi, M.Ando, H.Sakurai, Chem.Lett. 1986, 365.
24) For the antiperiplanar TS see ref. 22. For the synclinal TS see S.E.Denmark, B.R.Henke, E.Weber, J.Am.Chem.Soc. 1987, 109, 2532.
25) G.Palio, C.Nativi, M.Taddei, Tetrahedron:Asymmetry 1990, in press.
26) G.Palio, *"Tesi di Laurea"* (Thesis), University of Florence, 1988.
27) C.Nativi, M.Taddei, A.Mann, Tetrahedron, 1989, 45, 1131.
28) A.G.Brook, J.M.Duff, D.G.Anderson J.Am.Chem.Soc. 1970, 92, 7567.
29) S.J. Hathaway, L.A.Paquette, J.Org.Chem. 1981, 48, 3353. R.A.Daniels, L.A.Paquette, Organometallics 1982, 1, 1449.

30) G.L.Larson, E.Torres, J.Organometal. Chem. 1985, 293, 19. E.Torres, G.L.Larson, E.J.McGarvey, Tetrahedron Lett. 1988, 29, 1355.
31) B.F.Bonini, G.Mazzanti, P.Zani, G.Maccagnani, J.Chem.Soc., Chem.Commun. 1988, 365.
32) L.Coppi, "*Tesi di Laurea*", (Thesis), University of Florence 1986.
33) L.Coppi, A.Ricci, M.Taddei, Tetrahedron Lett. 1987, 28, 965.
34) L.Coppi, A.Mordini, M.Taddei, Tetrahedron Lett. 1987, 28, 969.
35) C.Nativi, A.Ricci, N.Ravidà, G.Seconi, M.Taddei, J.Org.Chem. 1990, in press.
36) See K.E.Konig, in "Asymmetric Synthesis" J.D.Morrison Ed., Academic Press, London, 1985, vol.5, 71.
37) T.H.Chan, D.Wang, Tetrahedron Lett. 1989, 23, 3041.
38) For other C-Centered Optically Active Silylated reagents see: M.E.Jung, K.T.Hogan, Tetrahedron Lett. 1988, 29, 6199. T.H.Chan, P.Pellon, J.Am.Chem.Soc. 1989, 111, 8737.
39) To our knowledge, the only C-Centered Optically Active Organosilane commercially available is Dimethylchlorosilylmethyl, 7,7-Dimethylnorpinane from Petrarch System Silanes and Silicones, Catalogue 1987, p.146.

Enantioselective Synthesis Using Organosilicon Compounds

T.H. Chan,* D. Wang, P. Pellon, S. Lamothe, Z.Y. Wie, L.H. Li, and L.M. Chen
DEPARTMENT OF CHEMISTRY, MCGILL UNIVERSITY, MONTREAL, QUEBEC, CANADA H3Y 2K6

1 INTRODUCTION

In the past several years, we and others have been interested in the use of organosilicon compounds for enantioselective synthesis. There are a number of approaches. The earlier attempts were made using chiral organosilicon compounds[1] where the chirality resided on silicon. A typical example was the work of Paquette[2] which showed that optically active allylmethylphenyl-α-naphthylsilane (1) reacted with the acetal 2 under Lewis acid conditions to give the product 3 but with very modest enantiomeric excess (ee).

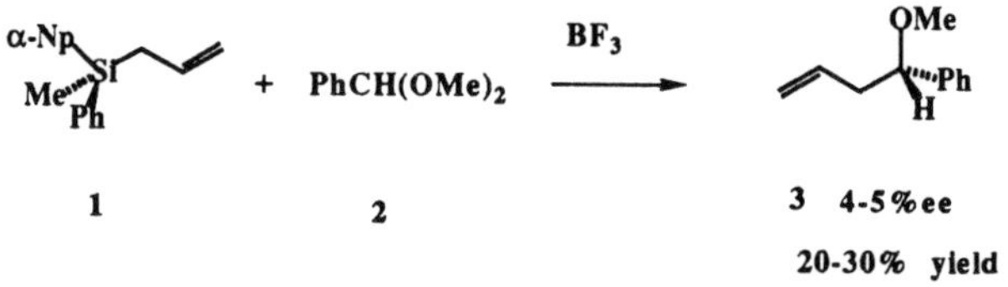

The alternative approach is to use organosilicon compounds with the chirality located at a site attached to, but removed, from silicon. An interesting example was provided by Jung[3] who used the C_2 binaphthyl compound 4 to reduce prochiral carbonyl compounds. The resultant alcohols showed modest ee.

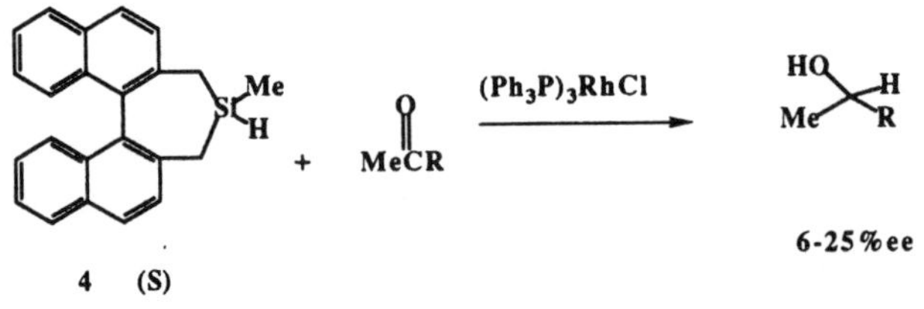

We ourselves have concentrated on using optically active organosilicon compounds which can be synthesized from readily available natural products. This chapter summarizes our research in this area. Emphasis will be placed on comparing the enantioselectivity of the same reaction using different chiral organosilicon compounds in order to gain an understanding of the underlying factors which control the selectivity.

2 LEWIS ACID MEDIATED ALLYLATION OF PROCHIRAL CARBONYL COMPOUNDS.

The reaction of carbonyl compounds with allylsilane under Lewis acid conditions to give homoallylic alcohols (Scheme 1), first described by Sakurai and Hosomi[4] in 1976, has been the subject of numerous studies. It is instructive to compare

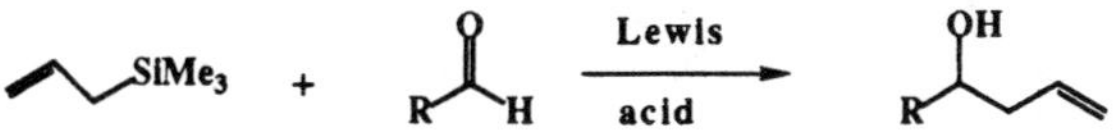

Scheme 1

the silane **1**, α-pinanyldimethylallylsilane (**5**) prepared by us[5] from (−)-β-pinene, and compound **6** prepared by Taddei.[6]

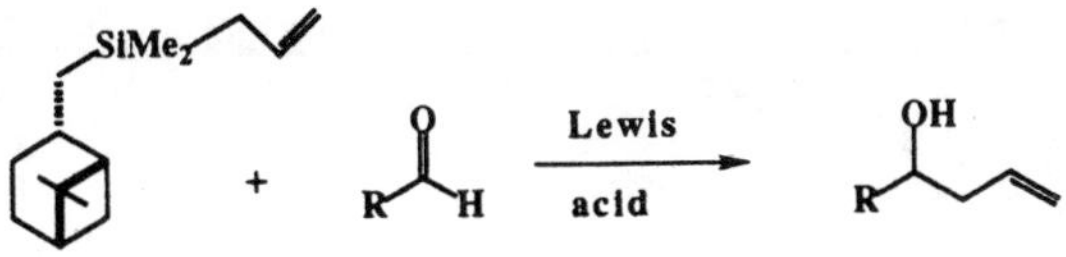

5   **7**   4-15%ee

Compound **1** was reported to be destroyed by Lewis acids (TiCl$_4$, SnCl$_4$) instead of undergoing the condensation reaction.[2] On the other hand, both **5** and **6** reacted under the usual Sakurai conditions to give the homoallylic alcohols **7** in good to excellent yields. Even though the ee of the products **7** remained modest, the results demonstrated nevertheless that a chiral auxiliary attached to silicon can indeed be used to induce enantioselectivity.

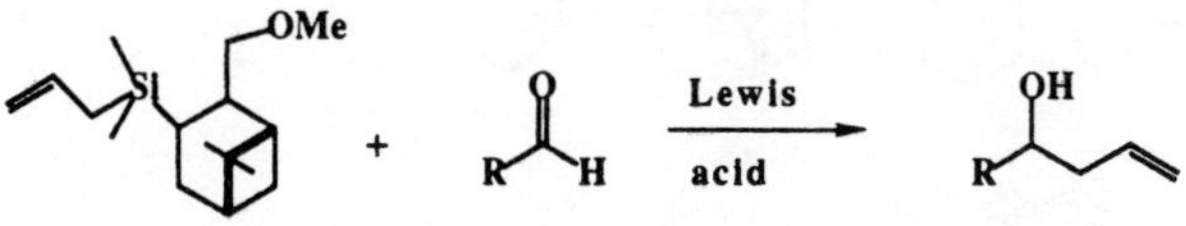

6   21-56%ee

The modest stereoselectivity obtained in these reactions is not totally unexpected if one considers the mechanism generally accepted for the reactions of allylsilanes with carbonyl compounds. From the work of Kumada[7] and Fleming[8], it has been concluded that the reaction proceeds through an antiperiplanar transition state **8**. That being the case, it would not be surprising that any chiral auxiliary, either at silicon or attached to silicon, would have little influence on the stereochemical outcome of the reaction. On the other hand, the work of Denmark[9] suggests that the synclinal transition state **9a** or **9b** may also be operative under certain conditions. It is possible that in **9b**, the silyl group may have a greater influence on the stereochemistry of the reaction. An approach to favour the synclinal transition state **9b** is to have ligands on the silyl moiety which can coordinate with the Lewis acid.

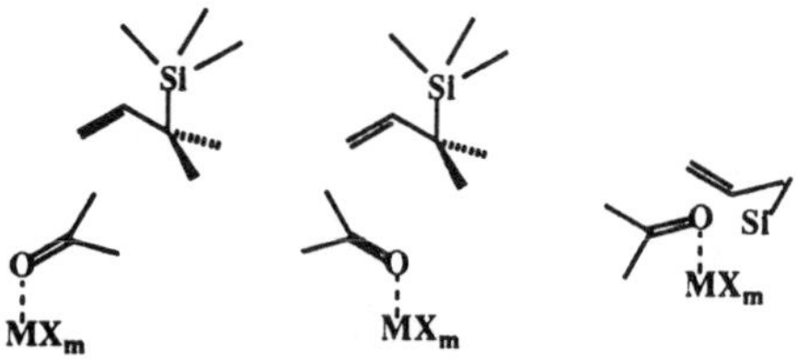

We have prepared a number of alkoxyallylsilanes **10** in which the alkoxy group was derived from readily available optically active alcohols.[10] These optically active alkoxyallylsilanes reacted with aldehydes and $BF_3 \cdot OEt_2$ to give the homoallylic alcohols **7** with ee in the range of 18-23%. While the enantioselectivity was still modest, it was an improvement over the pinanylsilane **5** system. Since the alkoxysilane **10c** was structurally similar to **5**, we attributed the improved stereoselectivity of the alkoxysilanes to the coordinating ability of the alkoxy group to the Lewis acid, thus relatively favouring the synclinal transition state **9b**.

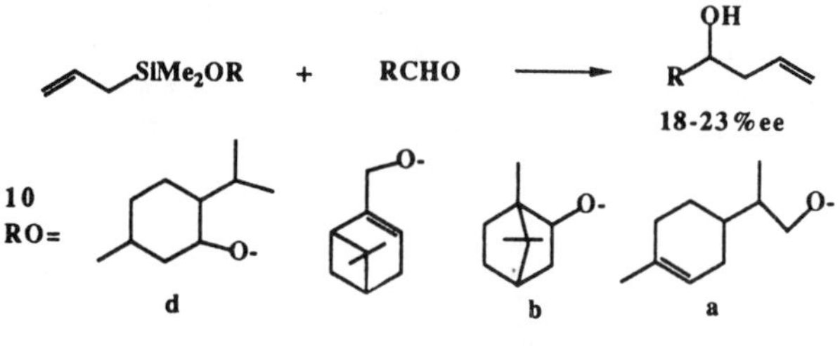

Better coordinating ligands were introduced onto the silyl group in the form of a series of pyrrolidinylmethyl-allylsilanes **11** prepared[11] from the bromomethylallylsilane **12**.

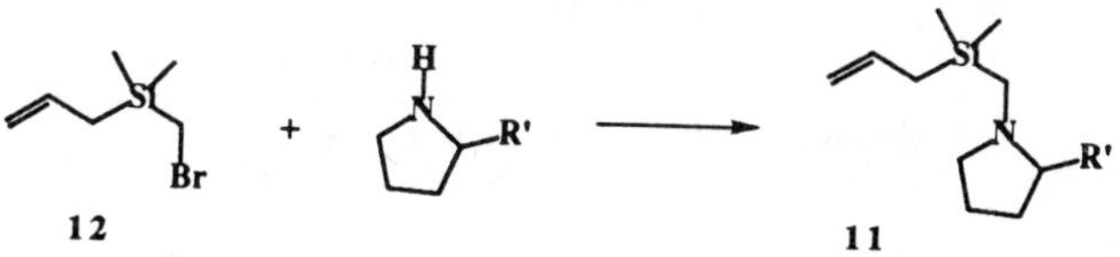

The ee of the homoallylic alcohol products **7** from the reaction of **11a-c** with aldehydes was much improved (up to 50%) relative to those from **5** and **10**.

11 a, R'=CH$_2$OMe
b, R'=CO$_2$Me
c, R'=CH$_2$CONHPh
d, R'=H

28-50% ee

(S)-

Coordination of the oxygen function was critical to the reaction. Compound **11d**, which lacks the oxygen function, failed to give any homoallylic alcohol product **7** under identical reaction conditions. Another interesting observation is that as the ratio of Lewis acid (e.g. TiCl$_4$) to aldehyde was increased from 1 to 10, the chemical yield of the product **7**, within a fixed period of time, was increased, but the optical yield was decreased. This was demonstrated for compounds **11b** as shown in Table 1.

This observation is consistent with the possibility that at lower Lewis acid concentrations, the Lewis acid coordinates with both the oxygen ligand in **11** and with the aldehyde, leading to more of the synclinal transition state **9c** as suggested by Denmark. At higher Lewis acid concentrations, the antiperiplanar transition state **8'** predominates with different molecules of Lewis acid coordinating separately to the aldehyde and to the oxygen ligand in **11**, thus accounting for the faster rate of reaction but lower ee of the product **7**. If this is indeed the correct explanation, further structural modification on the silyl moiety to enhance the synclinal transition state **9c** pathway can be envisaged. We are continuing to explore such modifications.

Table 1 Reaction of 11 with aldehydes under Lewis acid conditions

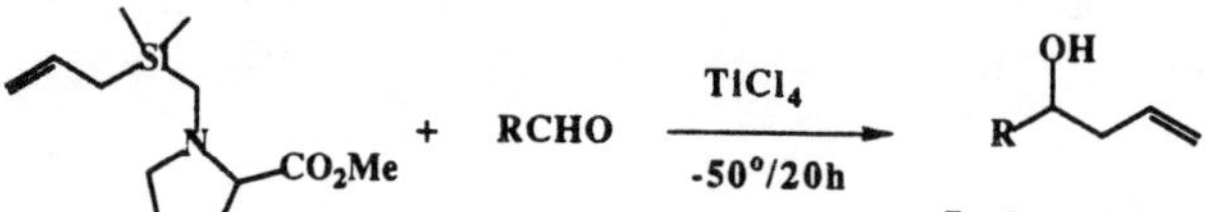

[Structure: allyl-Si-pyrrolidine-CO$_2$Me] + RCHO $\xrightarrow[-50°/20h]{TiCl_4}$ [Structure: R-CH(OH)-CH$_2$-CH=CH$_2$]

7, R=n-C$_8$H$_{17}$-

Ratio L.A./RCHO	chemical yield of 7, %	optical yield of 7, % ee	Abs. config
1:1	< 5	-	-
2:1	60	43	(S)
3:1	61	43	(S)
10:1	79	28	(S)

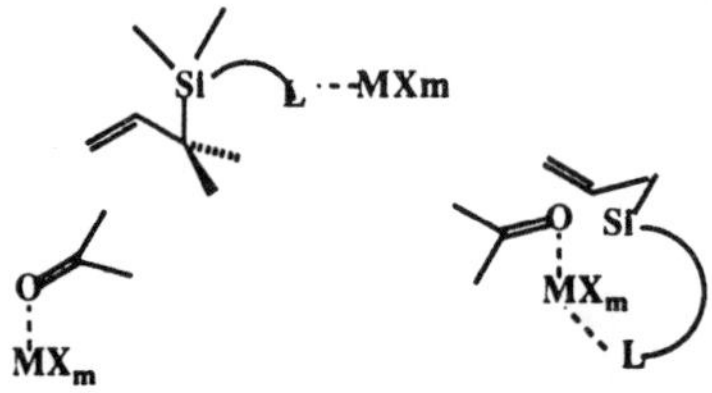

8'   9c

3 REACTIONS OF CHIRAL α-SILYLCARBANIONS.

α-Silylcarbanions are useful intermediates in organic synthesis. Recently we prepared the chiral organosilicon compound **13** from dimethyl(chloromethyl)benzylsilane (**14**) and (S)-(+)-2-(methoxymethyl)pyrrolidine (**15**) (Scheme 2).[12]

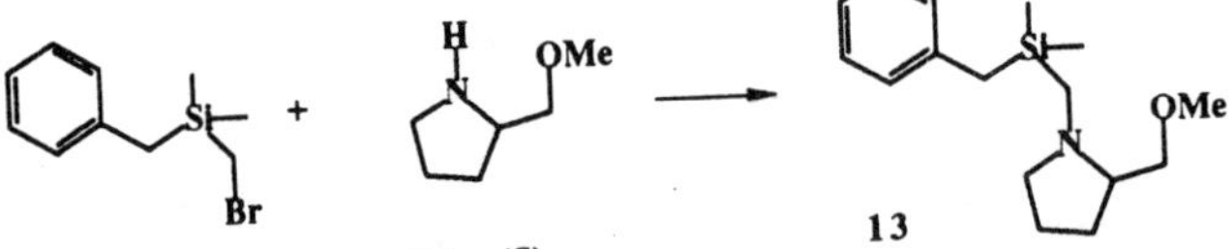

14   15 (S)-   13

Treatment of **13** with s-butyllithium gave the carbanion **16**, which on quenching with methyl iodide gave the alkylated product **17a** in good yield. The diastereoselectivity depended very much on the solvent used for the reaction. In THF, the diastereomeric excess (de) was found to be 50%, but in ether, compound **17a** was obtained as a single diastereomer (de>95%). Similar alkylation of **16** with other alkyl halides in ether gave the alkylated products **17** with equally high diastereoselectivity (Table 2).

Table 2

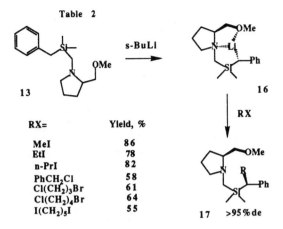

RX=	Yield, %
MeI	86
EtI	78
n-PrI	82
PhCH$_2$Cl	58
Cl(CH$_2$)$_3$Br	61
Cl(CH$_2$)$_4$Br	64
I(CH$_2$)$_5$I	55

Oxidative cleavage[13] of the carbon-silicon bond in **17a** (de>95%) with H$_2$O$_2$ and KHCO$_3$ gave (S)-(-)-phenylethanol of 98.5% ee. In one case (**17**, R=n-Pr), the arylcarbinol obtained was essentially optically pure according to capillary gas chromatography using a chiral column under conditions in which a lower limit of 0.4% of either of the enantiomers would have been detected.

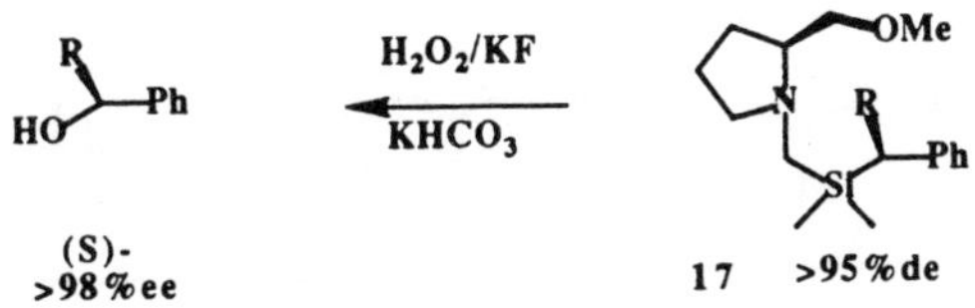

R=Me, 98.5% ee
R=Et, 99.0% ee
R=n-Pr, >99.5% ee

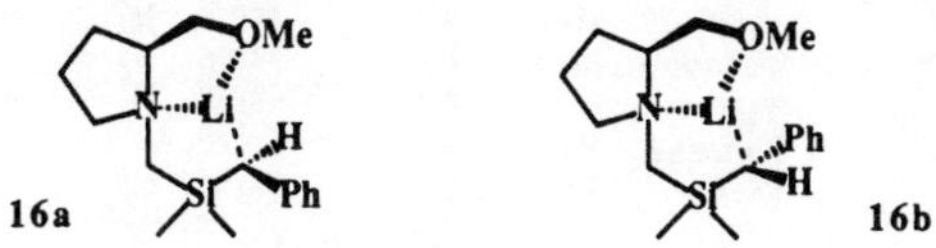

We attribute the stereochemical results in the following manner. Since it is well established that oxidative cleavage of the carbon-silicon bond occurs with retention of stereochemistry,[13] the alkylated compound **17** must have the (S)- configuration at the benzylic carbon as well. The carbanion **16** is most likely to have the lithium ion coordinated to both the nitrogen and the oxygen atoms of the pyrrolidine ligand as in **16a** or **16b**. To account for the stereochemical outcome of **17**, the alkyl halide can react with either **16a** with retention or **16b** with inversion. At the moment, we have no evidence to favour one pathway over the other.

In order to probe the factors controlling the stereoselectivity, and to extend the synthetic utility to other α-silylcarbanions, we have examined[14] carefully the reactions of the silylcinnamyl carbanions **18**. Methylation of the parent trimethylsilylcinnamyl carbanion **18a** in ether gave a 1:1 mixture of the α- and γ-products **19** and **20**. Replacement of one of the methyl groups on silicon by a pyrrolidinylmethyl group as in **18b** did not change significantly the α/γ ratio of methylation. This suggests that the pyrrolidine group by itself does not chelate significantly with the lithium ion to affect the regioselection of the reaction. A bidentate ligand, such as the O-methyl-(-)-ephedrine group in **18c** or (S)-(+)-2-(methoxymethyl)pyrrolidine in **18d** changed considerably the α/γ methylation ratio in ether strongly in favour of the α-isomer. Since we have established previously that α-regioselection in the alkylation of α-silylallyl anions is controlled by intramolecular chelation,[15] similar internal chelation likely exists in **18c** and **18d** as well. The stereoselection of the reaction depends however on the ligand used. Using ether as the reaction solvent, a much lower de was observed for **18c** than for **18d**. It is

$$Ph\diagdown Si_Y \xrightarrow{s\text{-BuLi}} Ph\diagdown Si_Y\text{Li} \xrightarrow{RX} Ph\diagdown Si_Y + Ph\diagdown Si_Y$$
  18   **19** R   **20** R

Table 3

	Solvent	RX	yield, %	α (de %, conf.)	γ (de %, conf.)
18a	ether	MeI		50	50
18b	ether	MeI		44	56
	THF	MeI		46	54
18c	ether	MeI	98	79(16, R)	29(10)
	THF	MeI	98	40(30, R)	60(0)
18d	ether	MeI	92	81(92, S)	19(66, R)
	THF	MeI	99	45(28, R)	55(98, R)
	Toluene	MeI	89	94(95, S)	6(80, R)
	THF	EtI	92	42(30, R)	58(98, R)
	Toluene	EtI	81	79(95, S)	20(90, R)
	ether	EtI	87	68(95, S)	19

interesting to note that for **18d**, diastereoselectivity was high for the α-isomer (de>95%) but only modest for the γ-isomer (de~50%). On the other hand, when the alkylation was carried out in THF, the regioselection was changed to favour slightly γ-alkylation. Furthermore, a low de was now observed for the α-isomer but a higher de for the γ-isomer. Even more interesting is the use of toluene as solvent. Not only was regioselection improved in favour of the α-isomer, the diastereoselection was also improved for both the α- and the γ-isomers. (Table 3)

From a synthetic point of view, since the α- and the γ-isomers could be separated by column chromatography, the present reaction offers a facile method for the synthesis of these chiral molecules. The carbon-silicon bond can likewise be cleaved under oxidative conditions to give the corresponding alcohols or carboxylic acids with equivalent optical purities.

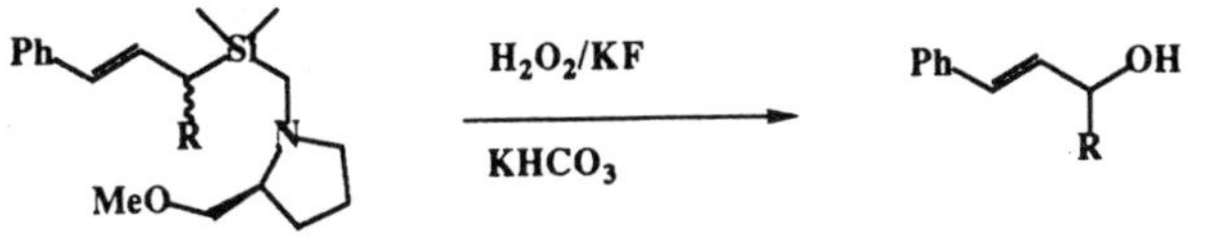

$$\text{19} \xrightarrow[\text{KHCO}_3]{\text{H}_2\text{O}_2/\text{KF}} \text{Ph} \diagup \text{OH}$$

R=Me, 94% ee
R=Et, 92% ee

$$\text{20} \xrightarrow[\text{MeCN/H}_2\text{O/CCl}_4]{\text{NaIO}_4/\text{RuCl}_3} \text{Ph} \diagup \text{CO}_2\text{H}$$

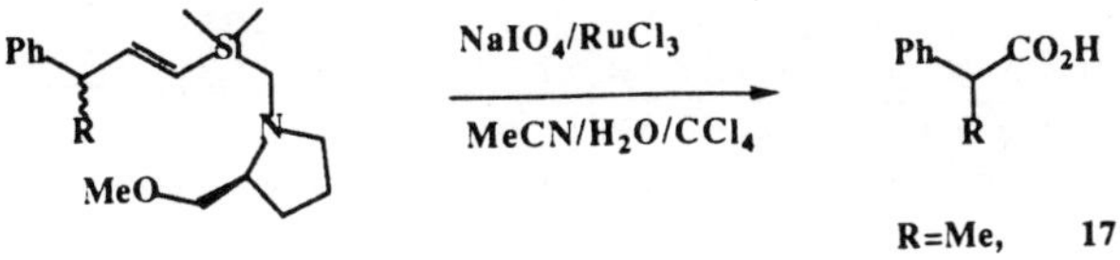

R=Me, 17% ee
R=i-Pr, 64% ee

4 ENANTIOSELECTIVE SYNTHESIS OF EPOXIDES *VIA* SHARPLESS EPOXIDATION OF ALKENYLSILANOLS.

The above reactions demonstrate the use of chiral organosilicon compounds for enantioselective synthesis. Equally useful in synthesis is the use of organosilicon compounds in catalytic enantioselective processes. The Sharpless epoxidation, since its discovery, has been hailed as one of the milestones in organic chemistry.[16] The reaction gives chiral epoxides with predictable stereochemistry and high enantiomeric purity. A limitation of the Sharpless reaction is the fact that the olefinic substrate must have an allylic hydroxy group. The hydroxy function is essential to assembling the substrate and reagents together and thus accelerating the reaction as well as conferring high stereoselectivity. For simple alkenes without the pendant hydroxy group, the Sharpless epoxidation is not useful. Recently, we proposed the use of alkenylsilanol as an allylic alcohol equivalent, and since the silyl group can be easily protodesilylated, it also serves as an alkene equivalent as well. The overall process is therefore an enantioselective synthesis of simple epoxides (Scheme 3).

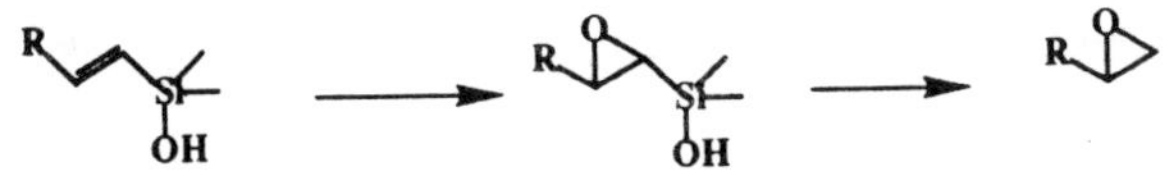

Scheme 3

The synthesis of optically active styrene oxide serves as a good example. The β-styryldimethylsilanol (**21**) was prepared by controlled $Ag_2O/AgNO_3$ oxidation of the silane **22** which was in turn obtained from β-styryllithium and dimethylchlorosilane. Epoxidation of **21** under the usual Sharpless conditions [Ti(O-i-Pr)$_4$, (+)-diethyl tartrate, t-BuOOH in CH_2Cl_2, -20°] gave the epoxysilanol **23** and the corresponding disiloxane **24**. Protodesilylation of both **23/24** gave (S)-styrene oxide in >85% ee. Several features of this reaction sequence are worthy of note. First of all, if **21** dimerised to the disiloxane **25** prior to epoxidation, then no epoxidation was observed. The hydroxy function on the silanol moiety was absolutely essential to the success of the reaction. Secondly, the absolute configuration of the epoxysilanol **23** should be the same as the styrene oxide obtained since we had demonstrated previously that the protodesilylation step proceeded with retention of stereochemistry.[18] The stereochemical course of the reaction of alkenylsilanol is the same as expected from the allylic alcohol model.[19] One can conclude therefore the same reaction mechanism operates in both cases.

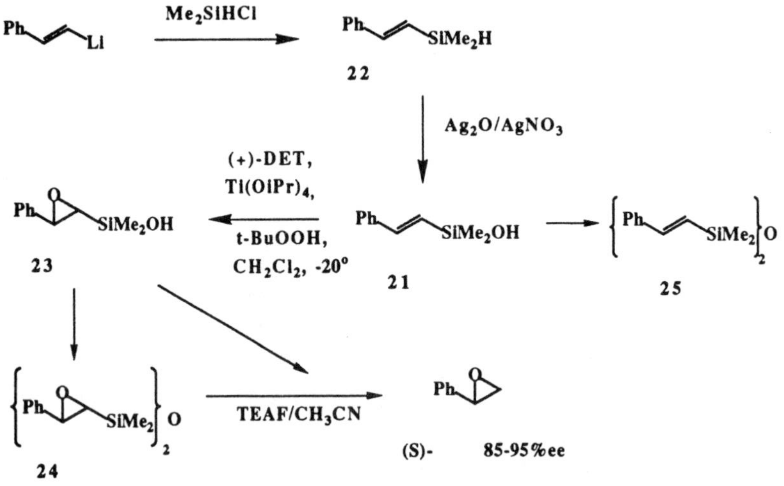

The same approach can be extended to other simple alkenylsilanols. For example, hydroalumination of 1-heptyne followed by quenching with dimethylchlorosilane gave the alkenylsilane **27** which on oxidation gave the silanol **28**. Sharpless epoxidation [(+)-DIPT] of **28** followed by protodesilylation gave (−)-1,2-epoxyheptane with 85% ee (Scheme 4).

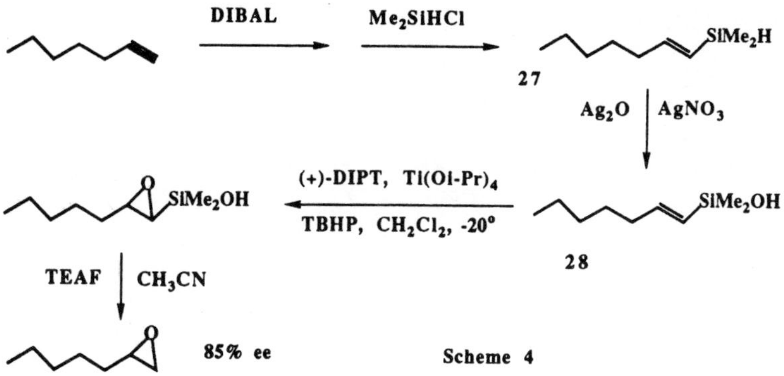

Scheme 4

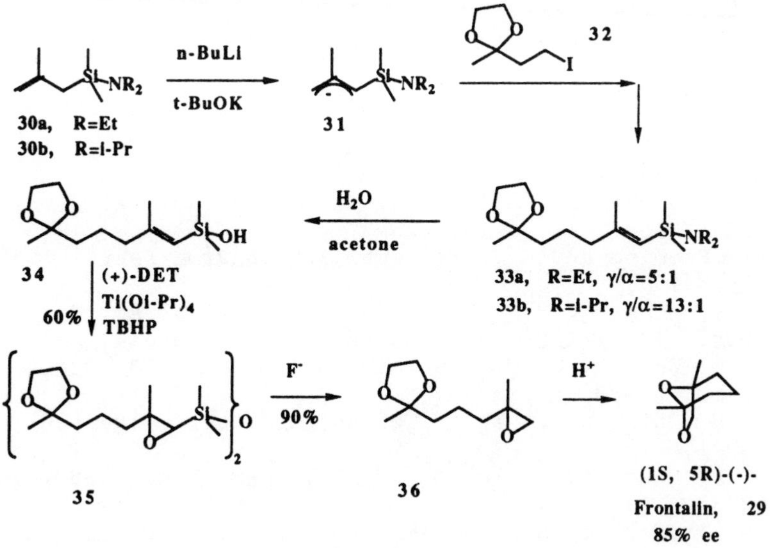

The synthetic utility of the reaction has been demonstrated[20] by the synthesis of (1S, 5R)-(−)-Frontalin (**29**), the pheromone isolated from the male western pine beetle.[21] The aminoallylsilane **30** was reacted with n-BuLi/KO-t-Bu to give the anion **31** which was alkylated with the halide **32** to give regioselectively[22] the γ-isomer **33**. The γ-selectivity was enhanced with the use of the more bulky di-i-propylamino group. Aqueous hydrolysis of **33** gave the silanol **34** which was epoxidised under Sharpless conditions [(+)-DET] to give the epoxysilane **35**. Fluoride ion mediated protodesilylation of **35** gave the epoxy compound **36** with 85% ee. Acid hydrolysis of the acetal followed by intramolecular cyclisation gave (1S, 5R)-(-)-Frontalin **(29)** with 85% ee.

5 CONCLUSION

Results obtained thus far demonstrate that organosilicon compounds can play a useful role in enantioselective synthesis. In view of the tremendous versatility of the silyl group in mediating many organic reactions, one can foresee continuing development in this area.

6 ACKNOWLEDGEMENT

Financial support of this research by NSERC and FCAR is gratefully acknowledged.

REFERENCES

1. A.G. Brook, J.M. Duff and D.G. Anderson, J. Am. Chem. Soc., 1970, 92, 7567.
2. S.J. Hathaway and L.A. Paquette, J. Org. Chem., 1983, 48, 3351.
3. M.E. Jung and K.T. Hogan, Tetrahedron Lett., 1988, 29, 6199.
4. A. Hosomi and H. Sakurai, Tetrahedron Lett., 1976, 1295.
5. D. Wang and T.H. Chan, Tetrahedron Lett., 1983, 24, 1573.
6. L. Coppi, A. Mordini, M. Taddei, Tetrahedron Lett., 1987, 28, 969.
7. T. Hagashi, M. Konishi, H. Ito and M. Kumada, J. Am. Chem. Soc., 1987, 109, 2512.
8. I. Fleming and N.V. Terrett, Tetrahedron Lett., 1983, 24, 4153.
9. S.E. Denmark, B.R. Henke and E. Weber, J. Am. Chem. Soc., 1987, 109, 2512 and references therein.
10. Z.W. Wei, D. Wang, J.S. Li and T.H. Chan, J. Org. Chem., 1989, 54, 5768.
11. T.H. Chan and D. Wang, Tetrahedron Lett., 1989 30, 3041.
12. T.H. Chan and P. Pellon, J. Am. Chem. Soc., 1989, 111, 8737.
13. K. Tamao, T. Takui and M. Kumada, J. Am. Chem. Soc., 1978, 100, 2268.
14. T.H. Chan and S. Lamothe, to be published.

15. T.H. Chan and R.F. Horvath, *J. Org. Chem.*, 1989, **54**, 317.
16. T. Katsuki and K.B. Sharpless, *J. Am. Chem. Soc.*, 1980, **102**, 5974.
17. T.H. Chan, L.M. Chen and D. Wang, *J. Chem. Soc., Chem. Commun.*, 1988, 1280.
18. T.H. Chan, P.W.K. Lau and M.P. Li, *Tetrahedron Lett.*, 1976, 2667.
19. K.B. Sharpless, *Chem. Tech.*, 1985, 692.
20. T.H. Chan, L.H. Li and D. Wang, to be published.
21. G.W. Kinzer, A.F. Fentiman, Jr., T.F. Page, R.L. Foltz and J.P. Vite, *Nature*, 1969, **221**, 447.
22. T.H. Chan and K. Koumaglo, *J. Organomet. Chem.*, 1985, **281**, 109.

β-Lactams from Allylsilanes

E.W. Colvin, M.A. Loreto, M. Monteith, and I. Tommasini
DEPARTMENT OF CHEMISTRY, UNIVERSITY OF GLASGOW,
GLASGOW G12 8QQ, UK

1 INTRODUCTION

Prior to 1970, β-lactam antibiotic research concentrated mainly on the penicillins and cephalosporins. Since that time, a dramatic expansion in the area has taken place with the discovery of a new generation of β-lactams,[1] showing antibiotic and/or β-lactamase inhibitory properties. A selection of some biologically important β-lactam types is shown in the Figure.

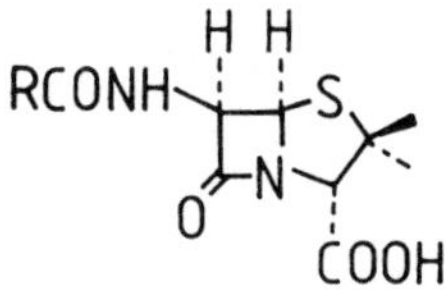

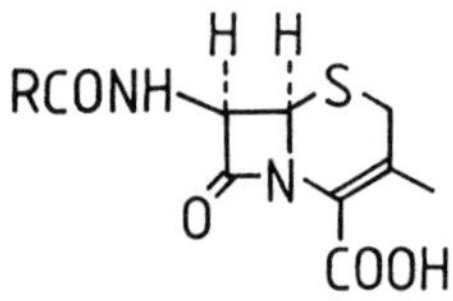

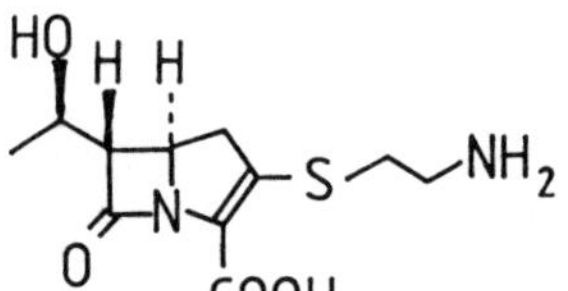

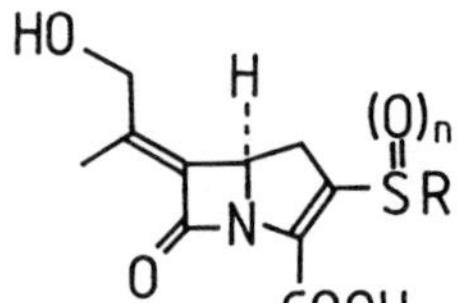

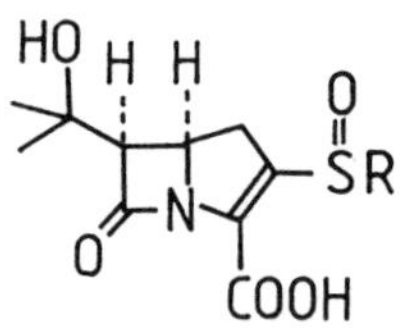

Figure   Selection of β-lactams

The top two structures are representative of the
familiar penicillins (1) and cephalosporins (2),
respectively. Thienamycin (3), the asparenomycins (4)
and carpetimycins (5), as carbapenems, represent one
group of this new generation of antibiotics and
β-lactamase inhibitors. With the discovery of these new
natural products, substantial efforts have been devoted
to the synthesis of β-lactams of varying structures.
Most of the successful syntheses of bicyclic β-lactams
involve the early creation of a monocyclic β-lactam,
followed by substituent elaboration and ultimate
cyclisation to form the second ring.

There are three general routes[2] to monocyclic β-
lactams: imine + ketene formal cycloaddition, cyclisation
of β-aminoësters via the metallated amine, and alkene +
isocyanate formal cycloaddition (Scheme 1).

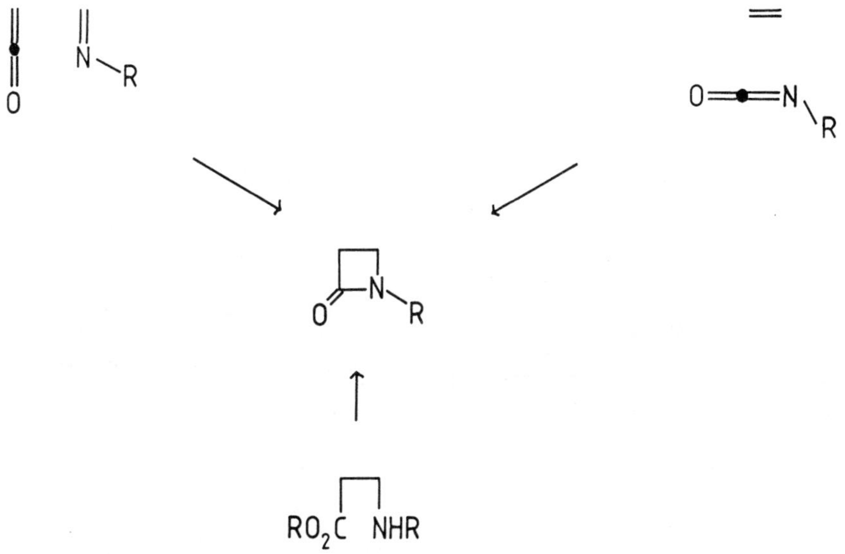

Scheme 1   Routes to monocyclic β-lactams

We have published our results[3] on a combination of
the first two routes. The topic of this lecture concerns
application of the third route, i.e., alkene + isocyanate
cycloaddition. The results presented are preliminary
details of an on-going study.

2 β-LACTAMS FROM ALLYLSILANES

The reaction of chlorosulphonyl isocyanate (CSI) with
functionalised alkenes, in particular with vinyl esters,
has proven of great utility in the preparation of monocyclic β-lactams. Two mechanisms for the reaction
between alkenes and CSI have been proposed (Scheme 2).

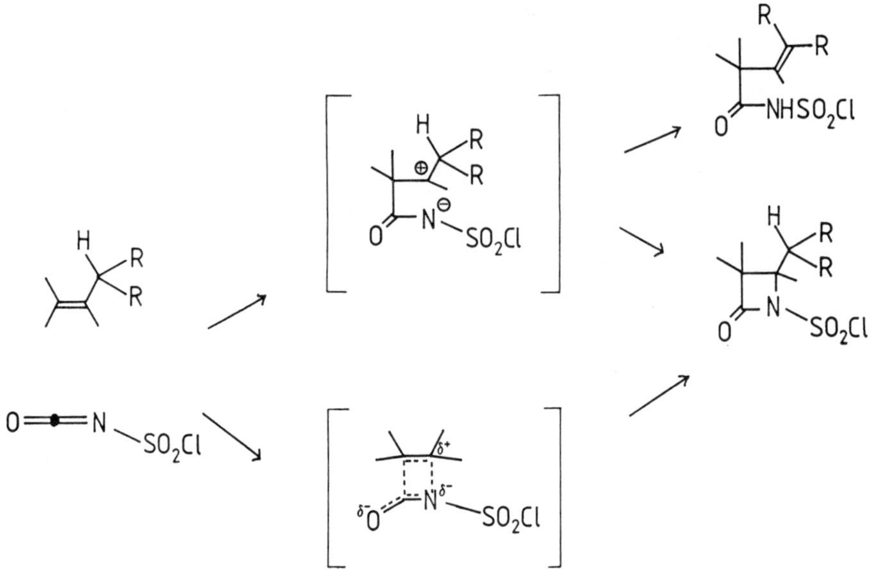

Scheme 2   Mechanisms of cycloaddition

Graf, the original investigator of this reaction, proposed[4] a two-step process, with an intermediate zwitterion, to account for the simultaneous formation in many cases of β-lactam and unsaturated amide (which could be the direct product of an 'ene' reaction) in ratios largely unaffected by change in reaction conditions. Moriconi, on the other hand, favoured[5] a near-concerted, thermally allowed [$_π2_s$ + $_π2_a$] cycloaddition, to account for the observation that simple (Z) and (E) alkenes underwent stereospecific cycloaddition. Alternatively, if the Graf two-step mechanism is correct, the lifetime of the zwitterionic intermediate must be significantly shorter than the rotational time about the developing carbonium ion.

We were intrigued by an early report[6] by Dunoguès on the reaction between allylsilanes and CSI. Allyltrimethylsilane was reported to produce the silyl imidate (6) directly (Scheme 3); however, using the dimethylallylsilane (7), an intermediate β-lactam (8) was detected by i.r. and ^{1}H n.m.r. spectroscopy; this unisolated intermediate was reported to rearrange in solution over a period of one hour at ambient temperature to the acyclic imidate ester (9), by an implied intramolecular silatropic shift. Indeed, Fleming has made good use of this process in a key step in a synthesis[7] of loganin aglycone.

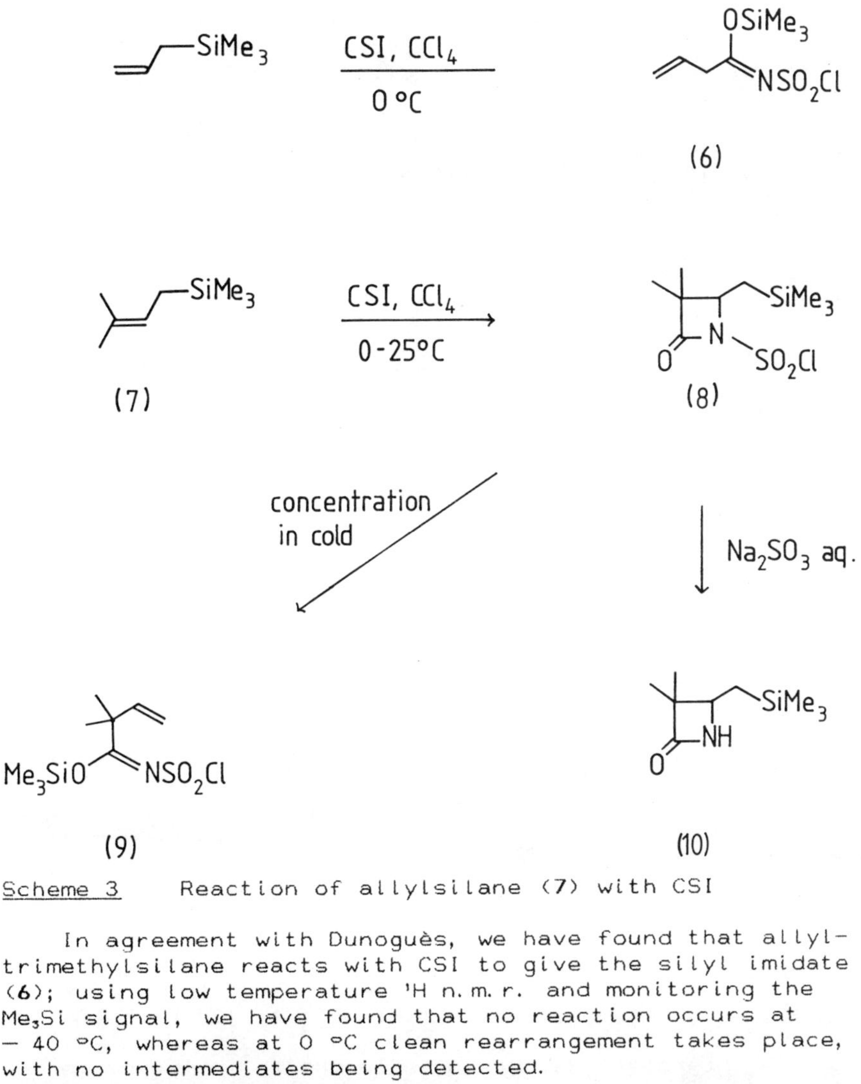

Scheme 3 Reaction of allylsilane (7) with CSI

In agreement with Dunoguès, we have found that allyl-trimethylsilane reacts with CSI to give the silyl imidate (6); using low temperature ^{1}H n.m.r. and monitoring the Me$_3$Si signal, we have found that no reaction occurs at −40 °C, whereas at 0 °C clean rearrangement takes place, with no intermediates being detected.

On the other hand, with the dimethylallylsilane (7) the intermediate N-chlorosulphonyl β-lactam (8) can be readily intercepted by *in situ* treatment with aqueous sodium sulphite[8] to afford 3,3-dimethyl-4-(trimethyl-silylmethyl)azetidin-2-one (10) as a white crystalline solid, m.p. 87-88 °C, in greater than 60% yield. Indeed, the intermediate N-chlorosulphonyl β-lactam (8) is stable in solution for at least 24 h at ambient temperature, but it does rearrange on concentration, even in the cold, to the imidate ester (9), suggesting a bimolecular pathway for this rearrangement. As a possible explanation for these differing results, it should be noted that Dunoguès operated at ≃ 3.5 M̄, Fleming at ≃ 2 M̄, whereas we used much more dilute, ≃ 0.2 M̄, conditions.

The butenylsilanes (11) and (12) were prepared (Scheme 4) as a mixture of geometric isomers of the terminal and of the internal silanes [ratio (g.l.c.) 3.73:1]. Reaction of this mixture of allylsilanes with CSI followed by cleavage gave the cis- and trans-β-lactams (13) and (14), accompanied by the lactam (15) and lactone (16) (as single diastereoisomers in the last two cases, assigned trans).

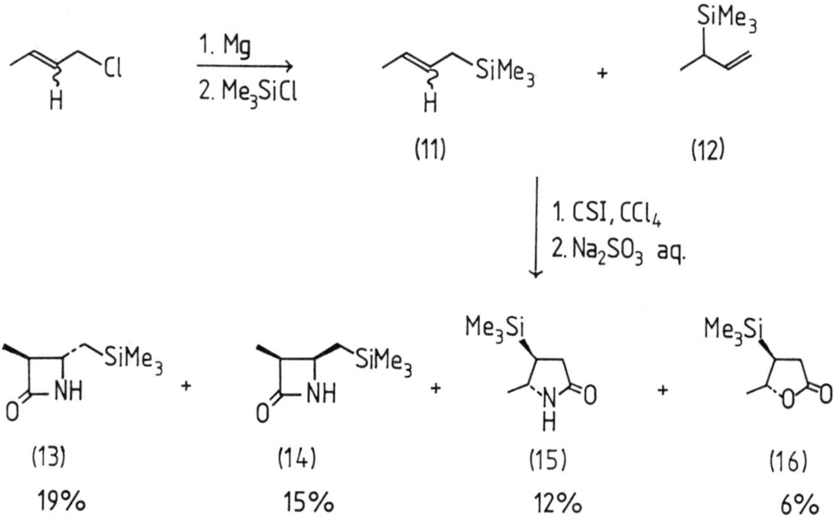

Scheme 4 Reaction of butenylsilanes (11/12) with CSI

We interpret these results in the following manner. The β-lactams have come from the terminal silanes; the lactam (15) and lactone (16), on the other hand, have come from the internal silane, possibly by initial attack (Scheme 5) to give the β-silyl carbonium ion, followed by a 1,2-migration of the silyl group (or possibly via a silyl-bridged intermediate), and cyclisation by attack by the delocalised anion through either nitrogen or oxygen.

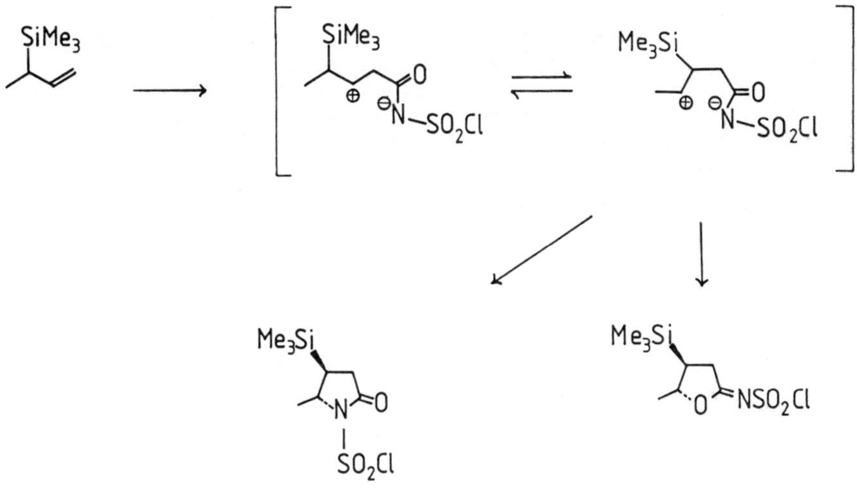

β-Lactams from Allylsilanes

Scheme 5 Formation of lactam and lactone

We are studying this reaction with a range of allylsilanes to establish its scope and utility. However, allylsilanes usefully functionalised in this context are relatively uncommon, and one is still left with saturated alkyl groups at C-3 and trimethylsilyl-methyl groups at C-4. We are therefore extending this study in two complementary directions, by varying the allyl substitution to provide more useful functionality at C-3, and by varying the silyl substitution to allow oxidative cleavage of the C-Si bond, resulting in the overall introduction of a hydroxymethyl or oxidatively related group at C-4.

3 β-LACTAMS FROM (ALLENYLMETHYL)SILANES

To generate more useful functionality at C-3, (allenylmethyl)silanes suggest themselves as ideally functionalised candidates: on cycloaddition (Scheme 6) introduction of alkylidene substitution would produce potential asparenomycin (4) and carpetimycin (5) precursors. The 3-alkylidene functionality can be transformed[9] into the carpetimycin side chain.

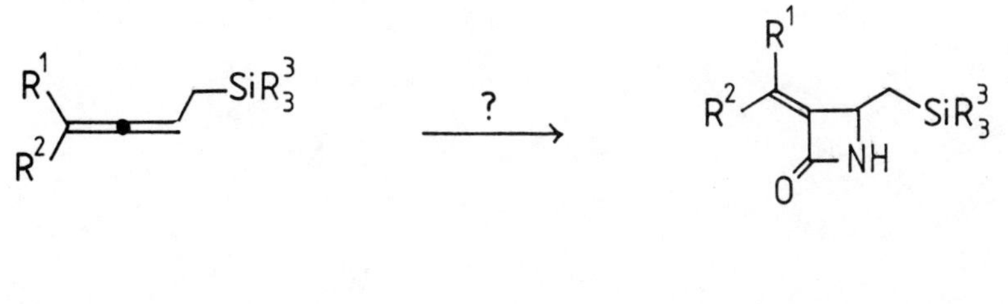

Scheme 6 Reaction of (allenylmethyl)silanes with CSI

(Allenylmethyl)trimethylsilanes are readily prepared[10] by reaction between [(trimethylsilyl)methyl]-copper(I) species and derivatives of propargylic alcohols. For example, the dimethylallene (**17**), prepared as shown (Scheme 7), reacted with CSI to give the crystalline 3-alkylidene β-lactam (**18**), m.p. 118-120 °C, in 22% yield.[11]

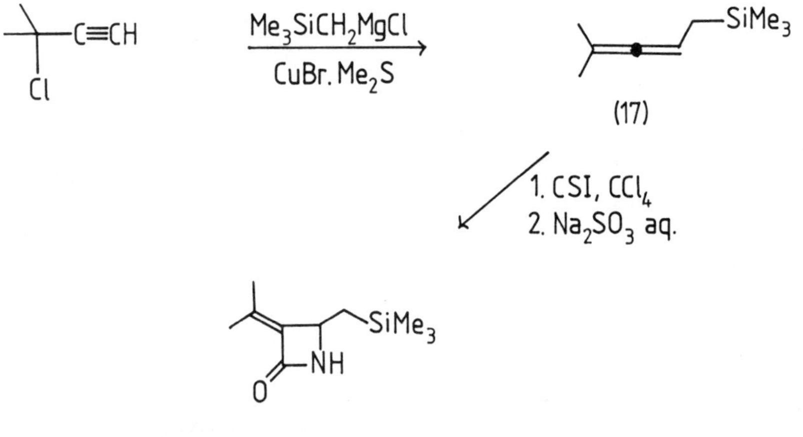

Scheme 7   Reaction of dimethylallene (**17**) with CSI

The monomethylallene (**19**) reacted similarly (Scheme 8), albeit in ≈10% yield, to give the β-lactams (**20**) and (**21**) as a 4:1 (E):(Z) mixture.

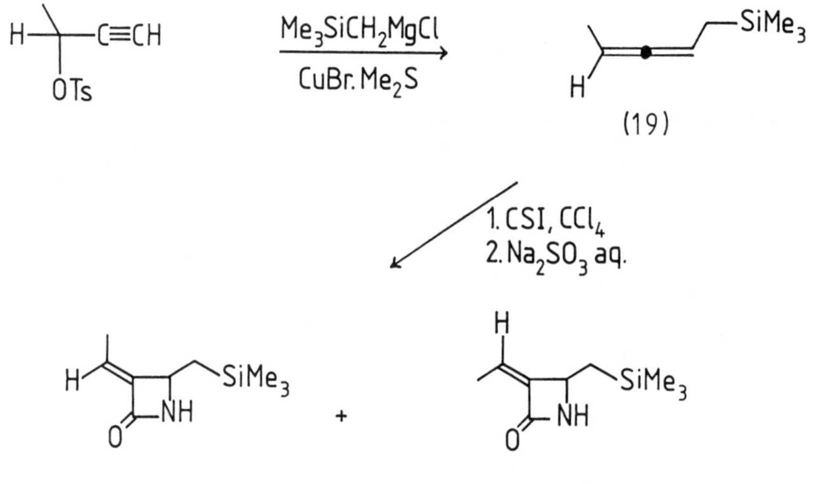

Scheme 8   Reaction of monomethylallene (**19**) with CSI

We have prepared the hydroxymethylallenylsilane (22) by a related protocol (Scheme 9). Reaction of chloroacetone with ethynylmagnesium bromide, followed by treatment of the chlorohydrin product with powdered KOH in ether, provided[12] the epoxybutyne (23). Reaction[13] of this with the cuprate shown gave the hydroxymethylallene (22).

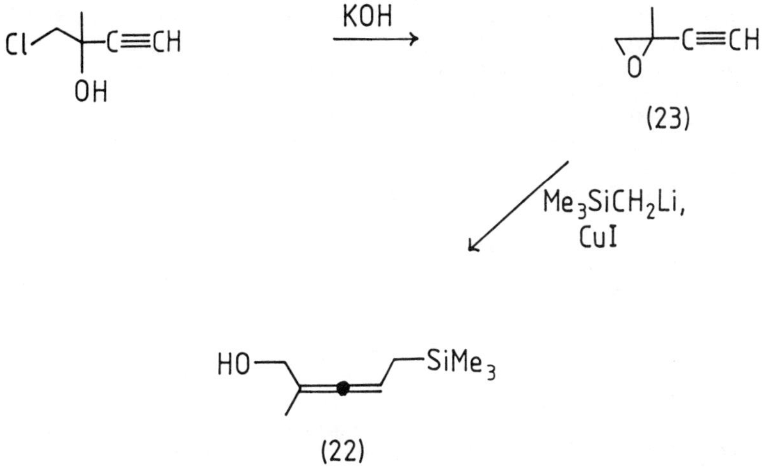

Scheme 9   Preparation of hydroxymethylallene (22)

Cycloaddition using the corresponding t-butyl-dimethylsilyl ether provided (Scheme 10) the β-lactams (24) and (25) in 21% yield as a crystalline 7:1 (E):(Z) mixture, i.e., the major isomer possessing the precise C-3 alkylidene functionality of the asparenomycins.

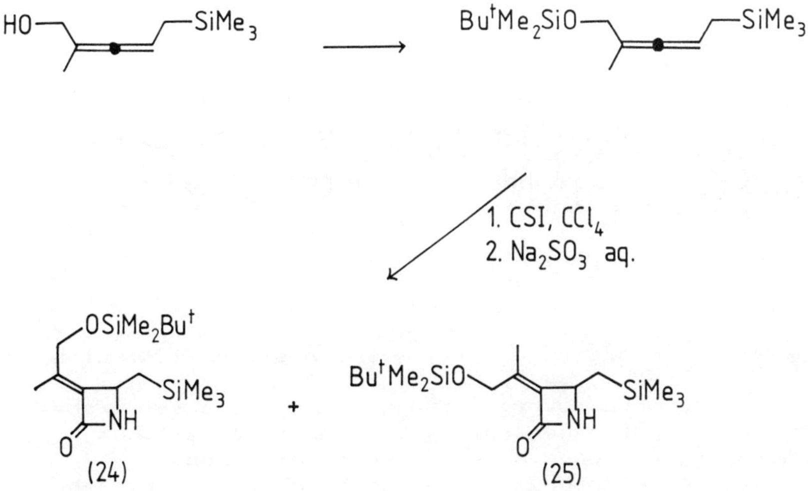

Scheme 10 Reaction of hydroxymethylallene (22) with CSI

4 β-LACTAMS FROM ALLYL(PHENYLDIMETHYL)SILANES

Oxidative cleavage of a carbon-silicon bond requires that the silane must carry at least one electronegative substituent, such as alkoxy or fluoro.[14] Either excess hydrogen peroxide or MCPBA may be used as oxidant, with fluoride ion often being a mandatory additive in what is believed to be an assisted rearrangement of a silyl peroxide. Phenyldimethylsilyl groups can be cleaved oxidatively to hydroxyl groups by a sequence of protiodesilylation using HF equivalents, followed by oxidation.[15] These cleavage conditions do not affect β-lactam sub-units.[16] Alternatively, such a substituent requirement can be fulfilled at an early stage, using alkoxysilane reagents, such as isopropoxydimethylsilyl[17] species, from the outset.

The allyl/vinyldisilane[18] (26) reacted smoothly with CSI to provide the crystalline <u>trans</u> β-lactam (27), m.p. 67-68 °C, in 55% yield (Scheme 11), suitably functionalised at C-3 for Peterson olefination [as in the reported[19] synthesis of (-)-asparenomycin C]. Further, this β-lactam underwent quantitative desilylation at C-3 on treatment with KF/MeCN, providing access to the otherwise unobtainable (by this protocol) 3-unsubstituted β-lactam (28). This provides a masked form of the synthetically useful 4-hydroxymethylazetidin-2-one.

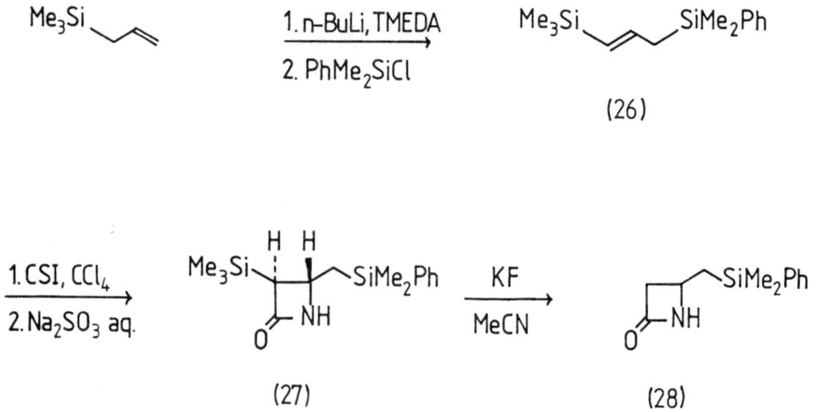

Scheme 11 Reaction of allyl/vinyldisilane (26) with CSI

The regiochemistry of the above cyclisation processes must be under the control of the β-effect, silicon encouraging the development of carbonium ions or partially developed such species β to it, and yet the silyl group is not lost. If a two-step, zwitterionic mechanism[4] holds in such cases, this is quite remarkable, since it has been clearly demonstrated[20] that electrophilic attack on allylsilanes normally leads to silyl loss with the formation of substituted products with a net double bond shift, *via* an intermediate β-silyl cation.

We plan to extend these potentially useful observations, in particular by constructing suitably functionalised homochiral allenes with either phenyldimethylsilylmethyl or isopropoxydimethylsilylmethyl substituents.

We thank the S.E.R.C. for financial support (studentships to M.M. and I.T.), and Professor Goré for experimental details for the preparation of (allenylmethyl)silanes.

REFERENCES

1. R. Southgate and S. Elson, Fortschr. Chem. Org. Naturst., 1985, 47, 1; A.G. Brown, Pure Appl. Chem., 1987, 59, 475.
2. G.A. Koppel, 'The Synthesis of the β-Lactam Function', ch. 2, in 'Small Ring Heterocycles,' ed. A. Hassner, John Wiley, New York, 1983.
3. E.W. Colvin, D. McGarry, and M.J. Nugent, Tetrahedron, 1988, 44, 4157.
4. R. Graf, Liebigs Ann. Chem., 1963, 661, 111.
5 E.J. Moriconi and W.C. Crawford, J. Org. Chem., 1968, 33, 370.
6. G. Déléris, J. Dunoguès, and R. Calas, J. Organometal. Chem., 1976, 116, C45; G. Déléris, J.P. Pillot, and J.C. Rayez, Tetrahedron, 1980, 36, 2215.
7. I. Fleming and B.-W. Au-Yeung, Tetrahedron, 1981, 37, Supplement No. 1, 13.
8. T. Durst and M.J. O'Sullivan, J. Org. Chem., 1970, 35, 2043.
9. J.D. Buynak and M. Narayana Rao, J. Org. Chem., 1986, 51, 1571.
10 M. Montury, B. Psaume, and J. Goré, Tetrahedron Lett., 1980, 21, 163.
11. 2-Methylpenta-2,3-diene reacts with CSI in similar yield but with the opposite regiochemistry; E.J. Moriconi and J.F. Kelly, J. Org. Chem., 1968, 33, 3036.
12. N.M. Klyueva and I.A. Rubtsov, Chem. Abstr., 1965, 63, 17875c.
13. H. Kleijn and P. Vermeer, J. Org. Chem., 1985, 50, 5143.
14. K. Tamao, T. Kakui, M. Akita, T. Iwahara, R. Kanatani, J. Yoshida, and M. Kumada, Tetrahedron, 1983, 39, 983.
15. I. Fleming, R. Henning, and H. Plaut, J. Chem. Soc., Chem. Commun., 1984, 29.
16. D.A. Burnett, J.C. Gallucci, and D.J. Hart, J. Org. Chem., 1985, 50, 5120.
17. K. Tamao and N. Ishida, Tetrahedron Lett., 1984, 25, 4245.
18. I. Fleming and J.A. Langley, J. Chem. Soc., Perkin Trans. I, 1981, 1421.
19. K. Okano, Y. Kyotani, H. Ishima, S. Kobayashi, and M. Ohno, J. Am. Chem. Soc., 1983, 105, 7186.
20. I. Fleming, J. Dunoguès, and R. Smithers, Organic Reactions, 1990, 37, 57.

Some Synthetic Applications of Organosilicon Reagents to Organic Synthesis

Kiitiro Utimoto
DEPARTMENT OF INDUSTRIAL CHEMISTRY, KYOTO UNIVERSITY, YOSHIDA, KYOTO 606, JAPAN

1 INTRODUCTION

Organosilicon reagents have found important roles in organic synthesis.[1,2] One of these is cyanotrimethylsilane (**1**) which is considered as a stabilized hydrogen cyanide. Reaction can be performed under aprotic conditions by the use of **1** in place of HCN. Typically **1** is used in preparation of synthetically useful nitriles: (1) addition to hetero double bonds such as C=O and C=N giving NC-C-OSiMe$_3$ and NC-C-NSiMe$_3$ moieties which serve as protected carbonyl or as the precursors of hydroxylamines and amino acids; (2) conjugated addition to enones to give γ-cyano ketones; (3) substitution of active halogen producing RCOCN, N-CN, P-CN, and S-CN compounds, respectively; (4) addition to oxiranes affording 3-trimethylsiloxy nitriles; (5) homologation of ketones being achieved via the adducts of **1** to a carbonyl group; (6) reaction with S$_N$1 active chlorides to form nitriles.[3-10] In addition to the above described ability of introducing cyanide functionality into various substrates, ambident nucleophilicity of **1** was reported to give either a nitrile or an isocyanide from both the substitution reactions with tertiary alkyl chlorides and the addition to oxiranes.[4,5] Although the selective use of the ambident reactivity of **1** was suggested by the above observations, relations between the ambident reactivity and the reaction conditions, solvents and catalysts, have not been completely clarified. This paper describes controlled utilization of ambident reactivity of **1** to effect either isocyanosilylation or cyanosilylation of oxiranes **2**. Selective cyanation of aziridines is also described.

2 ISOCYANIDES FROM OXIRANES

Treatment of oxirane **2a** with **1** in the presence of a catalytic amounts of Pd(CN)$_2$ in refluxing dichloromethane for 2 h afforded 2-(trimethylsiloxy)ethyl isocyanide (**3a**) quantitatively. Cyclohexene oxide (**2e**) afforded trans-2-(trimethylsiloxy)cyclohexyl isocyanide (**3e**) exclusively by Pd(CN)$_2$-catalyzed reaction with **1**.[5] In place of

Pd(CN)$_2$, SnCl$_2$ and Me$_3$Ga were also effective catalysts. Although the formation of isocyanide **3e** from **2e** has been reported under the catalytic action of ZnCl$_2$ or ZnI$_2$,[4] the Pd(CN)$_2$- or Me$_3$Ga-catalyzed reactions proceeded under mild conditions to give **3e** in better yield.[5] Nitriles could not be detected in any of the reaction mixture. Except for methyloxirane (**2b**), 2-trimethylsiloxy isocyanides **3** are prepared from various oxiranes **2** by regioselective attack of isocyanide on the more substituted carbon with inversion of configuration (Scheme 1).

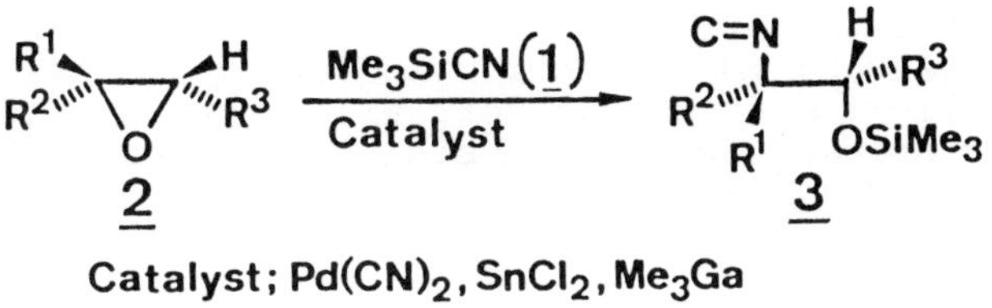

$$R^1R^2\text{-oxirane-}R^3H \xrightarrow{\text{Me}_3\text{SiCN (1)}}_{\text{Catalyst}} R^2R^1\text{C=N-}R^3\text{-OSiMe}_3$$

2 → **3**

Catalyst; Pd(CN)$_2$, SnCl$_2$, Me$_3$Ga

Scheme 1

Reaction on tertiary carbon atoms also proceeded stereospecifically. Results are shown in Table I.

Table I. Formation of Isocyanides **3** from Oxiranes **2**

Oxirane	R^1	R^2	R^3	Method[a]	Product[b] (yield %)	
2a	H	H	H	A	**3a** (99)	
2b	Me	H	H	A	**3b** (49),	**3k**[c] (49)
2b	Me	H	H	B	**3b** (41),	**3k** (41)
2b	Me	H	H	C	**3b** (49),	**3k** (49)
2c	Me	Me	H	A	**3c** (99)	
2c	Me	Me	H	B	**3c** (99)	
2c	Me	Me	H	C	**3c** (90)	
2d	H$_2$C=CH	H	H	A	**3d** (89)	
2d	H$_2$C=CH	H	H	B	**3d** (81)	
2d	H$_2$C=CH	H	H	C	**3d** (88)	
2e	H	-(CH$_2$)$_4$-		A	**3e** (99)	
2e	H	-(CH$_2$)$_4$-		B	**3e** (90)	
2e	H	-(CH$_2$)$_4$-		C	**3e** (100)	
2f	Me	-(CH$_2$)$_4$-		A	**3f** (99)	
2g	H	Et	Et	A	**3g** (84)	
2h	Et	H	Et	A	**3h** (82)	
2i	Et	Me	Me	A	**3i** (90)	
2j	Me	Et	Me	A	**3j** (95)	

[a] Method: A, Pd(CN)$_2$ (0.05 eq); B, SnCl$_2$ (0.05 eq); C, Me$_3$Ga (0.05 eq). [b] Homogeneity was determined by GLC as well as NMR. [c] Attack of isocyanide on the less substituted carbon gave CNCH$_2$CHMeOSiMe$_3$ (**3k**).

The following observations do not give any information about the formation of an active isocyanide species by the reaction of **1** with $Pd(CN)_2$, $SnCl_2$, or Me_3Ga: (1) treatment of oxiranes **2** with $Pd(CN)_2$ in refluxing dichloromethane did not give any product, (2) treatment of **1** with oxiranes **2** without any catalyst resulted in the recovery of the starting materials, and (3) Sn-CN species could not be detected by ^{119}Sn and ^{13}C NMR measurements in a mixture of $SnCl_2$ and **1**.

Regio- and stereoselective isocyanosilylation with **1** can be explained: (1) coordination of a Lewis acid to oxygen activates oxiranes **2** by stretching the bond between oxygen and more substituted carbon and (2) weakly nucleophilic **1** attacks the more electrophilic site of the activated oxiranes with inversion of configuration (Scheme 2).

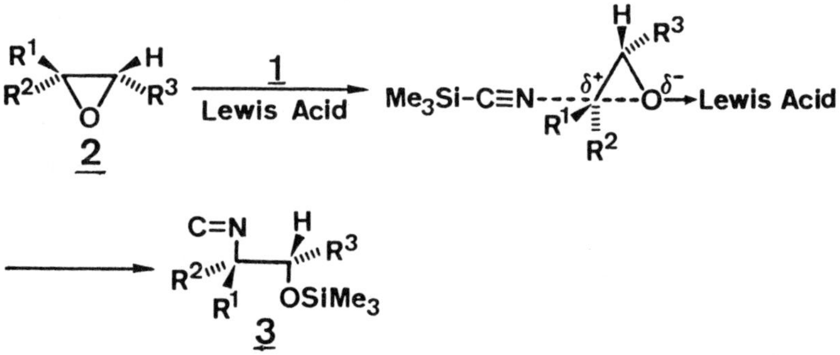

Scheme 2

The above described selective isocyanide formation was applied to both an hydroxymethyloxirane and an epoxy acid. Regio- and stereoselective opening of the oxirane ring would give either an β-amino α-hydroxy acid or an α-amino β-hydroxy acid; both are useful intermediates for organic synthesis. As shown in the following schemes, highly selective opening of the oxirane ring to give β-amino α-hydroxy acid exclusively (Scheme 3).[11]

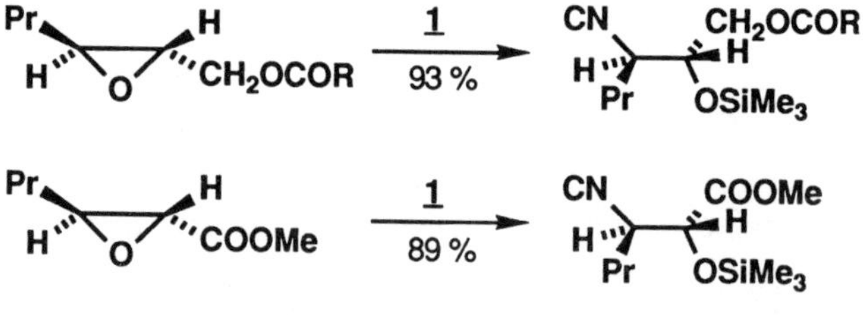

Scheme 3

As optically active starting materials are easily obtained from the corresponding allyl alcohols by Sharpless oxidation, optically active products are expected to be produced by the application of the above method.

3 NITRILES FROM OXIRANES BY ALUMINIUM CATALYST

In contrast to the above selective isocyanide formation under Lewis acid catalysis, the reaction of **1** with **2** in the presence of aluminium chloride or aluminium alkoxide afforded 3-trimethylsiloxy nitriles **4** exclusively.[5] Among the aluminium compounds examined, Al(OPr-i)$_3$ and (i-Bu)$_2$AlOPr-i afforded satisfactory results. Oxirane **2a** gave nitrile **4a** in quantitative yield, and methyloxirane (**2b**) afforded **4b** exclusively by regioselective attack of cyanide on the less substituted carbon of the oxirane ring. Results of the formation of nitriles **4** from various oxiranes **2** are summarized in Table II.

Table II. Formation of Nitriles **4** from Oxiranes **2**

Oxirane	R^1	R^2	R^3	Method[a]	Product[b] (yield %)
2a	H	H	H	D	**4a** (80)
2a	H	H	H	E or F	**4a** (100)
2b	Me	H	H	D	**4b** (82)
2b	Me	H	H	E or F	**4b** (100)
2c	Me	Me	H	D	**4c** (13), **3c** (21)
2c	Me	Me	H	E	**4c** (44), **3c** (41)
2c	Me	Me	H	F	**4c** (51), **3c** (6)
2e	H	-(CH$_2$)$_4$-		D	**4e** (79)
2e	H	-(CH$_2$)$_4$-		E	**4e** (90), **3e** (3)
2e	H	-(CH$_2$)$_4$-		F	**4e** (90)
2e	H	-(CH$_2$)$_4$-		G	**4e** (80)
2f	Me	-(CH$_2$)$_4$-		G	**4f** (34), **41**[c] (31)
2f	Me	-(CH$_2$)$_4$-		H	**4f** (67)[d], **41** (18)[d]
2g	H	Et	Et	G	**4g** (89)
2g	H	Et	Et	H	**4g** (100)[d]
2h	Et	H	Et	G	**4h** (87)
2h	Et	H	Et	H	**4h** (100)[d]
2i	Et	Me	Me	H	**4i** (100)[d]

[a] Method: D, Et$_2$AlCl (0.04 eq); E, i-Bu$_2$AlOPr-i (0.2 eq); F, Al(OPr-i)$_3$ (0.1 eq); G, i-Bu$_2$AlOPr-i (1.5 eq); H, Al(OPr-i)$_3$ (1.5 eq). [b] Homogeneity was determined by GLC as well as NMR. [c] Attack of cyanide on the more substituted carbon gave threo-1-methyl-2-(trimethylsiloxy)-1-cyclohexanecarbonitrile (**41**). [d] Including protodesilylated product.

In contrast to the formation of isocyanides 3 under the catalytic action of soft Lewis acids, such as $Pd(CN)_2$, $SnCl_2$, and Me_3Ga, the formation of a nitrile 4 from an oxirane 2 can be explained by the attack of Al-CN species to the less substituted carbon of the oxirane ring to give cyanoaluminated product that undergoes transmetallation with 1 affording a 3-trimethylsiloxy nitrile with regeneration of Al-CN species. The isolation of trimethylsilyl isopropoxide from the reaction mixture of 1 with aluminium isopropoxide suggests the formation of Al-CN species. Diethylaluminium cyanide seems to be produced by the reaction of 1 with diethylaluminium chloride. When 1 was treated with diethylaluminium chloride in heptane at 25°C for 3 h and the reaction mixture then distilled under reduced pressure, a viscous oil was obtained whose ^{13}C NMR is identical with that of authentic sample. Addition of **2b** to this oil afforded 3-hydroxybutanenitrile in 56% yield after hydrolytic workup. When chlorotrimethylsilane was added to the reaction mixture before hydrolytic workup, 3-(trimethylsiloxy)butanenitrile (**4b**) was obtained in 55% yield. Diethylaluminium cyanide exists as a polymer or a tetramer. A polymeric structure would allow isomerization of Al-CN to Al-NC, and a strongly nucleophilic Al-NC species should attack on the less substituted carbon of the oxirane ring with inversion of configuration (Scheme 4).

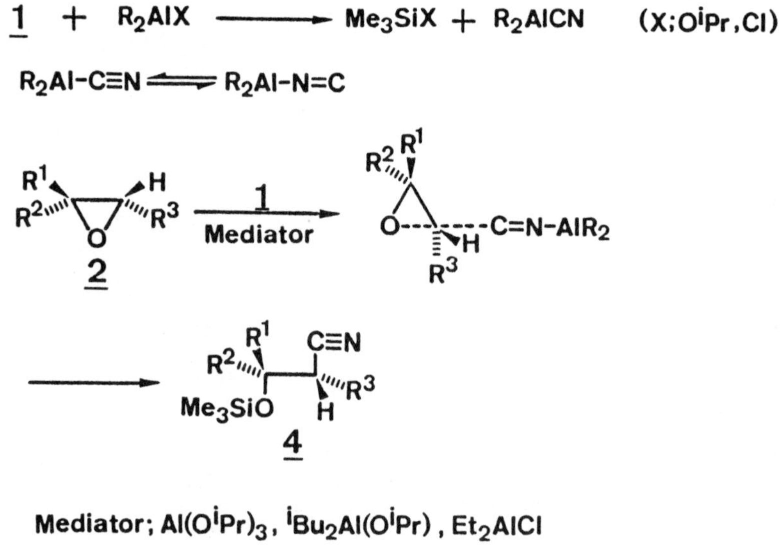
1 + R2AlX → Me3SiX + R2AlCN (X;OiPr,Cl)
R2Al-C≡N ⇌ R2Al-N=C
R1
R2
H
R3
O
2
1
Mediator
R2 R1
O
H
R3
C=N-AlR2
R1
C≡N
R2
R3
Me3SiO
H
4
Mediator; Al(OiPr)3, iBu2Al(OiPr), Et2AlCl

Scheme 4

Application of the above cyanide-forming reaction to 2-hydroxy-methyloxirane gives unsatisfactory results (Scheme 5).[11] Novel effective catalyst is needed to proceed highly selective reaction giving one product selectively.

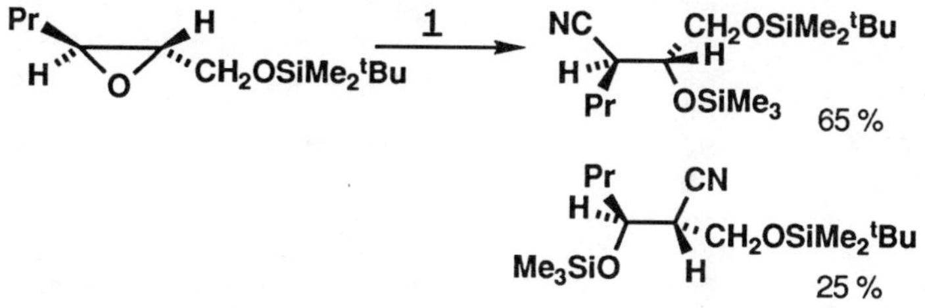

[Scheme showing reaction: Pr,H-oxirane-CH$_2$OSiMe$_2^t$Bu with reagent 1 gives NC/Pr/CH$_2$OSiMe$_2^t$Bu/OSiMe$_3$ product (65%) and Pr/CN/Me$_3$SiO/CH$_2$OSiMe$_2^t$Bu product (25%)]

Scheme 5

4 NITRILES FROM OXIRANES BY LANTHANOID CATALYST

As described in the previous section, selective formation of isocyanide from 2-hydroxymethyloxirane or 2-carboxyoxirane by the attack of isocyanide on the 3-position of the oxirane ring is performed by the use of a soft Lewis acid. Selective formation of nitrile from the same oxirane derivatives, on the other hand, has not afforded satisfactory results by the use of an aluminium mediator.

When a certain metal species works as a catalyst for the selective formation of nitrile without showing any catalytic activities as a Lewis acid, such a metal species could be a good catalyst for selective nitrile formation. Lanthanoid trichlorides have been found to be excellent mediators for the selective nitrile synthesis.[12,13]

A 1,2-dichloroethane solution of cyclohexene oxide (**2e**) is treated with cyanotrimethylsilane **1** in the presence of YbCl$_3$·6H$_2$O.[12,13] Reaction at room temperature does not give any adduct. When the reaction mixture is refluxed for several hours, however, trans-2-trimethylsiloxycyclohexane-carbonitrile **4e**[5] is formed in 80% yield. The catalytic properties of several lanthanoid salts were examined and the results are summarized in Table III.

These results indicate that lanthanide salts possess the desired catalytic activity with respect to both cyanotrimethylsilane (**1**) and an oxirane. Although YbCl$_3$ and Yb(fod)$_3$ gave **4e** in satisfactory yields, it is necessary to heat the reaction mixture to reflux even for

Table III. Lanthanoid-Catalyzed Reaction of **2e** with **1** affording **4e**

Catalyst[a]	Yield (%)[b]	Catalyst[c]	Yield (%)[b]
$ScCl_3 \cdot 6H_2O$	54	$ScCl_3$-BuLi	99
$Y(acac)_3$	92	YCl_3-BuLi	94
$CeCl_3$	11	$CeCl_3$-BuLi	75
$Pr(tfc)_3$	36		
$Eu(fod)_3$	66	$EuCl_3$-BuLi	84
$Eu(tfc)_3$	60		
$YbCl_3 \cdot 6H_2O$	80	$YbCl_3$-BuLi	99
$Yb(fod)_3$	85		

[a] A mixture of cyclohexene oxide (**2e**, 2.0 mmol), Me_3SiCN (**1**, 3.0 mmol), and catalyst (0.05-0.1 mmol) in 5 mL of 1,2-dichloroethane was heated at reflux for 5 h. [b] Isolated yield of trans-2-trimethylsiloxycyclohexanecarbonitrile (**4e**). [c] To a slurry of MCl_3 (0.1 mmol) in THF (7 mL), BuLi (0.33 mmol, hexane solution) was added. To the mixture, Me_3SiCN (**1**, 2.4 mmol) and cyclohexene oxide (**2e**, 2.0 mmol) was added and then the mixture was stirred at room temperature for 3 h.

highly reactive **2e**. Reaction of trans-1,2-diethyloxirane with **1** under $Yb(fod)_3$ catalysis gives the corresponding nitrile in 54% yield after refluxing for 18 h. Improvement of the catalytic activity was needed, and we attempted to accomplish this by the modification of the ligands.

A slurry of $YbCl_3$ (28 mg, 0.1 mmol)[14] in 7 mL of THF was treated with BuLi (0.33 mmol) at $-78°C$. After 15 min, cyclohexene oxide (**2e**, 2.0 mmol) and cyanotrimethylsilane (**1**, 2.4 mmol) was added and the resultant mixture was stirred at room temperature for 3 h. Glc analysis of the reaction mixture indicated almost quantitative formation of trimethylsiloxy nitrile **4e** which was isolated in 99% yield. The catalytic activity of some other organolanthanide species obtained analogously was examined and the results are shown in Table III. Although the species obtained from either $ScCl_3$ or $YbCl_3$ showed high reactivity, $YbCl_3$ was chosen for further studies.[15]

The ytterbium species obtained by the above method is effective for the regio- and stereoselective cleavage of various types of oxiranes; the cyano group is introduced at the less substituted carbon with backside displacement of the C-O bond. Examples are shown in Table IV.

Table IV. Yb(CN)$_3$ Catalyzed Reaction of Oxirane **2** with **1**

Oxirane	Condition Time (h)	Temp (°C)	Product	Yield (%)
n-Hex epoxide	3	25	n-Hex-CH(OSiMe$_3$)-CH$_2$CN	>99
isobutylene oxide	3	25	Me$_3$SiO-C(Me)$_2$-CH$_2$CN	93
1-methylcyclohexene oxide (methylenecyclohexane oxide)	3	25	1-(cyanomethyl)cyclohexyl OSiMe$_3$	90
cyclopentene oxide	6	25	trans-2-(OSiMe$_3$)cyclopentyl-CN	61
cyclohexene oxide	24	25	trans-2-(OSiMe$_3$)cyclohexyl-CN	92
butadiene monoxide (vinyl oxirane)	6	25	CH$_2$=CH-CH(OSiMe$_3$)-CH$_2$CN	69
epichlorohydrin	0.3	25	NC-CH$_2$-CH(OSiMe$_3$)-CH$_2$Cl	91
isopropyl glycidol (with OH)	2	45	Me$_3$SiO / NC / OH isopropyl product	81
trans-n-C$_5$H$_{11}$, n-C$_5$H$_{11}$ epoxide	12	50	anti product (Me$_3$SiO, CN)	82
cis-n-C$_5$H$_{11}$, n-C$_5$H$_{11}$ epoxide	7	65	syn product	63
n-C$_5$H$_{11}$, Me cis epoxide	6	25	two regioisomers 68:32	76
n-C$_5$H$_{11}$, Me trans epoxide	48	25	two regioisomers 71:29	42

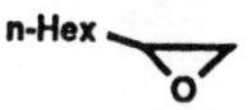

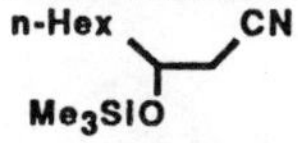

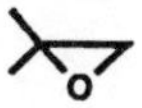

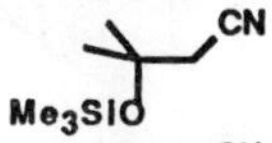

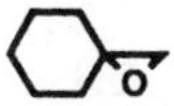

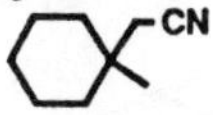

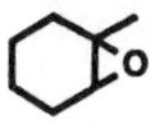

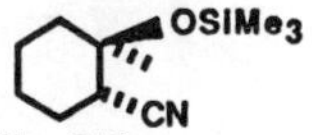

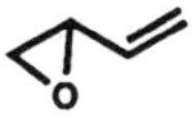

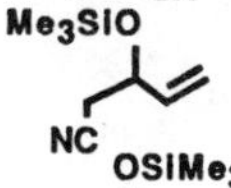

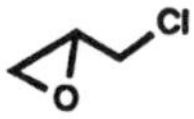

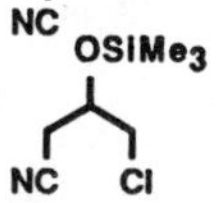

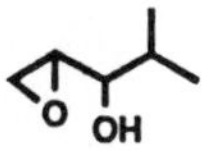

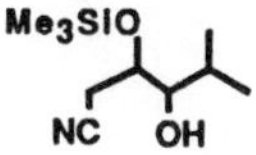

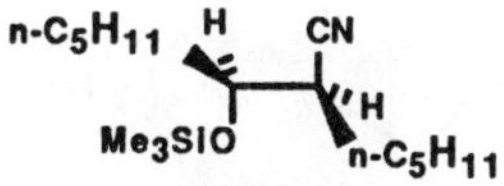

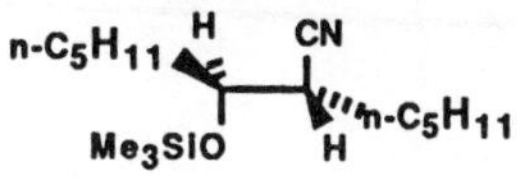

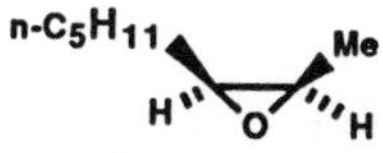

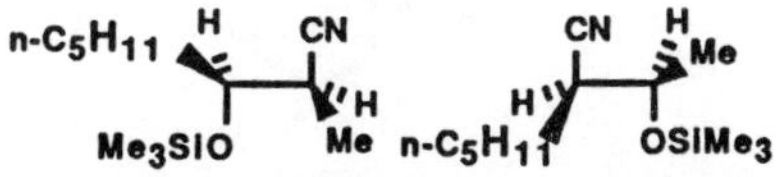

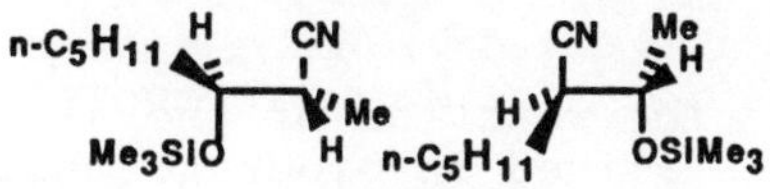

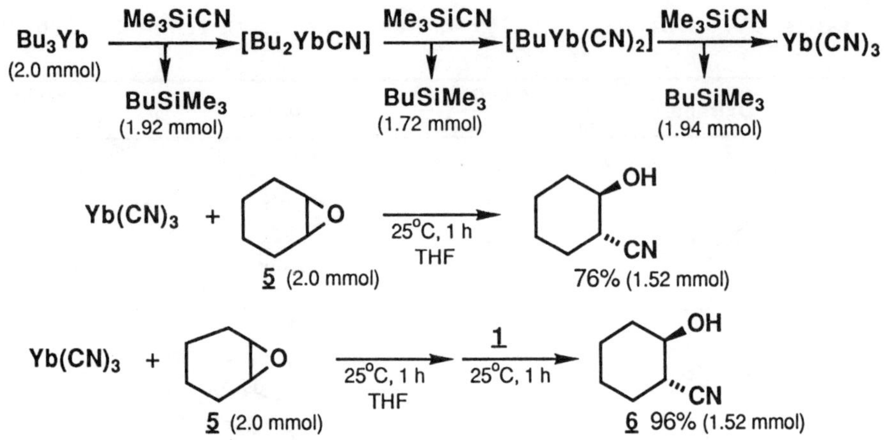

Scheme 6

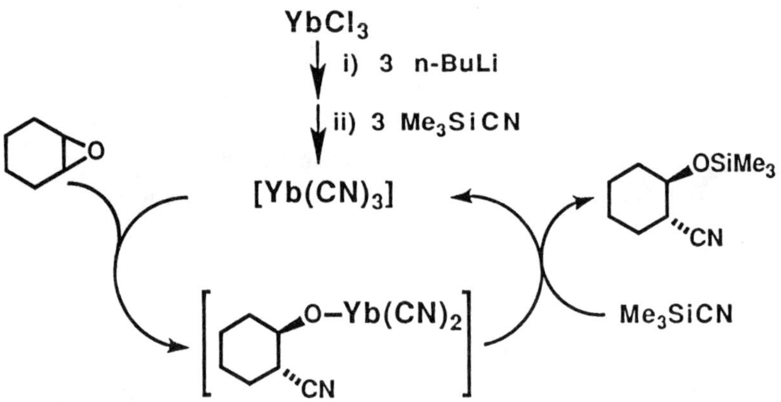

Bu_3Yb (2.0 mmol) →[Me_3SiCN]→ [Bu_2YbCN] →[Me_3SiCN]→ [$BuYb(CN)_2$] →[Me_3SiCN]→ $Yb(CN)_3$

↓ $BuSiMe_3$ (1.92 mmol)   ↓ $BuSiMe_3$ (1.72 mmol)   ↓ $BuSiMe_3$ (1.94 mmol)

$Yb(CN)_3$ + [cyclohexene oxide] **5** (2.0 mmol) →[25°C, 1 h, THF]→ [trans-2-cyanocyclohexanol] 76% (1.52 mmol)

$Yb(CN)_3$ + [cyclohexene oxide] **5** (2.0 mmol) →[25°C, 1 h, THF]→ →[**1**, 25°C, 1 h]→ **6** 96% (1.52 mmol)

Scheme 7

$YbCl_3$
↓ i) 3 n-BuLi
↓ ii) 3 Me_3SiCN
[$Yb(CN)_3$]

[cyclohexene oxide] → [cyclohexyl-O-$Yb(CN)_2$ with CN] → [cyclohexyl-OSiMe$_3$ with CN] + Me_3SiCN

The reaction of the alkylytterbium species with **1** was examined in order to obtain information about the structure of Yb catalyst. To a slurry of $YbCl_3$ (2.0 mmol) in THF maintained at -78°C, a hexane solution of BuLi (6.0 mmol) was added. The reaction mixture was stirred at -78°C for 15 min, and then 0°C for 30 min. Although the organometallic species present in the above reaction mixture has not been characterized, BuLi could not be detected by NMR analysis of the reaction mixture.[16] Therefore Bu_3Yb is assumed to be the main component. To this mixture, **1** was added three times (2.0

mmol, each; 6.0 mmol in all) at the interval of 20 min. Almost quantitative formation of butyltrimethylsilane was detected by glc analyses after each addition.[17] These results suggest that Bu$_3$Yb reacts smoothly with 3 eq of cyanotrimethylsilane to give Yb(CN)$_3$. The above mixture containing Yb(CN)$_3$ was treated with **2e** to give <u>trans</u>-2-hydroxy-cyclohexanecarbonitrile in 76% yield after hydrolytic work up. The addition of excess cyano-trimethylsilane (**1**) before hydrolytic work up afforded **4e** in 90% yield.

The above described formation of **6** by both catalytic and stoichiometric reactions suggests that Yb(CN)$_3$ is the active species to produce a nitrile from an oxirane. A plausible mechanism of the catalytic reaction could be depicted in Scheme 7.

5 NITRILES FROM AZIRIDINES BY LANTHANOID CATALYST

Optically pure α-amino acids are important chemical resources as chiral pool.[18] Since an optically pure aziridine derivative can be derived from an α-amino acid,[19] regio- and stereoselective nucleophilic ring opening reaction has been reported to give various important intermediates in organic synthesis.[20] Recently reported highly selective nucleophilic opening of oxirane cyanotrimethylsilane under lanthanide catalysis[12] prompted us to apply to selective opening of aziridine ring. The reaction of an optically pure 2-substituted aziridine with **1** in the presence of Yb(CN)$_3$ is described here, which occurs at the 3-position of the ring to give β-amino nitriles in an optically pure form.[21]

$$R\overset{\triangle}{\underset{\underset{Ts}{N}}{}} \xrightarrow[M(CN)_3]{Me_3SiCN} R\underset{TsNH}{\diagdown}CN$$

M: Yb, Y, Ce

Scheme 8

A THF (7 mL) solution of N-tosylcyclohexeneimine (**5**, 2.0 mmol) and cyanotrimethylsilane (**1**, 4.0 mmol) in the presence of Yb(CN)$_3$ (25 mol%, 0.5 mmol) was stirred at 65°C for 2.5 h, and then cooled mixture was treated with water. The mixture was extracted with ether and the ethereal solution was washed, dried over Na$_2$SO$_4$, and concentrated. Column chromatography (silica gel, hexane-ethyl acetate) gives <u>trans</u>-2-tosylaminocyclohexane-carbonitrile (**6**) in 90% yield. Reaction using other lanthanoid tricyanide, Y(CN)$_3$ or Ce(CN)$_3$, gave the same product in good yield. Optically pure (S)-3-tosylamino-4-phenyl-butanenitrile was obtained from (S)-N-tosyl-2-benzylaziridine in 84% yield.[22] Other N-tosylaziridines gave β-tosylamino nitrile in good yields. Results are shown in Table V.

Table V. Reaction of Aziridines with **1**

Aziridine	M(CN)$_3$	Time (h)	Product	Yield[b] (%)
cyclohexyl-NTs	Yb(CN)$_3$	2.5	cyclohexyl(NHTs)(CN)	90
	Y(CN)$_3$	2.5		98
	Ce(CN)$_3$	2.5		90
Ph–aziridine–Ts	Yb(CN)$_3$	3	Ph–CH(TsNH)–CH$_2$CN	85
	Y(CN)$_3$	3		80
n-Bu–aziridine–Ts	Yb(CN)$_3$	4	n-Bu–CH(TsNH)–CH$_2$CN	86
MeS(CH$_2$)$_2$–aziridine–Ts	Yb(CN)$_3$	2	MeS(CH$_2$)$_2$–CH(TsNH)–CH(H)CN	93
	Y(CN)$_3$	2		87
PhCH$_2$–aziridine–Ts	Yb(CN)$_3$	7	PhCH$_2$–CH(TsNH)–CH(H)CN	84

[a] A mixture of substrate (2.0 mmol), M(CN)$_3$ (0.5 mmol), Me$_3$SiCN (**1**; 4.0 mmol) in THF (7 mL) was stirred at 65 °C.
[b] Isolated yield.

As β-amino acid can be transformed in to the corresponding β-lactam with retention of configuration, the above described reaction opens a facile access to optically active β-lactams from easily accessible α-amino acids.

REFERENCES AND NOTES

1. E. Colvin, 'Silicon in Organic Synthesis', Butterworth, London, 1981.
2. W. P. Weber, 'Silicon Reagents for Organic Synthesis', Springer-Verlag, Berlin, 1983, Chapter 2.
3. K. Utimoto, Y. Wakabayashi, T. Horiie, M. Inoue, Y. Shishiyama, and H. Nozaki, Tetrahedron, **1983**, *39*, 967 and references cited therein.
4. P. G. Gassman, and T. L. Guggenheim, J. Am. Chem. Soc., **1982**, *104*, 1849; Org. Synth., **1985**, *64*, 39.
5. K. Imi, N. Yanagihara, and K. Utimoto, J. Org. Chem., **1987**, *52*, 1013 and references cited therein.
6. T. Shono, Y. Matsumura, K. Uchida, and F. Nakatani, Bull. Chem. Soc. Jpn., **1988**, *61*, 3029.
7. S. L. Buchwald, and S. J. LaMaire, Tetrahedron Lett., **1987**, *28*, 295.
8. N. Chatani, T. Takeyasu, N. Horiuchi, and T. Hanafusa, J. Org. Chem., **1988**, *53*, 3539.
9. H. Minamikawa, S. Hayakawa, T. Yamada, N. Iwasawa,

and K. Narasaka, Bull. Chem. Soc. Jpn., **1988**, <u>61</u>, 4379.
10. M. T. Reetz, M. W. Drewes, K. Harms, and W. Reif, Tetrahedron Lett., **1988**, <u>29</u>, 3295 and references cited therein.
11. K. Imi, S. Matsubara, and K. Utimoto, unpublished results.
12. S. Matsubara, H. Onishi, and K. Utimoto, to be published.
13. Catalytic use of some lanthanoide salts for the reaction of **1** with oxirane has been reported: A. Vougioukas and H. B. Kagan, Tetrahedron Lett., **1987**, <u>28</u>, 5513.
14. Experimental description of the preparation of $CeCl_3$ from its hydrate: T. V. Lee, J. A. Channon, C. Cregg, J. R. Porter, F. S. Roden, and H. T-L. Yeoh, Tetrahedron, **1989**, <u>45</u>, 5877. A weighed amount of $YbCl_3 \cdot 6H_2O$ is placed in a flask and heated up to $120°C$ under vacuum for 3 h, then dry argon is flushed in the flask. THF was added at once and the mixture was homogenized by ultrasonic irradiation.
15. Mainly for economic reasons.
16. ^{13}C NMR analyses of the reaction mixture indicated that carbon signals of Bu in BuLi disappeared and 3 new signals, assumed to be the Bu of Bu_3Yb, were observed.
17. Reaction of Bu-Yb species with **1** is complete within 15 min at room temperature.
18. G. M. Coppola and H. F. Schuster, 'Asymmetric Synthesis. Construction of Chiral Molecules Using Amino Acids', John Wiley & Sons, New York, 1987.
19. (a) W. Oppolzer and E. Flaskamp, Helv. Chim. Acta, **1977**, <u>60</u>, 204. (b) G. S. Bates and M. A. Varelas, Can. J. Chem., **1980**, <u>58</u>, 2562. (c) N. Yahiro, Chem. Lett., **1982**, 1479. (d) J. R. Pfister, Synthesis, **1984**, 969. (e) J. W. Kelly, N. L. Eskew and S. A. Evans, Jr., J. Org. Chem., **1986**, <u>51</u>, 95. (f) R. Haner, B. Olano, and D. Seebach, Helv.Chim.Acta, **1987**, <u>70</u>, 1676.
20. (a) D. Tanner, C. Birgersson, and H. K. Dhaliwal, Tetrahedron Lett., **1990**, <u>31</u>, 1903 and references cited therein. (b) J. E. Baldwin, R. M. Adlington, I. A. O'Neil, C. Shofield, A. C. Spivey, and J. B. Sweeney, J. Chem. Soc. Chem. Commun., **1989**, 1852. (c) B. B. Lohray, Y. Gao, and K. B. Sharpless, Tetrahedron Lett., **1989**, <u>30</u>, 2623. (d) K. Sato and A. P. Kozikowski, Tetrahedron Lett., **1989**, <u>30</u>, 4073. (e) J. Letgers, L. Thijs, and B. Zwangenburg, Tetrahedron Lett., **1989**, <u>30</u>, 4881. (f) A. Dureault, I. Tranchepain, and J.-C. Depezay, J. Org. Chem., **1989**, <u>54</u>, 5324. (g) D. Tanner and P. Somfai, Tetrahedron, **1988**, <u>44</u>, 619.
21. S. Matsubara, T. Kodama, and K. Utimoto, to be published.
22. Perfect retention of optical purity was determined by the HPLC analysis using Chiralcel OD (Daicel Chemical Industry).

Alkenylsilanes in the Synthesis of Nitrogen-containing Heterocycles

E. Lukevics and V. Dirnens
INSTITUTE OF ORGANIC SYNTHESIS, LATVIAN ACADEMY OF SCIENCES, RIGA, LATVIA

Introduction

Alkenylsilanes evoke interest as model compounds used to study the effects of silyl substituents on the reactivity of double bonds and serving as valuable synthons in organic synthesis. Because of (p-d)π interaction, vinylsilanes differ from their carbon counterparts in reactions of nucleophilic and electrophilic addition. Allylsilanes, in turn, are more nucleophilic than vinylsilanes and the corresponding alkenes, the separation of a multiple bond from the silyl substituent by two or more carbon atoms results in a reactivity comparable with that of the appropriate alkenes. Moreover, vinyl- and allylsilanes serve as convenient semiproducts in organic synthesis [1,2]. As many nitrogen-containing organosilicon compounds show marked biological activity [3], we summarized literature data and results of our own studies of addition reactions to unsaturated silanes leading to nitrogen-containing organosilicon heterocycles. Applicability of phase transfer catalysis [4] in the synthesis of silyl-containing heterocycles with the silicon atom attached to the heterocycle and capable of affecting its chemical properties has been appraised.

1. Addition of heterocyclic amines to alkenylsilanes

A convenient synthetic route to nitrogen-containing organosilicon compounds involves the amination of alkenylsilanes by heterocyclic amines in the presence of alkali metals, such as lithium, sodium, potassium or their amides. The reaction gives the products containing a nitrogen atom in β-position with respect to the silicon atom. In such a way, aziridine [5-13], azetidine [14], pyrrolidine [8,9,17], piperidine [8,9,15,17], piperazine [17], perhydroazepine [16,17], 3,5-dimethylpyrazole [19] have been attached to triorganylvinylsilane.

The effects of temperature, reagent ratio, catalyst and solvent concentration on reaction rate have been studied in the case of piperidine addition to triethylvinylsilane [10].

As revealed by the results of experiments, temperature elevation from 40 to 70°C increased the yield of the target-product from 35 to 70% during one hour. An excess of amine and particularly an increase in lithium piperidide concentration in the reaction mixture changed the reaction rate in the same direction. Solvent nature was essentially important. E.g., the rate of addition reaction in the presence of tetrahydrofuran was approximately twice as high as that in benzene or heptane.

As the reaction gets underway only in the presence of organic amides and variations in the amount of amides affect the reaction rate to a larger extent than an increase in the amount of piperidine, the following reaction mechanism has been proposed:

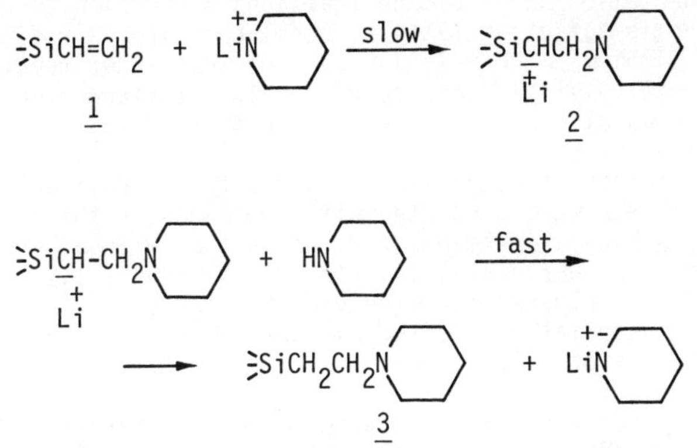

$$\equiv\text{SiCH=CH}_2 + \text{LiN}\bigcirc \xrightarrow{\text{slow}} \equiv\text{SiCHCH}_2\text{N}\bigcirc$$
$$\underline{1} \qquad\qquad\qquad\qquad \overset{+}{\underset{\text{Li}}{}} \quad \underline{2}$$

$$\equiv\text{SiCH-CH}_2\text{N}\bigcirc + \text{HN}\bigcirc \xrightarrow{\text{fast}}$$
$$\overset{+}{\underset{\text{Li}}{}}$$

$$\longrightarrow \equiv\text{SiCH}_2\text{CH}_2\text{N}\bigcirc + \text{LiN}\bigcirc$$
$$\underline{3}$$

Organolithium intermediates $\underline{2}$ have been successfully isolated [15] with subsequent hydrolysis to give trialkyl[2-(N-piperidyl)-ethyl]silane $\underline{3}$.

The influence of substituents on the reactivity of multiple bonds was evaluated in reactions of competitive addition of aziridine to vinylsilane bearing an alkoxyl, azacycloalkyl, phenyl and alkyl substituents at the silicon atom [20].

According to available experimental evidence, electron-withdrawing substituents $\overline{\text{CH}_2\text{-CH}_2\text{-N}}-$, $\text{C}_2\text{H}_5\text{O}-$ increase multiple bond reactivity in vinylsilanes to a greater extent than the phenyl group. The latter, in turn, enhances the reactivity of double bonds, compared with electron-donating groups. Substituents at the silicon atom with respect to their activating influence can be ranged in the following sequence:

$$\overline{\text{CH}_2\text{-CH}_2\text{-N}}- > \text{C}_2\text{H}_5\text{O}- > \text{C}_6\text{H}_5- > \text{CH}_3-, \text{C}_2\text{H}_5-$$

The yield of products resulting from the reaction of dimethyl-(2-furyl)- and dimethyl(2-thienyl)vinylsilanes with secondary heterocyclic amines in the presence of catalytic amounts of metallic lithium depends both on the nature of the heteroaromatic substituent at the silicon atom and on the amine [17]. The rate of addition at the double bond of thienyl-containing vinylsilane was increased in comparison with the furyl derivative. For instance, the yield of perhydroazepine addition amounts to 53 and 23%, respectively. A maximum yield reaching 65.5 and 71.4% for the furan and thiophene derivative, respectively, was found in the case of piperidine addition.

The reaction between dimethyl(2-thienyl)vinylsilane and pyrrolidine is characterized by complete elimination of the thiophene ring, whereas the reaction of dimethyl(2-furyl)vinyl silane under analogous conditions leads to furyl-containing β-substituted organosilicon amine in 24% yield. The breakage of the C_{Ar}-Si bond is also characteristic of methyldi(2-furyl)- and methyldi(2-thienyl)-vinylsilanes. According to PMR data, the reaction of addition at the double bond in THF in the presence of metallic lithium is accompanied by elimination of one or two heterocycles [17].

Dehydrocondensation and addition reactions occur simultaneously during interaction of organylvinylhydrosilanes with aziridine in the presence of lithium, the relative rate of the latter reaction increasing from diethylvinylsilane to diphenylvinylsilane. The addition reaction is also promoted by introduction of aziridine radical at the silicon atom [11].

In order to compare alkenylsilane reactivity in addition reactions, experiments were conducted with compounds of the general formula $R_3Si(CH_2)_nCH=CH_2$, where n=0,1,2, as well as with 3,3-dimethylbutene - a carbon analogue of trimethylvinylsilane [5]. The reaction proceeded fairly readily with vinylsilanes, the level of reactivity being essentially affected by substituents at the silicon atom. In the case of allylsilanes the reaction failed to occur in the presence of sodium or its amide. In some cases, the adducts were obtained only by using ethyleneimine amide as catalyst or by increasing the time of contact between the reagents [21]. This interaction is accompanied by $Si-C_{allyl}$ bond cleavage

$$R^1R^2R^3SiCH_2CH=CH_2 + \triangleright NH \longrightarrow \begin{cases} R^1R^2R^3SiN\triangleleft + CH_3CH=CH_2 \\ \underline{4} \quad R^1,R^2=CH_3; \; R^3=C_6H_5 \\ R^1R^2R^3SiCH_2CHCH_3 \\ \qquad\qquad\quad |N\triangle \\ \underline{5} \quad R^1,R^2,R^3=C_2H_5 \end{cases}$$

The nature of substituents at the silicon atom considerably affects the reaction pattern. For example, adduct 5 was obtained in 35% yield in the case of triethylallylsilane. If allylsilane contains at least one phenyl radical at the silicon atom, compound 5 fails to be formed, the reaction resulting in triorganylsilylaziridine 4 yield:10%. 2-Thienyl-dimethylallylsilane upon boiling for 18 hrs with aziridine in the presence of sodium gives dimethyldiaziridinyl-silane (yield: 90%) [13].

Trimethyl-γ-butenylsilane does not react with aziridine even at 100°C in the presence of sodium amide. 3,3-Dimethylbutene behaves similarly to γ-butenylsilane [5].

A dramatic difference in reactivity was clearly demonstrated by the reaction of addition of secondary heterocyclic amines to trimethylvinylsilane and its carbon analogue. Polarization of the double bond in vinylsilanes at the expense of electron-withdrawing silyl groups contributes to the addition of amines. A drop in the polarization of the multiple bond in γ-butenylsilane and the presence

of a weak electron-donating substituent in 3,3-dimethylbutene rules
out the occurrence of this reaction.

2. Cyclization of products resulting from the addition of
   N-halo derivatives of sulphamides and carbamates to
   unsaturated organosilicon compounds

An important synthetic route to 1-substituted 2-trialkylsilyl-
aziridines involves intramolecular cyclization of products resulting
from the addition of N,N-dihalosulphonamides and -carbamates to vi-
nyl- and allylsilanes.

N,N-Dichloro-p-toluenesulphonamide reacts with trimethylvinylsilane in the presence of atmospheric oxygen and Cu_2Cl_2 (used as
initiator) with subsequent treatment of the adduct with aqueous
sodium sulphite to give N-(2-chloro-2-trimethylsilylethyl)-p-tolue-
nesulphonamide (6) in 20% yield [22]. 1,2-Dichloroethyltrimethyl-
silane was obtained simultaneously in 62% yield (7). The reaction
occurring in argon atmosphere is characterized by increased yield
of adduct 6 (up to 53%) and decreased content of dichloroethylsilane
7 (20%) in the mixture.

Addition of dichloro-p-toluenesulphonamide to trimethylallylsilane in the presence of atmospheric oxygen without initiator,
i.e. under conditions favouring the ionic mechanism, proceeds according to Markovnikoff's rule. After treatment with sodium sulphite,
the only product isolated from the reaction mixture was N-(1-chloro-
methyl-3-trimethylsilylethyl)-p-toluenesulphonamide (8) [22]. The
reaction occuring in argon atmosphere in the presence of Cu_2Cl_2
yields organosilicon sulphonamide 8 as the sole product. Apparently
in this case, due to the high nucleophilicity of the multiple bond,
the addition occurs exclusively via the ionic mechanism.

1-p-Toluenesulphonyl-2-trimethylsilylaziridine and 1-p-toluene-
sulphonyl-2-trimethylsilylmethylaziridine were prepared by intramolecular alkylation of silicon-containing sulphonamides 6 and 8 under
liquid-solid phase-transfer catalysis conditions

$$Me_3SiCHClCH_2NHSO_2C_6H_4CH_3\text{-}p$$
6

$$Me_3SiCH_2CH(CH_2Cl)NHSO_2C_6H_4CH_3\text{-}p$$
8

hexane/NaOH
+
Oct_4NBr^-, 25°C

→ aziridine with SiMe$_3$, N-SO$_2C_6H_4CH_3$-p

→ aziridine with CH$_2$SiMe$_3$, N-SO$_2C_6H_4CH_3$-p

N,N-Dichlorourethane reacts with trimethylvinylsilane both in
inert atmosphere and in the presence of atmospheric oxygen [22-25].
Addition of catalytic amounts of Cu_2Cl_2 to the reaction mixture,
initiating radical addition of pseudohalogens to alkenes [26],
significantly accelerates the process.

The addition always occurs regiospecifically, contrary to Markovnikoff's rule. This equally applies to reactions of alkenylsilanes carrying a bromine atom or a methoxycarbonyl group in the α-position with respect to the silicon atom. Vinylsilanes react with N,N-dichlorocarbamates also in the presence of atmospheric oxygen without initiator of the radical reaction, suggesting the feasibility of ionic mechanism. The formation of adducts in this case, contrary to Markovnikoff's rule, is apparently due to the ability of the silicon atom to stabilize the positive charge in the β-position of çarbocation intermediate. The latter results from the addition of Cl^+ to vinylsilane, pseudohalogens serving as the source of these ions. Silicon-containing β-chloro-N-chlorocarbamates undergo reduction readily by aqueous sodium bisulphite or sulphite to give the corresponding β-chlorocarbamates **9** in good yield.

The pattern of interaction between trimethylallylsilane and dichlorourethane differs from that observed for vinylsilanes [22, 25]. The reaction mixture obtained by treating silane with N,N-dichlorocarbamate under conditions conducive to the radical reaction after treatment with aqueous sodium sulphite gives a mixture of ethyl ester of N-(2-chloro-3-trimethylsilylpropyl)carbamic acid (**10**) and ethyl ester of N-(1-chloro-3-trimethylsilyl-2-propyl)-carbamic acid (**11**) in the ratio 1:1 (^{1}H NMR spectroscopy data) and 40% total yield. Consequently, the addition of urethane to allylsilanes proceeds both in accordance with and contrary to Markovnikoff's rule. The mixture of products probably results from the attack of cation Cl^+ and radical ClNCOOEt on allylsilane.

Intramolecular alkylation of silicon-containing carbamates (**9,10** and **11**) under conditions of phase-transfer catalysis, as in the case of sulphonamides 6 and 8, leads to the corresponding 1-alkoxycarbonyl-2-trialkylsilylaziridine and 1-ethoxycarbonyl-2-trimethylsilylmethylaziridine

$$R_3SiCHCH_2NHCOOR^1 \quad\text{(Cl)}\quad \underset{9}{} \longrightarrow \underset{\underset{COOR^1}{|}}{\overset{SiR_3}{\triangledown_N}}$$

$$\left.\begin{array}{c} Me_3SiCH_2\overset{Cl}{C}HCH_2NHCOOEt \\ \underline{10} \\ + \\ Me_3SiCH_2\overset{CH_2Cl}{C}HNHCOOEt \\ \underline{11} \end{array}\right\} \xrightarrow[Oct_4NBr^-, 25°C]{hexane/NaOH} \underset{\underset{COOEt}{|}}{\overset{CH_2SiMe_3}{\triangledown_N}}$$

The example of methyl-N-(2-chloro-2-trimethylsilyl)-ethylcarbamate cyclization was applied to test the usefulness of ultrasound instead of phase-transfer catalyst. The reaction under PTC proceeds smoothly to afford 1-methoxycarbonyl-2-trimethylsilylaziridine (yield: 75%). The reaction occurring slowly in the absence of phase-transfer catalyst is considerably accelerated by ultrasonication

(100 W, 55 kHz), however after reaching a 45% yield the content of the target product begins to decline because of consecutive formation of 2-trimethylsilylazetidine [25,27].

α,α-Di-substituted silicon-containing carbamates obtained from α-bromo- and (α-methoxycarbonyl)vinyltrimethylsilanes undergo desilylation in the presence of solid bases and phase-transfer catalyst to give the corresponding disubstituted carbamates. Hence, the presence of an electron-withdrawing substituent in the α-position with respect to the silicon atom alters the direction of alkyl-N-(2-trialkylsilyl)ethylcarbamate transformation [22].

3. Synthesis of 2-triethylsilyl-1H-aziridinecarboxylic acid derivatives

The general procedure for the synthesis of 2-substituted 1H-aziridines involves base-induced cyclization of iodomethylates of products resulting from the addition of 1,1-dimethylhydrazine to acrylic acid derivatives [28,29]. Therefore, methyl-2-(triethylsilyl)acrylate 12 was applied as a starting silicon-containing compound, obtained by hydrosilylation of methyl ester of propiolic acid by triethylsilane in the presence of Speier's catalyst. In this case, as expected, methyl-*trans*-3-(triethylsilyl)acrylate (13) - a β-addition product is formed besides acrylate 12. The ratio of isomers 12 and 13 depends on reaction conditions. The same reaction occurring in THF predominantly yields a β-*trans*-addition product (12:13 = 30:70). Hydrosilylation of methylpropiolate in the absence of solvent gives a mixture of 12 and 13 in the ratio 70:30 (^{1}H NMR and GLC data).

As demonstrated by control experiments, 1,1-dimethylhydrazine reacting with a mixture of adducts (12 and 13) adds only to isomer 12 [30], therefore this reaction was performed without preliminary separation of the mixture

1,1-Dimethyl-2-(2'-methoxycarbonyl-2'-triethylsilyl)ethylhydrazine 14 was obtained in 60% yield by heating a mixture of acrylates (12 and 13) containing 70% of 12 together with 1,1-dimethylhydrazine in inert atmosphere (70°-80°C, 4 hrs). Compound 14 undergoes alkylation with methyl iodide in ether at room temperature to give 1,1,1-trimethyl-2-(2'-methoxycarbonyl-2'-triethylsilyl)ethylhydrazinium iodide 15 in almost quantitative yield. The conventional method for cyclization of 1,1,1-trimethyl-2-(2'-substituted)ethylhydrazinium salts to 2-substituted 1H-aziridines by treating them with sodium methylate in methanol [28,29] is inapplicable in the case of 15, because MeONa breaks the Si-C bond. For this reason we examined the possibility of 15 cyclization under solid-liquid phase-transfer catalysis (PTC) conditions providing satisfactory results upon intramolecular alkylation of alkyl-N-(2-chloro-2-trialkylsilyl)-ethylcarbamates occurring without rupture of the Si-C bond and leading to the formation of 1-alkoxycarbonyl-2-trialkylsilylaziridines [22-25]. Compound 15 in the two-phase system benzene-solid KOH in the presence of tetraoctylammonium bromide (TOAB) (15:KOH:TOAB= 1:1:0.1) at room temperature undergoes conversion to 2-methoxycarbonyl-2-triethylsilyl-1H-aziridine (16) in high yield (80%, GLC data). When NaOH is used instead of KOH, salt 15 upon continuous boiling in heptane (72h) or THF (48h) in the presence of TOAB converts to sodium salt of 2-triethylsilyl-1H-aziridine-2-carboxylic acid (17) in 85% yield. Compound 15 also undergoes conversion to 17 without

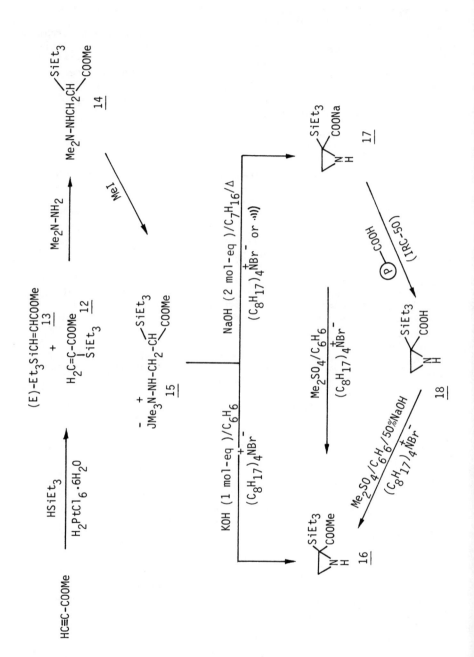

phase-transfer catalyst following ultrasonic irradiation of the reaction mixture. This allows one to decrease substantially both the reaction time (15 min, THF used as solvent) and temperature (approx. 0°C). Aqueous solution of salt 17 treated with diluted hydrochloric acid affords 2-triethylsilyl-1H-aziridine-2-carboxylic acid (18) in very low yield (2%). The use of strongly acidic sulphocationite instead of mineral acid appeared ineffective too. However, treatment of salt 17 with low-acidity cation-exchanger IRC-50 gives acid 18 in satisfactory yield (44%).

^{1}H, ^{29}Si and ^{15}N NMR spectra of 18 suggest that proton is located at the carboxyl group (under conditions of spectral registration) and no betaine form is present. The low basicity of the nitrogen atom in acid 18 is apparently due to the high strain in the three-membered cycle and to the steric shielding of the unshared nitrogen pair by the bulky triethylsilyl substituent. On the other hand, the strong electron-donating effect of the Et$_3$Si group is responsible for the decreased acidity of the COOH group in compound 18.

The structure of salt 17 and acid 18 was additionally confirmed by their transformation to 16 by alkylation with dimethylsulphate using known procedures [31,32] for the preparation of carboxylic acid esters under PTC conditions. Compound 17 undergoes alkylation in the two-phase solid-phase solid-liquid system benzene-solid salt 17 in the presence of TOAB at room temperature to give 16 in 55% yield (GLC data). Alkylation of acid 18 was also carried out in the two-phase system benzene/50% aq. NaOH (catalyst TOAB) at room temperature to give 16 in 90% yield (GLC data).

4. Reactions of vinylsilanes with ethoxycarbonylnitrene

The reaction of vinylsilanes with :NCOOEt was studied using vinyl-, α-bromovinyl- and (β-methoxycarbonyl)vinyl(trialkyl)silanes [30,33,34]. Among the available methods for the generation of ethoxycarbonylnitrene we chose base-induced elimination of p-nitrobenzosulphate anion from ethyl N-(p-nitrobenzosulphonyloxy)carbamate (19). The process was carried out under conditions of liquid-liquid phase-transfer catalysis (PTC). 1-Ethoxycarbonyl-2-trimethylsilyl-aziridine was formed as the main product in a two-phase system consisting of vinyltrimethylsilane, solvent, nitrene precursor 19 and triethylbenzylammonium chloride (TEBA) in dichloromethane and 1M aqueous solution of sodium bicarbonate at room temperature. 2-Trimethylsilyl-1H-aziridine probably resulting from saponification and decarboxylation was also detected from the reaction mixture:

$$R_3SiCH=CH_2 \quad \xrightarrow{19/CH_2Cl_2/1M\ NaHCO_3}_{catalyst} \quad R_3SiCH-CH_2 \atop \underset{\underset{COOEt}{|}}{\diagdown N \diagup} \quad + \quad R_3SiCH-CH_2 \atop \underset{\underset{H}{|}}{\diagdown N \diagup}$$

$$\qquad\qquad\qquad\qquad\qquad\qquad\qquad\qquad\qquad 20 \qquad\qquad\qquad 21$$

Application of saturated water solutions of sodium and potassium carbonates as bases decreased the yield of aziridine 20. Without phase-transfer catalyst the N-carboxy derivative 20 is formed in very low yield (1.5%, GLC data).

The reaction of α- and β-substituted vinylsilanes with ethoxycarbonylnitrene proceeds similarly to give the corresponding aziridines as the main products [30].

For example, α-bromovinyltriethylsilane leads to 1-ethoxycarbonyl-2-bromo-2-triethylsilylaziridine in 27% yield (GLC data). According to chromatomass spectroscopy data, the reaction mixture contains one more product (6%) identified as an isomer of aziridine. This may possibly be 4-bromo-4-triethylsilyl-2-ethoxy-1,3-oxazoline-2 resulting from rearrangement of silylaziridine or from 1,2-dipolar cycloaddition of ethoxycarbonylnitrene to silane. Methyl ester of 3-triethylsilylacrylic acid (E-form) reacts with :NCOOEt to give 1-ethoxycarbonyl-*trans*-2-triethylsilyl-3-methoxycarbonylaziridine in 22% yield [30]. The stereospecificity of this process indicates that the reacting species is singlet ethoxycarbonylnitrene [35].

The possibility of generating ethoxycarbonylnitrene from nitrene precursor 19 by α-elimination under liquid-solid PTC conditions was examined. Maximum yield of the target product was attained with potassium carbonate used as solid base [34]. Sodium carbonate and bicarbonate were less effective, whereas in the presence of calcium carbonate the reaction failed to occur. The yields of the corresponding silicon-containing aziridines were comparable with that gained by generating ethoxycarbonylnitrene under liquid-liquid PTC conditions.

The possibility of generating :NCOOEt in a two-phase liquid-solid system with ultrasound irradiation was investigated [34]. The corresponding aziridines 20 were formed in small amounts (5-15%) after continuous irradiation of a suspension of potassium carbonate in dichloromethane containing a nitrene precursor (19) and vinylsilane with aid of ultrasonic cell. An increase in ultrasound power attained by using an ultrasonic disintegrator allows one to decrease significantly the reaction time and increase product yield. Thus, the use of ultrasonication of high intensity (2000 W) instead of phase-transfer catalysis decreased reaction time more than 10-fold without affecting appreciably the yield of products.

5. Synthesis of silyl-substituted isoxazolines and their properties

The synthesis of 3-substituted 5-silylisoxazolines-2 is based on the ability of alkenylsilanes to undergo [2+3]-cycloaddition with nitrile oxides [36-45] generated by hydroxamic acid chlorides in the presence of bases or via dehydration of primary nitroalkanes with phenylisocyanate. Another method used for the synthesis of silylisoxazolines involves the reaction of silylated nitronic acid esters with unsaturated organosilicon compounds [46,47].

$$R^1CH_2NO_2 \xrightarrow[Et_3N, CCl_4]{Me_3SiCl} R^1C\underset{H}{=}N\underset{OSiMe_3}{\overset{O}{\diagup}} \xrightarrow{R^2_3SiCH=CH_2} \left[\begin{array}{c} R^2_3SiCH-CH_2 \\ O \diagdown \diagup CHR^1 \\ N \end{array} \right]$$

22

$R^1=H-; R^2=Me-, Et-, EtO-$
$R^1=Me-; R^2=Me-, Et-, EtO-$

$$R_3SiCH-CH_2 \atop O\diagdown_{N}\diagup CR^1_2 \xrightarrow{-Me_3SiOH} \overset{OSiMe_3}{23}$$

24

Primary nitroalkanes in the presence of trimethylchlorosilanes and equimolar amounts of triethylamine afford the appropriate silyl esters of nitronic acid 22. These esters react with silicon-containing olefins to give 3-unsubstituted 5-silylisoxazoline (in the case of O-trimethylsilyl ester of nitromethane) or 3-alkyl-5-silyl-substituted isoxazoline. The silicon-containing isoxazolidine intermediate 23 could not be isolated due to spontaneous elimination of trimethylsilanol [47].

The reaction with terminal alkenylsilanes occurs regioselectively to give invariably a product containing an oxygen atom at the carbon bearing the triorganylsilyl substituent.

Interestingly, alkenes with inactivated double bonds such as ethylvinyl ether, 1-hexene, cyclohexene do not react with silyl esters of nitronic acid or provide very low yields of the target product [48]. Triorganylsilyl group in vinylsilanes affects favourably the reactivity of the double bond because of its polarization.

The reaction of trimethylallylsilane with nitromethane O-silyl ester has very low yields (20%).

Methyl ester of 3-triethylsilyl-*trans*-acrylic acid (25) reacts regio- and stereospecifically with nitromethane O-silyl ester to give 4-triethylsilyl-*trans*-5-methoxycarbonylisoxazoline-2 [46]. Desilylation occurs concurrently under these reaction conditions, whereas the reaction between 25 and acetonitrile oxide produces two stereoisomers - 3-methyl-4-triethylsilyl-*trans*-5-methoxycarbonylisoxazoline-2 and 3-methyl-4-triethylsilyl-*cis*-5-methoxycarbonylisoxazoline-2 in the ratio 24:1.

The synthesis of 3-bromo-5-silyl-substituted isoxazolines, where a new functional group in position 3 enables further transformation to other nitrogen-containing organosilicon compounds, was attained by reacting alkenylsilanes with oxide of cyanogen bromide [49,50].

$$Br_2C=N\rightarrow OH \xrightarrow[EtOAc/TEBA]{K_2CO_3/H_2O} [BrC\equiv N\rightarrow O] \xrightarrow{Me_3M(CH_2)_nCH=CH_2}$$

26   27

$$\longrightarrow Me_3M(CH_2)_n\underset{O\underset{N}{\diagdown}}{\overset{CH-CH_2}{\diagup}}C-Br$$

a) n=0, M=Si
b) n=0, M=C
d) n=1, M=Si

28

Dibromoformaldoxime (26) in a two-phase system (aqueous solution of potassium carbonate/ethyl acetate) in the presence of phase-transfer catalyst (triethylbenzylammonium chloride) generated nitrile oxide 27 reacting regiospecifically with vinyl- and allylsilane as well as carbon analogue of trimethylvinylsilane - 3,3-dimethylbutene to give end products in good yield (50-70%) [50]. It should be pointed out that compound 28b is more thermostable than 3-bromo-5-trimethylsilylisoxazoline-2. For instance, isoxazoline 28a undergoes decomposition at 110°C, whereas 28b remains stable at 200°C during 16 h.

If substituted vinylsilanes have such substituents as bromine or 2,4-dinitrophenylthiol in α-position to the silicon atom, the reaction of nitrile oxide addition involves their elimination leading to the corresponding isoxazoles[36]. The reaction of trimethyl- and triethyl-α-bromovinylsilanes with oxide of cyanogen bromide in the presence of two equivalents of base also leads to 3-bromo-5-trimethyl(triethyl)silylisoxazoles.

Methyl ether of 3-triethylsilylacrylic acid reacts with nitrile oxide 27 to give two stereoisomers 3-bromo-4-triethylsilyl-trans-5-methoxycarbonylisoxazoline-2 and 3-bromo-4-triethylsilyl-cis-5-methoxycarbonylisoxazoline-2 in the ratio 24:1 [49]. Desilylation occurs concurrently under there reaction conditions.

The reaction of 5-triethoxysilylisoxazolines-2 with triethanolamine in toluene gives the corresponding 5-silatranylisoxazolines-2 [51].

3-Unsubstituted-5-trimethyl(triethyl)silylisoxazolines in acetonitrile in the presence of triethylamine at 80°C undergo ring opening which results in the corresponding silicon-containing 1,2-cyanohydrins, whereas in the two-phase system hexane/sodium hydroxide 5-trimethylsilylisoxazoline-2 affords 3-trimethylsilylacrylonitrile.

Reduction of 5-trimethylsilylmethylisoxazolines by hydrogen on Raney nickel catalyst or Pd/C involves N-O bond cleavage resulting in silicon-containing α,γ-amino alcohols [45]. The end-products without isolation were converted to unsaturated amines in the presence of titanium (IV) chloride. Reduction of silicon-containing isoxazolines by hydrogen on Raney's nickel was studied [52] in an attempt to retain the silyl group and the degree of oxidation at the carbon in position 3 of the isoxazoline cycle. The reaction performed in methanol in the presence of acetic acid or aluminium chloride involved ring opening at the N-O bond and subsequent deamination. Thus, 3-ethyl-5-triethylsilylisoxazoline yielded 1-triethylsilyl-1-hydroxypentanone-3.

REFERENCES

1. F.W. Colvin "Silicon in Organic Synthesis", London, Toronto, Butterworths, 1981.
2. W.P. Weber "Silicon Reagents for Organic Synthesis" Berlin, Heidelberg, New York Springer-Verlag, 1983.
3. M.G. Voronkov, G.I. Zelcans, E. Lukevics "Silicon and Life" Riga "Zinatne", 1978.
4. Yu. Goldberg, V. Dirnens, E. Lukevics, J.Organomet.Chem.Libr., 1988, 20, 211.
5. N.S. Nametkin, V.N. Perchenko, I.A. Grushevenko, Dokl.Akad. Nauk, 1964, 158, 404.
6. N.S. Nametkin, V.N. Perchenko, I.A. Grushevenko, L.G. Batalova, Proc. of Int.Symp. on Organosilicon Chem:, Prague, 1965, 323.
7. N.S. Nametkin, V.N. Perchenko, I.A. Grushevenko, G.L. Kamneva, Dokl.Akad.Nauk, 1966, 167, 106.
8. N.S. Nametkin, I.A. Grushevenko, V.N. Perchenko in: Organosilicon Compounds, Nauka, 1966, V.1, 47 (in Russian).
9. N.S. Nametkin, V.N. Perchenko, M.E. Kuzovkina, I.A. Grushevenko, Izv.Akad.Nauk SSSR,ser.khim., 1968, 1139.

10. N.S. Nametkin, V.N. Perchenko, I.A. Grushevenko, G.L. Kamneva, Izv.Akad.Nauk SSSR, ser.khim., 1968, 2074.
11. N.S. Nametkin, V.N. Perchenko, M.E. Kuzovkina, Dokl.Akad.Nauk, 1968, 182, 842.
12. I.A.Grushevenko,PhD Thesis,Moscow, 1967.
13. S.F. Thames, L.H. Edwards, J.Heterocycl.Chem., 1968, 5, 115.
14. L.J. Ledina, V.N. Perchenko, N.S. Nametkin, A.M. Krapivin, Dokl.Akad.Nauk, 1978, 243, 1200.
15. E. Lukevics, A.J. Pestunovich, M.G. Voronkov, Khim.Geterotsikl. Soed., 1969, 674.
16. E. Lukevics, A.J. Pestunovich, M.G. Voronkov, Khim.Geterotsikl. Soed., 1968, 949.
17. E. Lukevics, S. Germane, N.P. Erchak, O.A. Pudova, Khim.Farm. Zh., 1981, 15, 42.
18. E. Lukevics, E. Liepins, E.P. Popova, V.D. Shats, V.A. Belikov, Zh.Obshch.Khim., 1980, 50, 388.
19. V.D. Sheludyakov, N.A. Viktorov, V.F. Mironov, Zh.Obshch.Khim., 1972, 42, 364.
20. N.S. Nametkin, V.N. Perchenko, N.A. Grushevenko, M.E. Kuzovkina, Dokl.Akad.Nauk, 1969, 186, 1089.
21. S. Searles, M. Tammer, J.Amer.Chem.Soc., 1956, 78, 4917.
22. E. Lukevics, V.V. Dirnens, Y.S. Goldberg, E.E. Liepins, M.P. Gavars, I.J. Kalvins, M.V. Shymanska, Organometallics, 1985, 4, 1648.
23. V. Dirnens in Proc. of the 8-th Conf. of Young Scientists "Synthesis and Study of Biologically Active Compounds", Riga, 1984, 30.
24. Y.S. Goldberg, V. Dirnens, E. Lukevics in: New Methodological Principles in Organic Synthesis, IV-th All-Union Symp. on Organic Synthesis, Abstracts, Moscow, Nauka, 1984, 50 (in Russian).
25. E. Lukevics, V.V. Dirnens, Y.S. Goldberg, E.E. Liepins, I.J. Kalvins, M.V. Shymanska, J.Organomet.Chem., 1984, 268, C29.
26. V.I. Markov, V.A. Doroshenko, Zh.Org.Khim., 1972, 8, 1251.
27. E. Lukevics, V.N. Gevorgyan, V. Dirnens, L.M. Ignatovich, Y.S. Goldberg, M.V.Shymanska in: New Methodological Principles in Organic Synthesis, IN-th All-Union Symp. on Organic Synthesis, Abstracts, Moscow, Nauka, 1984, 34 (in Russian).
28. G.R. Harvey, J.Org.Chem., 1968, 33, 887.
29. S.A. Hiller, A.V. Eremeyev, I.J. Kalvins, E.E. Liepins, V.G. Semenikhina, Khim.Geterotsikl.Soed., 1975, 1625.
30. E. Lukevics, V.V. Dirnens, Yu.Sh. Goldberg, E.E. Liepins, J.Organomet.Chem., 1986, 316, 49.
31. E.V. Dehmlov, S.S. Dehmlov "Phase Transfer Catalysis" Weinheim, Verlag.Chem., 1980.
32. J.H. Wagenknecht, M.M. Baizer, J.L. Chruma, Synth.Commun., 1972, 2, 215.
33. Y.S. Goldberg, V. Dirnens, E. Liepins, I.J. Kalvins, M.V. Shymanska in: VI-th All-Union Conf. on the Chemistry and Appl. of Organic Compounds, Abstracts, Riga, 1986, 222.
34. V. Dirnens, Yu.S. Goldberg, E. Lukevics, Dokl.Akad.Nauk, 1988, 298, 116.
35. W. Lwowski in "Nitrenes" Ch.3, New York, Interscience Publishers, 1970.
36. A. Padwa, J.G. MacDonald, J.Org.Chem., 1983, 48, 3189.

37. I.G. Kolokoltseva, V.N. Chistokletov, A.A. Petrov, Zh.Obshch. Khim., 1970, 40, 2612.
38. I.G. Kolokoltseva, V.N. Chistokletov, A.A. Petrov, Zh.Obshch. Khim., 1970, 40, 2618.
39. I.G. Kolokoltseva, V.N. Chistokletov, M.D. Stadnichuk, A.A. Petrov, Zh.Obshch.Khim., 1968, 38, 1820.
40. Fr. 1,371,325; C.A. 1965, 62, 1689.
41. Fr. 84,686; C.A. 1965, 63, 4332.
42. G.A. Shwehgeimer, J.V. Arslanov, A. Baranski, Roczniki chem., 1972, 46, 2381.
43. A. Padwa, J.G. MacDonald, Tetrahedron Lett, 1982, 23, 3219.
44. R.F. Cunico, J.Organomet.Chem., 1981, 212, C51.
45. A. Hosomi, H. Shoji, H. Sakurai, Chem.Lett., 1985, 1049.
46. V. Dirnens, L.M. Ignatovich, E. Lukevics in Proc. IV-th All-Union Conf. on the Chem. of Nitrogen-containing Heterocycl. Comp., Novosibirsk, 1987, 92.
47. E. Lukevics, V. Dirnens, Latv.PSR Zin.akad.vestis, chem.ser. 1990, 235.
48. K. Torsell, O. Zeutlen, Acta chem.Scand., 1978, B32, 118.
49. E. Lukevics, V. Dirnens in VI-th All-Union Conf. on the Structure and Reactivity of Organosilicon Comp., Abstracts, Irkutsk, 1989, 11.
50. E. Lukevics, V. Dirnens, Latv.PSR Zin.akad.vestis, chem.ser, 1990, 236.
51. E. Lukevics, V. Dirnens in IV-th All-Union Conf. on the Structure and Reactivity of Organosilicon Comp., Abstracts, Irkutsk, 1989, 172.
52. V. Dirnens, Zh.G. Yuskovets, M.V. Shymanska, E. Lukevics in Chem. and Pract. Appl. of Organosilicon Comp.? Abstracts, Leningrad, Nauka, 1988, 45.

Palladium-catalyzed Insertion Reactions of Unsaturated Carbon Compounds into Silicon-Silicon Bonds of Polysilanes

Yoshihiko Ito
DEPARTMENT OF CHEMISTRY, KYOTO UNIVERSITY, KYOTO 606, JAPAN

Much interest has been focused on organosilicon compounds from the view-point of application in organic synthesis as well as synthesis of new silicon-containing functional materials. This paper describes new synthetic methods of organosilicon compounds by palladium(0) catalyzed insertion reactions of unsaturated carbon compounds into silicon-silicon bonds of polysilanes, which have been developed in our laboratory.

[1] Palladium Catalyzed Insertion Reactions of Isonitriles into the Silicon-Silicon Bonds of Polysilanes.

Insertion reaction of isonitriles into the silicon-silicon bond of disilanes was catalyzed by palladium(0) tetrakis(triphenylphosphine) complex or palladium(II) acetate to give the corresponding N-substituted bis-(organosilyl)imines in moderate yields.[1]

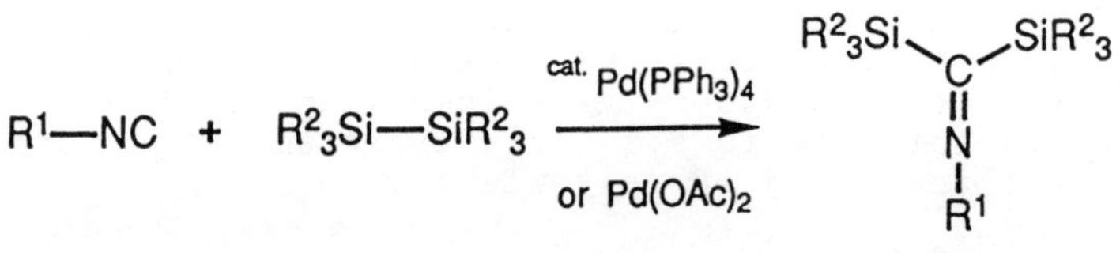

$$R^1-NC + R^2{}_3Si-SiR^2{}_3 \xrightarrow[\text{or Pd(OAc)}_2]{\text{cat. Pd(PPh}_3)_4} \underset{\underset{R^1}{\overset{\|}{N}}}{R^2{}_3Si\diagdown\underset{C}{}\diagup SiR^2{}_3}$$

R^1 = 2-tolyl, 2,6-xylyl, 2,6-di(isopropyl)phenyl, *cyclo*-hexyl   25 - 81 %

In general, N-substituted bis(organosilyl)imines thus prepared are yellow liquid which are thermally stable and distillable, but decomposed under air, turning dark-brown. N-Aryl bis(organosilyl)imines, especially N-(2,6-xylyl) and N-(2,6-diisopropylphenyl) bis(organosilyl)-imines, are more stable than the respective N-alkyl bis-(organosilyl)imines. Various isonitriles reacted with disilanes afford N-substituted bis(organosilyl)imines.

However, an exception was tert-alkyl isocyanides, such as tert-butyl isocyanide, 1-adamantyl isocyanide and 1,1,3,3-tetramethylbutyl isocyanide, which did not react with disilanes at all and remained unreacted in the reaction mixture. This finding was significantly important for the skeletal rearrangement of tetrasilanes and the bis-silylation of alkynes, which were both promoted or catalyzed by palladium(II) acetate in the presence of tert-alkyl isocyanide, as mentioned later.

The palladium catalyzed insertion of isonitrile into the silicon-silicon bond of disilane may be reasonably explained by a mechanism involving the following catalytic cycle (Scheme).

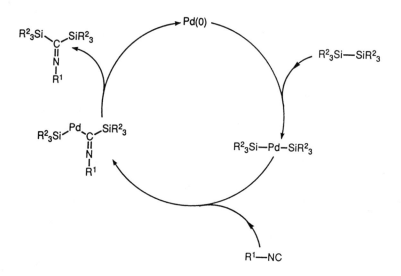

The palladium catalyzed insertion of isonitrile into the silicon-silicon linkage is extended to polysilanes, giving poly[sila(N-2,6-xylyl)imines], in which an isonitrile molecule is regularly inserted into each silicon-silicon bond of polysilanes.[2]

a :n=2 $R^1=R^2=R^3=R^4=Me$ 57%
b :n=2 $R^1=R^2=Me, R^3=R^4=Ph$ 78%
c :n=2 $R^1={}^tBu, R^2=R^3=R^4=Me$ 82%
d :n=3 $R^1=R^2=R^3=R^4=Me$ 55%
e :n=3 $R^1=R^2=Me, R^3=R^4=Ph$ 43%
f :n=4 $R^1=R^2=Me, R^3=R^4=Ph$ 40%
g :n=5 $R^1=R^2=R^3=R^4=Me$ 28%

For instance, decamethyltetrasilane was heated in toluene or dimethylformamide with an excess of 2,6-xylyl isocyanide in the presence of a catalytic amount of palladium(II) acetate to give 1,1,1,3,3,5,5,7,7,7-decamethyl-2,4,6-tris(2,6-xylylimino)-1,3,5,7-tetrasilaheptane in 57% isolated yield as a yellow crystalline solid, which was recrystallized from ethanol [mp 110-112 °C; IR 1544 cm^{-1}; UV(C_6H_{12}) λmax 407 nm (ε 540)]. The poly-insertions of 2,6-xylyl isocyanide into polysilanes up to hexasilane have been successfully achieved to afford the desired isolable crystalline poly[sila(N-2,6-xylyl)-imines].

There is no doubt that the palladium catalyzed poly-insertion of isonitriles into polysilanes proceeds by a stepwise mechanism. Indeed, a partial insertion of isonitriles into the silicon-silicon bonds of polysilanes was achieved by using a limited amount of isonitriles. Some examples are shown below.

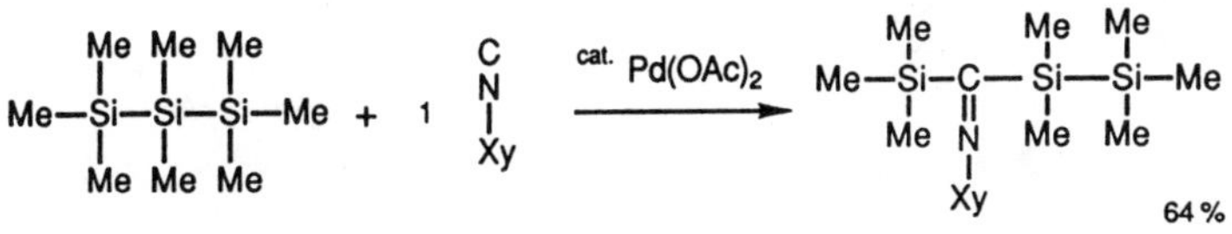

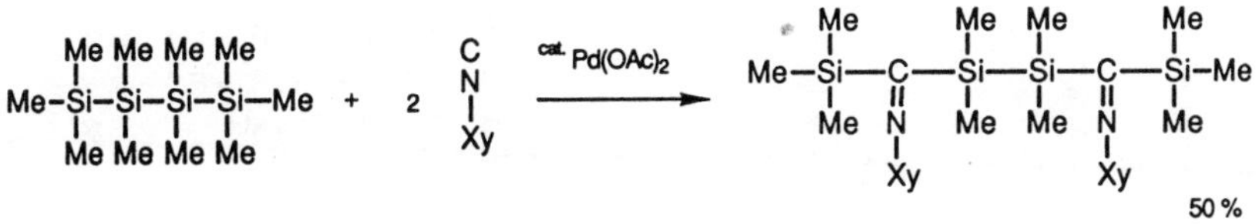

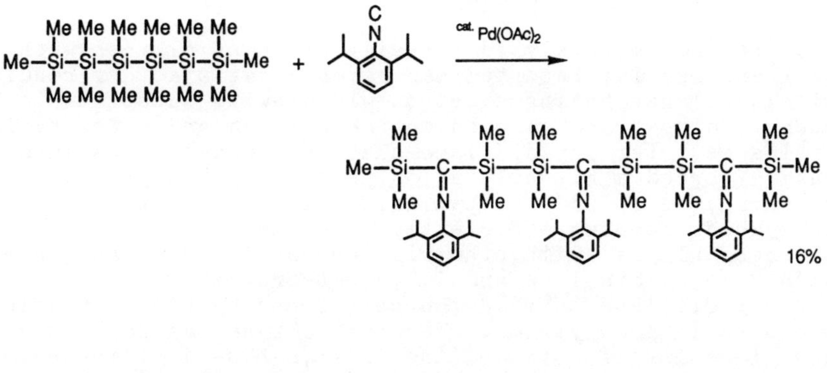

In the course of our study on the palladium catalyzed insertion of isonitrile into polysilanes, it was found that a novel skeletal rearrangement reaction of tetrasilane with aryl isocyanide took place to give 3,3-disilyl-2,4-disila-1-azacyclobutane derivative. The skeletal rearrangement is unique and interesting in that the product is completely reconstituted from the four fragments of tetrasilane and two fragments of isonitrile. The rearrangement product was formed in a moderate yield, when tetrasilane was refluxed in toluene with 1.5 molar equivalent of 2,6-xylyl isocyanide or 2,6-diisopropylphenyl isocyanide in the presence of 1,1,3,3-tetramethylbutyl isocyanide. The rearrangement reaction in the absence of 1,1,3,3-tetramethylbutyl isocyanide gave the product only in low yield. The role of the tert-alkyl isocyanide, which is not incorporated in the product, is not clear.[3]

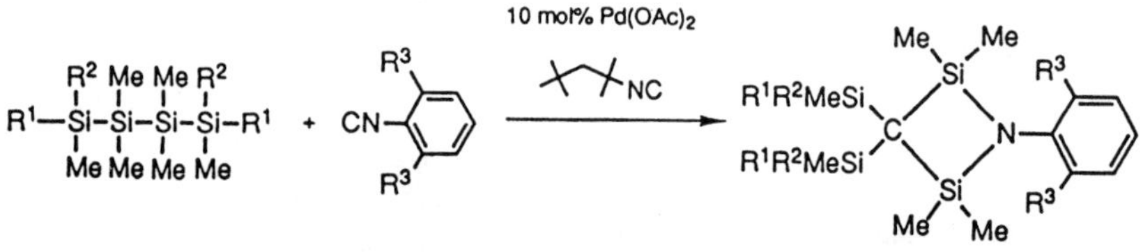

a: $R^1 = R^2 = Me, R^3 = Me$ — 40 %
b: $R^1 = R^2 = Me, R^3 = Pr^i$ — 45 %
c: $R^1 = Ph, R^2 = Me, R^3 = Me$ — 31%
d: $R^1 = Ph, R^2 = Me, R^3 = Pr^i$ — 62%
e: $R^1 = R^2 = Ph, R^3 = Me$ — 28 %
f: $R^1 = R^2 = Ph, R^3 = Pr^i$ — 48 %

[2] Bis-silylation of Alkynes Catalyzed by Palladium/tert-Alkyl Isocyanide Complex.

Palladium catalyzed bis-silylation of alkynes with disilanes so far reported has given a satisfactory result with disilanes having electron-withdrawing substituents such as alkoxy and halogen on the silicon and a few cyclic disilanes. Peralkyldisilanes such as hexamethyldisilane have afforded a bis-silylation product in low yield. Recently, we found that palladium(II) acetate/tert-alkyl isocyanide complex efficiently catalyzed the bis-silylation of alkynes with otherwise unreactive disilanes such as hexamethyldisilane and 1,1,2,2-tetramethyl-1,2-diphenyldisilane to give the corresponding bis-silylation products in good yields. Terminal alkynes as well as acetylene itself afforded the (Z)-1,2-bis-silylethenes. However, internal alkynes did not react with disilanes. In general, (Z)-isomers, which arose from *cis*-addition of the silicon-silicon bond to alkynes, were produced predominantly.[4]

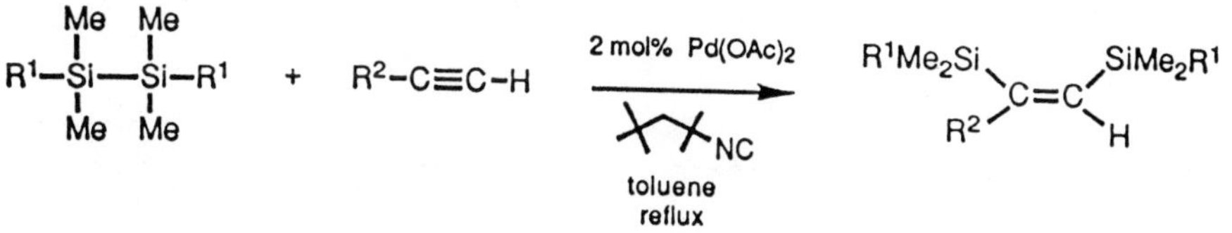

$$\text{R}^1\text{-}\underset{\underset{\text{Me}}{|}}{\overset{\overset{\text{Me}}{|}}{\text{Si}}}\text{-}\underset{\underset{\text{Me}}{|}}{\overset{\overset{\text{Me}}{|}}{\text{Si}}}\text{-R}^1 + \text{R}^2\text{-C}\equiv\text{C-H} \xrightarrow[\substack{\text{toluene}\\\text{reflux}}]{\text{2 mol\% Pd(OAc)}_2,\ \text{XXNC}} \underset{\text{R}^2}{\overset{\text{R}^1\text{Me}_2\text{Si}}{}}\text{C=C}\underset{\text{H}}{\overset{\text{SiMe}_2\text{R}^1}{}}$$

R^1	R^2	% Yield	Z : E
Me	Ph	82	96 : 4
Me	Hex^n	81	95 : 5
Ph	Hex^n	96	100 : 0
Ph	H	98	97 : 3

A significant feature of the palladium catalyst for the bis-silylation of alkynes is the use of excess tert-alkyl isocyanide as ligand. In the absence of tert-alkyl isocyanide, the bis-silylation did not occur at all.

The palladium catalyzed bis-silylation of alkynes was successfully applied to the reaction of octamethyltrisilane with phenylethyne giving a mixture of regioisomeric double bis-silylation products in high total yield.

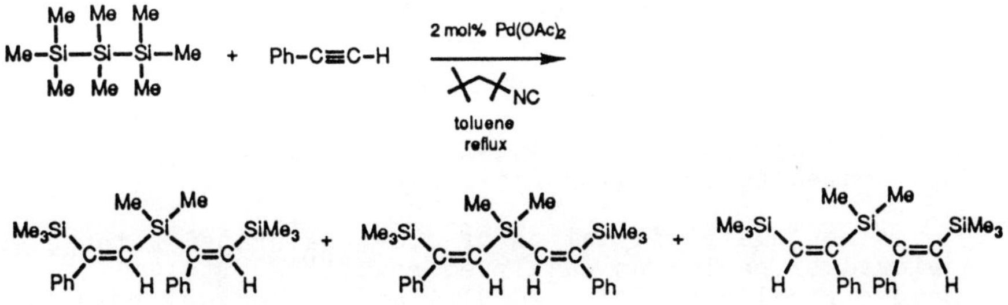

$$\text{Me-}\underset{\underset{\text{Me}}{|}}{\overset{\overset{\text{Me}}{|}}{\text{Si}}}\text{-}\underset{\underset{\text{Me}}{|}}{\overset{\overset{\text{Me}}{|}}{\text{Si}}}\text{-}\underset{\underset{\text{Me}}{|}}{\overset{\overset{\text{Me}}{|}}{\text{Si}}}\text{-Me} + \text{Ph-C}\equiv\text{C-H} \xrightarrow[\substack{\text{toluene}\\\text{reflux}}]{\text{2 mol\% Pd(OAc)}_2,\ \text{XXNC}}$$

products (three regioisomers shown) 95%, ≈ 4 : 2 : 3

A selective silylstannation of some alkynes with disilylstannanes was also catalyzed by palladium(0) tetrakis(triphenylphosphine) to give the corresponding (β-disilylalkenyl)stannanes, which may be derived from cis-addition of the silicon-tin bond to alkyne triple bond, adding the stannyl group at the internal carbon of alkynes.[5]

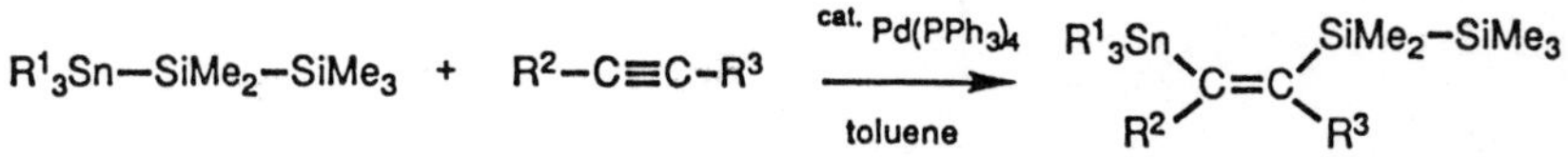

$R^1_3Sn-SiMe_2-SiMe_3$ + $R^2-C\equiv C-R^3$ $\xrightarrow[\text{toluene}]{\text{cat. Pd(PPh}_3)_4}$ $R^1_3Sn\diagdown_{R^2}C=C\diagup^{SiMe_2-SiMe_3}_{R^3}$

R^1	R^2	R^3	% Yield
Me	EtO$_2$C	H	63
Me	MeO$_2$C	MeO$_2$C	63
Me	Ph	H	42
Bu^n	MeO$_2$C	MeO$_2$C	55

Of interest is that (β-disilylalkenyl)stannanes thus prepared underwent a regioselective cycloaddition with phenylacetylene in the presence of palladium(0) tetrakis-(triphenylphosphine) catalyst to provide 3-sila-6-stanna-1,4-cyclohexadienes as a single isomer in high yields. Intermediacy of 1-sila-2-stannacyclobutene which might be formed via oxidative addition of the silicon-silicon bond of (β-disilylalkenyl)stannanes onto palladium(0) followed by an elimination of tetraalkylsilane, may be presumed.

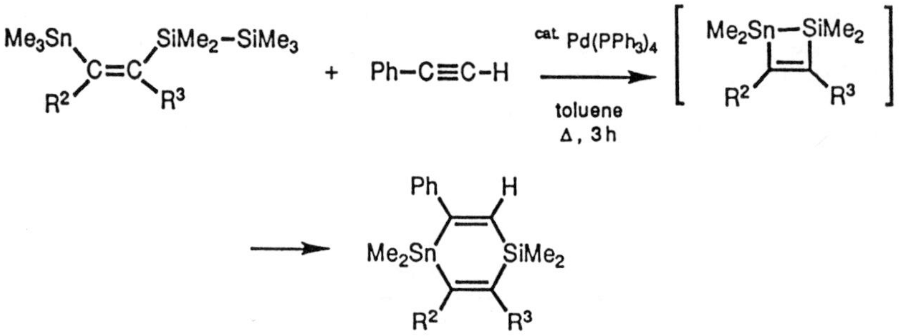

$Me_3Sn\diagdown_{R^2}C=C\diagup^{SiMe_2-SiMe_3}_{R^3}$ + $Ph-C\equiv C-H$ $\xrightarrow[\substack{\text{toluene}\\\Delta, 3h}]{\text{cat. Pd(PPh}_3)_4}$ $\left[\begin{array}{c} Me_2Sn-SiMe_2 \\ \diagdown_{R^2}\diagup^{R^3} \end{array} \right]$

$\longrightarrow$

Ph, H on Me$_2$Sn—SiMe$_2$ ring with R^2, R^3

[3] Palladium Catalyzed 1,4-Disilylation of α,β-Unsaturated Ketones with Disilanes.

A conjugate silylation of α,β-unsaturated ketones was achieved by palladium catalyzed reactions with the unsymmetrically substituted disilane, 1,1-dichloro-1-phenyl-2,2,2-trimethyldisilane added to some α,β-unsaturated ketones in the presence of palladium(0) tetrakis-(triphenylphosphine) complex with phenyldichlorosilyl group and trimethylsilyl group attacking at β-carbon and carbonyl oxygen, respectively. Treatment of the 1,4-disilylation product with an excess of methyllithium, which should generate lithium β-(phenyldimethylsilyl)-ketone enolate, followed by hydrolysis afforded the corresponding β-(phenyldimethylsilyl)ketone. The 1,4-disilylation product was also worked up with ethanol and triethylamine and isolated as β-(phenyldiethoxysilyl)-ketone.

The 1,4-disilylation of α,β-unsaturated ketones was not observed with other unsymmetrical disilanes $(MeO)_3SiSiMe_2X$ (X = Cl, F) or symmetrical disilanes $XMe_2SiSiMe_2X$ (X = F, Cl, Ph).[6]

The β-(phenyldimethylsilyl)ketones were readily converted to β-hydroxy ketones by fluorodephenylation with $HBF_4 \cdot Et_2O$ followed by oxidation of the resulting silicon-carbon bond with H_2O_2 in the presence of KF. The β-(phenyldiethoxysilyl)ketones were directly substituted with hydroxy group by the oxidation with H_2O_2. The oxidation of carbon-silicon bond has been known to proceed with retention of stereochemical configuration.

Alkylation of lithium β-(phenyldimethylsilyl)ketone enolates, which are generated *in situ* by treatment of the 1,4-disilylation product with methyllithium, took place at α-position with anti-stereochemistry (>20 : 1) to give 4-(phenyldimethylsilyl)-3-alkyl substituted ketones.

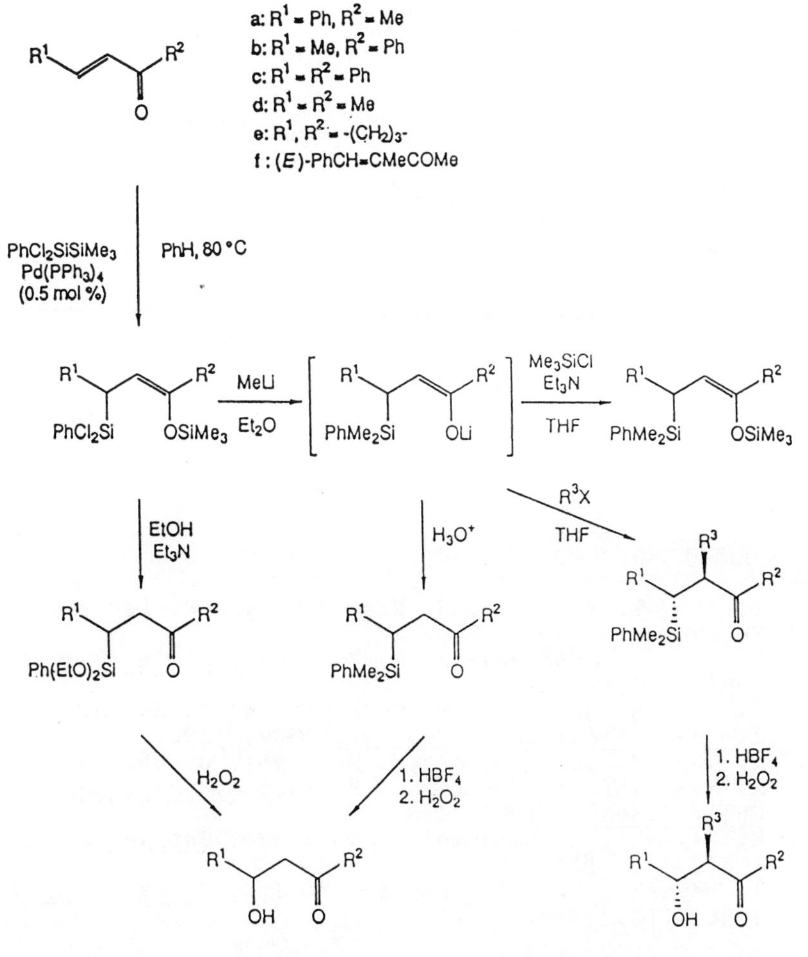

a: R^1 = Ph, R^2 = Me
b: R^1 = Me, R^2 = Ph
c: R^1 = R^2 = Ph
d: R^1 = R^2 = Me
e: R^1, R^2 = -(CH$_2$)$_3$-
f: (E)-PhCH=CMeCOMe

Catalytic asymmetric 1,4-disilylation of α,β-unsaturated ketones (E)-R^1CH=CHCOR^2 (R^1 = Me; R^2 = Me, Ph, 4-MeOC$_6$H$_4$: R^1 = i-Pr; r^2 = Ph: R^1 = Pr; R^2 = Me) was found to proceed with 1,1-dichloro-1-phenyl-2,2,2-trimethyldisilane in the presence of dichloro[(+)-2,2'-bis(diphenylphosphino)-1,1'-binaphthyl]palladium(II) as a catalyst (0.5 mol%), giving optically active β-(phenyldimethylsilyl)ketones of 74-92 %ee, after treatment with n-butyllithium followed by hydrolysis.[7]

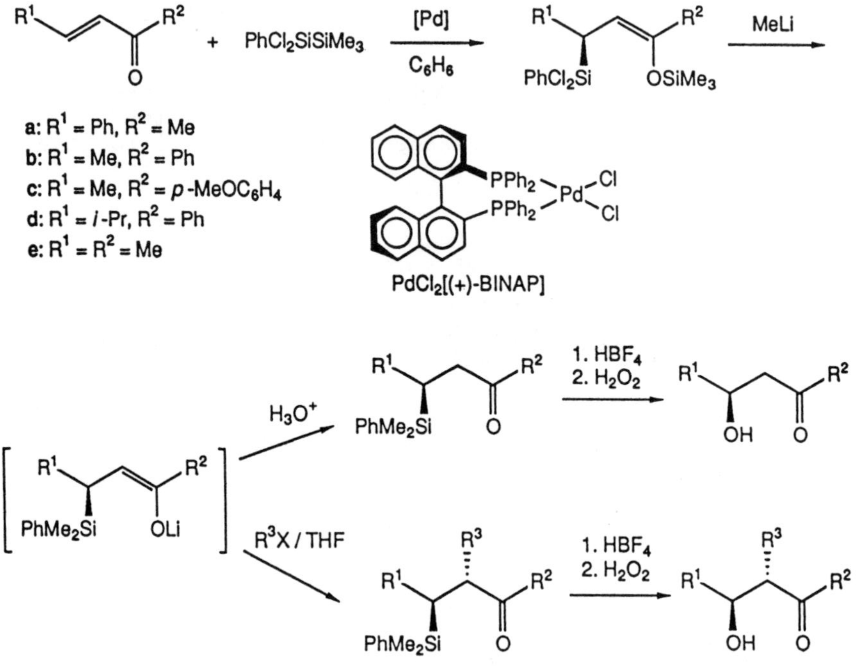

a: R^1 = Ph, R^2 = Me
b: R^1 = Me, R^2 = Ph
c: R^1 = Me, R^2 = p-MeOC$_6$H$_4$
d: R^1 = i-Pr, R^2 = Ph
e: R^1 = R^2 = Me

PdCl$_2$[(+)-BINAP]

REFERENCES

1. Y. Ito, T. Matsuura, S. Nishimura and M. Ishikawa, Tetrahedron Lett., 1986, 27, 3261.
2. Y. Ito, T. Matsuura and M. Murakami, J. Am. Chem. Soc., 1988, 110, 3692.
3. Y. Ito, M. Suginome, M. Murakami and M. Shiro, J. Chem. Soc., Chem. Commun., 1989, 1494.
4. Y. Ito, M. Suginome and M. Murakami, unpublished.
5. M. Murakami, Y. Morita and Y. Ito, J. Chem. Soc., Chem. Commun., 1990, 428.
6. T. Hayashi, Y. Matsumoto and Y. Ito, Tetrahedron Lett., 1988, 29, 4147.
7. T. Hayashi, Y. Matsumoto and Y. Ito, J. Am. Chem. Soc., 1988, 110, 5579.

Subject Index

Acylsilanes
  enantioselective microbial
    reduction, 224
(2-Adamantyl-3,3-di-*tert*-
    butyl-3-*H*
    silaphosphirene)
    tungsten pentacarbonyl,
  crystal structure, 124
Agostic interactions, 237,
    248
Alkenylsilanes
  addition of amines, 378
  anti-Markovnikov addition,
    381
  cycloaddition, [2 + 3];
    with nitrile oxides,
    386
  effect of substituents on
    additions to, 379
  reaction with
    ethoxycarbonyl nitrene,
    385-386
Alkenylsilanols
  Sharpless epoxidation,
    351-354
Alkynes
  bissilylation, Pd(0)
    catalysed, 394
  silylstannylation, Pd(0)
    catalysed, 395
Allenylsilanes
  cycloaddition with
    chlorosulphonyl
    isocyanate (CSI), 361
  optically active, 326-327
  synthesis, 362
Allylsilanes
  chemoselectivity using,
    332
  chiral, 344-355
  cycloadditions with CSI,
    357-360

Allylsilanes (continued)
  electrophilic substitution
    321
  enantioselectivity using,
    332, 338-340
  epoxidation, 321, 341, 351
  hydroboration, 321, 322,
    325
  isomerization
    photostationary states,
    136-137
  β-lactams from, 356-365
  Lewis acid catalysed
    aldehyde reactions,345-
    348
  nitrile oxide
    cycloaddition, 321
  osmylation, 321
  photochemical 1,3-silyl
    migration, 134-135
  protodesilylation, 329
  reaction with aldehydes,
    333, 335
    Felkin-Anh T.S, 338
    Hayashi-Kumada T.S, 337
    stereoselectivity, 336
  reaction with heterocyclic
    amines, 378
  reaction with iminium
    ions, 334
  reaction with lactones,334
  reaction with thioacetals,
    334
  Simmons-Smith reaction,
    321
  stereochemistry,
    conformational
    analysis, 321
  stereocontrol using, 321-
    331
  stereoselectivity using,
    332

Allylsilanes (continued)
  thermal 1,3-silyl
    migration, 134-135
    stereochemistry, 134,
      136
  urethane-cuprate synthesis
    of, 329
Azetidin-2-one
  3,3-dimethyl-4-
    (trimethylsilylmethyl),
    359

Benzocyclobutenes, 287
Binuclear metal alkyls
  (R=(Me$_3$Si)$_2$CH)
  aluminium, R$_2$AlAlR$_2$, 239
  gallium, R$_2$GaGaR$_2$, 239
  germanium, R$_2$GeGeR$_2$, 239,
    241
  indium R$_2$InInR$_2$, 239
  tin R$_2$SnSnR$_2$, 239
Biocatalysis, 224
Bioisosteric relationships,
  224
Bioorganosilicon chemistry,
  218-227
Biotransformation of
  organosilicon
  compounds, 218
Biphosphene, R$_2$P-PR$_2$
  [R=(Me$_3$Si)$_2$CH]
  P-P bond homolysis, 260
  structure, 260-261
  synthesis, 254
Bis(benzene-1,2-
  diolato)silicates, 189
1,2-Bis(dialkylchloro-
  silyl)ethane
  dehalogenation of, 52
1,2-Bis(dimethylchloro-
  silyl)ethane
  copolymerization with
    Me$_2$SiCl$_2$, 54
  dechlorination of, 52
  polymerization by Wurtz
    coupling, 54
Bis(dimethylchlorosilyl)-
  methane
  polymerization by Wurtz
    coupling, 53
Bis(trimethylsilyl)methane
  structure, 259

Bis(trimethylsilyl)methyl
  ligand, 231-250
Bis(trimethylsilyl)methyl
  metal complexes, 231-
  250
Bis-(tri-t-butyl)triazene
  reactions, 268
  synthesis, 268
  thermolysis, 269
Borylene (Me$_3$Si)$_2$CHB:
  formation and trapping,
    237-239
t-Butyl groups on silicon
  crowding in molecules,
    253, 262

Carbon-silicon bond
  oxidative cleavage, 364
  stereoselectively, 349
Carbonyl compounds
  reduction by anionic
    hydridosilicates, 187
Ceramic fabrication
  "catalysts", 41
  control of SiC to C
    content, 35
  densification, 29
  excess carbon, 29
  rule of mixtures, 29
Ceramic fibres, 28, 40
Ceramic matrix composites,
  28
Chiral organosilicon
  compounds, 218, 223,
    344-355, 398
Chlorosilanes
  acetolysis,
    steric substituent
      constants for, 209
  reaction with silanolates
    and alkoxides, 211
cis-1,1-Diadamantyl-2,3-
  dimethylsilirane
  photochemical
    decomposition, 102
  synthesis, 101
  thermal decomposition, 102
Cyanotrimethylsilane
  ambident
    nucleophilicity, 366
  reactions, 366-376
Cyclopentasilanes
  bond lengths, 279

Cyclopentasilanes (continued)
  pseudorotation, 279
Cyclopropenylsilanes
  precursors to low valent and multiply bonded silicon compounds, 285–294
  photochemical precursors, 287
  photoisomerism,
    products and mechanism, 287–291
  photolysis, 286
Cyclotetrasilanes
  bond lengths, 278
  conformation, 278
Cyclotrisilanes
  structures, 277

Decaborane
  reactions with diamines, 22
Decamethylsilicocene
  Au(I) complex, 312
  cyclic voltammetry, 309
  He(I) PES spectrum, 309
  mass spectrum, CI and EI, 308
  NMR, ^{29}Si, 308, 310
  oxidative addition to, 310, 314
  X-ray crystal structure, 309
Decamethyltetrasilane
  photolysis, 98
  pyrolysis, 89
    Arrhenius parameters, 91
    mechanisms, 94–95
    primary steps, 91
    with toluene and butadiene, 90
Decarborane-diamine polymers
  properties, 23
  pyrolysis, 23
    mechanism, 23
    to boron nitride, 23
Decarborane-diaminosilane polymers
  ceramic fibers from, 24
  formation and properties, 24

1,3-Dehydroadamantane
  hydrosilylation, 100–101
Dewar silabenzenes, 289
Di-*tert*-butylsilylene
  addition to aryl isonitriles, 131
  addition to 2,2'-bipyridyls, 129
  addition to nitriles, 123
  addition to phenyl isothiocyanate, 123
  addition to phosphaalkynes, 123
  addition to tri-*tert*-butylsilyl azide, 128
  formation, 122
2,2-Diadamantyl-1,3-diphenyl-1,1,3,3-tetramethyltrisilane
  photolysis, 100–101
1,1-Diadamantyl-2,2,2-triethyldisilane
  synthesis, 102
Diadamantyldiiodosilane
  dehalogenation, 101, 103, 105
  silylenoid from, 104
  synthesis, 102
2,2-Diadamantylhexamethyltrisilane
  photolysis, 100
1,1-Diadamantylsiliranes
  photochemical and thermal extrusion
    of diadamantylsilylene, 102–103
  synthesis, 101
1,1-Diadamantylsilirenes
  synthesis, 101
Diadamantylsilylene
  addition to *cis*-3-hexene, 102
  addition to 3-hexyne, 103–104
  addition to *trans*-3-hexene, 102
  calculated bond angle, 100
  insertion into triethylsilane, 103–104
Dialkyldiethynylsilanes,
  conducting polymers from, 11
  as optical switches, 14

Dialkyldiethynylsilanes (continued)
  iodine doping:bipolaron and soliton absorptions, 12, 13
  nonlinear optical properties, 14
Diborane
  reactions with silylamines and silazenes, 18
Dichlorobis(pentamethylcyclopentadienyl)silane reduction, 308
α,ω-Dichlorooligosilanes, hydrolytic polycondensation, 71
Diethynyldiphenylgermane, polymerization, 10
Diethynyldiphenylsilane,
  catalytic polymerization with $MoCl_5$, WCl_6, 11,
  structure of polymer, 11
  thermal polymerization, 9, 10
Dilithiobutadiyne, 10,
Dimesitylsilylene
  addition to tri-t-butylsilyl azide, 128
Dimethylallenylphenylsilane 135
Dimethylgermylene
  electronic absorption spectrum, 148
  extinction coefficient, 149
  mechanism for addition to 1,3-dienes, 107
  photochemical generation, 105
  PPh_3 complex, 148, 151
    electronic absorption spectrum, 151
    molar extinction coefficient, 151
    rate constants for reactions, 107, 152, 153
  reactivity with substituted 1,3-dienes, 106
  1,2-versus 1,4-addition to 1,3-dienes, 106

Dimethylphenyl(trimethylsilyl)germane
  photolysis, 105
  germene formation, 105
  rate constants for disappearance of transient, 107
Dimethylpropargylphenylsilane
  photoisomerization, 135
Dimethylsilylacetylene
  thermal isomerization, 4
Diphenylgermylene
  photochemical generation, 105
Diselenadisiletane, 316
1,4-Disila-1,3-butadiene-1,1,4,4-tetramethyl-2,3-bis(trimethylsilyl)
  formation and reactions, 173-174
1,2-Disila*closo*-dodecaborane(12)
  crystal structure, 25
  formation, 24
  properties, 26
1,3-Disilacyclobutane polymers, 50
1,2-Disilacyclobutanes
  synthesis, 51
Disilacylopropanes, 277
Disilanes
  Si-Si bond lengths, 272
  solid state conformations, 273
  X-ray and molecular mechanics, 273
Disilanyldiazo compounds
  photolysis, 171
Disilaoxiranes, 277
Disila[2.1.1]propellane
  attempted synthesis, 175-176
Disilathiiranes, 277
Disilazanes
  synthesis, 45
Disilenes
  structural parameters, 271
1,1-Disilylethane
  synthesis, 67
1,2-Disilylethane
  synthesis, 67

Subject Index

β-(Disilylalkenyl)stannanes,
  regioselective
    cycloaddition, 396
Disilylmethane
  synthesis, 63, 64
Disilylmethane
  feedstock for PECVD, 64
2,2-Disilylpropane
  synthesis, 67
Disyl groups, (Me$_3$Si)$_2$CH
    (see also homoleptic
    metal alkyls),
  bond lengths and angles,
    257
  structure by electron
    diffraction, 255
Dithiadisiletane, 316
Divinylsilylenes
  matrix isolation, 140-141
  photochemical generation,
    140-141
  transition energies and
    molecular orbitals,
    141-142

Ebelactone-a
  stereochemical
    analysis, 322-327
  strategy for
    synthesis, 322-327
Enantioselective enzymatic
  hydrolysis, 227
Enantioselective synthesis,
  344-355
Epoxides
  enantioselective
    synthesis, 352
  protodesilylation
    stereochemistry, 352
  ring opening with
    inversion of
    configuration, 370
Ethynylmethylphenylsilane,
  polymerization, 7

Fluorosilicates
  reactivity, 188

7-Germanorbornadiene
  7,7-dimethyl-1,4,5,6-
    tetraphenyl-2,3-dibenzo
  catalytic decomposition,
    154

7-Germanorbornadiene
    (continued)
  CIDNP, 147
  flash photolysis, 147
  matrix photolysis, 150
  molecular structure, 146
7-Germanorbornadienes
  photochemical
    decomposition
    mechanism of, 146
2-Germapropene
  rearrangement to
    dimethylgermylene, 109
Green body, 9, 29

Heptasila[7]paracyclophane
  synthesis, 180
  X-ray crystal structure,
    181
Hexa-t-butyldisilane, 265
Hexa-t-butyldisilazane, 264
Hexa-t-butyldisiloxane, 264
hexa-t-butylsulphane, 264
Hexa-t-butytlcyclo-
    trisilane
  photolysis, 122
Hexacoordinate silicon
  dynamic ^{1}H NMR, 193
  reactivity, 194
  X-ray crystal structure,
    193
Hexasilacyclooctyne
  permethyl
    addition of Mes$_2$Si, 177
    photochemistry, 177, 179
    synthesis, 177
Homoleptic metal alkyls
    (R=(Me$_3$Si)$_2$CH)
  lanthanides LnR$_3$, 243-244
    adduct formation, 248
    reaction, 243
  lithium (RLi)$_\infty$,.
  X-ray crystal structure,
    235
  reactivity, 245
  magnesium (MgR$_2$)$_\infty$
    neutron diffraction, 236
  manganese MNR$_2$,
    reactivity, 246
  rubidium [Rb(μ-R)-
    pmdeta]$_2$,
  X-ray structure, 235

Homoleptic metal alkyls
(continued)
  sodium (NaR)∞,
    X-ray crystal structure,
      235
  uranium UrR3,
    X-ray crystal structure,
      248
Hydridosilicates
  as SET reducing agents,
    187-188
  NMR ^{29}Si, 187-188
  preparation, 187
Hydrosilylation, 383
Hypervalent silicon, 185-195

Iridium complexes (see
  silene complexes)
Iron-silicon compounds, 275
Isocyanates
  reaction with
    pentacoordinate
    silicon, 191
Isonitriles
  from epoxides, 366
  regioselective formation,
    368
  stereoselective formation,
    368
Isothiocyanates
  reaction with
    pentacoordinate
    silicon, 191

Ketones, α,β-unsaturated
  catalytic asymmetric 1,4-
    disilylation, 396-398

β-Lactams
  synthesis, 356-365
  rearrangement, 358
Limits to Growth
  MIT report 1968, 81

Metabolic studies
  organosilicon compounds
    silanediols, 222
Methyl(trimethylsilyl)-
  germylene
  photochemical generation,
    105

Methyl(trimethylsilyl)-
  germylene (continued)
  dimerization, 110
  generation and potential
    rearrangements, 108
  reactions with 2,3-
    dimethylbutadiene, 109
(Methyl/silyl)methanes
  advantages over
    conventional
    feedstocks, 68
  feedstock for PECVD, 66, 67
Methylbis(trimethylsilyl)-
  germane
  pyrolysis, 108-109
  synthesis, 108
Methyldichlorosilane
  ammonolysis, 15
  polymerization of
    ammonolysis product, 15
Methyldisilane
  synthesis and properties,
    67
Methylphenylbis(trimethylsilyl)germane
  photolysis, 105
Methylphenyltrifluoro-
  silicate anion
  equilibrium constant for
    formation, 199
  formation in solution, 199
Methylsilane
  feedstock for PECVD, 63
Methyltris(trimethylsilyl)-
  silane
  photolysis, 97
  pyrolysis, 89
    Arrhenius parameters, 91
    mechanisms, 93-94
    primary steps, 91
    with toluene and
      butadiene, 90
Monolithic ceramics, 28

Nitriles
  from aziridines,
    lanthanoid catalysis,
      375
    regioselectivity, 375
    stereoselectivity, 375
  from epoxides, 369-371

NMR
  ^{13}C,
    Hammett plot di and tri-fluorosilicates, 202
NMR
  ^{29}Si, 72-74, 295-298, 308, 310, 312, 313, 318, 385
    disilanes, 275
    Hammett-type plots for $(Me_3SiO)_3SiX$, 208
    polysilapropylene. 44
Nucleophilic substitution in supersilyl compounds, 264
Nucleophilic substitution at silicon
  activation, 186
  stereochemical analysis for attack at trigonal bipyramidal silicates, 203
  stereochemistry, 185
  stereoelectronic effects, 215
  trajectory analysis by $\underline{X}$-ray crystallography, 203

Octamethyltrisilane
  pyrolysis, 89, 92
    mechanisms, 92
    with toluene and butadiene, 90
Olefin
  cis-trans thermal isomerization, effect of silyl-substituents, 4-6
Olefins
  oxygenated
    siloxanes from CO_2 laser reactions with silane, 166
Oligosilanes
  structures, 274
Optically active silanes, (see chiral organosilanes)
Organosilicon drugs, metabolism, 218
Oxiranes, (see epoxides)

Pauling bond order, 280
Pentacoordinate silicon
  kinetics of oxidation, 200
  oxidation of, 199
  stereochemistry of oxidation, 200
Pentacoordinate silicon compounds
  hydrides, anionic, 187-188
  hydrides, neutral, 189
  reactivity, 187-195
Pentamethylcyclopentadienyl ligand, 307-316
Pentasilacycloheptyne
  permethyl
    addition of Mes_2Si, 177
    crystal structure, 179
    photochemistry, 179
    synthesis and reactions, 177
Peterson olefination, 364
Phase transfer catalysis, 381, 383, 385, 387
Phenyl(trimethylsilyl)-germylene
  photochemical generation, 105
Phenylfluorosilanes
  activation to oxidation by F^-, 199
  oxidation, relative reactivity, 199
Polycarbosilanes
  preceramic polymer, 30
  TGA , 34
Polycarbosilanes
  direct synthesis, 43
  electrochemical synthesis, 42
  preceramic polymer, 41
Polyethylenepolysilylenes
  properties, 57, 58
Polyethylenetetramethyl-disilene
  characterization, 54-56
  NMR data, 56
  properties, 56-57
  synthesis and IR spectrum, 55
  thermal decomposition, 58

Polyoxymultisilylenes
  $-[O(Si)_m]_n-$
  by ring-opening
    polymerization
    anionic, 72
    cationic, 72
  morphology, 79
  NMR, ^{29}Si, 74
  spectral properties, 79
  thermal properties, 79
Polysila(N-2,6-xylyl)-
  imines, 393
Polysilaethers
  by cationic ring-opening
    polymerization, 74
  NMR, ^{29}Si, 72
Polysilahydrocarbons
  CO_2-laser pyrolysis, 58
  oxidative cleavage of
    Si-Si bond, 59
  structures, 50
Polysilanes
  electrochemical synthesis,
    42
  oxidation, 71
  Pd(0) catalysed isonitrile
    insertion, 393
  permethyl
    cyclic carbosilanes
      from, 95-96
Polysilanylsilenes
  from polysilanylbis-
    (diazo) alkanes, 172-
    173
Polysilapropylene
  synthesis, 43
  thermolysis, 44
Polysilazanes
  ceramic precursor, 32, 46
  conversion into silicon
    carbonitride, 17
  silacyclobutasilazane
    polymer, 32
  structure of polymer, 17
  TGA, 34
  treatment with
    $H_3B.S(CH_3)_2$, 20
    formation of borazine
      rings, 20
    network polymer from, 19
    pyrolysis of network
      polymer, 21

Polysiloxanes
  ceramic precursor, 35
  TGA, 34
Polysilylmethanes
  as feedstocks for PECVD,
    62
Polysilylmethanes
  structures, 63
Polyvinyltrimethylsilane,
  50
Preceramic polymers,
  27,40(see also,
    polycarbosilanes,
    polysilanes,
    polysilazanes,
    polysiloxanes and
    silicon acetylenes)
Prochiral carbonyl
  compounds
  allylation, 344-351
Pseudorotation, 279
Pyrrolidinylmethylallyl
  silanes
  in chiral synthesis, 347-
    351

Radicals, metal-centred
  $R=(Me_3Si)_2CH$
  AsE_2, 242-243
  GeR_3, 242-243
  PR_2, 242-243
  SnR_3, 242-243
Rearrangement
  dyotropic, 313
  sigmatropic, 313
Recycling, 81-85
Rochow synthesis
  by-products, 82
  chlorine recycling, 83
  yield, 83
Ruthenium complexes (see
  silene and silylene
  complexes)

Sharpless epoxidation, 351-
  354 341
Sila-1,3-butadiene,
  1,1-dimethyl-4-
    trimethylsilyl
  isomerization, 4
  preparation, 4

Subject Index

Sila-procyclidine,
  muscarinic antagonist, 223
Silacyclobut-2-ene,
  1,1-dimethyl-4-trimethylsilyl
  thermal equilibration, 3, 4
  thermal ring-opening, 4
Silacyclobutadiene
  decomposition, 284
  theoretical studies, 284
Silacyclobutadienes, substituted
  formation, 287
  photochemical ring opening, 288
  regiospecific *syn* addition, 288
  thermolysis, radical pathway, 288-291
Silacyclobutanes, 166
Silacyclopropenes
  formation, 317
  NMR, 318
Silaimines, 47
Silane
  CO_2 laser induced reactions, 158
  IR spectra of solid deposits from, 165
  mechanism, 165
  with fluorinated carbonyl compounds, 167, 168
  with oxygenated olefins, 163, 164
Silane-1,2-dichloro-1,2-difluoroethylene
  mechanism of CO_2 laser induced chemistry, 161-164
  products of CO_2 laser induced chemistry 162
Silane chlorotrifluoroethylene
  mechanism of CO_2 laser induced chemistry, 162
Silane-perhaloethene mixtures
  absorption spectra, 160
Silanols,
  as muscarinic agents, 219

Silanones
  formation, 190
Silanorbornadiene
  7,7-dimethyl-1,4,5,6-tetraphenyl-2,3-dibenzo-7-
  reaction with $PdCl_2(PPh_3)_2$, 154
7-Silanorbornadienes
  photochemical decomposition mechanism, 146
Silaselone, $(Me_5C_5)_2Si=Se$, 316
Silatetrahedranes
  possible intermediates in cyclopropenylsilane photolysis, 289-291
Silathiones
  stabilized, 192
Silatropic shift, 358
Silazanes
  intermediates in polymerization, 16
Silene complexes of iridium
  NMR spectra. ^1H; ^{13}C, 303
  synthesis, 303
  X-ray structure, 304
Silene complexes of ruthenium
  hydride migration in, 298, 300-303
  kinetic studies, 300-303
  NMR spectra; ^{29}Si, ^{13}C, 298
  pseudorotation, 299
  reaction with CO, 292
Silenes
  addition of alcohols, 138
  mechanism, 138-139
  copolymerization with methyl acrylate, 167
  dimerization and polymerization, 166
  photochemically generated cyclic, 138
  ultraviolet spectra, 171
  with oxygenated olefins, 166

Silicon
  oxidation state +2, 307

Silicon carbide, (see ceramics)
Silicon carbon bonds
  cleavage by $AlCl_3/Me_3SiCl$, 44
Silicon dust, reuse, 83
Silicon nitride-boron nitride composites formation, 21
Silicon-acetylene polymers $[-SiR_2-C{\equiv}C-]_n$, 8-14
  conversion to silicon carbide fibres, 9
  curing fibres by uv, 9
  from di-Grignard reagent, 8
  from dilithioacetylene, 9
  surface modificaton of melt-spun fibers, 9
Silicon-carbon bonds
  alcohols, by oxidation of, 197
  mechanism of oxidation, 197
Silicon-diacetylene polymers $[-SiR_2-C{\equiv}C-C{\equiv}C-]_n$, 9
  conversion to ceramic materials, 9-10
  crosslinking, 9
  formation from dilithiobutadiyne, 12
Silicon-olefin polymers $[-SiR_2-CH{=}CH-]_{n'}$, 7
  conversion to silicon carbide-carbon mixtures, 7
  from hydrosilylation of acetylenes, 7
Silicon-silicon bonds
  bond lengths, 271-280
  isonitrile insertion, 391
  Pd(0) catalysed insertion reactions, 391
Silicones
  annual production figures, 83
  recycling, 82
Siloxanes
  cleavage by HCl, kinetics, 210
  nucleophilic substitution stereochemistry, 214

Siloxanes (continued)
  reaction with Me_3SiOLi, 212-213
Silyl enol ethers
  photochemical 1,3-silyl migration, 137
Silyl SiH_3
  detection, 117
  generation, 116
  rate constants for detection, 119
Silyl transition metal complexes
  structures, 230, 275, 304
Silyl-substituted methanes
  bond lengths and angles, 258
Silylalkenes
  thermal isomerization via silyl-group migration, 5
Silylallene, 1-dimethylsilyl-3-trimethyl
  isomerization by 1,3-hydrogen shift, 3
  thermal equilibration, 3, 4
Silylaziridines
  basicity, 385
  synthesis, 381-386
Silylbutyrate, methyl-3-dimethylphenyl
  methylenation, 326
α-Silylcarbanions
  diastereoselectivity, 348-351
  regioselectivity, 348-351
  stereoselectivity, 348-351
Silylcarbenes
  rearrangements, 170
Silylcuprates
  addition to triple bonds, 328
  addition to unsaturated lactone, 324
  reaction with allenes, 362
  urethane-cuprate protocol, 329
Silylene complexes of ruthenium
  base free, 295
  donor adducts, 295
  NMR, ^{29}Si, 295, 297

Silylene SiH$_2$
  detection, 116
  generation, 116
  heat of formation of, 116
  rate constants for
    reactions, 117-118
Silylenes
  stable, 293
Silylenium ion
  (Me$_5$C$_5$)$_2$SiH$^+$ X$^-$, 311
β-Silylenolates
  alkylation, 322, 397
Silylidyne SiH
  generation and detection,
    114
  rate constants for
    reactions, 115
Silylisoxazolines
  properties, 386
  ring-opening, 388
  synthesis, 386
Silylphosphanes, 314
Sonication, 382, 385, 386
Speier's catalyst, 383
Spodoptera littoralis,
  biotransformations,
    organosilicon compounds
    in, 222
Stereocontrol
  using allylsilanes, 332-341
Stereocontrol, in organic
  synthesis, 321-331
Strain in molecules, 257
Substituent constants for
  silicon, 208-212
Supersilyl compounds
  aluminium(I) compounds,
    269
  bond angles, 269
  bond lengths, 263
  halides, reactions, 265
  nucleophilic substitution,
    264
  polyselanes, 266
  polysulphanes, 266
  radicals,
    preparation and use, 264
  silaneimine,
    preparation and
      structure, 267, 268
  synthesis, 264

Tetra-*tert*-butyldisilene
  addition to a
    3-thiazoline, 126
  addition to alkynes and
    olefins, 124
  addition to
    2,2'-bipyridyl, 129
  addition to
    diphenylisobenzfuran,
    125
  addition to isoquinoline,
    127
  addition to ketones, 125
  addition to N,N'-
    dicyclohexyl-
    1,4-diazabutadiene,
    129-130
    crystal structure of
      adduct, 130
  addition to N,N'-dicyclo-
    hexylcarbodiimide, 127
  addition to tri-*t*-butyl-
    silyl isocyanide, 128
  addition to tri-*t*-butyl-
    silyl cyanide, 125
  formation, 122
3-*H*-4-*H*-3,3,4,4-Tetra-
  tert-butyl-2-tri-
  tert-butylsilyl-3,4-
  disilaazete,
  crystal structure, 126
Tetracoordinate silanes
  oxidation,
    effect of R groups on
      silicon, 198
1,1,2,2-Tetrafluoro-1,2-
  disilacyclobutane, 51
1,1,2,2-Tetramethyl-1,2-
  disilacyclobutane
  addition reactions, 53
  polymerization, 53
  spectra, 53
  synthesis, 52
Tetrasiladecane
  2,2,5,5,6,6,9,9-
    octamethyl-2,5,6,9-
  IR spectrum of, 55
  structure, 57
Tetrasilane
  octamethyl-2,3-
    bis(trimethylsilyl)
    photolysis, 96-97

Tetrasilylmethane
  synthesis, 65, 66
trans-1,1-Diadamantyl-2,3-
  dimethylsilirane
  photochemical
    decomposition, 102
  synthesis, 101
  thermal decomposition, 102
Tri-t-butylcyclopropenium
  tetrafluoroborate
  reaction with
    polysilyllithium
    compounds, 286
Tri-t-butylsilyl
  compounds, (see
  supersilyl and
  specific compounds)
1,2,2-Trimethyl-2-
  silagermirane
  formation and
    rearrangements, 109
Trimethylsilyl-1H-aziridine
  carboxylic acid
    derivatives, 383, 385
Triplet silylenes
  approach to, 100
Tris(benzene-1,2-
  diolato)silicon, 194
Tris(trimethylsilyl)-
  borazine
  from (Me$_3$Si)$_2$NH.BH$_3$, 18
Tris(trimethylsilyl)methyl
  groups (see trisyl)
1,3,5-Trisila-1,4-
  pentadiene
  1,1,3,3,5,5-hexamethyl-
    2,4-bis(trimethylsilyl)
  formation and
    rearrangement, 174
2,4,5-Trisilabicyclo[1.1.1]
  pentane-2,2,4,4,5,6-
    hexamethyl-1,4-bis
    (trimethylsilyl)
  formation, 174
Trisilane
  2,2-diethylhexamethyl
    photolysis, 98
Trisila[3.1.1]propellane
  reactions, 175
  synthesis, 174
1,1,1-Trisilylethane
  synthesis, 66

Trisilylmethane
  feedstock for PECVD, 65
  synthesis, 64, 65
Trisyl groups
  bond lengths and bond
    angles, 255
  structure by electron
    diffraction, 254

Ultrasound (see sonication)

Vinylsilanes (see
  alkenylsilanes)

Yajima process, 28